Artificial Intelligence and Automation

ADVANCED SERIES ON ARTIFICIAL INTELLIGENCE – VOL. 3

ARTIFICIAL INTELLIGENCE AND AUTOMATION

Edited by

NIKOLAOS G BOURBAKIS

Binghamton University
Department of Electrical Engineering
Thomas J Watson School of Engineering
and Applied Science
Binghamton, New York, USA

World Scientific
Singapore • New Jersey • London • Hong Kong

Published by
World Scientific Publishing Co. Pte. Ltd.
P O Box 128, Farrer Road, Singapore 912805
USA office: Suite 1B, 1060 Main Street, River Edge, NJ 07661
UK office: 57 Shelton Street, Covent Garden, London WC2H 9HE

British Library Cataloguing-in-Publication Data
A catalogue record for this book is available from the British Library.

ARTIFICIAL INTELLIGENCE AND AUTOMATION
Advanced Series on Artificial Intelligence – Vol. 3

Copyright © 1998 by World Scientific Publishing Co. Pte. Ltd.

All rights reserved. This book, or parts thereof, may not be reproduced in any form or by any means, electronic or mechanical, including photocopying, recording or any information storage and retrieval system now known or to be invented, without written permission from the Publisher.

For photocopying of material in this volume, please pay a copying fee through the Copyright Clearance Center, Inc., 222 Rosewood Drive, Danvers, MA 01923, USA. In this case permission to photocopy is not required from the publisher.

ISBN 981-02-2637-3

This book is printed on acid-free paper.

Printed in Singapore by Uto-Print

PREFACE

This volume deals with several important issues in Artificial Intelligence and Automation. More specifically, it is divided into four main sections. Section 1 entitled "Issues in AI" includes five chapters. Chapter 1 deals with a new way of acquiring knowledge. Chapter 2 compares several knowledge representation schemes with the SPN KR scheme. Chapter 3 discusses the structures of word problems and their construction in natural language. Chapter 4 provides a statistical approach of resolving conflicts in inheritance reasoning. Chapter 5 presents an integration of low and high vision tasks for image understanding. The next section 2 entitled "AI Tools" includes six chapters. Chapter 6 presents the evolution of commercial AI tools, their barriers and objectives for the 1990s. Chapter 7 discusses the reengineering issue and AI tools for the market. Chapter 8 describes an intelligent tool for discovering data dependencies in relational databases. Chapter 9 presents a CASE based reasoning tool to assist traffic flow in airports. Chapter 10 provides an experimental study of a fuzzy expert system for financial applications. Chapter 11 describes an associative data parallel compilation model-tool for high performance knowledge retrieval and computing. Section 3 entitled "Automation", includes seven chapters. Chapter 12 discusses the software automation process. Chapter 13 describes how the software engineering gets benefit by using AI. Chapter 14 presents a knowledge based methodology for derivation of programs from specifications. Chapter 15 describes an automatic generation of functional models for parallel fault design error simulations. Chapter 16 provides an automated SPN methodology for visual reverse engineering of digital circuits. Chapter 17 discusses the impact of AI methodologies in VLSI Design Automation. Chapter 18 presents the automated acquisition of subcategorization of verbs, nouns and adjectives in natural language sentences. The final section 4 entitled "Planning" includes four chapters. Chapter 19 describes planning methodologies under different conditions and constraints. Chapter 20 uses learning to improve the path planning performance. Chapter 21 presents an incremental adaptation methodology to improve reactive behavior. Chapter 22 deals with an SPN-neural planning methodology applied on multiple robotic arms with constrainted placement.

CONTENTS

Preface v

Section 1: Issues in AI
CHAPTER 1
A NEW WAY TO ACQUIRE KNOWLEDGE
by H-Y Wang

Abstract 3
1. The Knowledge Acquisition and Developing on Resolving Problem 3
2. The Thinking Model of Solving a Problem in Social and Economic 4
3. The Structuring of the Expert Experience Geared to a Problem 5
4. The Structure of the Knowledge Base System Geared to a Problem 6
5. Acquiring Knowledge on Problem Resolving 7
6. Conclusion 9
   References 9

CHAPTER 2
AN SPN KNOWLEDGE REPRESENTATION SCHEME
by J. Gattiker and N. Bourbakis

Abstract 10
1. Introduction 10
2. Comparison of Knowledge Representation Methodologies 11
3. Petri Nets and Production Rules 13
   3.1. A Primitive for Representing Knowledge Atoms 13
   3.2. Representation of Knowledge 15
   3.3. Example 16
   3.4. Forward and Backward Chaining 16
4. Petri Nets and Logic 17
   4.1. Extension of Production Rules to Logic 17
   4.2. Representation of FOPC: Predicate Transition Nets 18
5. Petri Nets and Semantic Networks 19
6. Examples of Structured Computation Using Petri Nets 21
   6.1. Object Recognition 21
   6.2. Planning 23
   6.3. Predicate Calculus Circuit Simulation 23
7. Discussion 25
8. Conclusion 26
   References 26

CHAPTER 3
ON THE DEEP STRUCTURES OF WORD PROBLEMS AND THEIR CONSTRUCTION
by F. Gomez

    Abstract 28
1. Introduction 28
2. Schema-Based Understanding 30
3. The Deep Structure of Problems 32
4. The Deep Structure of Motion Problems 36
5. The Construction of the Deep Structure of Motion Problems 39
6. The Solution of Motion Problems 44
7. Conclusion 45
    References 46

CHAPTER 4
RESOLVING CONFLICTS IN INHERITANCE REASONING WITH STATISTICAL APPROACH
by C. W. Lee

    Abstract 48
1. Introduction 48
2. Interpretations of Defeasible Assertion 49
3. Evidential Probabilities 50
4. Model 51
5. Detecting the Existence of Rules 52
6. Inferences 54
    6.1. Specificity 54
    6.2. Generality 55
7. Algorithm 59
8. Examples 60
9. Conclusion 62
    References 62

CHAPTER 5
INTEGRATING HIGH AND LOW LEVEL COMPUTER VISION FOR SCENE UNDERSTANDING
by R. Malik and S. So

    Abstract 64
1. Introduction 64
2. Background 64
3. Problem Statement 65
4. Approach 65
5. Graph Extraction Algorithm 66
    5.1. Vertices Graph 67

    5.2. Regions Graph 68
    5.3. Exponent Regions Labeling 72
    5.4. Duality between Vertices Graph and Regions Graph 73
6. Graph Reduction Algorithms 73
7. Experiments and Results 76
8. Conclusion 77
    References 78

Section 2: AI Tools

CHAPTER 6
THE EVOLUTION OF COMMERCIAL AI TOOLS:
THE FIRST DECADE
by F. Hayes-Roth

1. Introduction 83
2. Setting the Stage 83
    2.1. The Commercial Milieu 83
    2.2. The Technical Milieu 85
3. Examples of Specific Advancements 86
    3.1. Support for the Task Specialization 86
    3.2. End-User Knowledge Programming and Maintenance 87
    3.3. Systems that Support Modularity and Embeddability 90
    3.4. Support for Co-operative Problem-Solving 91
4. Commercial and Technical Objectives for the 90s 93
    4.1. Current Technological Barriers and Objectives for the 90s 93
    4.2. Current Commercial Barriers and Objectives for the 90s 94
5. Conclusion 94
    References 95

CHAPTER 7
REENGINEERING: THE AI GENERATION — BILLIONS ON THE TABLE
by J. S. Minor Jr

    Abstract 96
1. Unit Continuous Optimization Manufacturing — Introduction 97
2. UCOM World Class Manufacturing Strategies 99
    2.1. Mass Customization 99
    2.2. Activity Based Costing 101
    2.3. Process Reengineering 101
    2.4. Concurrent Engineering 108
    2.5. Continuous Flow Manufacturing 109
    2.6. Strategies Review 116
3. UCOM Artificial Intelligence Review 116
    3.1. Senses 116
    3.2. Reasoning 120
    3.3. Neural Networks 121

    3.4. Hybrid 123
    3.5. Fuzzy Logic 123
    3.6. Genetic Algorithms 124
    3.7. Review 126
4. UCOM Simulation Review 126
    4.1. Review 127
5. UCOM Continuous Optimization Manufacturing MINOR I Model 127
    5.1. Engineering Redesign 128
    5.2. Engineering Impact 130
    5.3. Sales/Marketing Impact 130
    5.4. Plant Floor Impact 132
    5.5. Employee Impact 133
    5.6. Profitability Impact 134
    5.7. Vendor Impact 135
    5.8. Customer Impact 135
    5.9. Pilot Project 135
6. UCOM Continuous Optimization Manufacturing MINOR II Model ... 136
    6.1. Review 139
        References 141

CHAPTER 8
AN INTELLIGENT TOOL FOR DISCOVERING DATA DEPENDENCIES IN RELATIONAL DBS
by P. Gavaskar and F. Golshani

    Abstract 143
1. Introduction 143
    1.1. Overview of the Problem 145
    1.2. Preliminaries 146
    1.3. Organization of the Chapter 149
2. Previous Work 149
3. Knowledge Discovery in Databases 151
    3.1. Statistical Pattern Recognition 151
    3.2. Neural Networks 152
    3.3. Machine Learning 153
    3.4. Choosing a Method 155
4. Functional Dependency Algorithm 155
    4.1. Some Important Concepts 156
    4.2. Inference Rules for Keys 157
    4.3. Design of the Functional Dependencies Algorithm 158
5. Derivation of the Multivalued and Join Dependencies 168
    5.1. Some Fundamental Concepts 168
    5.2. Inference Rules for Multivalued Dependencies 170
    5.3. The Formal Framework 171
    5.4. The Algorithm 176
    5.5. Refinements 178

6. Realization of the Tool	182
6.1. The User Interface	182
6.2. Example	183
6.3. Performance Analysis	186
7. Conclusion	189
References	189

CHAPTER 9
A CASE-BASED REASONING (CBR) TOOL TO ASSIST TRAFFIC FLOW
by B. Das and S. Bayles

Abstract	192
1. Introduction	192
2. Background	193
3. Problem Analysis: The CBR Approach	195
4. The CBR Prototype Overview	196
4.1. Knowledge Engineering	196
4.2. Case Representation	198
4.3. Case Retrieval	199
5. Lessons Learned	201
6. Future Work	204
7. Conclusion	205
References	206

CHAPTER 10
A STUDY OF FINANCIAL EXPERT SYSTEM BASED ON FLOPS
by T. Kaneko and K. Takenaka

Abstract	207
1. Introduction	208
2. Expert System	210
2.1. Overview	210
2.2. Characteristics of an Expert System	211
2.3. Components	212
2.4. Applications	217
3. Fuzzy Expert System	220
3.1. Uncertainty in Expert Systems	220
3.2. Fuzzy Sets	220
3.3. Fuzzy Numbers	224
3.4. Fuzzy Inference	226
3.5. Fuzzy Expert System	227
3.6. Applications	228
4. FLOPS	229
4.1. Overview of FLOPS	229
4.2. Production System	230
4.3. FLOPS Anatomy	234

  4.4. Confidence Level 237
  4.5. Representation of Fuzziness in FLOPS 238
  4.6. Parallel FLOPS 241
5. Building a Financial Diagnosis System 243
  5.1. Overview 243
  5.2. Wall's Index Method 243
  5.3. Rules of Diagnosis 246
  5.4. Financial Data 247
  5.5. Process of Diagnosis 250
  5.6. Evaluation of the System 252
6. Conclusion 254
  References 255

CHAPTER 11
AN ASSOCIATIVE DATA PARALLEL COMPILATION MODEL FOR TIGHT INTEGRATION OF HIGH PERFORMANCE KNOWLEDGE RETRIEVAL AND COMPUTATION
by A. K. Bansal

  Abstract 257
1. Introduction 257
2. Background 261
  2.1. Preliminary Definitions and Notations 261
  2.2. Architecture for Associative Computing 262
  2.3. Associative Computing Paradigm 262
  2.4. Representing Data on Associative Architecture 263
  2.5. Associative Representation of Abstract Data 264
3. Overview of Logic Programming Concepts 264
  3.1. WAM — Conventional Execution Models 266
  3.2. Advantages of Associative Computation 267
4. An Algebra of Associative Computation 267
  4.1. Laws of Data Association 269
  4.2. Laws of Associative Search 269
  4.3. Laws of Associative Selection 270
  4.4. Laws of Data Parallel Computation 271
  4.5. Laws of Associative Update 271
5. Structure of the Model 272
  5.1. Associative Representation of Clause-heads 273
  5.2. Associative Representation of Global Bindings 274
  5.3. Storing and Manipulating Bags of Bindings 274
  5.4. Holding Temporary Values — A Data Parallel Version 274
  5.5. Handling Control Flow 274
  5.6. Handling Aliasing 275
6. Integrating Retrieval & Computation 275
  6.1. Alignment and Data Parallel Computation 275
  6.2. Alignment and Data Management 275
  6.3. Alignment and Aliasing 276

 6.4. Alignment and Associative Goal Reduction 276
 6.5. Deriving Unspecified Relations for Objects 277
7. The Model Behavior 277
 7.1. Handling Aliased Variables 278
 7.2. Handling Multiple Procedures 281
8. Performance Evaluation 281
9. Other Related Works 282
10. Conclusion 283
 References 283

Section 3: Automation
CHAPTER 12
SOFTWARE AUTOMATION: FROM SILLY TO INTELLIGENT
by J.-F. Xu, D.-X. Chen, Z.-J. Wang and L.-J. Dong

 Abstract 289
1. Introduction 289
2. Incorporating Machine Learning into Software Automation 290
 2.1. Inductive Program Synthesis 290
 2.2. Machine Learning in Inductive Problem Solving 292
 2.3. Analogical Program Derivation 295
3. Future Work 295
 3.1. Putting Experimental Systems on the Design Level into Practice 295
 3.2. Designing a New System 295
 3.3. Looking for Analogical Approach 295
 References 296

CHAPTER 13
SOFTWARE ENGINEERING USING ARTIFICIAL INTELLIGENCE:
THE KNOWLEDGE BASED SOFTWARE ASSISTANT
by D. White

 Abstract 297
1. Background 297
 1.1. Opportunity 297
 1.2. The Software Problem 298
 1.3. Expert Systems and Automatic Programming 299
2. An AI Approach 300
 2.1. Failings of Conventional Approaches 300
 2.2. Objectives 303
 2.3. Concept 303
 2.4. Approach 304
 2.5. Paradigm Scenario 307
 2.6. KBSA Program 309
3. Conclusion 313
 References 313

CHAPTER 14
KNOWLEDGE BASED DERIVATION OF PROGRAMS FROM SPECIFICATIONS
by T. Weight, J. Boyle, T. Harmer and F. Weil

Abstract 315
1. Introduction 315
2. The Software Development Process 316
   2.1. Idealized Software Development Process 317
   2.2. Practical Software Development Process 317
3. What is Program Transformation? 318
4. An Experiment in Program Transformation Applied to a
   Commercial Product 320
   4.1. Implementing Transcription to Hardware Specific Operations 320
   4.2. Implementing High Level Abstract Data Types 321
   4.3. Implementing Separation of Concerns 323
   4.4. Implementing High Level Specification Constructs 325
5. Conclusion and Future Work 328
   References 329
   Appendix 330

CHAPTER 15
AUTOMATIC FUNCTIONAL MODEL GENERATION FOR PARALLEL FAULT DESIGN ERROR SIMULATIONS
by S.-E. Chang and S. A. Szygenda

Abstract 348
1. Introduction 348
2. AFMG Domain Analysis 349
   2.1. Signal Modeling Phase 350
   2.2. Primitive Modeling Phase 350
   2.3. Functional Modeling Phase 353
   2.4. The Domain Independent Programming Knowledge 357
3. User Interface and Specification Tools 358
4. Behavioral Domain Model Generation — Automatic Primitive Modeling 363
5. Structural Domain Model Generation — Automatic Functional Modeling 369
6. Experiment Results 374
7. Conclusion 375
   References 376

CHAPTER 16
VISUAL REVERSE ENGINEERING USING SPNs FOR AUTOMATED DIAGNOSIS AND FUNCTIONAL SIMULATION OF DIGITAL CIRCUITS
by J. Gattiker and S. Mertoguno

Abstract 378

1. Introduction 378
2. The Detection Diagnosis Methodology 379
   2.1. Visual Reverse Engineering 379
3. Converging Functional and Structural Knowledge with SPNs 381
   3.1. Definitions and Primitive for SPN Knowledge Representation 383
   3.2. Mapping Graphs onto SPNs for Knowledge Processing 383
4. An Illustrative Example 383
5. Conclusion 390
   References 390

CHAPTER 17
THE IMPACT OF AI IN VLSI DESIGN AUTOMATION
by M. Mortazavi and N. Bourbakis

Abstract 391
1. Introduction 391
   1.1. VLSI Design Automation 392
   1.2. Top-Down Design Methodology for the DA Process 392
   1.3. Classification of AI Tools 396
2. Roles of AI Tools in VLSI Design Automation 398
   2.1. Expert Systems in Design Automation Process 398
   2.2. Rule Based Systems 398
   2.3. Heuristic Algorithms in VLSI Design Automation 399
   2.4. Learning Algorithms in VLSI Design Automation 399
3. AI Tools in Design Automation 400
   3.1. Synthesis Tool Systems 400
   3.2. Evaluation/Analysis Tools 405
   3.3. Design Enviornment/Management System Tools 406
4. Concluding Remarks 410
   References 411

CHAPTER 18
THE AUTOMATED ACQUISITION OF SUBCATEGORIZATIONS OF VERBS, NOUNS AND ADJECTIVES FROM SAMPLE SENTENCES
by F. Gomez

Abstract 415
1. Introduction 415
2. Overview of the Parser 416
3. The Acquisition of the Symantic Usages 421
4. Learning the Subcategorization of Verbs 423
5. Merging the Segment Usages into a Final Usage 424
6. Learning the Subcategorization of Nouns and Adjectives 426
7. Ambiguity and Other Problems 428
8. Related Work 429
9. Conclusion 430

xvi

References 431
Appendix 432

Section 4: Planning

CHAPTER 19
GENERAL METHOD FOR PLANNING AND RENDEZVOUS PROBLEMS
by K. I. Trovato

Abstract 439
1. Introduction 439
2. Framework 439
3. Algorithm Analysis/Accuracy 443
4. The Fire Exit Problem 444
5. The Robot Path Planning Problem 445
6. The Vehicle Maneuvering Problem 446
7. A Forward-Only Constraint 451
8. Use of the System for Complex Maneuvers and Larger Areas 452
9. A Radio Controlled Example 454
10. High Speed Vehicle Maneuvering 455
11. Planning the Coordination of Multiple Actors — Synergistic Planning .. 457
    11.1. Example Coordination Problem 459
    11.2. Coordination of Fleet Trucks 463
    11.3. Brief Analysis 463
    11.4. Other Applications 463
12. Invitation 464
13. Conclusion 464
    References 465

CHAPTER 20
LEARNING TO IMPROVE PATH PLANNING PERFORMANCE
by P. C. Chen

Abstract 466
1. Introduction 466
2. Related Work 468
3. Algorithmic Framework 469
4. General Analysis 472
5. A Specific Case Analysis 477
6. Particular Analysis 481
    6.1. Pessimistic Model 483
    6.2. Randomized Model 487
7. Application and Verification 489
    7.1. Pessimistic Model 489
    7.2. Randomized Model 490
8. Extension to Changing Environments 492
    8.1. Object-Attached Experience Abstraction 494

8.2. On-Demand Experience Repair	495
8.3. Other Repair Strategies	495
8.4. Solution Quality and Redundancy	496
8.5. Example	496
8.6. Computational Experience	498
9. Conclusion	499
References	500

CHAPTER 21
INCREMENTAL ADAPTATION AS A METHOD TO IMPROVE REACTIVE BEHAVIOR
by A. J. Hendriks and D. M. Lyons

Abstract	503
1. Introduction	503
2. Reactive Behavior	505
2.1. Pengi: Intelligence Through Interaction	505
2.2. The Subsumption Architecture	506
2.3. Design Reactive Machines	507
3. Incremental Adaptation	508
3.1. The Incremental Adaptation Methodology	509
4. Incremental Adaptation Architectures	511
4.1. THEO: Planning on Demand	511
4.2. DYNA: Learning Control Policies	512
4.3. ERE: Generation of Situated Control Rules for Action Selection	513
4.4. XFRM: Planning by Transformation	515
4.5. ADAPT: Planning and Reacting	516
5. Summary	519
References	520

CHAPTER 22
AN SPN-NEURAL PLANNING METHODOLOGY FOR COORDINATION OF MULTIPLE ROBOTIC ARMS WITH CONSTRAINED PLACEMENT
by N. Bourbakis and A. Tascillo

Abstract	522
1. Introduction	522
2. Notations and Definitions	523
3. SPN Planning Methodology	523
3.1. Why Stochastic Petri-Nets	523
3.2. SPN Model	524
3.3. Theoretical Aspects of the SPN Planning Model	524
3.4. SPN Planning in UD Using two Robotic Arms	525
4. An Illustrative Example of the SPN Planning with a Self-Organized Neural Network	528
5. Conclusion	536
References	536

CONTRIBUTORS

A. Bansal, Kent State University
S. Bayles, MITRE Corp.,USA
N. Bourbakis, Binghamton University, USA
J. Boyle, Motorola Corp., USA
S. E. Chang, University of Texas-Austin, USA
P. C. Chen, Sandia National Labs, USA
C. Daoxu, Nanjing University, China
B. Das, MITRE Corp.,USA
J. Gattiker, SUNY-B and Los Alamos Labs,USA
P. Gavaskar, Motorola Corp., USA
F. Golshani, Arizona State University, USA
F. Gomez, Central Florida University, USA
T. Harmer, Motorola Corp.,USA
F. Hayes-Roth, Cimflex Teknowledge Cop.,USA
A. J. Hendriks, Philips Labs, USA
X. Jiafu, Nanjing University, China
T. Kaneko, Kyushu Tokai University Japan
C. Lee, University of Connecticut, USA
D. Lijun, Nanjing University, China
D. M. Lyons, Philips Labs, USA
R. Malik, Stevens Institute, USA
S. Mertoguno, Binghamton University, USA
J. S. Minor Jr, IBM Corp.,USA
M. Mortazavi, Cadence Corp.,USA
S. So, Stevens Institute, USA
S. Szygenda, University of Texas-Austin, USA
K. Takenaka, HOYA Service Corp.,Japan
A. Tascillo, Ford Corp., USA
K. Trovato, Philips Research Labs, USA
H.-Y Wang, Xian Jiaotong University, China
T. Weight, Motorola Corp., USA
F. Weil, Motorola Corp., USA
D. White, AIRFORCE Rome Lab, USA
W. Zhijian, Nanjing University, China

CONTRIBUTORS

A. Bansal, Kent State University
S. Bayliss, MITRE Corp, USA
N. Bourbakis, Binghamton University, USA
J. Boyle, Motrola Corp., USA
S. K. Chang, University of Texas-Austin, USA
P. C. Chen, Sandia National Labs, USA
C. Daoxu, Nanjing University, China
B. Das, MITRE Corp, USA
E. Gauker, SUNY P and Los Alamos Labs, USA
R. Gevasier, Motorol Corp., USA
R. Grisham, Arizona State University, USA
R. Gomez, Central Florida University, USA
P. Harper, Motorola Corp., USA
R. Hayes-Roth, Cimflex Technologies Corp, USA
A. J. Hendriks, Phillps Labs, USA
X. hatu, Nanjing University, China
T. Kaneko, Kyushu Tokai University, Japan
C. Lee, University of Connecticut, USA
D. Lijun, Nanjing University, China
C. H. Lyons, Phillips Labs, USA
R. Mahs, Stevens Institute, USA
A. Metropono, Binghamton University, USA
J. S. Mhoor jr, IBM Corp, USA
M. Morasaxa, Cadence Corp, USA
A. So, Stevens Institute, USA
S. Szygenda, University of Texas-Austin, USA
K. takenaka, HOYA Service Corp, Japan
A. Tascillo, Ford Corp., USA
R. Trovato, Philips Research Labs, USA
H.-Y.Wang, Xian Jiaotong University, China
T. Weight, Motorola Corp., USA
P. Weil, Motorola Corp., USA
W. White, AIRFORCE Rome Lab, USA

SECTION : 1
ISSUES IN AI

A New Way to Acquire Knowledge
Knowledge Acquiring Geared to a Problem

HongYi   Wang

The School of Management, Xi'an Jiaotong University
Xi'an, Shaanxi Province 710049, *P. R. C.*

ABSTRACT

By the accumulation and reduction of facts and experience, new knowledge may be produced. But this knowledge acquireing is natural, no clear aimed for anything. In fact, when confronting a specific problem, one must, actively and aimly, probe into and search for new knowledge to solve the problem. We call this knowledge acquiring as knowledge acquiring geared to a problem. With informations in modern social sharp increasing, knowledge acquireing will, more and more, depend on this way. On the bachground of social and economic, this paper have prospected the development characteristic of knowledge, the thinking characteristic and thinking model of solving a problem in social and economic, the method of knowledge acquiring geared to a problm, the structural model of a expert system of policy analysis on knowledge acquiring geared to a problem

1. The knowledge acquisition and developing on resolving problem

It is usually considered that knowege is the understanding of objective realty raw[1], and the understanding is deepen gradually. On system view, each other of objective bodies is relative, therefore the development of knowledge is relative, conditional, dynamic, and we can divide the development process of knowledge as several phases: sensory—perception—experience—theory knowledge—common sense—old sense

Among the procoss, there is much experience which can not be transformed into theory knowledge, much theory knowledge can not be transformed into common sense, and in a certain condition the theory knowledge, or the common sense will become no practical value, old sense. From sensory to perception, through experience, finally to theory knowledge, the development procoss can be categorized two type: 1.) The development of knowledge is nature, for no goal. 2.) The development of knowledge is for a goal, for example, analysis and resolving a problem. In the second kind of knowledge development, the knowledge produced is often more effective, and more practical. We name the second knowledge development as knowledge acquiring geared to a problem. With social and economic informations sharply increasing, knowledge development will, more and more, depend on this way.

In social and economic, a problem is often with new characteristics and the solving of a problem often depends on the production of new knowledge. On the background of policy analyses in social and economic, the paper research the knowledge acquisition geared to a problem.

2. The Thinking model of solving a problem in social and economic

In social and economic, a social and economic state can be considered as a function as following: $Y = F(X, \text{effect}(X), \text{policy}(X), T)$

Y is a state set in social amd economic, a multiple vector. The X is a set of objective bodies with many characteristics in social and economic. Effect(X) is a set of effects on objective bodies by other objective bodies of the X. Policy(X) is a set of cotrol strategy, for example, goverment policy, or raw and rules. T is a time when the state is, which many objective bodies of the X, many effects on the X is relative to. When a policy is put on X, basing on the effects of each other of the X's objective bodies, Y will appear a state. F is a maping function from X set to Y set. The functions model have two concrete forms, one of which is the objective form people are still understanding, another one is a approximate form of the objective form.

When a policy is put on X set, Y have a actual value and there is expect value of Y in policy executor's brain. If the Y state do not equal the expect state (goal) to the policy, this, the gap between actual state Y and the expect state, is a problem. The posible cause of the problem is that new elements comes in the X, or the executor's understanding to the X and the effects is incomplete or with defeats. To clear the causes of the problem, one will analyze the X, the effects, and manage to improve the understanding to the X, effects which have always much room for improvement as time passes. We named the knowledge to be analyzed and improved asa doubtful point. In fact, when one faces a problem, he (or she) will, again and again , apply the infomations from the problem enviroment to his (or her) experience to deepen the understanding of the characteristic and the effects of the objective bodies in the problem and to improve his (her) experience until the cause of the problem is clear.

From analyses above, it is seen that a problem system consists of three parts: a problem space, a knowledge space and a objective body space. The problem space includes all problems in the field. The objective body space contains all objective bodies in the field. The knowledge space include all the knowledge about the objective body's characteristics and effects. Policy(X) is a kind of effect on the other bodies produced by goverment, which is a object body in social and economic. In social and economic field, problems always ceaselessly come out, so experience and theory knowledge is ceaselessly deepen, and this three space is also changing. The model of the policy problem analysis is also fit for other similar field.

3. The structuring of the expert experience geared to a problem

In the development process of knowledge, the phase from experience to theory kowledge is a important phase, in which person experience is transformed to theory knowledge all people can share. The essence of the phase is to structure person experience and represent it by a symble system (usally by a language). Therefore the structuring of expert experience is a key content of knowledge acquisition in artificial intelligence.

knowledge is the understanding of objective bodies raw[1], which includes the understanding of the characteristic of objective bodies, the effects on objective bodies by other bodies. The effects includes directed and indirected effects. The development and variation of objective bodies is all based on the effects. Indirect effect is formed by effect transfering from one objective bodies to another one. Expert experience is the understanding to indirect effects, but the transfer process of effects is not clear[2]. Therefore the core of knowledge is objective bodies, each other of which is relative in a field, so the characteristics, the effects of the objective bodies is relative, and they form an organic system on a speciality, a speciality knowledge system.

To make the effects clear and to represent it in knowledge base, we use RELATION to discrible the static quantitative relation of two object bodies which affect on each other, EFFECT only mean the dynamic process of the effect. In fact, we have structured knowledge as the characteristic of a objective body, relations of objective bodies, effects on objective bodies. To structure his experience by expert oneself, we may further structure the knowledge as follow:

Relation is structured as: relative objective bodies (body 1, body 2), relative direction (positive or negative), relative type (linear, exponential, power, logarithmic, ladder, ···), relative parameter (factor and coefficient).

Direct effect: the body affected, effective bodies, effect conditions, effect results, characteristics affected, effective characteristics.
Indirect effect: (same as direct effect).

4. The structure of the knowledge base system geared to a problem

On above analysis, the computer model of a policy analyse system on the knowledge is structured as Fig. 1:

```
┌──────────┐         ┌──────────┐
│knowledge │────────▶│reasoning │
│   base   │         │  engine  │
└──────────┘         └──────────┘
      │    ┌──────────┐    ▲
      │    │ control  │    │
      ├───▶│  model   │────┤
      │    └──────────┘    │
      ▼         │          ▼
┌──────────┐    │    ┌──────────┐
│ doubtful │    │    │ problem  │
│  point   │    │    │ function │
│ analysis │    │    │  model   │
└──────────┘    │    └──────────┘
                ▼
          ┌──────────┐
          │man—mechine│
          │ interface │
          └──────────┘
```

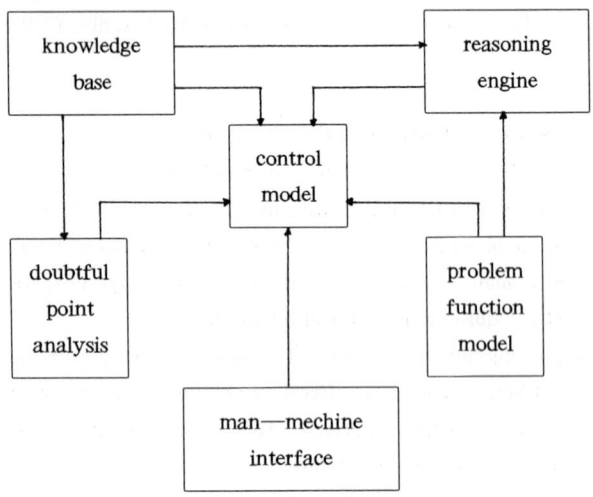

Figure 1.   The structure of knowledge manage system geared to a problem

The knowledge base in Fig. 1 contains knowledge about objective bodies and its characteristics, about relations among the bodies, about effects on the bodies. The bodies is a objective reality, a set of objective bodies studied in the problem field, the characteristic is the objective body's. An objective body and its characteristics is represented by frame in the base. Relations base on the objective bodies, represented by multiple. Effects base also on the objective bodies, represented by production rules. All the rules form a state network. On base of objective bodies, the base is a larger network and is a dynamic system. The structure of the base is depicted in Fig. 2.

The control model in Figure 1 cotrols the whole system work, accepts the task from man—mechine interface, give demands to man—mechine, calls reasoning engine to reason system state variation, calls problem function to calculate goal states, calls knowledge base to corect, improve, input the knowledge, calls doubtful point analysis to analyze, develop the knowledge.

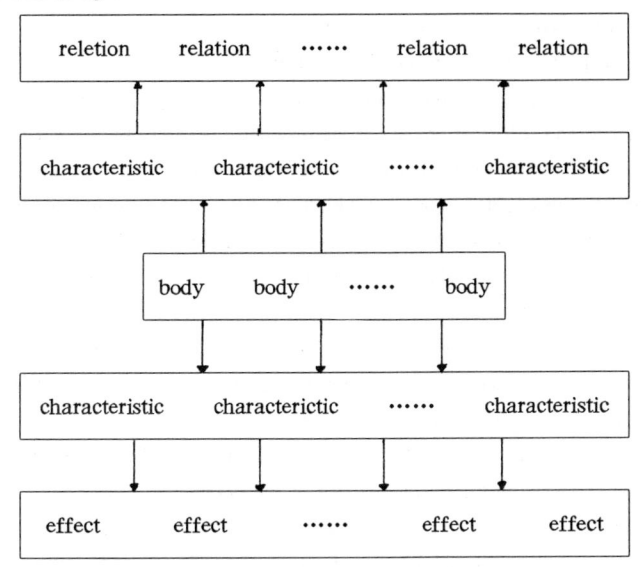

```
┌─────────────────────────────────────────────────────┐
│  reletion    relation   ······   relation   relation │
└────↑───────────↑──────────────────↑──────────↑──────┘
     │           │                  │          │
┌────┴───────────┴──────────────────┴──────────┴──────┐
│ characteristic  characterictic  ······  characteristic│
└────↑───────────↑──────────────────↑─────────────────┘
     │           │                  │
     ┌───────────┴──────────────────┐
     │  body    body   ······   body │
     └───┬───────┬──────────────┬────┘
         │       │              │
┌────────┴───────┴──────────────┴──────────────────────┐
│ characteristic  characterictic  ······  characteristic│
└────┬───────────┬──────────────────┬──────────┬───────┘
     │           │                  │          │
┌────┴───────────┴──────────────────┴──────────┴──────┐
│  effect     effect     ······    effect    effect   │
└─────────────────────────────────────────────────────┘
```

Figure 2.   The structure of knowledge base geared to a problem

The reason engine in Figure 1, on a policy(X), takes knowledge from knowledge base, reason the final state from the initial state X. The problem functin model in Figure 1 constructs Y's goal vector and caculates the value of a problem goal Y on relations from knowledge and the final states from reasoning engine. The man—mechine interface in Figure 1 completes displaying of the system questions and demands, and inputing of expert's answer and problem.

The doubtful point analysis model in Figure 1 locates the area of doubtful point, displays and analyzes the doubtful point in the imaginary forms such as chart, diagram ··· so as for expert to improve them.

5. Acquiring knowledge on problem resolving

As expert knowledge is stored by subconscious[3] and often difficult to be conveyed by words[4], so there are the difficulties of the understanding and representing of expert experience. Therefore the main task of acquireing knowledge is how to structure expert experience, how to improve knowledge in the knowledge base and how to convey expert experience by no words. By the structuring of knowledge on above, we have resolved knowledge representation. The clue of solving other problems above is as follow:

First, in the form of cure, chart, graph, diagram, and picture, the characteristic, relations and effects of objective bodies is displayed to expert, analyzed and judged by expert on his experience and informations about the problem. Second, by answering of the system prompt expert improves the knowledge or add new knowledge in the base. In this way, we can stir up expert's imagine thinking, impel expert to compare the knowledge with his experience, to structure and improve the knowledge in knowledge base and his experience. By calculating problem goal vector, expert can inspect, verify his experience and knowledge in the base. Finally, we not only develop the knowledge in the knowledge base and convey expert's experience to the knowledge base by no words, and also support the development of expert's experience in experts brain by structuring expert's experience. The pragram to do this way consists of tree step:

1) *Accept a problem and form a task*. The task can be divided two kind: one is only to calculate a goal vector value on knowledge existed in the knowledge base when a policy(S) is imitated. The task only applies the present knowledge to solve a problem. Another one is to analyze and solve a problem, solving of which needs developing new knowledge and deepening the understanding to the characteristics, relations, effects on the X. To form a task is to construct the Y vector, construct the problem function and input the informations about the problem including a policy(X) and for calculating the Y value.

2) *Imitate the executing of policy(X), calculate the Y vector value and locate the area of doubtful point*. As soon as task is formed, the control model will calls reason engine to reason the final state, and then calls problem function to calculate the Y vector. If the Y value is equal or close to the actual state by expert judging, the task ends. If not, control model calls doubtful point analysis to locate the area of the knowledge to be improved in help of expert's judging, and go to step 3.

3) *Analyze the doubtful points, improve the structural knowledge and add new knowledge in knowledge base*. In imaginary forms the doubtful points are displayed to expert, and in help of analysis tools in system, by expert's judging and answering system prompt, the knowledge in the base and expert's brain will be improved. When the anylysis is completed, go to step 2.

On this way, the system will be improved by a little and a little. The new or improving knowledge will be shared by other experts to resolve other problems.

6. Conclusion

1) Knowledge geared to a problem is an organic system on a special field. the core of knowledge is the objective bodies, the objective bodies studied in the special field. Being geared to a problem is helpful for systematizing the knowledge and arranging the kowledge base.
2) The development of knowledge is dynamic, little by little, from perception to experience and to theory knowledge. So the knowledge in the base should be arranged to a dynamic structure.
3) Structuring knowledge and then displaying it in imaginary forms is a direction to overcome the difficulties of the understanding expert's experience and conveying it by words, and being geared to a problem is helpful for doing this way.
4) The unification of knowledge development and acqusition is helpful for the knowledge acqusition in computer system, and also for expert's experience development in expert's brain.

References

[1] Atificial Intelligence Association of All—China Software Federation, An Atificial Intelligence Dictionary, Posts and Telecommunications Press, Pg. 203, 1992.
[2] Reif, F. , and J. Heller; "Knowledge Structure and Problem Solving in Physics. " in Educational Psychology, 17(2), 1982.
[3] Weiser, M. , and J. Shertz; "Programming Problem Representation in Novice and Expert Programmers," International Journal of Man— Machine Studies, 19 (1983).
[4] Kolodner, J. , Retrievaland Organizational Strategies in Conceptual Memory; A Computer Model, Erlbaum, Norwood, N. J. , 1984.

AN SPN KNOWLEDGE REPRESENTATION SCHEME

J.R. Gattiker and N.G.Bourbakis
Binghamton University, EE Dept., AAAI Lab
Binghamton, NY 13902

ABSTRACT

Stochastic Petri nets (SPNs) have traditionally been a powerful tool for modeling discrete parallel systems. Here a new methodology for knowledge representation based on Petri net structures is presented. Capabilities of the traditional predicate calculus and semantic network schemes for knowledge representation will be discussed, and the proposed methodology using Petri nets for equivalent representation of this knowledge is shown. Because of this equivalence, the proposed methodology can be used as a medium for conversion or convergence of knowledge forms. The capabilities of Petri nets exceed static representation, since Petri nets can also model functional, dynamic behavior. A Petri net is an active network with dynamics, and as such, if properly structured, can perform useful computation. In a knowledge representation system, this computation can be inferencing based on the encoded knowledge. This behavior is illustrated by examples showing an isomorphism to chaining in production rules, and to inferencing in logic.

1. INTRODUCTION

Classical knowledge representation techniques and inferencing have been the backbone of artificial intelligence. Storing knowledge with the goal for simulating human reason and/or behavior has taken many forms. Production rules are a format that structure knowledge as a set of rules to follow. The techniques of logic encoding include predicate calculus encoding, and inferencing and resolution. Semantic networks provide an additional refinement of content addressability to structure the directions of inference, and to provide hierarchy in the represented information. This paper will assume a familiarity with the concepts behind these knowledge representation methodologies, good references to these topics are Genesereth, Nilsson (1987), and Reichgelt (1991).

Petri nets are an established technique used mainly for modelling systems, and can represent parallelism and synchronization of entities in the system modelled. They have been widely applied and analyzed. Murata (1989) provides an excellent survey on the state of the art.

Considering the two areas of classical knowledge representation and Petri net modelling, both have strengths that are lacking in the other. Formalisms for knowledge encoding provide a good medium for representing *structural* knowledge. Although this distinction has some of the feeling of the declarative versus procedural knowledge controversy (discussed in the next section), this is intended to be a distinct issue - an observation that in general since logic is used for structured knowledge representation about a system, but does not embody the *function* of that system. Functional knowledge about a system is the strength of Petri nets. To make the distinction clearer, functional knowledge representation can be seen as modelling a system, where structural knowledge is seen as describing a system. To illustrate further with an example, structural knowledge about a car would be color, dimensions, and important attributes, where the functional knowledge would be a model of the active components of the system, the battery, the starter, the components of the engine, so that the representation is a model, which emulates the car. Of course, logic and semantic nets can be structured to reflect some idea of state and so be more functional in its character, but this is not the standard way of using these techniques.

One goal of this research is to provide a single knowledge representation methodology that can equally well represent both functional and structural knowledge, by exploring the ability of Petri net mechanisms to represent structural information. The knowledge representation techniques described above exemplify this, and so the exploration of implementing these knowledge forms using Petri nets has been undertaken.

A further goal of this work lies in the distinction between symbolic and connectionist ideals. Symbolic knowledge representation is essentially the manipulation of distinct symbols that stand for some entity in the world, and have very powerful application. Connectionist networks instead store knowledge in the behavior of many distributed nodes, and have desirable behavioral characteristics difficult to reproduce with other methodologies. The Petri net may be a way to span the chasm between these techniques by providing a way to encode symbolic knowledge into networks, which can then be manipulated through isomorphism with connectionist nets.

This paper explores encoding databases of production rules, logic, and semantic nets in the Petri net. The examples of the last section explore some of the issues in this encoding, and shows the main advantage of this: the implications of using an active network for knowledge encoding. Although the Petri nets in this paper are a formalism, the goal of this project is the generation of massively parallel databases for knowledge representation. This is precisely the nature of Petri nets.

2. COMPARISON of KNOWLEDGE REPRESENTATION METHODOLOGIES

This section will discuss the general characteristics of the broad categories of knowledge representation schemes, using comparison tables generated by Wah et al.(1990). First, in table 2.1, a comparison of local vs. distributed representations is made. This is of relevance to the methodology described in this paper, in that one of the goals of the Petri net knowledge representation is to help bridge the gap between distributed and local knowledge representation. Most knowledge representation falls into one of two camps: symbolic and connectionist, which can be generally construed as local and distributed, respectively. Table 2.1 describes some of the issues in the local/distributed tradeoff.

TABLE 2.1 Attributes of Local and Distributed Representation [Wah et al.(1990)]

Attribute	Local	Distributed
Storage technique	Each datum stored in dedicated hardware	Data represented over multiple units
Ease of understanding	Easy for humans to comprehend	Difficult for humans to interpret
Modification of stored data	Simple	More difficult
Fault tolerance	Loss of hardware results in loss of all stored data in that unit	Loss of small proportion of units does not seriously damage integrity of data

Table 2.2 presents a general comparison of the principal methodologies used in knowledge representation. The declarative/procedural distinction refers to a general issue in knowledge representation. As described in Reichgelt (1991):

> The procedural assertion position asserts that human knowledge is primarily knowing *how* [italics added], and that in order to capture knowledge in a computer, it is necessary to write computer programs consisting of a set of procedures, each of which represents some piece of knowledge. Thus, the knowledge that a person has, for example, about chess is identical to the procedures that he or she has for playing chess, and a procedural representation of this is a program that represents this knowledge as a set of internal procedures. The declarative position, on the other hand, believes that knowledge is primarily knowing *that* [italics added]. People know facts about the world. Consequently, knowledge should be represented explicitly and in a declarative format, rather than embedded inside the procedures of some program. Of course one needs procedures to use knowledge, for example, in order to derive implicit knowledge from stored knowledge. However, these procedures are assumed to be very general, and can be used for knowledge about different domains.

This paper is primarily concerned with representing declarative knowledge, since it is the declarative representation schemes of logic, semantic network, and production rules that will be examined. However, the goal of this exploration is to structure this knowledge in a Petri net so that the reasoning techniques operate implicitly. The reasoning techniques used in the knowledge methodologies discussed in this section are listed in Table 2.3. This implicit manipulation of the encoded knowledge is seen to move toward the long-term goal described in the introduction of incorporating the operation of symbolic methodologies and characteristics into connectionist networks, since perhaps the most distinguishing characteristic about a knowledge representation methodology is not the information it is capable of storing, but rather the possible ways that this encoded knowledge can be manipulated and the possible output of the methodology.

Table 2.2 Characterization of Knowledge Representation Schemes [Wah et al.(1990)]

Representation	Level of Representation	Characterization
Logic	Variable	Local/declarative
	Statement/relation	Local/declarative
	Program	Distributed/declarative
Production System	Variable	Local/declarative
	Statement/relation	Local/either
	Program	Distributed/declarative
Semantic Network	Node	Local/declarative
	Arc/relation	Local/declarative
	Network	Local/procedural
	Program	Distributed/declarative
Frames	Variable	Local/declarative
	Statement	Local/either
	Slot	Local/either
	Frame	Local/declarative
	Program	Distributed/declarative
Procedural	Variable	Local/either
	Statement	Local/procedural
	Program	Local/procedural
Connectionist	Connection strength	Local/declarative
	Propagation technique	Local/procedural
	Data and knowledge	Distributed/declarative

Table 2.3 Reasoning Techniques [Wah et al.(1990)]

Representation	Typical Reasoning Technique
Logic	Resolution (unification)
Production system	Forward/backward chaining
Semantic network	Spreading activation
Frames	Procedure attachments
Procedural	Control flow
Connectionist	Propagation of excitation

3. PETRI NETS and PRODUCTION RULES

Production rules, in their simplest form, are if/then clauses:
IF (condition clause) then (action clause).
The condition clause may be considered to be a logical combination of conditions, but the rules may easily be broken into clauses with only conjunction by splitting disjunctive clauses into separate clauses each with a different sub-clause. The rules are matched with atoms in the database, and when one is satisfied, its output is added to the database. Finally, a rule that leads to the goal is found (completing a forward chain), or in backward chaining a rule that leads to the conditions is found.

Representation of a discrete set of production rules as a Petri net begins with these analogies:
Places: Database atoms
Transitions: rules
Tokens: Instantiation of an atom
To elaborate on the distinction between the existence and instantiation of an atom, consider the basic rule format "If A then B". This rule defines (or implies definition of) a category of atoms, A and B. Instantiation is the actual assertion that these entities exist. The "If A then B" rule is to simply configure a Petri net so that there is a place A connected to a place B through a transition. The simplest representation has significant problems, including the problems of generating multiple redundant inference, and implementing negation. The next section describes a convenient representational primitive for use in constructing an Petri net system.

3.1 A PRIMITIVE FOR REPRESENTING KNOWLEDGE ATOMS

Our goal is to construct a simplified representation allowing shorthand illustrated in Fig. 1. This primitive is defined to solve the problems described above; it will guard against multiple inference, and in addition allow the representation of negative instantiation. This latter is a persistent problem of knowledge representation methodologies, namely that if a set of rules are designed to fire in the positive of a case, then the negative is not inherently represented, but instead must be explicitly provided for if desired. The problem really comes down to the fact that existence is a 3-valued entity: the uninstantiated concept of an atom, the positive occurrence of that atom and the negative occurrence. For example, if we consider an object, we know that it has the possibility of being blue. If it is blue, then we can imagine a token in a place to represent this, but if we know it is not blue this also should be represented. In other words there are three possible states: true, false, and unknown. In a representation with 3 states, consideration of contradiction is required, and must be addressed in the primitives.

A Petri net implementing the desired function of this primitive is shown in Fig. 2. This network module has inputs from external to two symmetric subsystems. One subsystem represents instantiation of the atom in the positive sense, the other represents instantiation in the negative. Once an input token arrives, it is multiplied into two tokens: one will go on to reflect the state of the module to external connections, the other remains internal to check for contradiction. If the atom is instantiated in both cases, a token will be passed to the contradiction output, and will lock the module by inhibiting the outputs, but does not effect tokens already distributed. The other path passes tokens through a delay to give priority to the detection of contradiction, then multiplies this to a place on every output. This means that each output will only be able to use each atom once, e.g. only one generation of a specific inference is possible. Although only one input is shown, each input leading to the atom uses a transition to distribute tokens to the two paths.

This Petri net system will be represented in shorthand as the rounded box in Fig.3, which uses the top portion of the box as inputs leading to the "true" or "positive" part, and the bottom of the box will represent the "false" or "negative" instantiation of the atom. The contradiction output is not shown; it is assumed that this will be handled by the system.

The question of how to handle the contradiction token output from the module is an application dependent decision. In the system described, contradiction will lead to a deadlock in the atom, removing it from further processing. Another option is to allow both to exist, and may be a way of representing uncertainty, a notion for future exploration. The viewpoint leading to the representation chosen is that contradiction is an important way of establishing database (in)consistency, detecting potential problems in the rules or in the input data. In the case of logic, using a *negative* assertion to achieve contradiction establish truth of the positive, so in this case contradiction is the desired result, and a completion condition. However, this is a design decision; in the system shown, if the contradiction indicator is

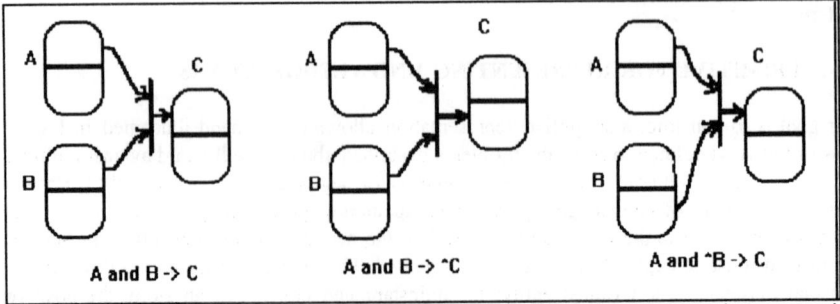

Figure 1: Examples of representing production rules with Petri Net primitives

A and B -> C   A and B -> ^C   A and ^B -> C

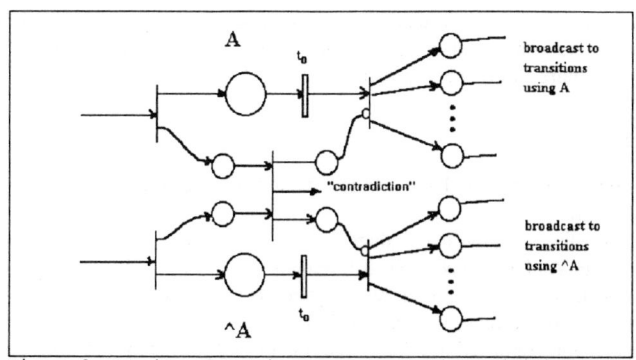

Figure 2: Details of Petri net representation primitive

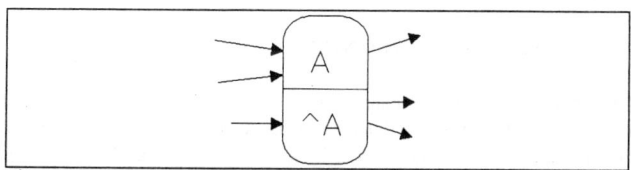

Figure 3: Petri net representational primitive, with positive and negative instantiation

ignored, processing will continue in a limited way: all of the output links remain active with their values prior contradiction, but the unit is locked and no additional data will be propagated. Note that each output place is expected to be used only once in a specific inference, so one token in each place is sufficient to satisfy all possible inferences using this atom. This mechanism as a whole prevents redundant operation of transitions, which corresponds to redundant inference.

3.2 REPRESENTATION OF KNOWLEDGE

As illustrated in Figs. 1-3, the coding of a rule in a Petri net format is simply the organization of the network so the tokens in the input places of the rule will cause a transition to fire. The transition represents the rule, and firing is applying that rule. The conditions of the rule are the inputs, and when they are satisfied, the transition fires, distributing tokens to the output places, representing the action of the rule. Since the database may be structured to use only conjunction in the condition clause (Horn clauses), a compound statement like "If (A and B) then C" is simply represented by a transition with two inputs, as shown in Fig. 1. Multiple actions are multiple outputs of the transition.

The methodology of this representation system will be to translate the rules into a Petri net as shown above. The Petri net is initialized to a marking according to the initial facts. Then, the system is run until either a place representing a goal state is satisfied, or the Petri net is dead, implying no more inference is possible.

3.3 EXAMPLE

The Petri net methodology will be illustrated by a simple example taken from Reichgelt (1991). The rules are:
1) If a car is a Ferrari, then it is Italian
2) If a car is an Alpha Romeo, then it is Italian
3) If a car is Italian, then it probably has good acceleration
4) If a car is a Volkswagen, then it is German
5) If a car has not been used in a long time, then it will probably have difficulty starting
6) If a car is Italian AND the weather is cold, AND it is not stored in a garage, then it will probably have difficulty starting

The Petri net for this set of rules is shown in Fig. 4. Each concept has been represented as a primitive logic element.

3.4 FORWARD AND BACKWARD CHAINING

The definition of the net to this point is analogous to forward chaining rules in a database. The advantage of backward chaining in a database is to avoid a possibly extreme number of "useless" inferences that may be generated through extraneous (to a particular context) rules. Backward chaining is a method of starting from the goal state, and branching back to possible sets of initial conditions. The goal of the Petri net system is to implement massively parallel active databases, which eliminates many drawbacks of forward chaining. However, backward chaining style operation may be generated by manipulating the rules themselves to generate

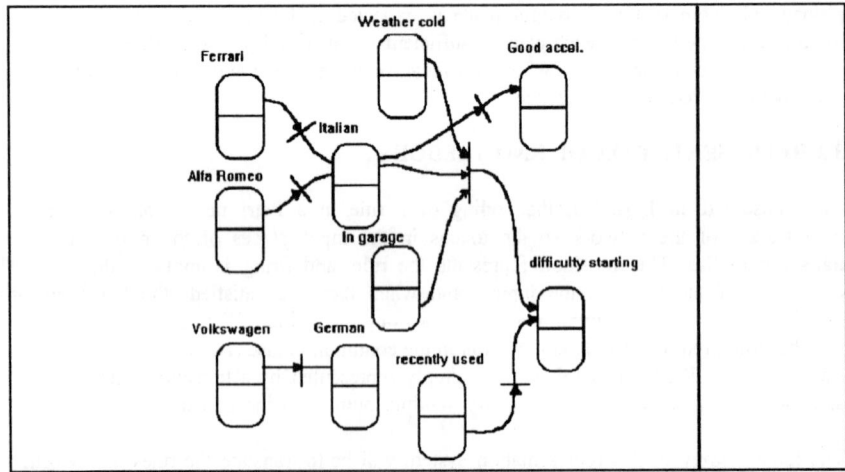

Figure 4: Petri Net implementation of production rule database

a set of "backwards" equivalent rules. This is reflecting the Petri net operation so that transitions remove consequents (actions) and distribute antecedents (conds.). Although this remains to be explored, a Petri net to implement similar but backwards function to the primitive detailed above is imaginable. A possible feature would be to use color tokens at branching points to represent alternative condition sets, in the case where there may be multiple initial conditions leading to a conclusion. To illustrate, in the above example, if the goal state is "difficulty starting", backward chaining would be used to generate the possible sets leading to this condition. In the example there are two possible main branches, namely rules 5 and 6. The backward chaining Petri net would take the double input to the conclusion now as a double output, each one of which will be given a different colored token. Then this color is propagated back. Each consecutive branch will add another dimension to the token color.

Another direction for exploration is the hybrid inferencing of specifying the pieces of information or evidence available, and completing the path which uses this data. This sort of searching and planning technique, explored by Calistri's (1991) Pathfinder program, uses an A* style searching technique to generate plans from incomplete information.

4. PETRI NETS and LOGIC

Logic encoding will be discussed with reference to the system of first-order predicate calculus.

4.1 EXTENSION OF PRODUCTION RULES TO LOGIC

A production rule:
    If A and B then C
is equivalent to the predicate calculus expression:
    $A \wedge B \rightarrow C$
for properties A, B and C. What has been implicit in production rule system described above is the instantiation of the properties, usually because the operation of the system regards one individual at a time. For example, a medical diagnosis tool has an implicit assumption that the information is regarding a specific patient, or a computer configuration expert system is concerned with one specific computer system at a time. To be more general, the proper predicate calculus expression of the above is:
    $(\forall x)[A(x) \wedge B(x) \rightarrow C(x)]$,
where x is a variable. In the case of the expert system, there is only one instantiation in the system, for example representing (as above) the current patient, or computer system. Expert systems may allow several mappings in the system at once, corresponding to the general predicate calculus instantiation. To allow this, colored tokens may be used to represent each instantiation.

The existential quantifier may be removed through the skolemization process, which is the replacement of the existential variable with a constant, or with function symbol for nested quantifiers. This would be indicated in the Petri net system as an initial marking in the system of a color-instantiated token in a particular place.

4.2 REPRESENTATION OF FOPC: PREDICATE TRANSITION NETS

Once skolemization is applied to a predicate calculus database, the resulting representation is a set of expressions in disjunctive normal form (DNF). Inference in DNF logic is a matter of matching parts of these expressions, possibly through substitutions, to generate inferences. Simple clauses of the form:
1) $P(x) \lor Q(x)$
2) $R(x) \lor S(x) \lor T(x)$,

may be decomposed, according to the possibilities of applying resolution, into the following rules:
1) If $\neg P(x)$ then $Q(x)$
   If $\neg Q(x)$ then $P(x)$
2) If $\neg R(x)$ and $\neg S(x)$ then $T(x)$
   If $\neg R(x)$ and $\neg T(x)$ then $S(x)$
   If $\neg S(x)$ and $\neg T(x)$ then $R(x)$.

These conjunctive implication clauses can then be implemented by our Petri net primitive, as shown in Fig. 5.

The next aspect of logic that will be approached is the inclusion of multiple variables in a predicate, for example the database statement:
3) $R(x) \lor Q(x,y)$

or:
3) $\char`\^R(x) \implies Q(x,y)$
   $\char`\^Q(x,y) \implies R(x)$

Encoding this statement requires more consideration than the simple properties above. This relation is universally quantified, so any instantiation of either variable is correct. This means regardless of the instantiation of the two arguments, the relation Q will be correct, and so the pairing is not critical. However, the intention of the predicate calculus database is that the instantiation will only be used if it is lead to by resolution (implication; inference). Also, the above example is too simple, e.g. consider the clauses:
4) $Q(x,y) \lor R(x) \lor S(y)$

or
5) $Q(x,y) \lor R(x,y) \lor R(y,z)$

Proper resolution of these clauses requires maintaining the <x,y> pairs.

Encoding complex forms of predicate calculus has been addressed by predicate transition (P/T) networks of Genrich and Lautenbach (1979), where the P/T networks have been shown to be functionally equivalent to predicate calculus. P/T nets have an additional functionality to Petri nets as defined so far. This functionality is manipulation of colored tokens in a much more complex way. In database statement 3, the Q relation requires the capability to consider the tokens as an ordered pair, <x,y>. It is straightforward to define the tokens as having color interpreted as an ordered pair, but the token must also be decomposable into its component parts for comparison with other clauses for resolution, e.g. the property R. In this case, the R place will deal with tokens of a single value, <x>. The burden is then on the transition to

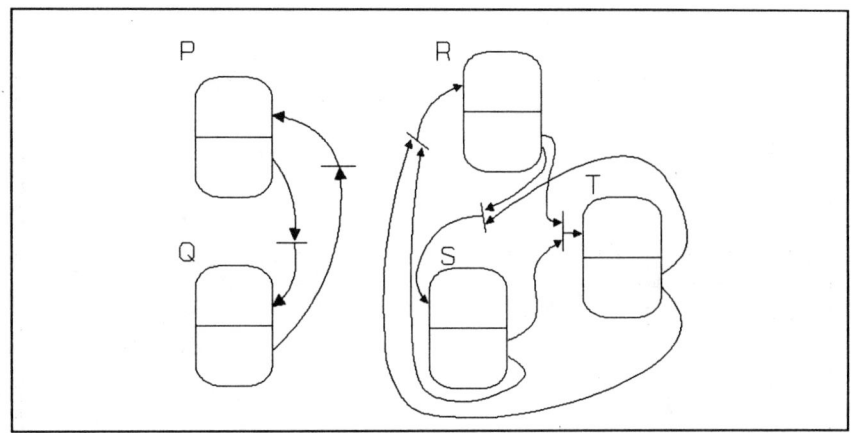

Figure 5: Petri net implementation of simple logic database, P(x) v Q(x), R(x) v S(x) v T(x)

make the correspondence, for example taking a pair <a,b> as input, and sending a token <a> as output. The convention for PR networks is to label the transition with the desired input/output substitution, for example in Fig. 6. Transitions are labelled to indicate the input/output function. In Fig. 6a, the transition takes any input from the input place, and distributes 2 to the output place. Fig. 6b show the operation on a relation, where the input place is populated by pairs <x,y>, and any input pair will be distributed to the output as a pair <x,x> (the second value of the input pair is not used). Fig. 6c shows a transition that links a place representing a relation to a place representing a property, and 6d illustrates multiple distribution.

The extension of these networks to the logical primitive unit (Fig. 2) is an obvious enhancement. The interior of the unit will operate as the P/T net, with the appropriate tuple for detection of contradiction and the distribution of tokens to the output.

In the P/T net, the elements of the token n-tuples may be either constants, or function constants, which is required for equivalence to first-order logic. The level of manipulation of logical construction will correspond to the level used in the transition. The operation of this is explored in the example of circuit function simulation, below.

5. PETRI NETS and SEMANTIC NETWORKS

There are obvious similarities between graph representations such as semantic networks, and Petri nets. Petri nets can easily implement association by spreading activation, the simplest function of semantic networks. By using color tokens and transitions, a network can be constructed to allow propagation of tokens while keeping track of their distance from the initially activated nodes. For example, the transition can consume tokens with an integer color

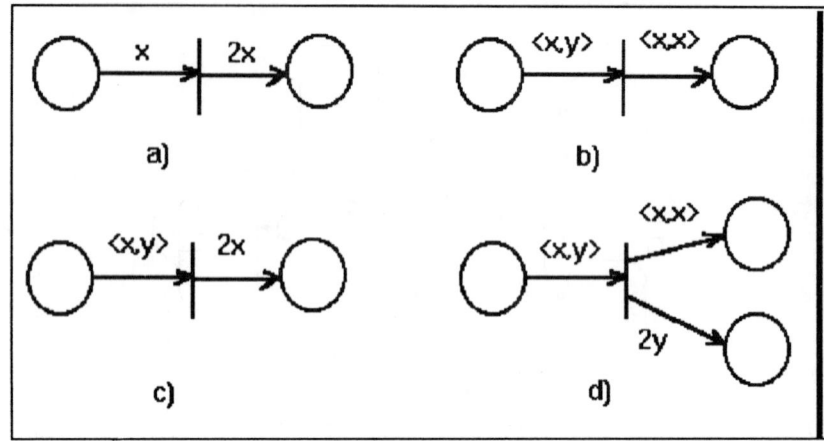

Figure 6: Examples of Predicate Transition (P/T) networks

i, and propagate tokens with the color $i+1$. Thus, distance can be established. Also, the number of separate paths to an item can be observed by the number of tokens in a place. If the goal is to detect relation between knowledge atoms, the search through spreading activation may take a form of two or multiple point constraint search, as in the Pathfinder system by Calistri (1991) mentioned above.

Although there is a clear correspondence between the Petri net and the knowledge representation style used in semantic networks, the goals generating a methodology for convergence are not clear. The semantic networks may store information in a precisely defined way, but the actual manipulation of this information in inference or computation is less defined; mostly they seem to embody the intuitive concept of associational representation. The main goal of the Petri net knowledge representation methodology is not only to provide a new way to structure knowledge, but most importantly to generate a causally active system that can perform useful computation: inference, planning, etc. The previous paragraph points out the correspondence between spreading activation and "closeness" determination between two knowledge atoms in a semantic network, and if this is the goal of the system definition ends here.

The main application of semantic networks may be to provide a heuristic for resolution of predicate-calculus style encoded knowledge. Considering resolution in a large database, the number of possible comparisons becomes very large, so resolution strategies indicate a subset of the possible inferences. In linear resolution, for example, the two clauses being compared must either be from the initial database, or one must be an "ancestor" inference of the other, thus keeping a constant line of descent. The tradeoff with a resolution strategy such as linear resolution is that because it greatly speeds inferencing by cutting down on the space of possibilities, it may also restrict away a critical inference necessary to generate the conclusion. Semantic networks may be seen as a similar methodology: they provide a map between concepts and so guide the inferences in a way to only try resolution between "related" clauses, where relatedness is the heuristic provided by the network.

In the case where the semantic network is to be used as a heuristic resolution strategy, the application to Petri net knowledge representation is clear. When the Petri net was generated from a predicate calculus database, as discussed in the last section, the network was structured to allow the flow of the tokens through the network to generate any possible inference. Application of a semantic network is to restrict the network topology, allowing fewer inferences, but implicitly introducing knowledge in the form of preferred inferences.

Semantic networks are also a way to indicate hierarchy, or default reasoning. This is a structuring of the network of general categories to specific categories. Using this in determining characteristics of a node by using default characteristics of the general type can be performed by spreading activation. More complicated methods of assigning defaults we consider to belong more properly to the frame-based knowledge representation methodologies, which will be considered in future exploration.

6. EXAMPLES of STRUCTURED COMPUTATION USING PETRI NETS

6.1 OBJECT RECOGNITION

This example will show the use of the nets for object recognition based on high-level object features using identification rules as a framework. The basic problem is: given an input of primitives extracted from an image, recognize high-level constructs. In this case, the primitive constructs will be:
 - line segments,
and the primitive properties will be:
 - parallel
 - adjacent
 - equal (length).
We can describe objects using these primitives:
 triangle:
  3 line segments, each adjacent to 2 others.
 isosceles triangle:
  3 line segments, each adjacent to 2 others, two equal,
  ==> triangle with two sides equal
 equilateral triangle:
  3 line segments, each adjacent to 2 others, all equal
  ==> isosceles with third side equal
 parallelogram:
  4 line segments, each adjacent to exactly 2 others, each equal to 1 other, each parallel to one other.
 rectangle:
  4 line segments, each adjacent to exactly 2 others, each equal to 1 other, each parallel to one other, all adjacent at right angles (or, one adjacent at right angles)

==> parallelogram with adjacent right angle
square:
    4 line segments, each adjacent to exactly 2 others, all equal, each parallel to one other.
==> rectangle with adjacent sides equal

The network for recognizing a triangle is shown in Fig. 7. Note that this network structure utilizes two pieces of information not in the original formulation, relating to definition of the primitives. First is the fact that adjacent and equal are symmetric properties, which is taken into account by the input transitions, which generates the symmetric relation when the fact is stored. This means that every possible configuration of an object will be generated, e.g. if a triangle <a,b,c> is deduced, then so will <a,c,b>, <b,c,a>, etc. This could also have been implemented by parallel transitions covering the possible combinations. The second information is the hierarchical nature of the categories, which was noted in the definitions. A net for recognizing the 4-sided figures is also shown in Fig. 7, where the primitive nodes (adjacent etc.) have been duplicated for ease of representation. It seems likely that useful image recognition can be built from a small set of relational primitives like these, without the necessity of quantitative calculation.

Note that a set of production rules, as well as a set of predicate calculus expressions can be abstracted from this net.

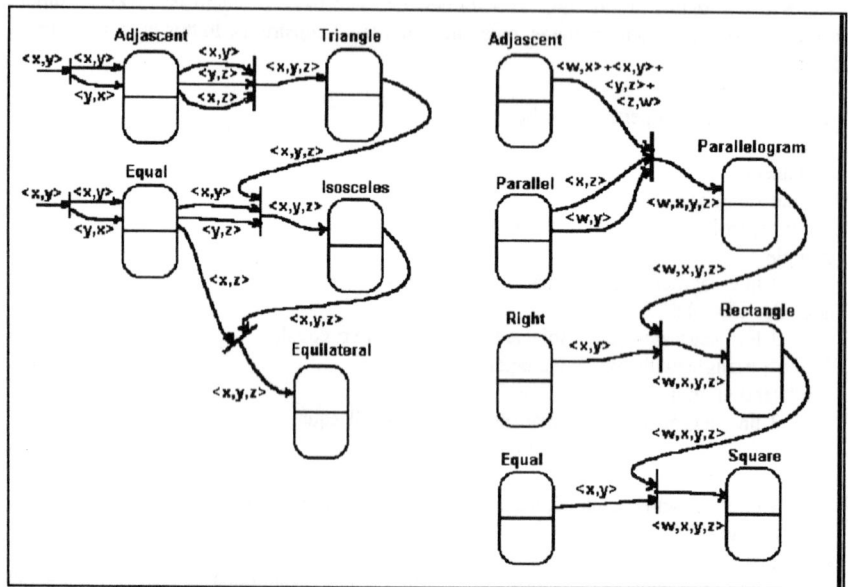

Figure 7: Recognition of 3- and 4-sided figures from primitive features using PT nets

6.2 PLANNING

This section will discuss previous work in this area: Bourbakis and Tascillo (1994). In this work, a method for planning in a system where the constraints have been represented in a stochastic Petri net has been discussed. This is based on the observation that the paths from an initial Petri net marking to a final marking are represented by the reachability tree. Thus, path planning is equivalent to deciding which paths to follow down the reachability tree. In the example, a Petri net was generated that represented the constraints of *two* robot grippers operating simultaneously in a "blocks-world" environment. An ART neural network was used to characterize the decision to be made at the stochastic distribution sites (the stochastic Petri net is one with multiple outputs from a node, so that it must be decided which transition will fire). In this way, the behavior of the system in examples was characterized based on the current network state. When the stochastic distribution sites were governed by these rules, the Petri net simulation provided good performance on new problems.

6.3 PREDICATE CALCULUS CIRCUIT SIMULATION

An example of predicate calculus applied to circuit characterization and simulation from Genesereth and Nilsson (1987) will be used. First, the input/output behavior of building blocks is defined. Then, the connections of the circuit are defined, and finally the database is given specific instantiation of the inputs to the circuit. Through inference, the final state of the circuit is generated. The example circuit (a full adder) is shown in Fig. 8. The primitive blocks making up the adder are AND, OR, XOR, and the usually implicit CONN, or connection. The inputs and outputs of a block are numbered from the top down.

The logical primitive definitions are:
  Adder(x) - x is a full adder
  Andg(x) - x is an AND gate
  Org(x)   - x is an OR gate
  Xorg(x) - x is an XOR gate
  I(x,y)   - input x of (block) y
  O(x,y)   - output x of (block) y
  V(x,y)   - signal at entity x is y
  Conn(x,y) - connection between entity x and y
          note that entities should be I() or O().

The logic specifying functionality of the primitives is:

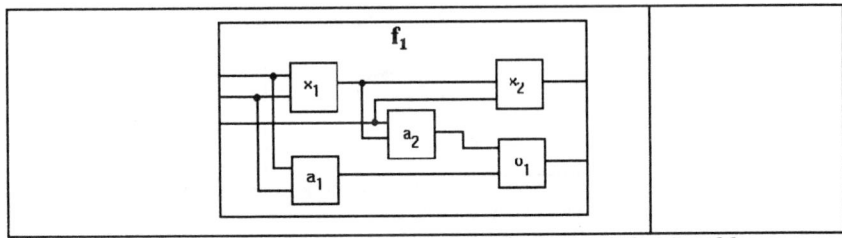

Figure 8: Primitive logic blocks making up a full adder

$\forall x$ (Andg(x) ^ V(I(1,x),1) ^ V(I(2,x),1) ==> V(O(1,x),1))
$\forall x \forall y$ (Andg(x) ^ V(I(y,x),0) ==> V(O(1,x),0))
$\forall x \forall y$ (Org(x) ^ V(I(y,x),1) ==> V(O(1,x),1))
$\forall x$ (Org(x) ^ V(I(1,x),0) ^ V(I(2,x),0) ==> V(O(1,x),0))
$\forall x \forall y$ (Xorg(x) ^ V(I(1,x),y) ^ V(I(2,x),y) ==> V(O(1,x),0))
$\forall x$ (Xorg(x) ^ V(I(1,x),1) ^ V(I(2,x),0) ==> V(O(1,x),1))
$\forall x$ (Xorg(x) ^ V(I(1,x),0) ^ V(I(2,x),1) ==> V(O(1,x),1))
$\forall x \forall y \forall z$ (Conn(x,y) ^ V(x,z) ==> V(y,z))

The problem specific database (circuit description) is:
Adder(F1), Xorg(X1), Xorg(X2), Andg(A1), Andg(A2), Org(O1)
Conn(I(1,F1),I(1,X1))   Conn(I(2,F1),I(2,X1))
Conn(I(1,F1),I(1,A1))   Conn(I(2,F1),I(2,A1))
Conn(I(3,F1),I(2,X2))   Conn(I(3,F1),I(1,A2))
Conn(O(1,X1),I(1,X2))   Conn(O(1,X1),I(2,A2))
Conn(O(1,A2),I(1,O1))   Conn(I(1,A1),I(2,O1))
Conn(O(1,X2),O(1,F1))   Conn(O(1,O1),O(2,F1))

A network which demonstrates implementation of this logic is shown in Fig. 9. This shows only the implementation of one Andg rule, and the Conn rule, but is sufficient to illustrate construction. This example demonstrates one additional of the network that is required for operation: the place V (value) may contain colored tokens of the form <O(block,rank),value>. In the transition representing the Andg rule, the variable in this transition is the block identifier within the first quantity, giving a wildcard rule <O(x,1),1>. This means that only those doubles instantiating V with the form given in the transition rule will be operated on. However, in the Conn rule, the entire first quantity is a variable, indicating that any instantiation of the first quantity in that position will be allowed. This capability of the transitions is required to allow general inference.

The example network whose construction is illustrated by Fig. 9 is not what people will envision when using Petri nets to model circuit behavior. Specific limitations of the database

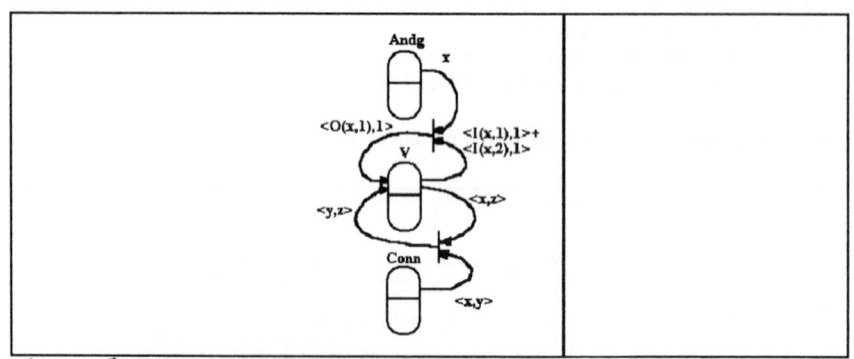

Figure 9: Demonstration of P/T network implementation for the circuit characterization example

structure in the example include the limitation that this is for feed-forward circuits only, e.g. each connector can change only from uninstantiated to instantiated. Change between values will yield contradiction in the database. Note that although the positive/negative instantiation primitive is not used because of its absence in the framework of logic, the circuit could be implemented using the positive/negative places in V to represent the on/off values of logic, making place V a function rather than a relation.

This example shows an optimally generic medium for simulating logic circuits of any configuration (within limitation discussed above). In this case, the network structure contains only the information about the generic behavior of circuit building blocks, and in logical form contains no redundancy. Parallel simulation both encodes circuit specifics in the structure of the network, and produces redundancy (function is duplicated for e.g. every AND gate). A method for implementing this is to consider the initial database instantiation of circuit-specific information as places, and so each block edge is modelled by a knowledge atom. For example, AND gate A1 will have 3 parts:
I(A1,1) - Input 1 of A1,
I(A1,2) - Input 2 of A1, and
O(A1,1) - Output 1 of A1.
These parts are connected in the circuit according to the connection rule (not pursued here), and the Andg function rules, as illustrated in Fig. 10. This illustrates a general tradeoff in the representation of knowledge, between specific/parallel/redundant, and general/serial/concise.

7. DISCUSSION

Since this paper represents some initial work in developing a comprehensive Petri net knowledge representation methodology, some comment on the insights gained from the project so far are the most important results.

The concept behind representing structural knowledge in Petri nets is to allow not only the representation, but also the inference and manipulation of that knowledge as well as the original form. Production rules are perhaps the most suited to this representation. The action

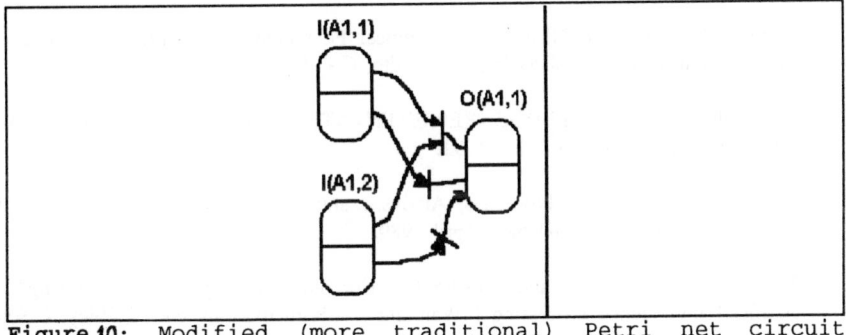

Figure 10: Modified (more traditional) Petri net circuit simulation structure

of the transitions and knowledge atom places is a natural fit to the chaining action of the rules. Extensions of production rules, such as the inclusion of truth or certainty values in the application of the rules seem to be possible through enhancement of the transition function, but are yet to be explored. The more complex task of encoding general logic so that the inferencing capability is not impacted is somewhat more difficult. The framework above shows that databases of simple functions and implication can be simply transferred to color Petri nets. However the inclusion of relations provides a difficulty that was surmounted by the extensions of predicate transition nets. Even these much richer building blocks may have difficulty encoding complex nested relations. Further examination of this encoding methodology will be undertaken, with the goal of reducing all of the functionality of general predicate calculus to colored Petri nets with simple transition functions.

The examples illustrate that this methodology is not just a formal representation format, but examines the notion that these building block can form a realizable distributed system to perform useful computation with the complexity of logical inference.

8. CONCLUSION

These analyses of conversion of knowledge forms into Petri nets is the first step in providing a comprehensive knowledge representation language, the step of encoding structural knowledge. These forms are defined as structural knowledge because the descriptive character of the knowledge.

The second step in this methodology is the integration of this system with more traditional Petri nets used to represent functional knowledge. Functional knowledge can be defined as the knowledge describing changing state. Petri nets have been largely a modeling tool, and so this aspect of representing knowledge has been explored. Concluding this effort will be integrating the structural and functional representations methodologies with a query system in a functional database.

REFERENCES

1. Genesereth, M.R., and Nilsson, N.J., Logical Foundations of Artificial Intelligence, Morgan Kaufmann Publishers Inc., Palo Alto, CA, 1987.

2. Reichgelt, H., KNOWLEDGE REPRESENTATION: An AI Perspective, Ablex Publishing Corp., Norwood, NJ, 1991.

3. Murata, T., "Petri Nets: Properties, Analysis, and Applications", Proceedings of the IEEE, vol.77, no.4, pp. 541-580, April 1989.

4. Wah, B., Lowrie, M.B., and Guo-Jie Li, "Computers for Symbolic Processing", in Artificial Intelligence Processing B.Wah and C.V.Ramamoorthy eds., J.Wiley & Sons, 1990.

5. Calistri-Yeh, R.J., "An A* Approach to Robust Plan Recognition", in Applications of Learning & Planning Methods, N.G.Bourbakis ed., World Scientific Publishing Co, Teaneck, NJ, 1991.

6. Genrich, H.J., and Lautenbach, K., "The Analysis of Distributed Systems by Means of Predicate/Transition-Nets", Semantics of Concurrent Computation: Lecture Notes in Computer Science 70, pp. 123-146, Springer, Berlin, 1979.

7. Bourbakis, N.G., and Tascillo, A., "An SPN Planning Methodology with Neural Networks", Proceedings of the IEEE International Conference on Neural Networks, June 26-July 2, 1994, Orlando, FL, IEEE Press.

8. Bourbakis N.G.,"A neural knowledge base using SPNs", IJAIT 1996.

ON THE DEEP STRUCTURES OF WORD PROBLEMS AND THEIR CONSTRUCTION

FERNANDO GOMEZ
Department of Computer Science
University of Central Florida
Orlando, Fl 32816
USA

ABSTRACT

The relation between problem-solving and comprehension is studied. It is shown that word problems have a deep structure, which needs to be revealed if these problems are to be solved in a principled manner. It is also indicated that the comprehension of a word problem is not a frame-based process, but rather is a constructive process by means of which the deep structure is built from the sentences describing the problem, in a way similar to the construction of a house from the building materials. This constructive process is done by means of integration rules, which build the deep structure on the basis of the semantics of the logical forms of the sentences and on the deep structure under construction.[1]

1 Introduction

This paper deals with an aspect of the problem of automating the comprehension of expository texts, namely understanding word problems. It will be shown that problems have a *deep structure* similar to the deep structure of a sentence, and that the task of the comprehender is to reveal and build this deep structure. It will be indicated that the deep structure built from the natural language description of the problem has in it all elements necessary to answer a simple information retrieval question, an arithmetic question or an algebra question. The idea of deep structure of a problem is not entirely new. It can be traced back to an earlier paper by Paige and Simon [13] in which they study the behavior of subjects solving word algebra problems and compare it to Bobrow's STUDENT program. They distinguish between "direct" and "auxiliary" translations of natural language expressions. By "direct" translation the authors refer to the direct mapping of an English sentence into an algebraic expression. For instance, the sentence *Three times a number plus 18 is equal to 78* will be directly mapped into $3x + 18 = 78$. This is the behavior that characterizes

---

[1]This paper is an extended abstract of a technical report with the same title.

STUDENT and some of the subjects they studied. However, some kinds of problems require "auxiliary" representations such as the construction of diagrams that capture the physical situation described by the problem. Later in the paper, they say:

> Let us propose a hypothesis. We suppose that the subject constructs an *internal representation* of the problem situation from the problem statement. We do not insist that this representation be "visual" in any literal sense, but we do require that it contain in implicit form the same relations that are implicit in the diagram (p. 273).

The authors clearly establish that some subjects use internal representations that are functionally equivalent to diagrammatic representations depicting the physical situation described by the problem. The notion of internal representation of a problem, now called *deep structure* of a problem, is further discussed in chapters 2 and 3 of *Human Problem-Solving* [11]. In chapter 3, the authors assert:

> In contrast [... to the theories of deep structure sketched by linguists], the internal structures we shall postulate for problem solving situations generally constitute large, complex, interrelated contexts that do not factor out in any simple way into components that are isomorphic with single sentences.

It is clear that the authors maintain that their notion of internal structure goes beyond the meaning of individual sentences. The internal structure the authors postulate for problem solving is the notion of *problem-space*. One of the main points of this paper, however, will be that the internal representation of a problem stated in natural language is a structure that *mediates* between the English description of the problem and the *problem-space* representation, and that it cannot be identified with the *problem-space* representation of a problem. In our opinion, the *problem-space* representation derives from the deep representation, which is semantically based. The *problem-space* representation underlies many problem-solving methods: logic, hill-climbing, means-end analysis, generate-and-test, etc. In a *problem-space* representation, the semantic content of the terms that make up the problem has been removed. A river is not any longer a river in a *problem-space* representation, but just a token. Paraphrasing Hilbert in his description of the formalist program for the foundations of mathematics, one may say that the meaning of the terms in the *problem-space* have been reduced to stains of ink. Even in a domain as clearly semantic as medical diagnosis [3], many problem-solving methods have been also syntactical. In most expert systems, terms such as "fever," "inflammation," etc., are just tokens with no meaning for the problem-solver. The mapping of a problem stated in natural language into any of these *problem-space* methods is not only a hard task, but one that in most cases is only possible if one knows a solution prior to constructing the *problem-space* representation. It is also extremely interesting that different people provide very diverse representations within a given method; e.g. different axioms in the case of logic or different operators in the case of means-ends analysis. This, in my opinion, indicates that the underlying structure of the problem cannot be identified with the *problem-space* representation. It is the internal representation what makes possible the construction of the *problem-space*.

The other main point of this paper is that the comprehension of a word problem is not a script-based process, but is a process consisting of constructing the deep structure of the problem from the logical forms of the sentences in a way similar to the construction of a house from the building materials. It is a constructive process, not filling holes in a pre-established structure. This paper is organized as follows. Section 2 argues that the comprehension of word problems is not a schema-based process. Section 3 provides examples and representations of deep structure based on the concept of *classification*. Section 4 explains the *deep structure* of motion problems based on the notion of *subevent* and other temporal links. Section 5 shows how the *deep structure* of motion problems can be constructed from the logical form of the sentence, by using integration rules. Section 6 outlines how the solution of the problem can be obtained from the *deep structure*. Finally, section 7 gives our conclusions.

2 Schema-Based Understanding

The AI paradigm for understanding problems stated in natural language has been the same as that for understanding narratives, namely frames, scripts or schemas. Schema-based understanding is a top-down process with two main components: schema recognition and schema instantiation. Once a relevant schema is recognized, understanding proceeds by instantiating the slots of the schema with the concepts in the sentence. Several programs have been built using the notion of schema-based understanding [12, 4, 14]. There are serious limitations with this approach to understanding problems. First of all, the recognition of the relevant schema and the actual instantiation of the slots are problems for which solutions have been found only in the simplest situations. Schema-based understanding can be considered - with serious reservations - as a model of the understanding of a problem by an expert. The programs that embody schema-based understanding have compiled in them a tremendous amount of knowledge, not only about how to integrate different parts of the sentence into the schema, but also about the final representation of the problem on the basis of just a few initial clues in the first sentence of the problem. It is certainly the case that a good student who has solved a considerable number of word problems can recognize a type of problem by reading the first sentence of a problem. Yet, it is also true that the same student can be taken aback by the formulation of a problem, realizing only in the last sentences that he/she is mistaken about the type of problem. Certainly, experts are not just a compiled mass of knowledge, but they are able to deal with cases that deviate from stereotypical situations. This is true not only of problem-solving, but also of comprehension.

Although the understanding of some of these problems can be done by identifying a relevant schema and instantiating it, the comprehension of most problems can not be schema-based. There are several reasons for this. First of all, the clue identifying the relevant schema may occur too late in the problem to be of any value. Consider the following problem describing an overtake situation:

> Leesburg is on the route from Orlando to Tampa. The distance from Orlando to Leesburg is 200 miles. The distance from Leesburg to Tampa is 100 miles. Peter left Orlando to go to Tampa by car at 55 mph. Mary left Leesburg to go to Tampa at the same time. She drove at 25 mph. Will Peter overtake Mary before she gets to Tampa ?

Note that until the question is posed, there has not been a good way to find out what schema this is. Perhaps, it could have been suspected in the fifth sentence, but that is too late to make any sense of the first four sentences. Redundant information is another aspect of problem comprehension which would be very hard to accommodate in a schema-based system. Children are frequently given problems with irrelevant information in order to test their problem-solving skills. These "noisy" problems pose serious difficulties for the children. The reason for this is that these problems deviate from stereotypical problems. If a child has learned a rigid schema which is good for at most three or four problems, the child is unable to match redundant or irrelevant information to the schema.

Our research on the comprehension of expository texts has centered on the investigation of understanding in the face of limited knowledge. In [5, 6], we have proposed a knowledge-lean model of comprehension. The comprehender of our model is a layman who reads the texts with limited knowledge about their content. Since the reader is acquiring new concepts, the schema-based understanding model can barely be applied in this context. Instead, comprehension is a bottom-up process by means of which new concepts are built. In our model, concepts are constructed by means of *formation* rules, which build concepts from the logical form of the sentence. Once these concepts are built, a process, called *recognition*, is invoked in order to check if the concepts that have been built are already in long-term memory (LTM). Those concepts that remain unrecognized are passed to an *integration* phase, which is responsible for integrating them in LTM. The front end of SNOWY, the system that incorporates these ideas, is a general purpose parser, called WUP (word usage parser), based on the notion of syntactic usage of a word. WUP produces a shallow parse of a sentence, leaving the attachment of prepositions and relatives, the determination of the verbal concept and thematic roles to the *interpretation* phase of SNOWY. The *formation* phase is also a general purpose algorithm producing an output that is the same for a variety of tasks in the comprehension of expository texts (e.g. acquiring knowledge for an expert system [7], understanding an elementary scientific text, solving a word problem, etc.). The *recognition* algorithm is also the same for all these tasks. The *integration* algorithm is the only one that changes with a given task. These changes are not major ones, and are mainly due to the different way in which events, e.g.

Peter went to the theater, and general atemporal facts, e.g. *All whales live in the sea* are integrated.

3 The Deep Structure of Problems

Consider the following problem:

> Problem C1: All students who went to the theater went to the game. Twenty students went to the theater. Did at least twenty students go to the game?

In spite of its simplicity, this problem can hardly be solved by using production rules that would match the surface terms of the sentences. The problem has an underlying structure that needs to be revealed if it is to be solved in a principled manner. The underlying structure of this problem is one of *classification* in a sense explained below. Consider the first sentence of problem C1 *All students who went to the theater went to the game*. In [8], we have shown how to represent restrictive relative clauses as classification hierarchies. The clause *Students who went to the theater* is represented, or formed, by creating a concept with a dummy name, say x1, but containing a slot, called *cf*, that identifies the concept by providing necessary and sufficient conditions. The representation of this concept is:

```
X1
 (cf(is-a(student) actor(x1(q(all))) primitive(ptrans)
     theme(x1(q(all))) to-loc(theater-a)))
```

We will use the primitive *ptrans* throughout the paper to mean the transfer from a location to another location by an animate being [15]. The content of the *cf* slot says that *x1* is a subset of students who went to a theater. The precise meaning of the *cf* representation in FOPC is:

$$\forall(x)(X1(x) \iff Student(x) and R1(x))$$

where R1 is the relation "x went to the theater." During *formation*, the name *X1* replaces the restrictive relative clause in the main clause, which, in our case, yields *All x1 went to the game*. This in turn is represented (formed) by building a relation predicated from all X1 as indicated in Figure 1. Concepts denoting entities (student, book, x1) are called *object-structures*. Conceptual relations denoting actions or events are represented in the *object-structures* by indicating the relation followed by the *theme* (if any) and a pointer to the representation of the action structure. In Figure 1, *a1* is an action structure. The conceptual relation is fully represented in *a1*. The action structures can be connected to other action structures. The letter

```
              student
                │ is-a
                x1
[cf (is-a (student)) (actor (x1 (q(all))) (prim (ptrans) theme (x1 (q (all)))
     (to-loc (theater-a)))]
  (cardinality (20))
  (ptrans (game-a a1))
                │
                ▼
             a1
             args (x1 x1 game-a)
             prim (ptrans)
             actor (x1 (q (all)))
             theme (x1 (q (all)))
             to-loc (game-a)
```

Figure 1: Deep structure of problem C1.

q is the quantifier of the concept immediately following it. Concepts that are not followed by a quantifier are individuals or constants, e.g. theater-a, game-a, Orlando, etc. The scope of the quantifiers is from left to right. (See [8] for an explanation of the meaning of these structures using FOPC.)

Existentially quantified sentences, e.g., *Some people like music* and also sentences whose *subject* or *actor* has a numeral in it are also represented as classification hierarchies. Hence, the sentence *Twenty students went to the theater* is represented by creating a class, say X2, whose characteristic features are "is-a student and went to the theater," and the cardinality of the class is twenty. Formally:

```
X2
  (cf(is-a(student) actor(x2(q(all))) primitive(ptrans)
      theme(x2(q(all))) to-loc(theater-a)))
  (cardinality(20))
```

When X2 is going to be integrated in memory, the *recognizer* algorithm is activated and realizes that X2 already exists in memory, namely the concept X1 created above. Hence, the result of integrating X2 in memory is simply to insert the slot *cardinality* with the value twenty in the concept X1. Now, the answer to the question *Did at least twenty students go to the game?* is obtained by descending to the concept *X1* from the concept *student* and noticing that the quantifier of *X1* is a universal quantifier and the cardinality of the class *X1* is greater than or equal to twenty. Note that if the first sentence of the problem had been *Twenty students went to the theater*, the

y1
(cardinality (100))
[cf (is-a (people))
    (actor (y1 (q (all))))
    (prim (ptrans)) (theme (y1 (q (all))))
    (to-loc (theater-a)))]
(pay (dollar a1))

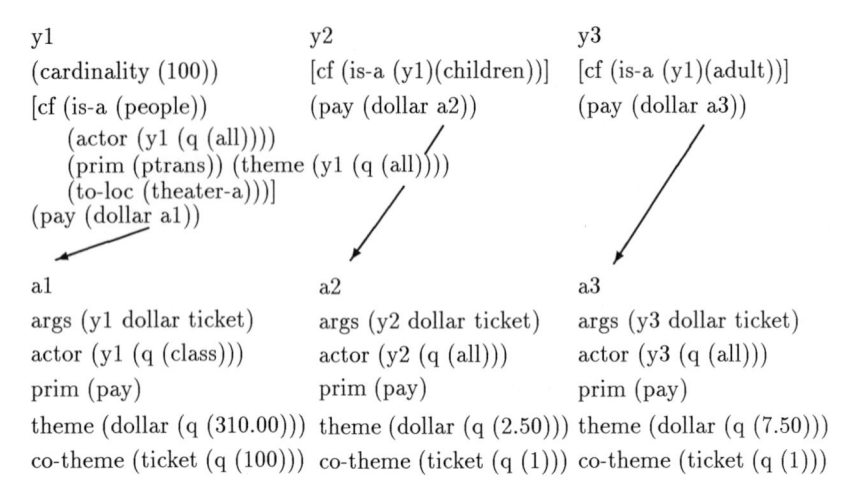

y2
[cf (is-a (y1)(children))]
(pay (dollar a2))

y3
[cf (is-a (y1)(adult))]
(pay (dollar a3))

a1
args (y1 dollar ticket)
actor (y1 (q (class)))
prim (pay)
theme (dollar (q (310.00)))
co-theme (ticket (q (100)))

a2
args (y2 dollar ticket)
actor (y2 (q (all)))
prim (pay)
theme (dollar (q (2.50)))
co-theme (ticket (q (1)))

a3
args (y3 dollar ticket)
actor (y3 (q (all)))
prim (pay)
theme (dollar (q (7.50)))
co-theme (ticket (q (1)))

Figure 2: Structure definitions of problem C2.

structures built would have been the same. Classification is the deep structure of many word algebra problems. Consider the problem below that is one or two levels more complex than the one just explained, but it has a similar deep structure.

> Problem C2: A group of 100 people consisting of children and adults went to the theater. They paid a total of $310.00 for 100 tickets. Each child paid $2.50 per ticket, and each adult paid $7.50 per ticket. How many children went to the theater?

The representation built for this problem is depicted in Figure 2 and 3. *Y1* represents the concept "people who went to the theater." Note the quantifier *class* in the relation *pay*, structure *a1*, under the concept *Y1*. This is a collective quantifier. Its meaning in this sentence is that they paid as a group or class $310.00. Note that "all" or "100" are wrong ways to quantify *Y1* in the relation "pay." *Y2* represents the concept "children who went to the theater" and *Y3* represents the concept "adults who went to the theater." Figure 3 depicts the classification relations between these concepts, providing the deep structure of the problem. The answer to the question is obtained by the following algorithm.

An answer to a question of the form *How many x relation?* is obtained by searching for the quantifier of x if x is the *theme* of the relation and by searching for the *cardinality* of x if x is the *actor* of the relation. Let us consider the latter case. If x has a *cardinality* slot, the value of that slot is the answer. If x does not have it, the cardinality slot of the superconcept of x is examined. Then, the algorithm tries to obtain the answer by finding the cardinality of the brothers of x and by applying some arithmetic. That would be the case if the cardinality of the brothers of x is known.

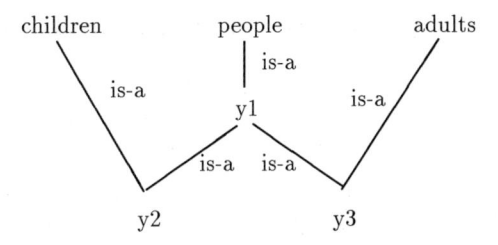

```
    children        people         adults
         \          |  is-a        /
          is-a      y1      is-a
           \       /   \       /
            \   is-a   is-a   /
             \  /        \   /
             y2           y3
```

Figure 3: Deep structure of problem C2.

A simple subtraction would solve the problem. In this case, the algorithm tries to solve the problem by setting up an equation of two unknowns. The algorithm asks the information retrieval component to find the algebraic relations that exist between a superconcept and its subconcepts. One of these relations is that the sum of the cardinality of the subconcepts is equal to the cardinality of their superconcept. For the problem we are discussing, this will result in producing the equation "Y2 + Y3 = 100." The information retrieval prints that equation with the following warning "I am assuming that Y2 and Y3 are the only subclasses of Y1 and that they are disjoint classes." The second equation that the information retrieval would produce is based on the following general rule. Let arg1 denote an argument of a relation:

```
If (a) the argument, say arg1, of a relation in a concept,
        say x, is quantified with a class quantifier, and
   (b) the same relation is predicated of all subconcepts of x,
   (c) and the quantifier of arg1 is ''all'' in every
        subconcept of x, then:
   (d) an arithmetic or algebraic relation exists between the
        superconcept of x and its subconcepts.
```

Once this relation is established, a procedure that matches the other arguments in the relation produces the final equation that in our case is "2.50 * Y2 + 7.50 * Y3 = 310.00." Note that this rule is independent of the relation "pay." For instance, similar relations will be established for problem C3 below, which is clearly *isomorphic* to the one just explained because the two have the same *deep structure*.

C3: One hundred people consisting of faculty and students ate 300 cakes. Each faculty member ate 2 cakes and each student ate 4 cakes. How many cakes did the students eat?

4 The Deep Structure of Motion Problems

We proceed, now, to analyze the deep structure of motion problems. Children at school are told that they must construct a graphical representation of these problems prior to trying their actual solution. For instance, consider the following problem:

> Problem M1: Peter traveled from a to b through a1. He used a train and a plane. The train going at 30 mph took 2 hours longer than the plane. The plane traveled at 150 mph. He traveled a total of 1400 miles. How long did the train take?

One possible graphical representation of problem M1:

```
train: 30 mph;duration = 2 + x   plane:150 mph;duration = x
<---------------------------a1----------------------------->
a--------------------- 1400 mi ---------------------------b
```

There is a considerable body of evidence indicating that many children and also adults have trouble in solving these problems because they do not understand their linguistic formulation. By asking them to build the graphical representation, they are forced to fully comprehend the description of the problem. Of course, what the diagrammatic representation does is to make apparent the deep interrelations that define the problem [10]. In other words, the graphical representation is one way to expose the deep structure underlying the problem. For instance by looking at the graphical representation above, one can *see* that it is about a motion problem describing a trip consisting of two subtrips. A question such as *How far is a from b?* can be directly answered by looking at the figure. The answer to the question *How long did the trip by train take?* is also facilitated by the graphical representation, because it shows clearly that the distance traveled by the train plus the distance traveled by the plane is equal to 1400 miles.

The diagram is a kind of mental model of the representation of the problem. The analysis of word algebra problems about motion revealed to us that if a program is to solve these problems in a principled manner, it needs to construct a representation that will be at least homomorphic to the graphical representation. The representation built by our system is a propositional representation that is homomorphic to the graphical representation. The deep structure corresponding to problem M1 is depicted in Figure 4.

The representation in Figure 4 says that problem M1 is an instance of a problem of a trip in stages, having three events. Event1 represents the event of going from place *a* to place *b*. Train-a and train-b are internally generated names that stand for instances of trains.

Event2 and event3 are linked to event1 by the relation *subevent*. Event e1 is a subevent of e2 iff the occurrence of e1 is a necessary condition for the occurrence of e2 and e1 occurs *in* the time interval of e2. The concept of *subevent* can be defined

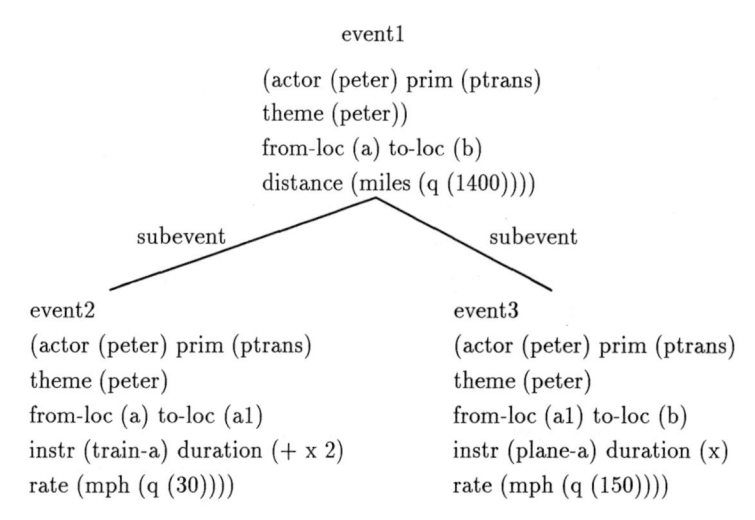

```
                        event1
                (actor (peter) prim (ptrans)
                theme (peter))
                from-loc (a) to-loc (b)
                distance (miles (q (1400))))
         subevent                      subevent

event2                             event3
(actor (peter) prim (ptrans)       (actor (peter) prim (ptrans)
theme (peter)                      theme (peter)
from-loc (a) to-loc (a1)           from-loc (a1) to-loc (b)
instr (train-a) duration (+ x 2)   instr (plane-a) duration (x)
rate (mph (q (30))))               rate (mph (q (150))))
```

Figure 4: Deep structure of problem M1.

formally using Allen's temporal logic [1] as follows. First, we need the following Allen's definitions for intervals. STARTS(t1,t2): time interval t1 shares the same beginning as t2, but ends before t2 ends; FINISHES(t1,t2): t1 shares the same end as t2, but begins after t2 begins; DURING(t1,t2): t1 is fully contained within t2; We need also the predicate IN, meaning that one interval is wholly contained in another, as follows:

$\forall (t1, t2)(IN(t1, t2) \iff DURING(t1, t2) \vee STARTS(t1, t2) \vee FINISHES(t1, t2))$

Another of Allen's predicates is that of OCCUR. This predicate takes an event, e, and a time interval, t, and is true if the event happened over the time interval t and there is no subinterval of t, say $t\ prime$, over which e is true. This is captured in the axiom:

$\forall (e, t, t')(OCCUR(e, t) \wedge IN(t', t) \implies \neg OCCUR(e, t'))$

Then, a formal definition for the concept of *subevent* is:

$\forall (e1, e2)(Subevent(e1, e2) \iff \exists t1 \exists t2(OCCUR(e1, t1) \wedge OCCUR(e2, t2) \wedge NECESSARY(e1, e2) \wedge IN(t1, t2)))$

The predicate *necessary* is an essential element of the definition of *subevent*, since potentially infinite events occur in the time interval of the event, say, *Peter drove to*

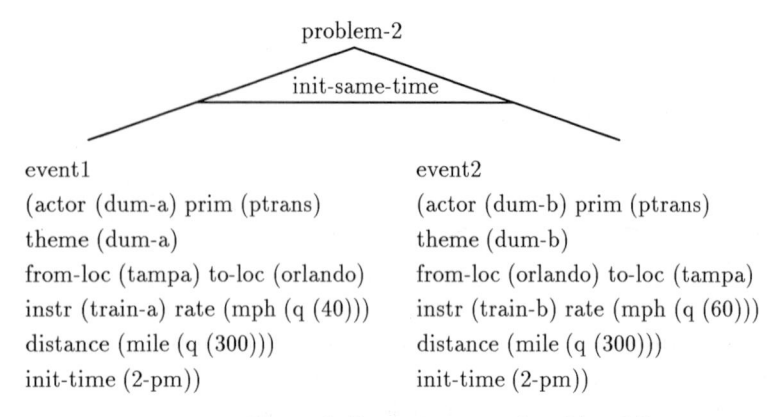

```
                    problem-2
                   /    |    \
                  / init-same-time \
                 /                  \
event1                              event2
(actor (dum-a) prim (ptrans)        (actor (dum-b) prim (ptrans)
theme (dum-a)                       theme (dum-b)
from-loc (tampa) to-loc (orlando)   from-loc (orlando) to-loc (tampa)
instr (train-a) rate (mph (q (40))) instr (train-b) rate (mph (q (60)))
distance (mile (q (300)))           distance (mile (q (300)))
init-time (2-pm))                   init-time (2-pm))
```

Figure 5: Deep structure of problem M2

Tampa. But only just a few of them are preconditions for the event *Peter drove to Tampa* to occur. Some of those could be *Peter opened the door of the car*, *Peter entered the car*, etc. The meaning of *necessary* in this definition is logic necessity. That is, "event e1 is a necessary condition for event e2" means that e2 does not occur unless e1 occurs, or $\neg occur(e1) \implies \neg occur(e2)$, or, by contraposition, $occur(e2) \implies occur(e1)$. Thus, if one considers the representation of the event *Peter went from a to b through a1* depicted in Figure 4, event *event2 Peter went from a to a1* and event *event3 Peter went from a1 to b* are necessary conditions for the event in the root node to occur. This is so because, although Peter could have gone from *a* to *b* through many other routes, once the event *Peter went from a to b through a1* took place, events *event2* and *event3* become necessary conditions for event *event1* to occur.

The deep structure is not only a computationally efficient representation, but, more importantly, one that is required if a solution is to be found, unless a subject uses rote memorization for the solution of a problem. Of course, this does not mean that a subject needs to be aware of the deep structure of a problem when he/she is solving it. Note that the representation that has been built does not classify the type of problem as being an overtake problem, round trip problem, etc. Consider the problem:

> M2: At 2 pm, a train left Tampa to go to Orlando at 40 mph. Another train left Orlando to go to Tampa at the same time at 60 mph. The distance from Tampa to Orlando is 300 miles. When will the two trains meet?

The representation built for this problem and depicted in Figure 5, consists of two events connected by the relation *init-same-time* meaning that the two events initiated at the same time. This relation does not indicate, in any way, which event is going to end or finish first.

5 The Construction of the Deep Structure of Motion Problems

The deep structure of a motion problem is built by using *integration* rules similar to those explained for the classification problems. As for those rules, these rules access two data bases, one formed by the logical form of the sentence, and another one consisting of the deep structure under construction. These rules encode common sense knowledge about the physical constraints in the logical forms and in the representation being built. In the description of a problem, initial sentences set up a context and subsequent ones need to fit in this context. This may be illustrated with the following sentence: *Peter traveled a distance of 1200 miles from a to b.* This sentence produces the following initial representation (we use the diagrammatic representation):

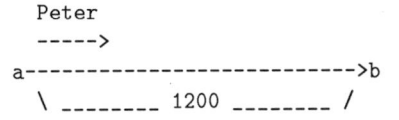

```
    Peter
    ----->
a--------------------------->b
 \ _____ 1200 _____ /
```

Subsequent sentences need to fit in this initial representation, which already imposes constraints on their semantics. For instance, suppose that the next sentence is *Peter returned to c at 20 mph.* This sentence does not make any sense because there is not a location called "c" in the initial model, and it cannot be integrated in the representation. If the next sentence is *He continued from b to c at a speed of 30 mph,* "b" is found in the model and this representation can be produced:

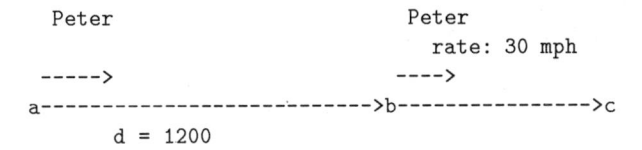

```
    Peter                       Peter
                                rate: 30 mph
    ----->                      ---->
a--------------------------->b---------------->c
         d = 1200
```

The representation also deals with cases of ambiguity. For instance, suppose that the next sentence is *It took Peter two hours.* This sentence is ambiguous only in the context of the last diagram, because it has two segments to which the duration can be attached. However, the sentence is not ambiguous in the first representation that consists only of one segment. There are also rhetoric and discourse constraints imposed by the initial sentences of a problem [2].

Two types of integration rules are needed: those that *merge* an incoming event into an event already in the model and those that *reorganize* the events by connecting them with temporal links. We use DS to refer to the deep structure being built and I-EVENT to the event being integrated. The first integration rule says:

```
R1
If the DS is empty, then insert the I-EVENT in the model.
```

Then, if the first sentence read is *A train traveled from a to b at 24 mph*, the DS will contain:

```
(event1
   (actor(dum-a) primitive (ptrans) theme(dum-a) from-loc (a)
    to-loc(b) instrument (train-a) rate  (mph (q(24))))
```

Terms like dum-a, dum-b stand for a unknown *actor, theme, from-loc*, etc. We refer to them as dummy terms. Had the first sentence read been *Peter traveled 1400 miles*, then after applying rule R1, the DS would contain:

```
(event1
  actor(Peter) primitive (ptrans) theme(Peter)
  distance (miles(q(1400))))
```

Suppose that the first sentence read is *Peter traveled from a to b*. This sentence will produce the following representation in the DS:

```
(event1
 (actor (Peter) primitive (ptrans) theme (Peter)
  from-loc (a) to-loc (b))
```

The following rule merges information from the I-EVENT into the events in the model.

R2
If the I-EVENT and an event in the DS match then
   a. remove the event from the DS
   b. create a new event by forming the union of the two events.
   c. integrate the union event in the DS.

Two events, say e1 and e2, *match* if all slots that appear in both events have identical fillers.

$\quad$ Match(e1,e2) $\iff \forall(x)\forall(y)$ (Slot(x,e1) and Slot(y,e2) and Equal(x,y) $\implies$ Equal(filler(x), filler(y))

Suppose that the DS consists only of the following event:

```
(event1
 (actor(Peter) primitive (ptrans) theme (Peter)
  from-loc(Tampa) to-loc(Orlando)))
```

Now, the next sentence is: *He traveled by train*. The logical form produced for this sentence is:

```
(actor (Peter) primitive (ptrans) theme (Peter)
  instrument (train-a))
```

Rule R2 will fire since the event in the DS and this one will *match*, because every slot that appears in both events has the same filler. The union of the two events will result in the event:

```
(event1
 (actor(Peter) primitive (ptrans) theme (Peter)
  from-loc(Tampa) to-loc(Orlando) instrument (train-a)))
```

Since rule R2 has removed the only event in the model from it, the DS is empty. As a consequence, the event above will be integrated in the DS by rule R1. Let us assume that the first two sentences of a problem are *Peter went from Tampa to Orlando. He used a train and a plane.* The representation built for the first sentence is:

```
(event1
 (actor(Peter) primitive (ptrans)
  theme(Peter) from-loc (Tampa) to-loc(Orlando)))
```

The parser output for the second sentence is:

```
1.a (subject ((pron he)) verb ((main-verb use) (tense sp))
     object ((udet a) (noun train)))
1.b (subject ((pron he)) verb ((main-verb use) (tense sp))
     object ((udet a) (noun plane)))
```

Interpretation rules resolve the referent of "he" and create unique names for the indefinite noun groups "a train" and "a plane." This produces:

```
2.a (subject (Peter) verb (used) object (train-a))
2.b (subject (Peter) verb (used) object (plane-a))
```

Now, it becomes necessary to determine the meaning of "used." One of the interpretation rules for finding the meaning of "use" in a sentence consisting only of an *subject* and an *object* says: Find the most recent logical form, say LF1, containing an action primitive and an *actor* that matches the *subject* of the sentence with "use." If the rule succeeds, as in this case, the subject of the verb "use" becomes the *actor*, the verb "use" is replaced with the primitive of LF1, and the *object* slot is changed to an *instrument* slot. This rule will produce:

```
3.a (actor (peter) primitive (ptrans) instrument (train-a))
3.b (actor (peter) primitive (ptrans) instrument (plane-a))
```

The logical form 3.a is integrated using rule R1. However, R1 does not fire in integrating logical form 3.b. This logical form is integrated using one of the rules that inserts events in the DS and reorganizes it. In this case, the rule is:

```
                        event1
              (subevents (event2 event3))
              (actor (peter) prim (ptrans) theme (peter)
              from-loc (tampa) to-loc (orlando))
         subevent                           subevent
event2                                event3
(actor (peter) prim (ptrans)          (actor (peter) prim (ptrans)
theme (peter)                         theme (peter)
from-loc (tampa) to-loc (dum-a)       from-loc (dum-a) to-loc (orlando)
instr (train-a))                      instr (plane-a))
```

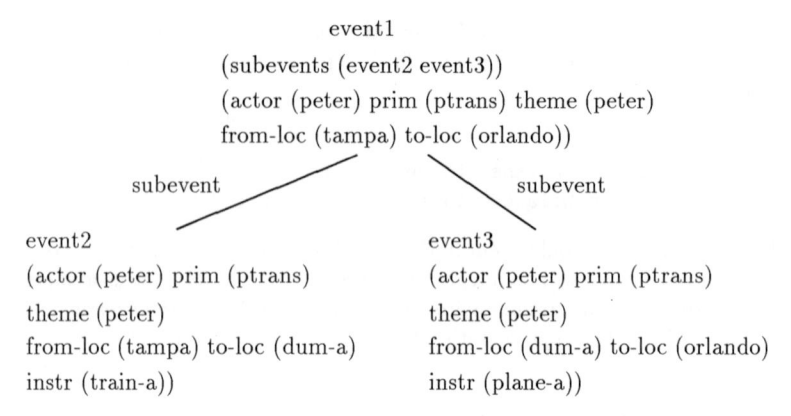

Figure 6: Structure built for "Peter went from Tampa to Orlando. He used a train and a plane."

R3
```
If the I-EVENT and an event in the DS match, say event e1,
except for the instrument slots then:
a. create a new event in the DS.
b. connect the I-EVENT and e1 with a subevent link.
```

The application of this rule results in the structure depicted in Figure 6. The rule is encoding the piece of common sense knowledge that says that if somebody has traveled using two vehicles, then he/she has made a trip in stages. The rule produces the correct hierarchy with an event at the top level and two subevents connected to it. The integration rules are independent of the surface form of the sentences. One can see that the rules above will produce the same deep structure if the sentence is *Peter traveled from Tampa to Orlando using a train and a plane.* Consider the following more complex example:

> Problem M3: (1) Peter went from Tampa to Orlando using a train and a plane. (2) He traveled by train from Tampa to Leesburg and (3) traveled by plain from Leesburg to Orlando. (4) He also went from Orlando to Miami.

Sentence (1) will create the structure depicted in Figure 6. The integration of sentences (2) and (3) will result in inserting "Leesburg" into the slots *to-loc* and *from-loc* of event2 and event3. This is done by using rule R2. Slots with fillers dum-a, dum-b, etc. are ignored by the matching algorithm. The integration of sentence (4) is done by the following subevent rule.

R4

If the I-EVENT and an event in the DS, say e1, have the same
actor and the same primitive and the from-loc slot of the
I-EVENT is the same as the to-loc slot of e1, then:
a. create an superevent, say e2, whose from-loc is the from-loc
   of e1 and the to-loc is the to-loc of the I-EVENT
b. one subevent of e2 is e1
c. the other subevent of e2 is the I-EVENT

This rule is applied in a top-down fashion starting with the root node of the hierarchy of events and descending. The application of this rule stops as soon as an event in the DS matches. The structure created is depicted in Figure 7. If the sentences describing problem M3 are followed by:

> The trip from Tampa to Miami took 7 hours. The trip from Tampa to Leesburg took 2 hours, and the trip from Leesburg to Orlando, 3 hours.

These sentences will be integrated by rule R2. The definite description "the trip" in the sentence *The trip from Tampa to Miami took 2 hours* is handled by applying rule R2 to the logic form below:

```
(actor(dum-a) primitive (ptrans) from-loc(Tampa) to-loc(Miami)
   duration(hour(q(2))))
```

Note that only one event in the DS matches. However, rule R2 will find more than one match if the sentence were *The trip took 2 hours*. In this case, the definite description cannot be resolved, and an assumption needs to be made about its reference.

If none of the rules fires for an event, that event is inserted in the DS by a catch-all rule, which inserts the event in the DS without connecting it to any of the events already in the DS. Hence, we have the rule:

R4
If no rule fires, then insert that event in the DS without
connecting it to any of the other events in the DS.

Consider the sentences *Peter traveled from Tampa to Orlando. Mary left from Tampa toward Orlando at 2 p.m.* Let us call e1 and e2 the events produced by the first and the second sentence, respectively. Event e1 will be integrated by rule R1, while event e2 will be integrated by rule R4. No link will connect those two events in the DS. Suppose that the next sentence is *Peter left Tampa at 2 p.m.* In this case, rule R2 will fire modifying event e1 that will become:

```
(actor(peter) primitive(ptrans) init-time(2-pm) from-loc(Tampa)
   to-loc(Orlando))
```

```
                    event1
                    (subevents (event2 event3))
                    (prim (ptrans) actor (peter) theme (peter)
                    from-loc (tampa) to-loc (miami))
              subevent    ╱         ╲    subevent

event2                              event3
(subevents (event2.1 event2.2))     (prim (ptrans) actor (peter)
(prim (ptrans) actor (peter)        theme (peter)
theme (peter)                       from-loc (orlando) to-loc (miami))
from-loc (tampa) to-loc (orlando))
     subevent  ╱         ╲  subevent

event2.1                            event2.2
(prim (ptrans) actor (peter)        (prim (ptrans) actor (peter)
theme (peter)                       theme (peter)
from-loc (tampa) to-loc (leesburg)  from-loc (leesburg) to-loc (orlando)
instr(train-a))                     instr (plane-a))
```

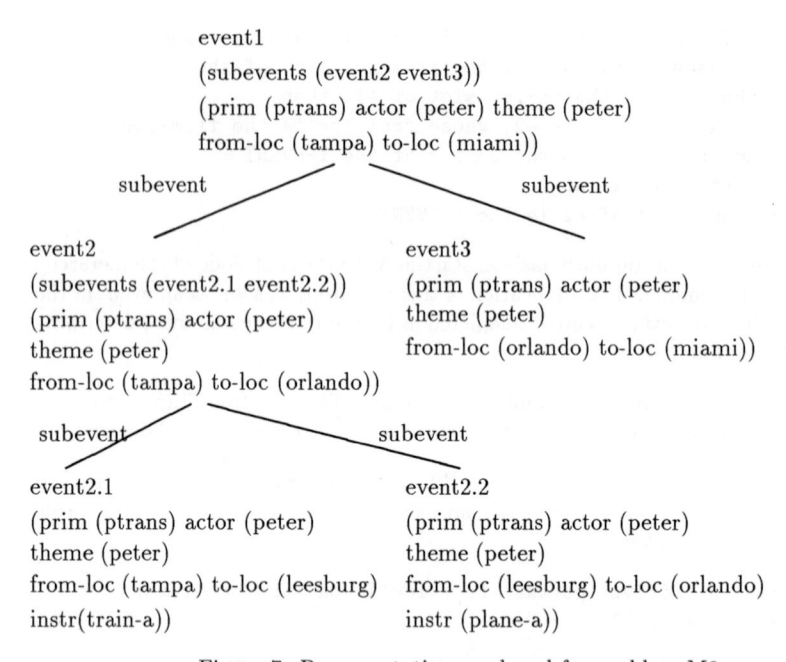

Figure 7: Representation produced for problem M3

The slot *init-time* recording the time in which the event starts has been built by rule R2. As the reader may remember, when rule R2 merges knowledge into an event, it removes that event from the model and activates the rules that connect the events in the DS in order to integrate that event. In this case, the following rule integrates the event in the DS:

R5
If the I-EVENT and an event in the DS have an init-time
slot with the same filler, then:
connect the two events with an init-same-time link.

The representation produced for this last example is depicted in Figure 8.

6 The Solution of Motion Problems

The solution of motion problems represented in a hierarchy of subevents are solved in a way similar to problems whose deep structure is organized using an *is-a* hierarchy. Each type of question has an associated hierarchy of problem-solving methods that will provide an answer to it. For instance, consider the question *How long did Peter*

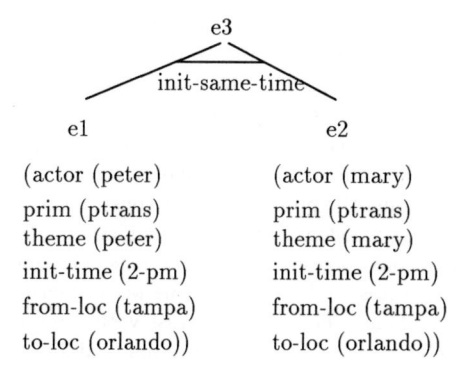

```
                        e3
                     ╱──┴──╲
                ╱ init-same-time ╲
              e1                    e2
      (actor (peter)          (actor (mary)
      prim (ptrans)           prim (ptrans)
      theme (peter)           theme (mary)
      init-time (2-pm)        init-time (2-pm)
      from-loc (tampa)        from-loc (tampa)
      to-loc (orlando))       to-loc (orlando))
```

Figure 8: The representation produced for "Peter traveled from Tampa to Orlando. Mary left from Tampa toward Orlando at 2.p.m. Peter left Tampa at 2 p.m."

take going from A to B. The first problem-solving method in the hierarchy tries to find an answer to the question by accessing the appropriate slot in the right event in the DS. This is simple knowledge retrieval. If this fails, a slightly more complex problem solving method that tries to apply some arithmetic knowledge is activated. This problem-solver method checks to see if the events in the DS are organized by the *subevent* relation. Let us assume that the DS consists of event2 and event3 which are subevents of event1. The problem-solver method tries to achieve the subgoal: duration(event1) = duration(event2) + duration(event3). This subgoal is attempted by directly accessing the "duration" slot in those events, that is, by activating the information retrieval method. If the arithmetic problem-solver fails because the duration of event2 or the duration of event3 is unknown, then the problem is tried by using some algebra, that is, by invoking the algebraic problem-solver method. The subgoal activated is distance(event1) = distance(event2) + distance(event3), which is equal to rate(event1) * duration (event1) = rate(event2) * duration(event2) + rate(event3) * duration(event3). The execution of this subgoal is again tried by activating the information retrieval method, which fills the content of "rate" and "duration." Each slot acts like a question to the information retrieval method, e.g., find the content of "rate" in event1, etc.

7 Conclusion

All the ideas described in this paper have been implemented. The program that deals with motion problems contains only about fourteen general organization rules in order to integrate events describing motion problems. Besides these rules, there are some integration rules which are specific to some lexical terms. For instance, the verb "return" has a special rule to integrate it. Expressions such as "upstream" and "downstream" also trigger specific integration rules.

In the body of the paper, we say that the understanding of a word problem is similar to the construction of a house from scratch. The house in our metaphor corresponds to what we have called the *deep structure* of a problem, and the building materials are the events describing the problem. In the construction of a house, there are minor actions like putting a wooden frame in a window, and major actions like setting a wall. The minor actions correspond to the actions performed by our *merge* rules, and the major actions correspond to the actions performed by our event insertion and organization rules. This is clearly in contrast with the schema-based understanding paradigm in which the house is already pre-built, and the only things left to do are minor actions.

Where do problem-spaces come from? This is one of the most fundamental questions in the field of human problem-solving. It is generally accepted that if a problem-space is provided for a problem, the solution follows without much difficulty since it reduces to search, which is a well-understood technique to which massive parallel methods, to use the favorite phrase, can be applied. However, the solution of many problems hinges not as much on *how* to search, but on *what* and *where* to search, that is, on the problem-space itself. In this paper, we have gathered some evidence showing how the problem-space for some kinds of word problems is based on and derives from the *deep structure* of the problem.

References

[1] J. Allen. (1984). Towards a General Theory of Action and Time. *Artificial Intelligence*, **23**, pp. 123-154.

[2] J. Batali. (1991). *Automatic Acquisition and Use of Some of the Knowledge in Physics Texts*. Ph.D Dissertation, Massachusetts Institute of Technology, Artificial Intelligence Laboratory.

[3] F. Gomez. (1981). *On General and Expert-Based Methods in Problem-Solving*. Ph.D Dissertation, The Ohio State University, Department of Computer Science.

[4] F. Gomez. (1982). Towards a Theory of Comprehension of Declarative Contexts. Proceedings of the 20th Meeting of the Association of Computational Linguistics, pp. 36-43.

[5] F. Gomez. (1985). A Model of Comprehension of Elementary Scientific Texts. Proceedings of the Workshop on Theoretical Approaches to Natural Language Understanding. Halifax, Nova Scotia, pp. 70-81.

[6] F. Gomez and C. Segami. (1989). The Recognition and Classification of Concepts in Understanding Scientific Texts. *Journal of Experimental and Theoretical Artificial Intelligence*, **1**, pp. 51-77.

[7] F. Gomez and C. Segami. (1990). Knowledge Acquisition from Natural Language for Expert Systems Based on Classification Problem-Solving Methods. *Knowledge Acquisition*, **2**, pp. 107-128.

[8] F. Gomez and C. Segami, (1991). Classification-Based Reasoning. *IEEE Transactions on Systems, Man and Cybernetics*, **1**(3), pp. 398-405.

[9] F. Gomez. (1992). Representing Biological Systems as Temporal Event Hierarchies. Technical Report, Dept. of Computer Science, UCF, Orlando, Florida.

[10] J. Larkin and H. Simon, (1987) Why a Diagram is Sometimes Worth Ten Thousand Words, *Cognitive Science*, **11**, pp. 65-99.

[11] A. Newell and H. Simon. (1972). *Human Problem-Solving*. Englewood Cliffs, N.J., Prentice-Hall.

[12] G. S. Novak. (1976). Computer Understanding of Physics Problems Stated in Natural Language. Journal of Computational Linguistics, Microfiche 53.

[13] J. M. Paige and H. Simon. (1979) Cognitive Processes in Solving Word Algebra Problems. In H. Simon, *Models of Thought*, Yale University Press.

[14] C. H. Rapp. (1986). Algebra Reader: An Expert Algebra Word Problem Reader. TR-86-30-6, Computer Science Department, Oregon State University.

[15] R. Schank. (1975) *Conceptual Information Processing*. North Holland, Amsterdam.

RESOLVING CONFLICTS IN INHERITANCE REASONING WITH STATISTICAL APPROACH

Changhwan Lee
Department of Computer Science and Engineering
University of Connecticut, Storrs, CT 06269

Abstract

Inheritance reasoning is a classical mechanism used in common sense reasoning in artificial intelligence. Although many inheritance reasoners have been proposed in artificial intelligence literature, most previous works are missing clear semantics which is why these reasoners sometimes provide anomalous conclusions. In this paper, we describe a set-oriented inheritance reasoner and propose a method of resolving conflicts with clear semantics of defeasible rules. The semantics of default rule are provided by statistical analysis, and likelihood of rule is computed based on the evidence in the past. Two basic rules, specificity and generality, are defined to resolve conflicts effectively in the process of reasoning. We show that the mutual tradeoff between specificity and generality can prevent many anomalous results occurring in traditional inheritance reasoners. An algorithm is provided, and some typical examples are given to show how the specificity/generality rules resolve conflicts effectively in inheritance reasoning.

1 Introduction

Inheritance reasoning appears in various forms in the literature of artificial intelligence and is becoming more important because inheritance hierarchies are simple, natural, and useful. Most inheritance reasoners have been characterized by algorithms which operate on an inheritance network, and some of them present translations of the inheritance network into some nonmonotonic logics, such as Reiter's default logic [6], Moor's autoepistemic logic [10], and circumscriptive theories [8] [9]. Although many inheritance reasoners have been proposed in artificial intelligence literature, most previous works are missing clear semantics which is the central reason why these reasoners sometimes provide anomalous conclusions.

In this paper, instead of converting network structure to certain theories, we describe an inheritance reasoner and propose a method of resolving conflicts with clear semantics of defeasible rules. The semantics of default rule are provided by statistical analysis, and likelihood of rule is computed based on the evidence in the past. The $\mathcal{X}^2$ method is used to decide the existence of defeasible rules between predicate classes. In order to get presumably appropriate probabilities for each rule, probabilities are defined based on the observations of the occurrences in the past, and the probabilities are represented as intervals instead of single point values. These statistically induced degrees of beliefs have an advantage over the subjective probabilities since they are based on objective information about the world, information which could in principle be obtained through experience/observation.

Two basic rules, *specificity* and *generality*, are defined to resolve conflicts effectively in the process of reasoning. First, like many other inheritance reasoners, the model uses a specificity rule as a tool to resolve conflicts. However, specificity is interpreted differently from the viewpoint of set cardinality. Second, generality rule, a complementary criteria of specificity, is defined based on the degree of biasedness of the observations in predicate classes. As a complementary rule of specificity, the generality rule is defined as a measure of the degree of typicality of instances of each class. The basic idea behind the generality rule is that the set of instances in the class must correctly represent the general characteristic of the class they belong to. We studied two conditions of which the instances in a class must avoid in order to maintain the minimum degree of generality: immature class and biased class.

We begin, in Section 2, by providing a description of the semantics of the defeasible assertions in this work. Section 3 introduces evidential probabilities and explicit negation. In Section 4, we give a formal definition of the reasoning model. Section 5 explains a statistical method for detecting the existence of rules. Section 6 shows how inheritance reasoning methods in the model can resolve conflicts, and Section 7 provides an abstract algorithm for the reasoner. Section 8 gives some examples of the reasoning and shows how conflicts in these examples can be resolved. Finally, conclusions are given in Section 9.

2 Interpretations of Defeasible Assertion

Exception handling has been identified as the major issue of many inheritance reasoners. For example, the defeasible inference that Clyde is gray since he is an elephant and elephants are typically gray. As has been discussed in the literature, such defeasible assertions can not be modeled as universally quantified assertions since there are some elephants that are not gray. However, even though these exceptional cases falsify a universal quantifier, they do not invalidate the entire defeasible assertions. Without a well defined semantics for the defeasible assertions, it is very difficult, maybe even impossible, to develop an inheritance reasoner which will provide justifiable inferences in all cases. It is clearly impossible for an inheritance reasoner to generate correct or reasonable answers in all cases. However, even if we may occasionally be wrong we would still like to have some global justification for all of the inferences that the system generates.

The proposed system interprets the defeasible assertions as being statistical assertions. In order to determine the degree of belief of assertion, a good approach is to take advantage of known facts. A fundamental assumption of this probability is, "The probability of an indicative conditional of the form *if A is the case then B is* is that the probability of *if A then B* should be equal to the ratio of the probability of *A and B* to the probability of *A*(ratio of conjunction of antecedent and consequent to antecedent)." For example, in assertion "Most of birds fly", we interpret this assertion based on statistical evidence. In other words, among the instances of birds, how many of them can fly. Roughly speaking, if there are X objects with property α, and Y of these have property β, then for any term t, $P(\beta(t)|\alpha(t)) = Y/X$, unless we have some additional information about the term t. For formulas like "Fly(Clyde)" where Clyde is known to be a bird, if 90% of all birds fly and Clyde is known to be a bird, the inductive assumption would attach a degree of belief of 0.9 to the formula "Fly(Clyde)" by assuming that Clyde was a randomly selected bird.

Statistical semantics was introduced before by Bacchus [1]. He introduced statistical majority probabilities. In his study, if more than half the objects among L_1 satisfy L_2, a rule $L_1 \rightarrow L_2$, L_1 is usually L_2, is defined. It is possible that the probabilities of some rules coincidentally become larger than a constant ϵ and thus become defeasible rules. Another problem occurs when the set cardinality of the precedent class is very small. When the cardinality of a class domain is very small, the probability of getting an anomalous conclusion is very high. Geffner has studied another probabilistic reasoning, called ϵ-semantics reasoning [7]. He has considered probabilistic versions of inheritance reasoning. However, he used probabilities infinitesimally close to 1 and 0. Obviously, there are very rare cases in which the properties are related via infinitesimal probabilities.

3 Evidential Probabilities

To make the degree of belief of default assertions more precise, we adopt an explicit negation system. Most reasoning systems regard unknown states as false states for the sake of efficiency. We claim that *false* states are different from *unknown* states. Various methods for implicit negation have been proposed such as unique-name assumption, domain closure, closed world assumption, and predicate completion [5]. The main issue common in these strategies is the efficiency of representing knowledge, sacrificing expressive power for the sake of efficiency. In the proposed approach, we use *explicit negation* to differentiate unknown from false. Notice that, as we mentioned earlier, the semantics of the system is totally based on current evidence. Therefore, if we do not differentiate false from unknown, we start with incorrect evidence.

Each rule in the system comes with its corresponding probability, which is generated automatically. In most current reasoning systems, an expert assigns his/her subjective probability to each rule. One of the big problems in this approach is the preciseness of the value of the probability assigned. There is also an issue of whether it is reasonable to describe probability by a single point rather than a range. While an agent might agree that the probability of an event lies within a given range, say between 1/3 and 1/4, he might not be prepared to say that it is precisely 0.287. In the proposed system, the probabilities are represented as intervals instead of point values. Introducing unknown states allows us to define the likelihood of a proposition A as a subinterval of the unit interval [0,1]. The lower bound of this interval is the degree of support of the proposition, S(A), and the upper bound is its degree of plausibility, Pl(A). The likelihood of an assertion A is written as [S(A),Pl(A)]. The support of A is meant to describe a lower bound on the degree of belief of an agent that A is actually the case. The corresponding upper bound is called plausibility. Intuitively, we view the interval [S(A),Pl(A)] as providing lower and upper bounds on the "likelihood" of A. The detailed method of calculating these probabilities is explained in the following section. In summary, the main features of the system are listed in the following.

- The existence of rules between predicates is decided by statistical analysis.
- Probabilities are based on evidence available in the current knowledge base.
- We use probability intervals to represent the likelihood of rules.
- Specificity and generality of rule are being used to resolve conflicts.

4 Model

This section introduces the basic framework of the reasoner. The system consists of three components: a set of objects, a set of predicates, and a set of edges. We have constant symbols which represent the members of the object, predicate symbols, and two types of edges. When an object belongs to a predicate, there is a strong edge from the object to the predicate. Similarly, if there is a strong or defeasible rule between two predicates, there is a strong edge or defeasible edge, respectively, between two predicates. Reasoner Γ is being represented as follows.

$$\Gamma = <\mathcal{D},\ \mathcal{P},\ \mathcal{E}> \qquad (1)$$

where $\mathcal{D}$ shows the domain of universe, $\mathcal{P}$ shows the set of predicates, and $\mathcal{E}$ are the edges of the network. As we mentioned, there are two types of edges: strong edges and defeasible edges. A strong edge will have the form $x \Rightarrow y$, or $x \not\Rightarrow y$, where y is a class. If x is an object, such an assertion would be interpreted as an ordinary atomic statement: for instance, they are analogous to $y(x)$ or $\neg y(x)$ in logic. They might represent statements like "Tweety is a bird" or "Tom isn't a Bird." If x is a class, these assertions would be interpreted as generic statements. For example, $x \rightarrow y$ and $x \not\rightarrow y$ might represent the statements "Birds fly" and "Penguins don't fly," respectively. In addition, as we mentioned earlier, the edges are accompanied by probabilities such that the general form of an edge is $P_1 \rightarrow P_2:[S(P_1 \rightarrow P_2), Pl(P_1 \rightarrow P_2)]$. Values $S(P_1 \rightarrow P_2)$ and $Pl(P_1 \rightarrow P_2)$ represent support probability and plausibility probability, respectively.[1] Especially, when $S(P_1 \rightarrow P_2) = Pl(P_1 \rightarrow P_2) = 1$, we say $P_1 \Rightarrow P_2$.

To differentiate unknown state from false state, we adopt the following definitions. Each predicate has three subsets: P^+, P^-, and P^*. P^+ contains objects known to have properties of P, P^- contains objects known to have properties of $\neg P$, and P^* contains objects which are inconclusive about P. Graphical notations of these sets are shown in Figure 1. A semantic structure for a network is an assignment of a triple of sets of objects (P^+, P^-, P^*) for each node in $\mathcal{P}$. The support and plausibility are given as

$$S(P \rightarrow Q) = \min\{Pr(Q|P)\} = \frac{\min\{Pr(Q \wedge P)\}}{\max\{Pr(P)\}}$$

$$Pl(P \rightarrow Q) = \max\{Pr(Q|P)\} = \frac{\max\{Pr(Q \wedge P)\}}{\min\{Pr(P)\}}$$

Therefore, the support and plausibility are formally defined in the following.

Definition 1

$$S(P \rightarrow Q) = \frac{\|P^+ \cap Q^+\|}{\|P^+\| + \|P^* \cap Q^-\| + \|P^* \cap Q^*\|}, \qquad (2)$$

$$Pl(P \rightarrow Q) = \frac{\|(P^+ \cup P^*) \cap (Q^+ \cup Q^*)\|}{\|P^+\| + \|P^* \cap Q^+\| + \|P^* \cap Q^*\|} \qquad (3)$$

---

[1] In this paper, we use $Pr(P_1 \rightarrow P_2)$ instead of $Pr(P_2 \mid P_1)$.

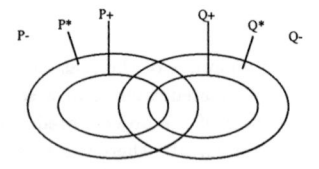

Figure 1: Set notation for probability definition

We can easily check that the support and plausibility satisfy the following properties.

$$0 \leq S(P \rightarrow Q) \leq Pl(P \rightarrow Q) \leq 1$$
$$S(P \rightarrow Q) + Pl(P \rightarrow Q) \leq 1$$
$$Pl(P \rightarrow Q) = 1 - S(P \not\rightarrow Q)$$
$$S(P \Rightarrow Q) = Pl(P \Rightarrow Q) = 1$$

5 Detecting the Existence of Rules

One of the most prominent features of the proposed reasoner is the separation of detecting a rule and calculating the probability of the rule. In this section, we show how the system detects the existence of rules and how their corresponding probabilities are computed. Suppose there are two predicate A, B. We will see how the existence of the defeasible rule between A and B is detected and how the probability of the detected rule is calculated.

We shall examine the chi-square($\mathcal{X}^2$) method of testing the hypothesis that the two predicates are dependent on each other. In most reasoners, the existence of a defeasible rule is decided (1) by an expert or rational agent, or (2) if the ratio of consequent and antecedent is greater than a certain constant, they define it as a defeasible rule. We use a different approach to define dependency. First we differentiate the existence of the defeasible rule from the probability of the rule. That is, even though the success probability between two predicates is high, it is still possible that they do not have a causal relationship. In the proposed approach, the existence of defeasible rule is decided by statistical analysis. A Chi-square test is usually used to test the hypothesis that the observations agree or disagree with the theoretical frequencies [11]. In our case, it can be used to test whether the domain sets of two predicates are inter-dependent or not. The statistic we will use is

$$\mathcal{X}^2 = \sum_{i,j \in \{+,-,*\}} \frac{(f_{ij} - F_{ij})^2}{F_{ij}} \qquad (4)$$

where f_{ij} is the observed frequencies, and F_{ij} is the theoretical frequencies. For example, in Figure 2, the theoretical frequency of $A^+ \cap B^+$ is given as $250 \cdot \frac{200}{1000} = 50$. Other theoretical

	B$^+$		B$^-$		B*		
A$^+$	180	(50)	15	(119.4)	5	(30.6)	200
A$^-$	65	(190)	550	(453.7)	145	(116.3)	760
A*	5	(10)	32	(23.9)	3	(6.1)	40
	250		597		153		1,000

Figure 2: $\mathcal{X}^2$ table for detecting defeasible rule

frequencies can be calculated in a similar way. To determine the dependency between A and B, the total $\mathcal{X}^2$ value of Figure 2 is computed as 566.58 and is compared with the value of $\mathcal{X}^2_{[4,0.95]}$, where $\mathcal{X}^2_{[4,0.95]}$ is the $\mathcal{X}^2$ value with degree of freedom being 4 and level of significance begin 0.95. As the value of $\mathcal{X}^2_{[4,0.95]}$ is given as 9.49, we can conclude that A is not independent of B. The other issue concerning the detecting rule is how to decide the direction of the detected rules. We employ the following strategy in the system.

$$\begin{cases} P \to Q & : \quad P^+ \cap Q^+ \geq P^+ \cap Q^- \\ Q \to P & : \quad otherwise \end{cases}$$

According to Definition 1, the probability interval of the above rule is given as follows.

$$[S(A \to B), Pl(A \to B)] = \left[\frac{180}{200 + 32 + 3}, \frac{180 + 5 + 5 + 3}{200 + 5 + 3} \right] = [0.76, \ 0.92].$$

Sometimes, there are cases that the predicate A is a subset of the predicate B, which means there exists a strong rule between A and B. In this case, A$^+\cap$ B$^-$ and A$^+\cap$ B* in the above table are empty, and we can easily figure out that the system produces a strong rule between A and B.

5.1 Semantics of Rules

The semantics for each defeasible or strict rule are given in the following. Suppose c denotes an instance and P, Q denote predicates.

1. c $\Rightarrow$ P is true iff P$^+$(c)
2. c $\not\Rightarrow$ P is true iff P$^-$(c)
3. P $\to$ Q : [S(P $\to$ Q), Pl(P $\to$ Q)], iff $\mathcal{X}$(P,Q) > $\mathcal{X}^2_{[4,0.95]}$ and P$^+\cap$ Q$^+$ $\geq$ P$^+\cap$ Q$^-$
4. P $\not\to$ Q : [S(P $\not\to$ Q), Pl(P $\not\to$Q)], iff $\mathcal{X}$(P,Q) > $\mathcal{X}^2_{[4,0.95]}$ and P$^+\cap$ Q$^+$ < P$^+\cap$ Q$^-$
5. P $\Rightarrow$ Q is true iff P $\to$ Q and P$^+$ $\subseteq$ Q$^+$
6. P $\not\Rightarrow$ Q is true iff P $\not\to$ Q and P$^+\cap$ Q$^+$ = $\emptyset$

5.2 Inheritance Reasoner As a Network

Because the reasoner is modeled as a network, network-oriented terminology is defined and will be used to explain the algorithm later in this paper. *Paths* are sequences of edges in a hierarchy network, and are subject to two restrictions. First, a negative link can occur in a path, if at all, only at the very end: a → p ↛ q is a path, but a ↛ p → q isn't. Second, an object can occur only as the initial node of a path: p → a ↛ q isn't a path.

Let us denote $*$ for $\{\Rightarrow, \not\Rightarrow, \rightarrow, \text{or} \not\rightarrow\}$, $\overset{+}{*}$ for $\{\Rightarrow \text{or} \rightarrow\}$, and $\overset{-}{*}$ be $\{\not\Rightarrow \text{or} \not\rightarrow\}$. Similarly, ▷ is denoted to be $\{\rightarrow, \text{or} \not\rightarrow\}$ and ⊵ is denoted to be $\{\Rightarrow, \text{or} \not\Rightarrow\}$. Given an inheritance hierarchy, we say that x_n is *reachable from* x_0 if there is some path $x_0 \overset{+}{*} x_1 \overset{+}{*} \cdots \overset{+}{*} x_{n-1} * x_n$. The path $x_0 \overset{+}{*} x_1 \overset{+}{*} \cdots \overset{+}{*} x_{n-1} * x_n$ is defined to be *supported* by the network. If the final edge of a path is positive, we say that it is a *positive path*. Similarly, we can define a *negative path*. Ambiguity arises when two paths conflict. Formally, an inheritance hierarchy is *ambiguous* if there are some nodes such that both $x_0 \overset{+}{*} x_1 \overset{+}{*} \cdots \overset{+}{*} x_{n-1} \overset{+}{*} x_n$ and $x_0 \overset{+}{*} x_1 \overset{+}{*} \cdots \overset{+}{*} x_{n-1} \overset{-}{*} x_n$ are supported by the hierarchy. According to Touretzky's terminology [14], the network is heterogeneous(i.e. it allows both strict and defeasible assertions), nonmonotonic(i.e. permits exceptions to inherited properties), bipolar(i.e. contains both positive and negative edges), and credulous(i.e. draws as many conclusions as possible) system.

6 Inferences

Inferences are performed based on the hierarchical structure generated from the data. Inheritance reasoning in this case can be regarded as a classification problem given an available data set. Our claim in this system is that the most fundamental tools in classification type reasoning are specificity and generality. Specificity has already been proposed in many inheritance reasoning systems while generality is a new inference tool introduced in this paper.

6.1 Specificity

Many inheritance reasoners use specificity as a tool for selecting natural preference criterion. This preference criterion is based on simple intuition: the more knowledge that is used to generate the degree of belief the better is the degree of belief. Sentence α represents more knowledge than β if $\alpha \Rightarrow \beta$ is deducible from the knowledge base. For instance, the fact that "Royal_Elephant(Clyde) ⇒ Elephant(Clyde)" indicates that the knowledge Royal_Elephant(Clyde) should be preferred when including a degree of belief in Gray_Thing(Clyde). The idea of specificity was first given by Touretzky in the form of inferential distance [14]. Since then, many variations of specificity have appeared in the literature. The basic idea behind specificity-based algorithms including shortest distance is that the more specific the information is, the more precise the result is.

The system also adopts specificity as one of the tools for resolving conflicts. While the definition of specificity in other inheritance reasoners is based on network path or standard nonmonotonic logic, we are interpreting the specificity in terms of set relationships among classes. Suppose we have two conflicting reference classes, say A, B, and class A is a subset of class B(A is more specific than B), we choose A as the reference class. This is basically the same as the way in which the conventional inheritance reasoners solve conflicts. Now, the question is what if A has nothing to do with B ? In this case, B is irrelevant to class A and we have conflicts. In the system, it creates a derived class which is the conjunction of A and B, and that derived class becomes the target reference class, which is obviously more specific than either A or B. By doing this, to get the most specific reference class, the system does specification using set conjunction as many times as it can.

However, specificity alone can not solve all conflict problems. Let us pay a second visit to above example. Assume that the class Elephant has only 3 instances, and all these instances are known to not gray. It is not safe saying that Elephants usually are not gray. Where does this problem come from ? The reason is that the size of the class Elephant is so small that it can not represent correctly the characteristic of the class Elephant in the real world. The smaller the number of instances for a class is, the less likely the instances can represent the property of the real class correctly. Thus, if the reference class, whether original or derived, does not have enough instances, it becomes an immature class, and thus be prohibited to be a reference class because it does not have enough statistical information. The following section investigates this problem in more detail.

6.2 *Generality*

Generality is a complementary criteria of specificity. The basic idea behind the generality concept is that the more information we have, the more correct estimates we can have. Speaking in terms of class, when we have more sample instances, we have higher quality guesses. The instances of the class must have the generality which can represent the general characteristic of the class they belong to. We propose two conditions which the samples in a class must satisfy to maintain the minimal degree of generality of a class. In case a class has a very small number of samples, it is quite possible that these samples are not correctly representing the general characteristic of the class, and lose their generality. Even though the cardinality is large enough to avoid the immature class problem, if the sample is not evenly collected from its subpredicates, we have biased class problem. These requirements are discussed in more detail in the following.

6.2.1 *Immature Class*

For example, suppose we have a rule "Americans are tall," "Connecticut residents are not tall," and "Jim is Connecticut resident." If the instances of Connecticut about property *tall* are very few, we can not conclude that Jim is not tall based on the Connecticut class. When the cardinality of Connecticut class is very small, it is quite possible that this small sample does not represent correctly the characteristic of Connecticut residents. Then, how large should the class size be ? We need a quantitative study for deciding the cardinality of classes.

We should consider two kinds of predicates, leaf nodes and intermediate nodes, since each of these has different characteristics of minimum cardinality. First, for leaf predicates, each class, say A, is divided into three subgroups(A^+, A^-, and A^*). Among them, we can consider a set $A^+ \cup A^-$ as a sample of class A. Considering the predicate A as a variable, we can easily see that the distribution of A follows a binary distribution. Statistically speaking, the problem can be regarded as selecting an appropriate sample size in estimating the proportion of A. In this type of predicate, we have dichotomous values, 'yes' or 'no', and thus we only need to estimate the proportion of the first value.

We will consider the following situation such that restricting to an acceptable level the probability that the difference between population mean P of predicate A and sample mean p of A is greater than a specified value. Let n denote the sample size of the population. Note that p is defined as $\frac{\|A^+\|}{\|A^+ \cup A^-\|}$, and n is denoted as $\|A^+ \cup A^-\|$ where $\|\cdot\|$ means cardinality function. N is the cardinality of domain of the universe, which is denoted as $\|\mathcal{D}\|$.

If the permissible error in the estimate of the population value of the mean is d and the degree of assurance desired is $1 - \alpha$, then the following inequality holds.

$$Pr(|P - p| > d) \leq \alpha \tag{5}$$

According to *the central limit theorem* [3], p follows the normal distribution, i.e., $p \sim N(P, (1-f)\sigma^2(i)/n)$. Using t-distribution analysis, the above equation is rearranged to conclude

$$n \geq N[1 + N(\frac{d}{t_\alpha s})^2]^{-1}.$$

In particular, when we have a very large number of objects in our domain, the approximate size of a sample can be decided by the following.

$$n > \frac{t_\alpha^2 p(1-p)}{d^2}. \tag{6}$$

Here we say that n is the minimum cardinality of A to avoid the immature class. If the class cardinality is less than the above number, the corresponding class is considered to be an immature class. One drawback of this approach is that when the size of the original population is not large enough, it needs a correct value of the size of the population, which we have to estimate.

Secondly, we will investigate the intermediate predicates. Intermediate predicates are defined to have a set of subpredicates. If there is a strong edge from predicate P to predicate Q, we define the predicate P as the subpredicate of Q. For an intermediate predicate Q and a set of subpredicates of Q, say P_1, P_2, ... P_k, these subpredicates partition the domain of universe into k disjoint subsets. In this case, predicate Q is called the parent predicate. If we consider each of these subpredicates as a category value of discrete variable, the distribution of predicate Q follows a multinomial distribution. For a parent predicate, computing the minimal cardinality involves a multinomial distribution, which means that we need to consider the statistical distributions of all subpredicates simultaneously. To estimate the minimal cardinality of the intermediate predicate, we need a correct value of the size of each subpredicate. Suppose $\|P_j'^+\|$ and $\|P_j'^-\|$ represents the cardinality of P_j^+ and P_j^-, respectively, in the real population. $Q_j'^+$ and $Q_j'^-$ are defined in a similar way. Let w_j and W_j be defined, respectively, as $w_j = \|P_j^+\|/\|Q_j^+\|$ and $W_j = \|P_j'^+\|/\|Q_j'^+\|$. The proportion of each subgroup w_j can be represented simultaneously, for a predicate Q and jth value assignment in Q. The question is how to derive the smallest cardinality n for a random sample from a multinomial population so that the probability would be at least $1 - \alpha$ and that all of the estimated proportions would simultaneously be within

```
for n=1 to N
    for j=1 to k
        z_j = d_j√n/√(p_j(1-p_j))
        α_j = 2(1 - Φ(z_j))
    end for
    if Σα_j < α then
        return n
end for
return "no value of n found"
```

Figure 3: Algorithm computing n for intermediate classes

specified distances of the true population proportions. This constraint can be expressed by the following equation.

$$Pr(\bigcap_{j=1}^{k} |W_j - w_j| > d_j) \leq \alpha \tag{7}$$

For an individual parameter P_j, the probability that the sample value in the table lies outside the specified interval is

$$\alpha = Pr(|Z_j| \geq d_j\sqrt{n}/\sqrt{w_j(1-w_j)}) = 2(1 - \Phi(z_j)) \tag{8}$$

where Z_j is a standard normal random variable, Φ is the cumulative standard normal distribution, and $z_j = d_j\sqrt{n}/\sqrt{w_j(1-w_j)}$. The algorithm is provided in Figure 3 to calculate the minimal cardinality of intermediate predicates.

6.2.2 Biased Class

Even though a class has a large number of instances, when the sample among subpredicates are not proportionally distributed, it may cause another problem. For instance, suppose class America has subpredicates including California, Connecticut etc. If the samples selected from each of the subpredicates are not evenly distributed, like 80% from Connecticut and 10% from California, obviously the characteristic of class America has biased values, which in turn leads the reasoner to draw anomalous conclusions.

Biased class problems have a somewhat different characteristic from the immature class problems. If we have a very large database which has a large number of instances, the immature class problem would have been eliminated. On the other hand, having large number of instances does not mean even distribution of samples across the subpredicates. Sampling must recognize the different natures of the subpredicates and must be done according to the characteristic of subpredicates. In this section, we present a framework that can be used to measure the degree of biasedness of the parent predicate with respect to its subpredicates and how such a measurement can be used to estimate the trustworthiness of the parent predicate.

When a predicate consists of a number of subpredicates, the simple random sampling technique alone can not always represent the characteristic correctly. If the accumulated samples are not selected proportionally across the subpredicates, we might have a biased parent predicate. Statistically speaking, a subpredicate is the division of the population to be sampled into a number of blocks, each of which is to be sampled separately. Notice that one class may have different sets of disjunctive exhaustive subpredicates. If the average values of subpredicates are significantly different from each other and cardinalities are not proportional to the size of the subpredicates in the population, then we have a biased class.

The optimum cardinality of each subpredicate is to use sampling fractions, or rates, which are directly proportional to the size of each subpredicate. However, a drawback to optimal sampling is that it is practically almost impossible to employ an optimal sampling fraction. Optimal allocation to one set of subpredicates can not be optimal for other subpredicates. Another thing is that correct sampling can only be achieved if one has prior knowledge of the sizes of subpredicates in the target population as we have seen in the above formulas. Therefore, it is worth noting the effect of inaccuracy in subpredicate cardinalities. Since optimal fractions would rarely be used in practice, we should use an approximate number of the size of each subpredicate. In practice, we do not know the mean values for the subgroups of the original population and thus we may not be able to compute the degree of biasedness directly by comparing the mean values for the population subgroups and the mean values for the sample subgroups. The suggested approach is first to set the limit d of the bias to a predetermined value (e.g., 0.1) and then compute the confidence limit in the following way.

For an intermediate predicate Q and a set of subpredicates of Q, say $P_1, P_2, \ldots P_k$, these subpredicates partition the domain of universe in k disjoint subsets. Suppose we reason about predicate R, and $p_{(j,R)}, j = 1, \cdots, k$ denotes the average value of subpredicate P_j with respect to R. As shown earlier, we note that $w_j = \|P_j^+\|/\|Q_j^+\|$ and $W_j = \|P_j'^+\|/\|Q_j'^+\|$. Then the mean value of the parent predicate Q with respect to the value of R is given as

$$p_Q = \sum_{j=1}^{k} w_j p_{(j,R)} . \qquad (9)$$

According to [3], the overall distribution of predicate Q follows the t-distribution, and thus the estimate of the overall variance, S_Q^2, is then

$$S_Q^2 = \frac{1}{N^2} \sum_{j=1}^{k} \|P_j'^+\| (\|P_j'^+\| - \|P_j^+\|) \frac{\sigma_Q^2}{\|P_j^+\|} . \qquad (10)$$

As the variance of the original population is not known in practice, the general practice is to substitute it with the variance of the sample. After the substitution, the above equation becomes

$$S_Q^2 = \frac{1}{N^2} \sum_{j=1}^{k} \|P_j'^+\| (\|P_j'^+\| - \|P_j^+\|) \frac{S_j^2}{\|P_j^+\|} \qquad (11)$$

where S_j^2 is the computed variance of predicate P_j with respect to R. The substitution entails us to use the t-distribution from which the confidence limit is computed as $p_Q \pm t_\alpha S_Q$. Finally, it can be rearranged as

$$\alpha = t^{-1}(d/S_Q) \qquad (12)$$

where $1 - \alpha$ is said to be the confidence level. Now the confidence level of each intermediate predicate can be computed and if the user provides a certain degree of certainty as a parameter, intermediate predicates whose certainty level are lower than that will be prohibited to be reference classes.

7 Algorithm

We will now define the reasoning algorithm and discuss the basic ideas which motivated this particular method. Before we describe the algorithm, a set of rules are specified in the following. The probability for a strict rule is omitted for the sake of simplicity. By definition, strict rules($\Rightarrow$) contain the probability of [1,1]. If we use only [**AX0**]-[**AX5**], we can do monotonic deductive inference because rules [**AX0**]-[**AX5**] are sound. However, monotonic deductive inference alone cannot do a large amount of useful inheritance reasoning.

[**AX0**] If $\{c \Rightarrow P_1, P_1 \Rightarrow P_2\}$, $c \Rightarrow P_2$.
[**AX1**] If $\{c \Rightarrow P_1, P_1 \not\Rightarrow P_2\}$, $c \not\Rightarrow P_2$.
[**AX2**] If $\{P_1 \Rightarrow P_2, P_2 \Rightarrow P_3\}$, $P_1 \Rightarrow P_3$.
[**AX3**] If $\{P_1 \Rightarrow P_2, P_2 \not\Rightarrow P_3\}$, $P_1 \not\Rightarrow P_3$.
[**AX4**] If $\{P_1 \rightarrow P_2:[\alpha,\beta], P_2 \Rightarrow P_3\}$, $P_1 \rightarrow P_3:[\alpha,\beta]$.
[**AX5**] If $\{P_1 \rightarrow P_2:[\alpha,\beta], P_2 \not\Rightarrow P_3\}$, $P_1 \not\rightarrow P_3:[\alpha,\beta]$.
[**AX6**] If $\{P_1 \Rightarrow P_2, P_2 \rightarrow P_3\}$, $\{P_1 \rightarrow P_2, P_2 \rightarrow P_3\}$, $\{P_1 \Rightarrow P_2, P_2 \not\rightarrow P_3\}$, or $\{P_1 \rightarrow P_2, P_2 \not\rightarrow P_3\}$,
    1. $P_1 \rightarrow P_3:[S(P_1 \rightarrow P_3), Pl(P_1 \rightarrow P_3)]$,
       if $\mathcal{X}^2(P_1,P_3) > \mathcal{X}^2_{[4,0.95]}$ and $P_1^+ \cap P_3^+ \geq P_1^+ \cap P_3^-$
    2. $P_1 \not\rightarrow P_3:[S(P_1 \not\rightarrow P_3), Pl(P_1 \not\rightarrow P_3)]$,
       if $\mathcal{X}^2(P_1,P_3) > \mathcal{X}^2_{[4,0.95]}$ and $P_1^+ \cap P_3^+ < P_1^+ \cap P_3^-$
    3. otherwise, there is no edge between P_1 and P_3.

The algorithm takes network information and a focus node s as input, and generates conclusions with probability for all of the destination nodes reachable from the focus node. As we have mentioned, immature or biased class is apt to generate anomalous results. Therefore, these classes are deleted in advance from the network. The algorithm then proceeds by first finding a set of nodes(N_s) which the focus node belongs to, and delete nodes, within N_s, which s doesn't belong to. Now for each node P in N_s, calculate all possible paths reachable from P using [**AX0**]-[**AX6**]. The approach we present in the algorithm has the characteristic of credulous extension. Credulous extension of an inheritance hierarchy Γ with respect to a node s is a maximal unambiguous a-connected subhierarchy of Γ with respect to a. When there is a path $s \Rightarrow E$ or $s \Rightarrow \cdots \Rightarrow E$, then this path substitutes all paths from s to E. If there are two positive(negative) paths which share the same intermediate and end node, delete the redundant paths(resp. negative). Given two paths

$$s \Rightarrow A_1 \overset{+}{*} A_2 \overset{+}{*} \cdots A_k \overset{+}{*} E \tag{A}$$

and

$$s \Rightarrow B_1 \overset{+}{*} B_2 \overset{+}{*} \cdots B_l \overset{-}{*} E, \tag{B}$$

If there is an edge $A_i \Rightarrow B_j$, $i = 2, \ldots, k$, $j = 2, \ldots, l$, then
    delete path (B);
else
    create a derived class $D = (A_1 \wedge A_2 \wedge \cdots \wedge A_k \wedge B_1 \wedge B_2 \wedge \cdots \wedge B_l)$;
    If D is not immature class nor biased class then
        delete both paths;
        create new path $s \Rightarrow (A_1 \wedge A_2 \wedge \cdots \wedge A_k \wedge B_1 \wedge B_2 \wedge \cdots \wedge B_l) \triangleright E$;
    else
        delete both paths;
        create new path $s \Rightarrow ((A_1 \wedge A_2 \wedge \cdots \wedge A_k) \vee (B_1 \wedge B_2 \wedge \cdots \wedge B_l)) \triangleright E$;
    end if
end if

Figure 4: Algorithm for resolving conflicts

repeat the procedure described in Figure 4 until there is only one path left for each destination node.

Now only one path is assigned to each destination node in the network. Furthermore, we have temporary node sets, which are formulas closed under union and conjunction. To calculate the probabilities of each path which are defined based on these temporary nodes, probability definitions in Definition 1 must be extended. For predicates A and B, the following set operations can be easily understood to accommodate the extended formula.

1. $(A \cap B)^+ =_{def} (A^+ \cap B^+)$

2. $(A \cap B)^- =_{def} (A^- \cup B^-)$

3. $(A \cap B)^* =_{def} (A^+ \cap B^*) \cup (A^* \cap B^+) \cup (A^* \cap B^*)$

4. $(A \cup B)^+ =_{def} A^+ \cup B^+$

5. $(A \cup B)^- =_{def} A^- \cap B^-$

6. $(A \cup B)^* =_{def} (A^- \cap B^*) \cup (A^* \cap B^-) \cup (A^* \cap B^*)$

8 Examples

This section shows a couple of examples which show how the specificity or generality works in these cases. Nixon's diamond, shown in Figure 5, has been largely used to show the ambiguity of inheritance reasoning. Nixon is a Republican and at the same time a Quaker. Since $\mathcal{X}^2(R,P) = 1161$ is greater than $\mathcal{X}^2_{[4,0.95]} = 9.49$ and $R^+ \cap P^+(130) < R^+ \cap P^-(320)$, there is a rule "R(Republican) is usually not P(Pacifist)" based on the semantics of rules. Similarly, a rule "Q(Quaker) is usually P(Pacifist)" exists since $\mathcal{X}^2(Q,P) = 923$ is greater

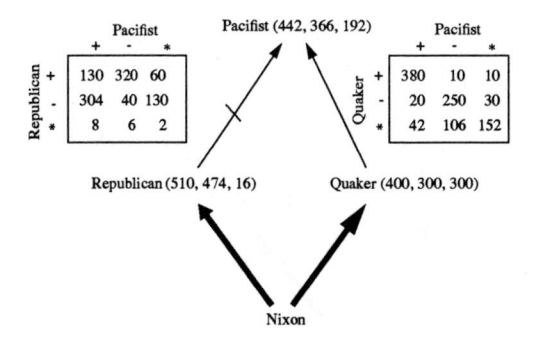

```
           Pacifist        Pacifist (442, 366, 192)      Pacifist
           +  -  *                                      +   -   *
    ┌─┬─────────────┐        ↗    ↖              ┌─┬──────────────┐
Rep │+│ 130 320  60 │       /      \        Quak │+│ 380  10   10 │
ubl │-│ 304  40 130 │      /        \       er   │-│  20 250   30 │
ica │*│   8   6   2 │                             │*│  42 106  152 │
n   └─┴─────────────┘                             └─┴──────────────┘

        Republican (510, 474, 16)       Quaker (400, 300, 300)
                    ↖                    ↗
                      ↖                ↗
                        Nixon
```

Figure 5: Nixon's diamond

than 9.49 and $Q^+ \cap P^+(380) > Q^+ \cap P^-(10)$. Therefore, Republicans are usually not-Pacifist while Quakers are Pacifists. In this case, skeptical reasoning does not generate any conclusion because conflict occurs at pacifist, while credulous reasoning provides two extensions: Nixon⇒Quaker→Pacifist and Nixon⇒Republican↛Pacifist, and does not give any preference among these extensions. In the proposed approach, the diagram will be transformed into one of the following paths.

1. Nixon→(Quaker ∩ Republican)→Pacifist, if
    - $\mathcal{X}^2$(Quaker ∩ Republican, Pacifist) $> \mathcal{X}^2_{[4,0.95]}$.
    - (Quaker)$^+$ ∩ Republican$^+$ ∩ Pacifist)$^+$ ≥ (Quaker)$^+$ ∩ Republican$^+$ ∩ Pacifist)$^-$

2. Nixon→(Quaker ∩ Republican)↛Pacifist, if
    - $\mathcal{X}^2$(Quaker ∩ Republican, Pacifist) $> \mathcal{X}^2_{[4,0.95]}$.
    - (Quaker)$^+$ ∩ Republican$^+$ ∩ Pacifist)$^+$ < (Quaker)$^+$ ∩ Republican$^+$ ∩ Pacifist)$^-$

3. There is no edge between Nixon and Pacifist, if
    - $\mathcal{X}^2$(Quaker ∩ Republican, Pacifist) $\leq \mathcal{X}^2_{[4,0.95]}$.

We present another version of the Nixon example in Figure 6 to show the problem of immature class. As we have seen in Figure 5, "R(Republican) is usually not P(Pacifist)." Similarly, "Q(Quaker) is usually P(Pacifist)" because $\mathcal{X}^2(Q,P) > 9.49$ and $Q^+ \cap P^+(2) > Q^+ \cap P^-(1)$. However, the set for Quaker is (3,6,991), which means only nine elements are in the class Quaker as samples. In this case, because it does not satisfy the minimum cardinality requirement, we just disregard the class Quaker because it is an immature class. Therefore, the system concludes that Nixon is not Pacifist.

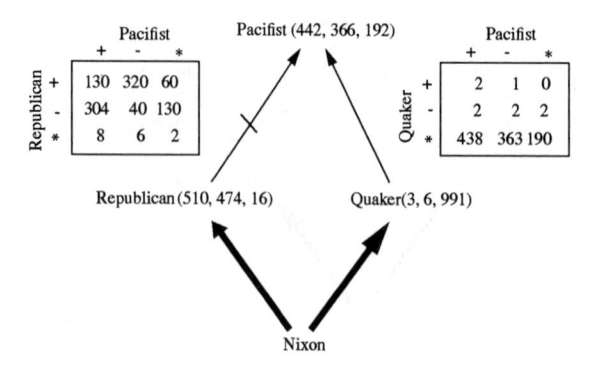

Figure 6: Nixon's diamond with immature class

9 Conclusion

In this paper, we proposed a set-oriented statistical model for inheritance reasoner. The $\mathcal{X}^2$ method using contingency table is used to decide the existence of rules. Due to the introduction of unknown state, likelihood of each rule is given as an interval. Among the features of the general inheritance reasoner, we focused on the way of resolving conflicts. Two basic inference tools were introduced and explained to resolve conflicts in inheritance reasoning. Specificity, one of the most common inference tool, is interpreted differently from the viewpoint of set cardinality. As a complementary rule of specificity, the generality rule is defined as a measure of the degree of generality of the instances of each class. Generality plays the role of complementing the drawback of the specificity. The tradeoff between specificity and generality can prevent anomalous results caused by small cardinality and uneven distribution of instances.

References

1. F. Bacchus. A Modest, but semantically Well-Defined, Inheritance Reasoner, *Proceedings of 11th IJCAI* 1989.

2. F. Bacchus, *Representing and Reasoning about Probabilistic Knowledge*, MIT Press, 1990.

3. V. Barnett, *Sample Survey Principles and Methods*, New York:Oxford University Press, 1991.

4. R. Cox, Probability, frequency, and reasonable expectation, *American Journal of Physics* 14:1, 1946, pp. 1-13.

5. E. Davis, *Representation of Commonsense Knowledge*, Morgan Kaufman Publisher. 1990.

6. D. W. Etherington, *Reasoning with Incomplete Information*, Morgan Kaufmann, Los Altos, CA, 1988.

7. H. Geffner, *Default Reasoning:Causal and Conditional Theories*, MIT Press, 1992.
8. B. A. Haugh, Tractable Theories of Multiple Defeasible Inheritance in Ordinary Nonmonotonic Logics, *Proceedings AAAI-88*, St. Paul, MN, 1988
9. T. Krishnaprasad, M. Kifer and D. S. Warren, On the Declarative Semantics of Inheritance Networks, in *Proceedings IJCAI-89*, Detroit, MI 1093-1103.
10. H. Przymusinska and M. Gelfond, Inheritance Hierarchies and Autoepistemic Logic, Technical Report, Computer Science Department, University of Texas at El Paso, TX, 1988.
11. H. Reichenbach, *Theory of Probability*, University of California Press, Berkeley and Los Angeles, CA, 1949.
12. G. Shafer, *A Mathematical theory of Evidence*, Princeton University Press, 1976.
13. L. A. Stein, Resolving Ambiguity in Nonmonotonic Inheritance Hierarchy, *Artificial Intelligence*, **55**, pp. 259-310, 1992.
14. D. S. Touretzky, J. F. Horty, and R. H Thomason, A clash of Intuitions:The current state of Nonmonotonic multiple inheritance systems, in *Proceedings IJCAI-87*, pp. 476-482, 1987.

INTEGRATING HIGH AND LOW LEVEL COMPUTER VISION FOR SCENE UNDERSTANDING

Raashid Malik and Siu So
Department of Electrical Engineering and Computer Science
Stevens Institute of Technology, Hoboken, NJ 0030
Tel:(201)216-5623, Fax:(201) 216-8246

ABSTRACT

Object recognition involves "high level" reasoning on information extracted from "low level" grey scales images stored as matrices in computers. Analyses and operations directly on image matrices are referred to as low level processing whereas operations on information extracted from these images (such as object boundaries, corners, shadows etc) are referred to as higher level processing.

The information that is extracted from these images is often not without ambiguity. Reasoning about the shape and structure of an object can lead to serious errors if the boundaries and corners in a scene are ignored or misclassified. Labeling schemes for polyhedral objects, such as the Huffman-Clowes-Waltz (HCW) method, ignore the possibilities of errors in vertex identification and may therefore lead to scene "misunderstanding". The HCW method is a powerful scheme for establishing consistencies in vertex labeling in line sketch images. Such line sketch images are usually derived from grey scale image matrices. Insufficient contrast amongst surfaces in the scene may lead to failures by edge detection processes to mark and identify "lines" in the line sketch. This in turn will lead to problems with the HCW method.

The research presented in this paper is concerned with utilizing high level methods to assist in enhancing lower level methods for vertex and edge detection. Several algorithms are presented to extract and represent images where lines have been located. Another class of algorithms is introduced which reduces these representations to line (or primal) sketches of the original images. The results of experiments that use these methods are also described.

I. INTRODUCTION

Important information about an object in a scene is retained in an outline of that object. Even more information about an object is captured in a line sketch of the object. Humans, including babies, can easily recognize objects from simple line drawings. This phenomenon has convinced many vision researchers that the information in a line sketch is often sufficient for object recognition tasks. Line drawings however are typically generated by human artists in rendering scenes containing objects. Electronically captured images of such scenes are however grey scale images usually represented in computer systems as matrices of integers (also known as pixels). Automating object recognition tasks on a computer requires working with such (typically large) matrices of pixels. An initial objective, prior to undertaking the recognition task, is to convert the image matrix into a suitable line sketch. The line sketch is a much simpler representation of the scene and may thus be easier to work with in developing recognition algorithms. An objective of this paper is to present a class of methods for transforming grey level images into line sketches. We concentrate on scenes containing polyhedral objects, in which case the sketch will only contain straight lines.

II. BACKGROUND

Most well known methods [1][2] for transforming grey level images into line sketches involve two steps. The first step is concerned with identifying all pixels in the image which are on, or close to, a significant change in grey level intensity. This stage is referred to as edge detection/thresholding. The second step involves isolating and consolidating groups of edge detected pixels into their boundary lines. This stage is referred to as line or boundary extraction. This standard scheme is outlined in Fig. 1a.

The scheme used in this paper is however different. This new scheme also involves two steps. The first step is concerned with locating the orientation in the image at which significant changes in grey intensity occur. This step is known as line location and uses projections of the image to find these lines. The second step in this scheme involves determining the exact extent of all line segments, which implies detecting the vertices in the image. This scheme is outlined in Fig. 1b. An important difference in the two schemes is that in the new scheme the second processing stage (the vertex detector) utilizes the original grey level image in addition to the line list in making decision concerning the line sketch.

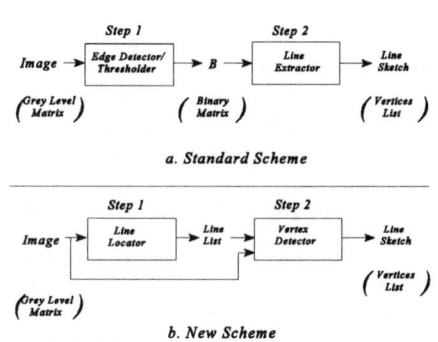

Fig. 1 Standard and new schemes in extracting a line sketch from a grey level image.

Line locators have been described in previous work [3][4]. In this paper we concentrate on the second step in this scheme which is the extraction of the line sketch (Fig. 1 step 2).

III. PROBLEM STATEMENT

A grey level image matrix $\mathcal{I}(x,y)$ is available. The column index x ranges from 1 to H and the row index y from 1 to V. The number of pixels in this image, m (=H×V), is a measure of the resolution in the image. This matrix has been processed using projection methods (which are akin to the Hough Transform) to locate lines or directions along which significant changes in grey intensity are indicated. Fig. 2a shows an example of an original image $\mathcal{I}$ and Fig. 2b the lines located in this image. Our objective in this paper is to reduce Fig. 2b to the line sketch shown in Fig. 2c. Notice that the transformation from 2b to 2c involves the removal or elimination of vertices and/or line segments in Fig. 2b. The problem therefore is to devise algorithms for efficiently converting the information in the original image (Fig. 2a) and the line list (Fig. 2b) into a line or primal sketch[8] (Fig. 2c).

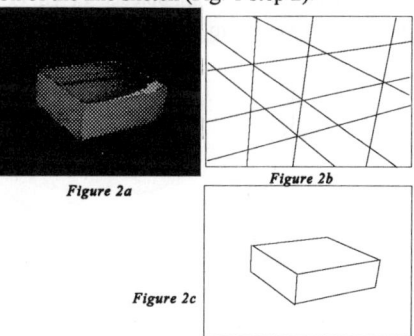

Fig. 2 The processing stages in obtaining a line sketch of an object in a scene. a) Grey level image, b) a representation of the output of the Line Locator (Fig. 1b step 1), c) a representation of output of the Vertex Detector (Fig. 1b step 2).

IV. APPROACH

The approach taken in this paper is from a graph theoretic perspective. The located line list may be viewed as a graph and the resulting line sketch may be viewed as a transformation from one graph to another. Since the resulting graph is in fact a subgraph of the original graph, the problem is really to determine an appropriate **graph reduction algorithm**.

In Section V we present a number of **graph extraction algorithms** for representing the located lines in suitable ways. Two representations have been found to be useful. These are the Vertices Graph--a graph whose nodes are the line intersections and the Regions Graph--the nodes of which are the spaces between the lines. The weights on the links in these graphs are extracted directly from the grey scale image matrix $\mathcal{I}(x,y)$. The computational complexity, as a function of the number of lines, n, and the image resolution, m, is also evaluated for these graph extraction algorithms.

The graph reduction algorithms are discussed in Section VI. These algorithms reduce the graphs extracted using the procedures in Section V to generate subgraphs that are in fact representations of the desired line sketch. The elimination of nodes and links in the graph depends on various decision criteria based on pixel variation in the image. Links are essentially removed based on the brightness variation in the image and on "high level" understanding of the significance of the removal of a link. Link removals may result in sketches of impossible or physically unrealizable objects. The reduction algorithm must incorporate this knowledge in order to generate "correct" line sketches.

V. GRAPH EXTRACTION ALGORITHM

Let $\mathcal{L}_0$ be the set of lines extracted by a Line Locator (Fig. 1b new scheme) from an image $\mathcal{I}$ which contains polyhedral objects. $\mathcal{L}_0 = \{l_1, l_2, ..., l_i, ..., l_n\}$, where l_i's are parameters of the ith line in the image in which a total of n lines were located.

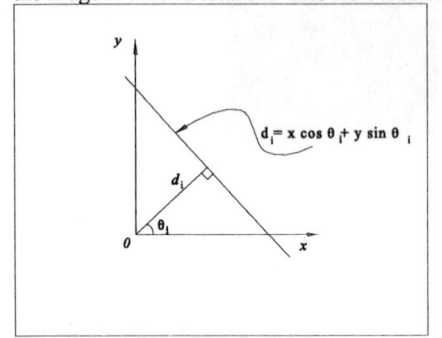

Fig. 2 Line equation represented in parametric form, where x and y are image coordinate axes and 0 the coordinate origin.

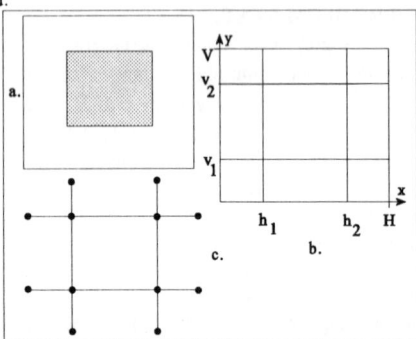

Fig. 3 a. Image of a square object on a white background. **b.** Extracted lines:$\mathcal{L}_0 = \{(v_1,0), (h_1, \pi/2), (v_2,0), (h_2, \pi/2)\}$. **c.** Resulting vertices graph. The points are the nodes and lines between them the links.

The line equations are written in parametric form for representation purposes (see Fig. 2):
$$d_i = x \cos \theta_i + y \sin \theta_i$$
Therefore, $\quad l_i = (d_i, \theta_i), \quad 1 \leq i \leq n$.

In some of the algorithms that follow the image frame lines are also required, so we define an enhanced line set $\mathcal{L}$.

$\mathcal{L}=\mathcal{L}_0 \cup \{(1,0), (1,\pi/2), (H,0), (V,\pi/2)\}$, where $(1,0), (1,\pi/2), (H,0), (V,\pi/2)$ corresponds to the image frame lines x=1, y=1, x=H and y=V. The example in Fig. 3a shows the image of an object and Fig. 3b a representation of the extracted lines for that image.

5.1 Vertices Graph

The vertices graph is a graph:
$$\mathcal{G}_V=(\mathcal{N}_V, \mathcal{L}_V),$$
where the nodes of the vertices graph, $\mathcal{N}_V$, are the intersections among the lines in $\mathcal{L}$ and the links, $\mathcal{L}_V$, are the line segments connecting the intersection points.

Since the line sketch image will have fewer lines and vertices than the image resulting from the line locator, the Vertices Graph is the appropriate data representation model to eliminate vertices and edges. The node set $\mathcal{N}_V$ identifies clearly which vertices must be checked and $\mathcal{L}_V$ contains the edge set that must be examined.

Let (x_{ij}, y_{ij}) be the intersection point of line $l_i \in \mathcal{L}_0$ and line $l_j \in \mathcal{L}$ such that $i \neq j$. Then the set of nodes $\mathcal{N}_V = \{(x_{ij}, y_{ij}) \mid 1 \leq x_{ij} \leq H$ and $1 \leq x_{ij} \leq V\}$, i.e., the set of intersection points which lie within the image frame. The number of elements in $\mathcal{N}_V$, $|\mathcal{N}_V| \leq n(n-1)/2 + 2n$, where $n(n-1)/2$ is the maximum number of intersection amongst n lines and $2n$ is the number of end points (frame intersections).

The set of links $\mathcal{L}_V = \{\{(x_{ij}, y_{ij}), (x_{ik}, y_{ik})\} \mid 1 \leq i \leq n, 1 \leq j \leq n+4, 1 \leq k \leq n+4, i \neq j, i \neq k, j \neq k$ and (x_{ij}, y_{ij}), (x_{ik}, y_{ik}) are neighbors on line l_i and elements of $\mathcal{N}_V\}$.

5.1.1 Vertices Graph Extraction (VGE Algorithm)

Given the line sets $\mathcal{L}$ and $\mathcal{L}_0$, the node and link sets, $\mathcal{N}_V$ and $\mathcal{L}_V$ may be obtained using the following algorithm.
0. Initialize $\mathcal{N}_V$ and $\mathcal{L}_V$ to be empty sets.
1. For each line l_i in $\mathcal{L}_0$, Begin
2.   For each line l_j in $\mathcal{L}$, $i \neq j$, Begin
3.   Compute the legitimate intersection points with each line l_j in $\mathcal{L}$, $i \neq j$. (Legitimate intersection points have to satisfy the condition that $1 \leq x_{ij} \leq H$ and $1 \leq y_{ij} \leq V$.)
4.   Let $\mathcal{N}_i$ be the set of intersection points on line l_i resulting from step 3. Add $\mathcal{N}_i$ to $\mathcal{N}_V$.
5.   End
6.   If θ_i (the orientation of l_i) is not $\pi/2$, then
         sort the elements of $\mathcal{N}_i$ by their x-coordinates using radix sort
     else   sort $\mathcal{N}_i$ on y-coordinates using radix sort
     Name this sorted set of $\mathcal{N}_i$ vertices as $\mathcal{S}_i$.
7.   For each pair of consecutive vertices in $\mathcal{S}_i$, add to $\mathcal{L}_V$.
8. End
9. Return $\mathcal{N}_V$ and $\mathcal{L}_V$ as the node and link sets of $\mathcal{G}_V$

5.1.2 Complexity of Obtaining the Vertices Graph

The double FOR loops in steps 1 and 2 always need $n(n+4)$ times to execute, where n is the number of lines in the image. Steps 3 and 4 will run in $n(n+4)$ times regardless of the number of intersection points, which is $O(n^2)$. Total execution time in step 6 consists of the sorting in step 6 and the FOR loop in step 1. The net effect will be $O(n^2)$, where $O(n)$ is for the sorting on step 6 and another $O(n)$ is a result of the outer loop on step 1. Step 7 is similar to step 4 which is $O(n)$. Therefore, the complexity of this algorithm has running time of $O(n^2)$.

5.2 Regions Graph

The regions graph is a graph:

$$\mathcal{G}_R = (\mathcal{N}_R, \mathcal{L}_R),$$

where the nodes of the regions graph, $\mathcal{N}_R$, are the regions obtained by successively segmenting the image frame. The links, $\mathcal{L}_R$, are the adjacency information about the regions in $\mathcal{N}_R$.

The regions graph is related to the dual of the vertices graph. The nodes in $\mathcal{N}_R$ represent subregions of the image as shown in Fig.s 4a, b and c. If two regions share a common line segment in the image then the corresponding nodes in $\mathcal{N}_R$ are connected by the link from $\mathcal{L}_R$. The regions may be obtained by successively segmenting the image as shown in Fig. 6.

The utility of the Regions Graph arises from the need to identify the parts of the image $\mathcal{I}$ which straddle a particular edge. This information is required for deciding if an edge really is an edge or not.

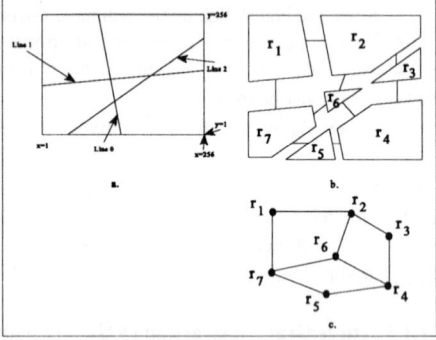

Fig. 4 a. An image with 3 extracted lines. b. Separation of the image into regions. c. The resulting regions graph.

5.2.1 Intersections of Line and Region

Intersection of a line l_i (defined by $d_i = x \cos \theta_i + y \sin \theta_i$) and region r_j (defined by N_j vertices $\{(x_{j1}, y_{j1}), (x_{j2}, y_{j2}), \ldots, (x_{jN_j}, y_{jN_j})\}$) is established by evaluating

$d_i - x_{jk} \cos \theta_i + y_{jk} \sin \theta_i$ and checking for a sign change in successive evaluations. A change in sign indicates that successive vertices lie on opposite sides of the line.

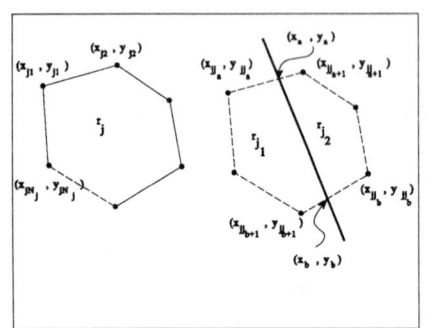

Fig. 5 a. A region r_j with N_j vertices. b. After splitting r_j, r_{j_1} and r_{j_2} emerges.

5.2.2 Region Splitting

Suppose the intersection points of line l_i and region r_j are (x_a, y_a) and (x_b, y_b) which occur after vertices (x_{jj_a}, y_{jj_a}) and (x_{jj_b}, y_{jj_b}) respectively (see Fig. 5). Then the two subregions r_{j_1} and r_{j_2} are

$r_{j_1} = \{(x_{j1}, y_{j1}), (x_{j2}, y_{j2}), \ldots, (x_{ja}, y_{ja}), (x_a, y_a), (x_b, y_b), (x_{jb+1}, y_{jb+1}), \ldots, (x_{jN_j}, y_{jN_j})\}$ and
$r_{j_2} = \{(x_b, y_b), (x_a, y_a), (x_{ja+1}, y_{ja+1}), \ldots, (x_{jb}, y_{jb})\}$.

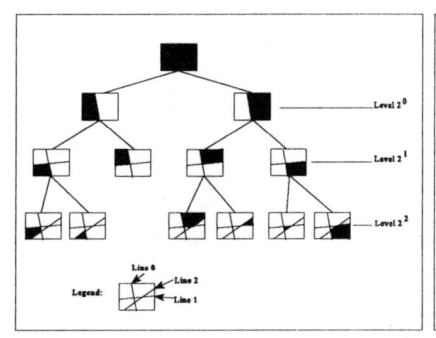

Fig. 6 Process of segmenting the image by the extracted lines. The segmented regions are the leaf nodes of the binary tree.

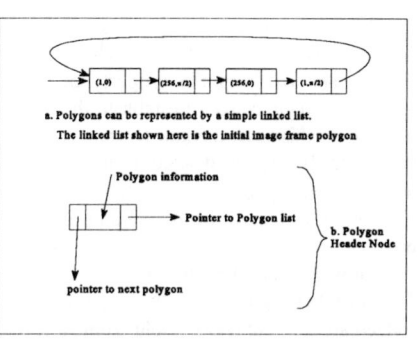

Fig. 7 Data Structure of the Regions Graph at segmentation stage.

Although the process of segmenting the image results in a binary tree data structure, a more efficient data structure for storing the regions graph is a simple linked list(Fig. 7a). A region or polygon are represented by a simple circular linked list of its vertices (in clockwise order). The polygons are connected by a list of polygon header nodes(Fig. 7b). Following this convention, a region r_j is then defined by N_j lines $\{l_{j1}, l_{j2}, \ldots, l_{jN_j}\}$.

5.2.3 Adjacency Properties of Regions Graph:

Two regions r_a and r_b are adjacent if exactly two consecutive vertices of each polygon are common to both r_a and r_b regions.

This property is used in the algorithm below for obtaining the adjacency information among the regions. Two theorems from plane geometry are needed for the algorithm extracting the regions graph. They are stated without proof as follows:

A line intersects the boundary of a convex region in at most two points. A line therefore splits a convex region in at most two regions.

The two regions resulting from the splitting of a convex region by an intersecting line are also convex.

5.2.4 Link Routing

Link routing is the process of maintaining the adjacency relationships reflected in the set, $\mathcal{L}_R$, among the regions, $\mathcal{N}_R$, at region splitting. The scheme can be explained by the following example. Suppose the current regions are $\mathcal{N}_R = \{r_4, r_5, r_6, r_7\}$ as shown in Fig. 8a with $\mathcal{L}_R = \{\{r_4, r_5\}, \{r_4, r_6\}, \{r_5, r_7\}, \{r_6, r_7\}\}$.

When line l_3 is processed, all the regions intersected by l_3 are put onto a stack, S, with three elements in each allocation. The contents of S are:

Old Region	New Regions
r_5	r_{51}, r_{52}
r_6	r_{61}, r_{62}
r_7	r_{71}, r_{72}

Now consider the first link $\{r_4, r_5\}$ in $\mathscr{L}_R$. Since only r_5 is on S, r_4 has to check against r_{51} and r_{52} using the property of adjacency of regions to re-establish the adjacency. In this case, the link $\{r_4, r_{51}\}$ is inserted and $\{r_4, r_5\}$ is deleted from $\mathscr{L}_R$ because r_4 and r_{51} are adjacent to each other, otherwise $\{r_4, r_{52}\}$ would have been inserted.

Now consider the third link $\{r_5, r_7\}$ in $\mathscr{L}_R$. Since both r_5 and r_7 are in S, r_{51} and r_{71}, which are new subregions split from r_5 and r_7 respectively, are checked for adjacency using the property of adjacency of regions, similar to the first link mentioned above. As a result, the links $\{r_{51}, r_{52}\}$, $\{r_{71}, r_{72}\}$, $\{r_{51}, r_{71}\}$ and $\{r_{52}, r_{72}\}$ are inserted while $\{r_5, r_7\}$ is deleted from $\mathscr{L}_R$. The whole process is presented in an algorithmic form as follows:

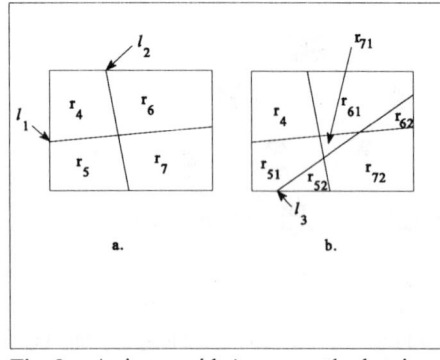

Fig. 8 a. An image with 4 segmented subregions. b. A third line l_3 further segments the image.

LINK_ROUTING($\mathscr{L}_R$, $\mathscr{S}$)
1. FOR each link of the form $\{r_a, r_b\}$ in $\mathscr{L}_R$
2.   IF both r_a and r_b are NOT in old region of $\mathscr{S}$ then
3.   process next link in $\mathscr{L}_R$
4.   ELSE IF both r_a and r_b are in old region of $\mathscr{S}$ then
5.   IF the new regions r_{a1} and r_{b1} in $\mathscr{S}$ from r_a and r_b are adjacent to each other
6.   insert $\{r_{a1}, r_{b1}\}$, $\{r_{a1}, r_{a2}\}$, $\{r_{a2}, r_{b2}\}$, $\{r_{b1}, r_{b2}\}$ into $\mathscr{L}_R$
7.   ELSE
8.   insert $\{r_{a1}, r_{b2}\}$, $\{r_{a1}, r_{a2}\}$, $\{r_{a2}, r_{b1}\}$, $\{r_{b1}, r_{b2}\}$ into $\mathscr{L}_R$
9.   ENDIF
10.   delete $\{r_a, r_b\}$ from L_R
11.   ELSE IF only r_a in old region of $\mathscr{S}$ but NOT r_b
12.   IF the new region of $\mathscr{S} r_{a1}$ and r_b are adjacent to each other
13.   insert $\{r_{a1}, r_b\}$, $\{r_{a1}, r_{a2}\}$ into $\mathscr{L}_R$
14.   ELSE
15.   insert $\{r_{a2}, r_b\}$, $\{r_{a1}, r_{a2}\}$ into $\mathscr{L}_R$
16.   ENDIF
17.   delete $\{r_a, r_b\}$ from $\mathscr{L}_R$
18.   ENDIF
19. ENDFOR

5.2.5 Algorithm for Obtaining the Regions Graph: (Regions Graph Extraction, RGE)
0. Initialize $\mathscr{N}_R$ to contain one region which is the image frame and $\mathscr{L}_R$ to ϕ
1. For each line l_i in $\mathscr{L}_0$, Begin

2.   For each region r_j (or polygon) currently in $\mathcal{N}_R$, Begin
3.   if l_i intersects r_j then
4.   Let r_{j_1} and r_{j_2} be the regions resulted from the splitting of r_j
5.   Push r_j, r_{j_1} and r_{j_2} onto stack S mentioned in Link Routing
6.   Replace r_j by r_{j_1} and r_{j_2} (the split regions) in $\mathcal{N}_R$
7.   End
8.   End
9.   Maintain $\mathcal{L}_R$ using the method LINK_ROUTING
10. End

Note that the link set $\mathcal{L}_R$ of the Regions Graph is identical to the link set $\mathcal{L}_V$ in the Vertices Graph.

5.2.6 Complexity of Obtaining the Regions Graph

The complexity is based on the maximum possible number of regions at each iteration at step 2 and 9 in RGE, which is the bracketed term with the summation sign, as obtained in Fig. 9. At step 1 of LINK_ROUTING, the number of line segments got generated at each iteration at the worst case is i^2. At steps 2, 4 and 11, the worst case behavior depends on the size of stack S that is $i+1$ at each iteration of i. The timing for steps 9 through 11 is $\Sigma i^2(i+1)$ for line i. The number of times executed by the outer FOR loop and step 2 through step 8 and step 9 the link routing in RGE is as following, where n is the number of lines:

$$[(\sum_{i=1}^{1} i) + 1 + \sum_{i=1}^{1} i^3] + [(\sum_{i=1}^{2} i) + 1 + \sum_{i=1}^{2} i^3] + \ldots + [(\sum_{i=1}^{n} i) + 1 + \sum_{i=1}^{n} i^3]$$

The number of square bracketed terms is n. The formula is rewritten to be $n(n+1)(n+2)/6 + n + n^2(n+1)^2/4$. The combined timing for all n lines is $n^2(n+1)^2/4 + n(n+1)(2n+1)/6$. Therefore the algorithm is $O(n^4)$.

5.2.7 Labeling of the Regions in the Image (Region Labeling, RL)

The regions graph defines the regions and the adjacency of the regions in an image given the extracted lines. However the members or pixels inside a region, have not been explicitly identified. The following algorithm will label all the pixels in the image once the regions are identified using the regions graph extraction (RGE) algorithm:

5.2.8 An Algorithm of Regions Labeling (RL)

1. For all regions r_j in $\mathcal{N}_R$, Begin
2.   For all x from 1 to H, Begin
3.   For all y from 1 to V, Begin
4.   If $(x,y) \in r_j$ then $\mathcal{R}(x,y) = j$

Number of Lines	Max. # of Regions	Max. # Line Segments		Min. # of Regions	Min. # Line Segments	
0	1	0		1	0	
1	2	1		2	1	
2	4	4		3	2	
3	7	9		4	3	
4	11	16		5	4	
⋮	⋮	⋮		⋮	⋮	
i	$(\Sigma i)+1$	i^2		$i+1$	i	

Fig. 9 Maximum and minimum number of regions based on number of lines successively splitting the sub-regions obtained.

5.   End
6.   End
7. End
8. return $\mathcal{R}$ as the matrix containing the labels of the regions.

5.2.9 Complexity Analysis of Regions Labeling
The complexity of this algorithm is $O(n^2)$ at step 1, since the number of regions $\leq n^2$ and $O(m)$ at step 2 and 3. The combined complexity is $O(n^2m)$. [Note m=H×V]

5.3 Exponent Regions Labeling (ERL)
A more powerful labeling algorithm that doesn't require the use of the RGE is exponent regions labeling (ERL) algorithm. Exponent regions labeling is a process of segmenting the given image $\mathcal{I}$ such that adjacency information is preserved in the segment labels. This process first initializes a matrix to zeros. It then keeps adding a number, which is a power of 2, to the element at (x,y) if (x,y) is not on the same side of $(0,0)$ relative to the processing line.

5.3.1 Exponent Regions Labeling Algorithm
0. Initialize a matrix $\mathcal{E}$ of size m=H×V (same size as the image $\mathcal{I}$) to zeros.
1. For each line l_i in $\mathcal{L}_0$, i=1,2, ..., N Begin
2.   For 1≤X≤H, Begin
3.   For 0≤y≤V, Begin
4.   If (x,y) is left of l_i, add 2^{i-1} to $\mathcal{E}(x,y)$.
5.   End
6.   End
7. End
8. return $\mathcal{E}$ as the exponent regions labels

5.3.2 Complexity of Exponent Regions Labeling
The complexity of obtaining $\mathcal{E}$ is mn, where m is the size of the grey level matrix and n is the number of extracted lines. Therefore the complexity is $O(mn)$.

5.3.3 Property of Adjacency of Exponent Regions Labeling
The adjacency information is determined by the hamming distance amongst the labels in $\mathcal{E}$. If the labels on two pixels have a hamming distance of 0, they belong to the same region. If two pixels have labels with a hamming distance of 1, they are in adjacent regions.

Based on $\mathcal{E}$, the regions graph can be constructed using this property of adjacency of exponent labeling. Figure 10 illustrates the adjacency relationships and labels as results of exponent labeling.

5.3.4 Algorithm for obtaining the Regions Graph Using ERL: (Regions Graph Extraction, RGE)

After labeling each pixels of $\mathcal{I}$, the labels are sorted by radix sort to produce a sorted array of labels, S_L. This array S_L contains duplicates because pixels in the same regions have the same label. Duplicates are then eliminated from S_L and the result is in $\mathcal{N}_R$, the node set of the regions graph. Extracting $\mathcal{N}_R$ is on the order of $O(mn)$ because the number of times each pixel being processed is proportional to the sorting time and elimination of duplicates.

Given $\mathcal{N}_R$ the following algorithm obtains $\mathcal{L}_R$, the links between the nodes in $\mathcal{N}_R$:

0. Initialize $\mathcal{L}_R$ to empty
1. For $i=1$ to $|\mathcal{N}_R|$, Begin
2.   For $j=i+1$ to $|\mathcal{N}_R|$, Begin
3. l_i and l_j=binary label of r_i and r_j
4.   compute $C(l_i \oplus l_j)$ which is the counting of number of 1 bits in the binary Exclusive OR operation
5.   if $C(l_i \oplus l_j)=1$ then
6.   add the link $\{r_i, r_j\}$ to $\mathcal{L}_R$.
7.   End
8.   End
9. End

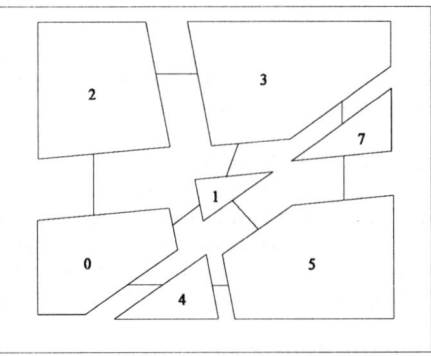

Fig. 10 Final result of the region graph by exponent labeling. Note the hamming distance between 3 and 7 is one, hence their corresponding region are adjacent.

As a result of the For loops at line 2 and 3, the complexity is $O(m(m+1)/2)$, where m is the number of nodes in $\mathcal{N}_R$. From the analysis on RGE, at the worst case, the number of regions or nodes with n given lines is $O(n^2)$. Therefore, the overall complexity of this algorithm is $O(m^2n^2)$.

5.4 Duality between Vertices Graph and Regions Graph

The nodes set $\mathcal{N}_V$ and $\mathcal{N}_R$ are different in the Vertices and Regions Graphs however the link sets $\mathcal{L}_R$ and $\mathcal{L}_V$ are identical.

VI. GRAPH REDUCTION ALGORITHMS (GRA)

In the previous section algorithms were presented to extract the graphs, $\mathcal{G}_V$ and $\mathcal{G}_R$, to represent the underlying relationships between the line segments in $\mathcal{L}_V$ (or $\mathcal{L}_R$), and the regions, $\mathcal{N}_R$ or vertices $\mathcal{N}_V$. In order to obtain the line sketch, $\mathcal{G}_V$ (or $\mathcal{G}_R$) has to be reduced. This results in eliminating pseudo vertices and line segments from $\mathcal{G}_V$ or $\mathcal{G}_R$. The approaches described in this section first associate weights to the links of the graph then consider whether each link in $\mathcal{L}_R$ should be kept or eliminated. The results of the algorithms described in this section are targeted line sketch images.

6.1 Link Weights

Weights, w_b, can be associated with links $\mathscr{L}_R$ in $\mathscr{G}_R$ to indicate the strength of the line segment. The weights can be ordered ascendingly such that $w_{1_1} \leq w_{1_2} \leq \ldots \leq w_{1_i} \leq \tau \leq w_{1_{i+1}} \leq \ldots \leq w_{1_z}$ where τ is a threshold which separates the order into two sequences. From the beginning of the sequence up to but not including τ, where all the w's are less than τ, are regarded as line segments that do not belong to the final line sketch.

6.2 Analysis of Threshold

The threshold, τ, is a number that should satisfy the following relation: $\min(w_i) \leq \tau \leq \max(w_i)$. A formula found to be useful in selecting a threshold τ is: $\overline{w} + k\sigma$. Here $\overline{w}$ is the average value of all the weights and σ is the standard deviation among the weights and k is some constant. k=-0.2 gave very accurate results in most of our experiments. However, due to variations such as texture and lighting conditions, some images may require slightly different k value.

6.3.1 Algorithm GRA.0

Exponent labeling is used for the extraction of subregion pixels from the image matrix. The average grey values, A_i, of each region in $\mathscr{N}_R$, is obtained by dividing the summation of the pixel values in region r_i by the number of pixels in r_i. A can be obtained in $O(mn^2)$ time because of m pixels and n^2 regions.

One obvious choice of the link weights is :
> the absolute value of difference between the average grey values of the adjacent regions separated by the line segment, i.e. the weight $w_{ab} = |A_a - A_b|$ where A_a is the average pixel value in region r_a and A_b is the same for r_b.

The graph reduction algorithm based on pruning the links in $\mathscr{L}_R$--the link set in $\mathscr{G}_R$--is as follows:

1. For each link e_i in $\mathscr{L}_R$, Begin
2.   weight $w_i = |A_a - A_b|$ where regions r_{i_a} and r_{i_b} are on either side of link e_i
3. End
4. set $\overline{w} = \frac{1}{k}\sum_{i=1}^{k} w_i$ and $\sigma_w = \frac{1}{k}\sum_{i=1}^{k} (w_i - \overline{w})^2$ where $k = |\mathscr{L}_R|$
5. $\tau = \overline{w} + k\sigma$
6. For each link e_i in $\mathscr{L}_R$, Begin
7.   if $w_i < \tau$, then remove e_i from $\mathscr{L}_R$
8. End
9. return $\mathscr{L}_R$ as the set containing the line sketch

6.3.1.1 Complexity Analysis of GRA.0

Steps 1 through 3, step 4 and steps 6 through 9 will run in $O(n^2)$, where n is the number of extracted lines, due to $\mathscr{L}_R$ has $O(n^2)$ links (line segments).

6.3.2 Algorithm GRA.1

The contribution of smaller regions to the weights and to threshold may be exaggerated in

GRA.0. One way of reducing this effect is to start by merging these small regions with larger regions and recomputing the line weights. GRA.1 merges the regions and reassign weights to the remaining lines whenever a line segment is eliminated.

// This algorithm uses steps 1 through 5 in algorithm GRA.0.
6. sort the links $\mathscr{L}_R$ according to the weights w_i
7. For each link e_i in $\mathscr{L}_R$, Begin
8.   If $w_i < \tau$ then
9.   Begin
10.   reduce e_i from $\mathscr{L}_R$
11.   designate a new region r_{i_3} to be the new merged region
            of r_{i_1} and r_{i_2} separated by e_i
12.   compute the new pixel average of r_{i_3}, $A_{i_3} = \dfrac{c_{i_1} P_{i_1} + c_{i_2} P_{i_2}}{c_{i_1} + c_{i_2}}$
13.   For each link e_j currently in $\mathscr{L}_R$, Begin
14.   if the left, r_{j_1}, or right, r_{j_2}, region of e_j are affected by removal
                of e_i
15.   then adjust the weight w_j of e_j by the new pixel average A_{i_3}
16.   End
17.   End
18.   recompute a new threshold τ from the remaining w_i
19. End
20. return $\mathscr{L}_R$ as the set containing the line sketch

6.3.2.1 Complexity Analysis Algorithm GRA.1

The outer FOR loop from step 6 through 19 runs in $O(n^2)$ time, where n is the number of extracted lines. The inner FOR loop at step 13 and step 18 together with the outer FOR loop together behaves like $O(n^2(n^2-1)/2)$ because of the adjustment only applies to the links that are not yet visited. The combined complexity for algorithm GRA.1 is $O(n^4)$.

6.3.3 Algorithm GRA.2
Another choice in weight w of the links is:
> *Increase or decrease in variance among the grey levels*
> *when the regions are considered as one consolidated region.*

This choice of w is based on the observation that: when two essentially different regions are considered as one, the variance within the grey levels tends to increase dramatically. The threshold for this algorithm is simply 1 because either there is an increase or a decrease. Justification for this method is presented in the Appendix.

1. For each link e_i in $\mathscr{L}_R$, Begin
2.   Compute the variance be $\sigma_{i_1}^2$ for the left region of e_i

3. Compute the variance be $\sigma_{i_r}^2$ for the right region of e_i
4. Let σ_i^2 be the combined variance of the left and right regions of e_i
5. weight $w_i = max(\sigma_{i_l}^2, \sigma_{i_r}^2)$
6. if $w_i \geq \tau$,
7. then there exists a line segment represented by link e_i
8. else eliminate e_i from $\mathscr{L}_R$
9. End
10. return $\mathscr{L}_R$ as the set containing the line sketch

6.3.3.1 Complexity Analysis of Algorithm GRA.2

Compared to GRA.0, this algorithm computes the square of individual pixels, which is also an $O(mn^2)$ operation, in addition to the average of the pixels. Similar to algorithm GRA.0, the determination of a line segment e requires computing the variance

$$\sigma_X^2 = var(X) = \Sigma(x-\mu)^2 = \frac{1}{n}\Sigma(x^2) - \mu^2 \quad where \quad \mu = \frac{1}{n}\Sigma x$$

of the two regions separated by e and the new combined standard deviation, which is the form of:

$$\sigma_{X+Y}^2 = var(X+Y) = \frac{1}{n_x+n_y}\Sigma(x^2+y^2) - \left(\frac{xn_x+yn_y}{n_x+n_y}\right)^2$$

where n_X and n_Y are the number of pixels in the region X and Y respectively. The running time is $O(n^2)$.

VII. EXPERIMENTS AND RESULTS

The graph reduction algorithms were tested on the image shown in Fig. 2. The results depended critically on the threshold τ, which is $\bar{w} + k\sigma_w$. This threshold of course depended on the weight derived from the image but also on the scale factor k. Algorithm GRA.0 and GRA.1 produce results like either Fig. 11 or Fig. 12 depending on the value selected for k. That is either has too many or too few line segments removed.

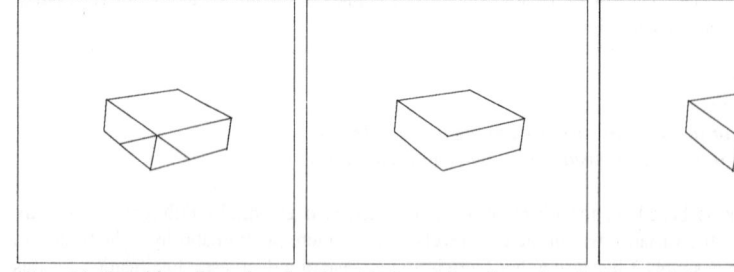

Fig. 11 Resulting line sketch when τ set too low.

Fig. 12 Resulting line sketch when τ set too high.

Fig. 13 The perfect line sketch of the image of a wooden block.

The order of visiting the edges in algorithm GRA.1 affects the result severely. The order, in which the edges are visited in a descending order in weight, produce a better result than an ascending order. Dangling edges were removed automatically because the average pixel values on the left side and the right side were equal. There is no obvious difference in performance between algorithm GRA.0 and GRA.1 although GRA.1 uses a more sophisticated region merging algorithm.

Algorithm GRA.2 only produces one result, i.e., Fig. 11. It is understandable because the face with two extra lines has a texture where the grey levels, as a result the standard deviations, varies tremendously. However, with appropriate feedback algorithm, those two lines will be removed as they will produce unrealizeable object and they are the ones closest to being removed.

VIII. CONCLUSION

We have described methods of extracting a line sketch from an image, which has been preprocessed to give all possible line locations of edges in the image. The utility of the methods described depend, in part, on the preprocessing stage. An extension of this work would be combining the two methods and testing the resulting methods.

APPENDIX A: Mathematical Justification to GRA.2

This section is based on the Gaussian modeling of pixels in an image described in Malik and So [3]. With an edge-based algorithm like GRA.2, only two regions, for instance A and B, are considered at an edge. Only two hypotheses H_0 and H_1 exists and their a priori probabilities are assumed equally likely for simplicity: $P(H_0) = P(H_1) = \frac{1}{2}$. Here H_0 and H_1 represent the situations where no line segment is detected (referred as the merge case) and a line segment exists (referred as the split case) between the regions A and B respectively. Suppose n_A and n_B are the number of pixels in regions A and B respectively. μ_A and μ_B are the mean values of the pixels of regions A and B. σ_A and σ_B are the standard deviations of the pixels in regions A and B. The following are the probability density function for hypotheses H_0 and H_1:

$$p_x((x_1, x_2, \ldots, x_{n_A+n_B}) | H_0) = \frac{1}{(2\pi)^{\frac{n_A+n_B}{2}} \sigma_{A+B}^{n_A+n_B}} e^{-\frac{1}{2\sigma_{A+B}^2} \sum_{i=1}^{n_A+n_B}(x_i - \mu_{A+B})^2} \quad \text{A.1}$$

$$p_x((x_1, x_2, \ldots, x_{n_A}, x_{n_A+1}, \ldots, x_{n_A+n_B}) | H_1) = \frac{1}{(2\pi)^{\frac{n_A+n_B}{2}} \sigma_A^{n_A} \sigma_B^{n_B}} e^{-\frac{1}{2\sigma_A^2}\left(\sum_{i=1}^{n_A}(x_i - \mu_A)^2\right) - \frac{1}{2\sigma_B^2}\left(\sum_{i=n_A+1}^{n_A+n_B}(x_i - \mu_B)^2\right)} \quad \text{A.2}$$

The decision rule which maximizes the probability of a correct decision or minimize the probability of an error decision when both hypotheses are equally likely is:

Select H_0 if $p_x(x|H_0) \geq p_x(x|H_1)$ else select H_1.

Since the mean and variance are a prior unknown they must be estimated from sample set. The estimate of the variance $\hat{\sigma} = \frac{1}{n}\sum_{i=1}^{n}(x_i - \mu)^2$. Therefore, in A.1 and A.2, the exponent term can be reduced to $\frac{(n_A+n_B)}{2}$, where the decision rule can be rewritten as: Select H_1 if $\sigma_{A+B}^{n_A} \sigma_{A+B}^{n_B} > \sigma_A^{n_A} \sigma_B^{n_B}$

else H_0.
Four scenarios exists:

case 1: $\sigma_{A+B} > \sigma_A$ and $\sigma_{A+B} > \sigma_A$, case 2: $\sigma_{A+B} < \sigma_A$ and $\sigma_{A+B} > \sigma_A$,
case 3: $\sigma_{A+B} > \sigma_A$ and $\sigma_{A+B} < \sigma_A$, case 4: $\sigma_{A+B} < \sigma_A$ and $\sigma_{A+B} < \sigma_A$.
Case 1 always selects H_1. Similarly case 4 always selects H_0. Case 2 and 3 are equivalent cases. They do not conclusively predict the selection of H_0 or H_1 because whether the final result,

$$\sigma_{A+B}^{n_A}\sigma_{A+B}^{n_B} > \sigma_A^{n_A}\sigma_B^{n_B}$$ or not, depends on the values of σ_A, σ_B and σ_{A+B}.

The term in the decision will easier to be evaluated if rewritten as:

$$\sigma_{A+B} > \sigma_A^{\frac{N_A}{N_A+N_B}} \sigma_B^{\frac{N_A}{N_A+N_B}}$$

and the right hand side always falls between σ_A and σ_B

$$\sigma_A < \sigma_A^{\frac{N_A}{N_A+N_B}} \sigma_B^{\frac{N_B}{N_A+N_B}} < \sigma_B$$

if $\sigma_A < \sigma_B$. The

proof follows:
$\sigma_A^{N_A} < \sigma_B^{N_A}$ and $\sigma_A^{N_B} < \sigma_B^{N_B}$ implies $\sigma_A^{N_A}\sigma_A^{N_B} < \sigma_A^{N_A}\sigma_B^{N_B} < \sigma_B^{N_A}\sigma_B^{N_B}$ and therefore

$$\sigma_A^{N_A+N_B} < \sigma_A^{N_A}\sigma_B^{N_B} < \sigma_B^{N_A+N_B}$$

The decision rule used in GRA.2 is a slight variation of this model and treats case 2 or 3 as case 4:
    Select H_1 (edge exists between the regions) if
        combined variance is larger than the maximum of individual
        region's variance, which is $\sigma_{A+B} > max(\sigma_A, \sigma_B)$
    else select H_0.

VIII. REFERENCES

[1] Burns, J. B., A. R. Hanson, and E. M. Riseman, "Extracting Straight Lines," *IEEE Transactions on Pattern Analysis and Machine Intelligence*, Vol. PAMI-8, 1986, pp.425-455.

[2] Duda, R., and P. Hart, "Use of Hough Transformation to Detect Lines and Curves in Pictures," *Communications of the ACM*, Vol. 25, 1972, pp. 11-15.

[3] Malik, R. and Sinha, P., "Edges from Projections," *Proc. IS&T/SPIE 1993 International Symposium on Electronic Imaging Science & Technology*, (Nonlinear Image Processing IV), San Jose, CA, Feb. 1993.

[4] Malik, R. and So, S., "Vertices and Corners: A Maximum Likelihood Approach", SPIE, *Intell. Robots and Comp. Vision XII*, vol 2055, Sept. 1993

[5] Huffman, D. A. "Impossible Objects as Nonsense Sentences," *Machine Intelligence*, vol. 6, B. Meltzer and D.Michie, Eds. Edinburgh, Scotland: Edinburgh University Press, 1971, pp. 295-323.

[6] Clowes, M. B., "On Seeing Things," *Artificial Intelligence*, vol. 2, 1971, pp. 79-116.

[7] Waltz, D., "Understanding Line Drawings of Scenes with Shadows," *Psychology of Computer*

Vision, Patrick H. Winston(ed.), McGraw-Hill, New York, 1970, pp. 19-92.

[8] Leavers, V., *Shape Detection in Computer Vision Using the Hough Transform*, Springer-Verlag, 1992.

SECTION : 2
AI TOOLS

THE EVOLUTION OF COMMERCIAL AI TOOLS: THE FIRST DECADE*

F. HAYES-ROTH

Cimflex Teknowledge Corporation, Palo Alto, CA 94303

1. Introduction

Considerable progress has occurred in the development of commercial Artificial Intelligence tools over the past decade. During that time many improvements have been made in the quality of those tools. The key areas of improvement are:

1. New tools are more modular and easier to embed in existing applications.
2. New tools have many features that make them easier to use not only by novice but also experienced users.
3. New tools have features that facilitate the co-operation between tools relative to some common problems.
4. New tools support problem decomposition and task specialization in terms of both their representation and their control.

However, there are still many technical and commercial barriers that must be crossed before widespread use of such tools is possible. The goal of this paper is to assess the evolution of AI tools over the past ten years from both a technical and a commercial perspective. The paper will conclude with a discussion of trends in the development of AI tools over the next decade.

2. Setting The Stage

The advances that took place during the past decade were shaped by the technical and commercial milieu that existed in the early 1980's. In this section, a brief discussion of both the technical and commercial bases which has influenced the development of AI tools over the past decade is briefly presented.

2.1. The commercial milieu

In the early 1980's, AI technology was perceived by many as being simply a collection of unrelated methods and techniques best employed in research environments

---

*This work was edited by Dr. Robert G. Reynolds to whom the author expresses his great appreciation.

by skilled practitioners. In order to be successful commercially, the first generation of tool developers had to demonstrate that these techniques can be used to simplify business problems and their solutions. To be successful, developers of these tools needed to alter the perceptions of potential buyers and users. Buyers had to be convinced that the tool would solve their problem and must be a "safe" purchase. Users had to be convinced that the tool could be integrated into their existing operation and that they had the skills to use the tool effectively themselves. Both buyers and users needed to be convinced that there would be continued technical support and improvement of the product.

The first tool that was able to make this type of impact on commercial buyers and users was designed to capture and activate the existing problem solving expertise of business personnel. These tools allowed businesses to view specific problem solving knowledge as a commodity and an asset for the first time. These systems appealed to a wide variety of buyers. They included expert system developers, system integrators, corporate management, and end-user consumers.

Several factors were important in influencing the choice of a system.

1. How well did the features of the proposed tool match up with the requirements of the buyer?
2. How did the price/performance ratio for the product compare with that of others running on similar platforms?
3. The perceived viability of the product in the competitive market.
4. The extent to which the product was supported by the vendor.
5. The consistency of the product with organizational standards.
6. Experiences of others who has actually used the system to solve similar problems.

In the early 1980's, there were few suppliers of commercial AI tools but this number had swelled to over 300 by 1986. This rapid increase in the number of competing firms was due to several factors. Firstly, there were few barriers to entry into the market place. Secondly, the large initial diversity of products meant that few competed head to head. Thirdly, there was little brand identification on the part of buyers so that new companies with strong products could make an immediate impact. Fourthly, the presence of an inexperienced user community meant that there were few initial standards that all had to meet.

However, as the end of the eighties approached, companies began to consider alternatives to knowledge-based AI tools. These alternatives often entailed less perceived risk and supported the "business-as-usual" approach. The AI industry was forced to compete in a shrinking market, often head to head with more traditional products such as conventional programming systems, user-friendly generic applications, decision support systems, and CASE systems, among others. This resulted in a marked downturn in the number of tool suppliers at less than 100.

In addition to competitive pressure, AI companies had to become more dependent on others to supply portions of their products. At the beginning of the decade, the chief suppliers of the industry were hardware and networking vendors along with vendors of operating systems and programming languages. Towards the end of the decade there arose an increased dependence on component software vendors. These vendors provided software such as user interface kits, databases and query links, documentation and hypertext media, renderers and geometric reasoners, and instrumentation packages. The establishment of such a network of alliances was often difficult for startup companies to accomplish.

2.2. The technical milieu

Knowledge-based system development in the early 1980's was done predominantly in LISP. Knowledge was represented most frequently in terms of rules and frames. Reasoning was performed using methods to support unification and various inference strategies. The most popular inference strategy was based upon the principle of modus ponens. These systems saw their most successful early applications in the areas of diagnosis and system configuration. These knowledge-based system shells supported the separation of control knowledge from descriptive knowledge as well as objects and object-oriented graphics.

It became clear early on that for knowledge-based systems to be truly competitive with other technologies certain deficiencies had to be addressed. These deficiencies were:

1. The knowledge engineering process itself was difficult and time consuming. It therefore became important to develop standard frameworks for doing this. In addition, these frameworks could support the portability of a solution from one application to another.
2. In practice, expert systems seldom stood alone. Therefore it was important to develop standard embedding and integration interfaces.
3. User interfaces were expensive to build and difficult to modify. The development of more modular interfaces that could be easily modified and reconfigured was necessary.
4. The needs of the computing infrastructure in which these knowledge-based systems were embedded were constantly shifting. One particular direction was the need to generate knowledge servers that could support co-operative problem-solving applications.
5. The knowledge maintenance activity was often performed by the knowledge engineers themselves. Over the long run, the longevity of a knowledge-based system required that the users perform that activity. This required the development of representations that allowed end-users to easily perform this task.

During the 1980's work was done on each of these problems. The progress that was made relative to each deficiency is described below:

1. In order to address problem 1 above, standard Knowledg: Engineering(KE) shells were developed for tasks that had been commonly occurred.

2. Several innovations were developed in order to tackle the problems of embedding and integration. Firstly, languages other than Lisp were used to implement KE shells in order to make them more portable to a wide variety of application environments. Secondly, SQL and other database interfaces were provided. Thirdly, management systems to control the configuration and use of user interfaces were produced.
3. The need for more modular knowledge processing systems was addressed in several ways. One approach was to develop standard knowledge-based function modules. This allowed KE shells to be reimplemented as a collection of functional modules and stored as a library of modules.
4. In order to develop knowledge servers that supported co-operative work it was necessary to recognize and exploit coarse-grained parallelism in a distributed problem-solving environment. Inference engines that could provide inference-on-demand to processes active within a distributed system were produced. Also, control schemes were generated that supported concurrency as well as means to collect and deliberate on results from distributed sources.
5. In order to allow end-users to accomplish the knowledge maintainence task more effectively, new form-filling interfaces were introduced. In addition, simplified and standardized case frames were adopted for performing specific knowledge maintainence tasks.

Taken together, these advances facilitated the development of products with the following features. These products often contained task-specific tools that provided specialized representation and control formalisms to simplify the knowledge engineering process. They also supported end-user knowledge programming and maintainence activities. Internally, they were highly modular at a functional level and their interfaces were built from collections of functional components. This fine-grained modularity supported the representation of these systems as libraries of components. As such, these new systems were easier to embed in existing application environments. Lastly, coarse grained modularity was exploited in the production of systems that supported co-operative work in distributed environments. In the next section, examples of specific systems that can serve to illustrate these advancements will be presented.

3. Examples of Specific Advancements

In this section, examples of existing systems that illustrate some of the above properties are presented. The specific properties discussed here include support for task specialization, end-user programming and maintainence support, support for system modularity and embeddability, and support for co-operative problem solving.

3.1. *Support for task specialization*

Task specialization concerns the refinement of a knowledge engineering tool to fit the requirements of a specific task, such as system diagnosis or configuration. There are several advantages to doing this. Since knowledge engineering tools and methods are general, this facilitates a more domain-specific form of problem solving. This prestructuring simplifies knowledge-based design because it allows knowledge-base concepts and utilities that support a given task to be collected together. This also supports their reuse for future instances of a task. This approach allows the knowledge engineer to focus the acquisition of knowledge around the specific problem solving tasks that must be supported.

"Task analysis" is synonymous with the concept of "domain analysis" in software engineering. It involves the identification and description of the basic objects of interest, their relations, and the problem-solving methods that can be applied to modify and reason about objects and their relationships. Tasks can be expressed in terms of these objects and relationships. Control plans can be developed to actualize the performance of a task. Task-specific knowledge acquisition and maintenance aids can be developed, that will expedite the knowledge engineering process.

Enterprise/Dx(tm) is an example of a knowledge-base system designed specifically to support problems in large-scale diagnosis. The product supports multiple views of the diagnosis activity in terms of symptoms, failures, tests, and repairs. Enterprise/Dx is a suitable approach when the diagnosis problem is complex or when experts rather than knowledge engineers are required. This approach supports the generation of a line of related diagnostic products in the knowledge base or allows a single system to be used in different consultation frameworks.

Figure 1 illustrates the cycle of diagnostic activities supported by Enterprise/Dx. A task is selected from an agenda of possible diagnostic activities. In this case, test error code is selected. The error code obtained is used to provide evidence for a failure. This information is then propagated through a belief network with subtypes and causal links in order to identify the possible causes. Then the belief network is updated and another task is selected.

Enterprise/Dx can be used in variety of problem solving scenarios. First, it supports the classic question-and-answer consultation process. It also supports the integration of an application with a larger suite of software, or the complete embedding of an application within a system. It allows a knowledge maintainer to augment existing failure networks and associate new tests, repairs, and other supporting constructs with them. The maintainer also has the ability to select certain policies to be maintained in producing a solution. Such policies relate to identifying single or multiple faults as well as whether the closed world assumption is to be followed. The entire system was built using an OOPS class library that supports the generation of knowledge engineering systems.

3.2. End-user knowledge programming and maitenance

In this section a tool, the Requirements Manager(RM), that supports techniques for practitioners to produce their own knowledge bases is described. In order to do this, the system must support the forms and syntax needed to state goals, means

The Diagnostic Cycle

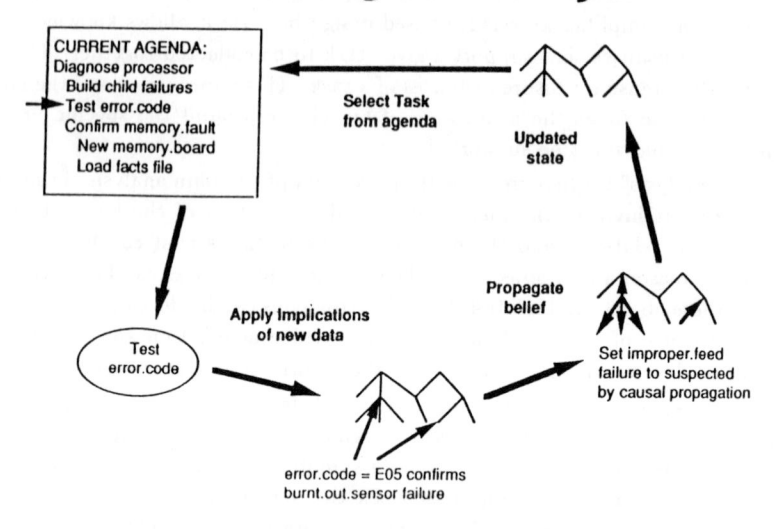

Fig. 1.

to achieve them, and constraints to be followed in their achievement. There are a number of good reasons for doing this. First, general knowledge engineering tools and methods require programming as well as knowledge engineering skills. Most organizations want in-house experts to monitor and update the knowledge base as opposed to programmers or knowledge engineers. This is due to the fact that, updating is a vital and urgent organizational activity that is most effectively carried out by those using the system.

In order for a system to support end-user knowledge programming, a particular knowledge-based design and problem-solving organization are selected. A language that allows end-users to specify and edit knowledge, in this framework, is provided. Other related functions such as knowledge base access, update, and configuration are provided.

The RM is a computer system that is intended to support concurrent engineering and decision-making by teams of specialists who co-operate in real time on the development and realization of complex product designs [2]. It enables them to define, maintain, and verify the requirements, performance measures, specifications, policies, and constraints that influence their decisions. The "Planners Workbench" in the early 1980s that supported rapid replanning, and the "Libra" system in the mid-1980s that supported extensive organization memory needs for product development teams were precursors of RM.

In a typical RM scenario, a chief engineer or system engineer uses RM to tailor and produce an initial requirements database that supports the flowdown of high-level requirements to specific product specifications. Other engineers involved in the development process can then augment the existing requirements to include additional information concerning design, manufacturing, and service constraints. RM can be used by team member to check their evolving design against recent specification and to address areas of non-compliance. RM maintains a history of these changes, their rationales, and evaluation of the results that can be used to generate management reports or possibly be reused in future projects.

Figure 2 illustrates an example of application of RM in the electronics development domain. Note that the RM is integrated in with other contributors to the development process. The RM keeps a requirements knowledge base that contains product and process attributes, methods to calculate the values for dependant variables, information about goal interactions, and interdependancies, as well as the changes made to the specification and the rationale for doing so. It is able to acquire design data and activate certain design and evaluation tools to monitor the extent to which the current design supports the current set of requirements.

Example Application of Requirements Manager to Electronic Product Development

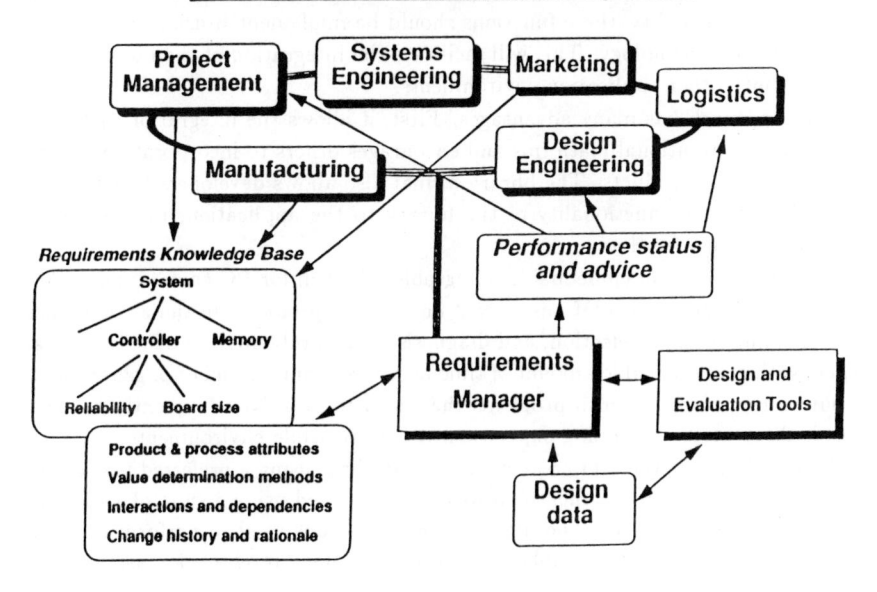

Fig. 2.

These design tools can be viewed as intelligent assistants for the support of tasks that require great user initiative and decision-making freedom. In order to accomplish this, an embedded expert system shell-M.4(tm) — is used to provide full knowledge representation and reasoning capabilities. The user is able to specify requirements for a given domain model using a standard syntax and to check these requirements against the current design. In this sense the requirements actively shape the design and a mismatch between the design and the specifications can trigger a user-defined response.

The integration of RM into the development environment is supported by an SQL protocol for distributed data management, an interface to preferred applications such as word processing software (e.g. Microsoft Word[1](tm)), and a high-bandwidth human interface. It is implemented on a PC platform in Smalltalk/Objectworks for MS-Windows and is also on SUN SPARC UNIX environment.

3.3. *Systems that support modularity and embeddability*

For a system to be both modular and embeddable it is important that the knowledge processing(KP) functions be separated from everything else, including the knowledge base. Each knowledge-processing function should possess a standardized interface to allow easy configuration with other KP modules. These modules can be grouped together into classes of related functions. In order to facilitate embeddability and portability, these functions should be implemented using conventional object-oriented technology. This will facilitate the integration of the library of KP functions into other application environments.

This approach has many advantages. First, it allows the integration of KP activities into conventional platforms and encourages others to incorporate KP capabilities into their products. The library format also allows developers to selectively tailor the size and functionality of the library to the application, thereby giving them more design freedom.

M.4 Windows is an embeddable, integrable KE shell for PC environments. M.4 implements judgmental solutions to well-understood problems in the areas of analysis, planning, design, selection, and diagnosis. Such problems can be solved by an expert within a reasonable amount of time and have a finite number of generic sorts of outcomes. These are also problems that need to be solved frequently by users with little or no programming expertise in PC compatible environment.

In M.4 Windows the user creates a knowledge base using a preferred text editor. The knowledge system can then be tested and debugged using a suite of knowledge system development tools. The library of KP functions used to perform inference and other tasks can then be embedded in a C language program for use with the

---

[1] Microsoft and Word are trademarks of Microsoft Corporation. ABE, Enterprise/DX, and M.4 are trademarks of Cimflex Teknowledge Corporation.

knowledge base. This supports the seamless integration with other application programs and the ability to customize the application to specific end-user requests. The KP functions supported by the M.4 library are rule-based programming, backward and forward chaining, pattern match variables and general symbolic unification, symbolic list processing, computation of certainty factors, retraction of knowledge, explanation of results, and use of meta-facts and meta-propositions.

3.4. Support for co-operative problem-solving

Co-operative systems are composed of coarse-grained, loosely coupled concurrent servers that interact via request/reply protocols. They can be seen as aggregations of semi-autonomous systems running on heterogeneous platforms and containing both conventional and knowledge processing modules. Such environment can provide several advantages to users. First, they can amplify the value of a legacy system by providing it with access to knowledge-based reasoning capabilities. Second, they facilitate the independent development of subsystems for problem-specific applications by exploiting various processor and platform technologies.

Various techniques can be used to support the development of co-operative problem solving environment: the use of module-oriented programming to generalize OOP to support recursive system development; the use of frameworks to support high-level system design; the use of high-level messages with abstractly defined structures to mediate the interactions between modules; and the use of common virtual machine services to generate platform transparency between applications.

ABE(tm) is an environment for building co-operative intelligent systems[1, 3, 4]. It is a distillation of many years of experience in building multi-module intelligent systems at Teknowledge, CMU, Stanford, ISI, and for various customers. ABE provides a suite of high-level tools that directly supports the following: modular software architectures and system structures, perspicuous modelling, component reuse, process heterogeneity, rapid prototyping and evolutionary development, knowledge-based components, system deployment in the field, distributed and concurrent processes, extensibility, and team development methodologies.

ABE is a toolkit for building customizable and recursive modular application architectures based on Module-Oriented Programming (MOP). This is a generalization and specialization of OOPS using modules, intermodule events, data communications, and supporting process and resource scheduling. ABE supports an extensible set of "frameworks" for developing application systems. These application modules and their data types can be stored for future reuse. ABE supports a number of graphical tools for building and validating complex applications, including time-critical ones. These tools are integrated with a number of supporting tools and components in C, C_{++}, and Commonlisp. A schematic description of the technology incorporated into the structure of ABE is given in Fig. 3.

In a typical ABE scenario, an application architect (AA) incrementally models the application in a top-down fashion by creating dummy modules. These modules

ABE Technology Map

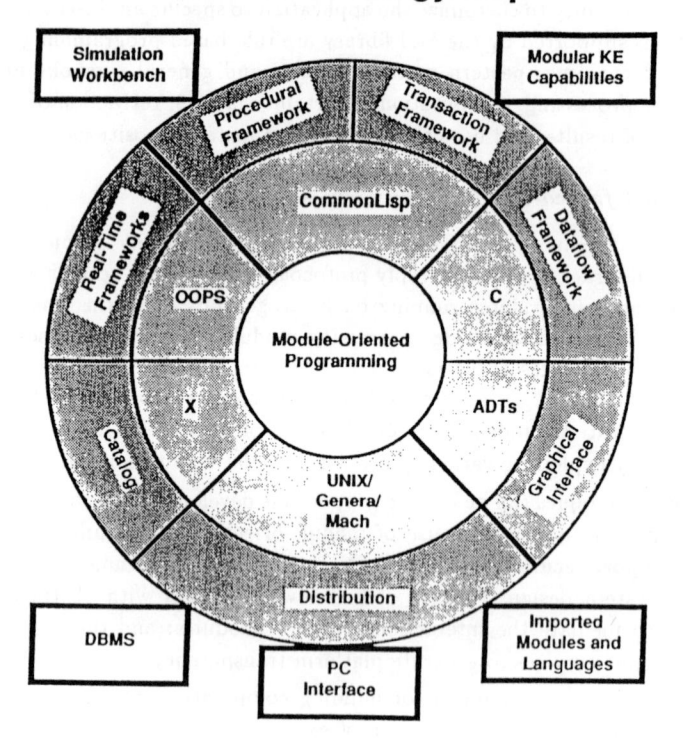

Fig. 3.

can be decomposed and composed along with their inter-connections. Modules are implemented by application developers (ADs). A module can be implemented in several ways. A dummy module's description can be elaborated by adding detailed code in an incremental manner in order to achieve a fully implemented version. The module can also be replaced with a fully elaborated one from an existing library of modules. A final way to implement the module is to import a component from outside the system.

The AA can also model the delivery hardware configuration, assign modules to processors, simulate the execution of the application in that environment, and collect statistics on its performance. The application can be modified by the AA via the modification of individual modules; the rearrangment of modules; the addition, deletion, or replacement of modules; the modification of the delivery hardware configuration; and the reassignment of modules to processors in the delivery configuration. Thus, the ABE toolkit can be used by an application developer to produce an application system configured for a particular delivery environment.

4. Commercial and Technical Objectives for The 90s

While a number of advances in the development of AI tools over the past decade have been made, much remains to be done. In this section, a number of technological and commercial barriers that will be the chief obstacles to the development of AI tools in the 1990s will be discussed.

4.1. *Current technological barriers and objectives for the 1990s*

There are many technical obstacles to the development of large-scale AI tools in the coming decade. Some of these obstacles are:

1. Large size of the problem space and the sparseness of acceptable solutions will require more heuristic search strategies.
2. The lack of formal principles to specify reusable knowledge.
3. The lack of a large-scale knowledge codification industry to support the development of large-scale knowledge bases which are required to support more complex tasks.
4. The lack of standard problems and performance benchmarks that can be used to assess the utility of an approach.
5. The lack of semantic integration into existing database technology makes it more difficult to ask the right questions and retrieve the appropriate information.
6. The need to integrate knowledge-based tools with systems that support real-time control.
7. The need to marry various reasoning paradigms together such as symbolic, numeric, and geometric reasoning.
8. The need to develop global initiatives for the development and integration of product and process description standards across the AI community.

Taken as a whole, the characteristics described above will lead to the development of certain classes of tools. Some examples are:

1. Compilers that convert declarative knowledge into pragmatic knowledge;
2. Modular, inter-operable KP components with standardized tasks, ontologies, and knowledge representations;
3. Standard metrics and tools to measure them on benchmark problems;
4. Tools to support the large-scale codification of knowledge across many domains;
5. Tools that support the unification of knowledge bases and databases by separating out issues of representation, ontology, and control;
6. Real-time reasoning engines and hybrid controllers that combine algorithmic and declarative approaches;
7. Tools that support the use of hybrid representation and reasoning systems such as the joining of symbolic reasoning with that of numeric and geometric reasoning.

4.2. *Current commercial barriers and objectives for the 1990s*

There are several commercial obstacles to the development of large-scale AI tools in the next decade. Firstly, many of the large-scale problems that AI tools can address are very domain specific. Therefore, prices will need to be high for such products in order to compensate for the low volume of anticipated sales. Secondly, there are few solution techniques in AI that are generic enough to counteract this trend towards specialization. Thirdly, with increased domain specialization will come with an increased emphasis on knowledge acquisition and maintenance, which is a difficult problem on a large scale. Finally, the lack of standards, even for well-known types of knowledge processing tasks, will tend to make tool generation for such large-scale problems difficult.

In order to combat these difficulties several things can be done. One approach is to form consortia among those firms with similar large-scale problems. These consortia can be used to provide standard frameworks for knowledge codification and problem-solving methods for particular large-scale problems. In addition, standard process and product models should be developed for problems that are currently within the realm of AI tools. It is also important to identify problems in new, diverse commercial niches that can be addressed by current technology. As users in new domains become familiar with AI tools for smaller problems, they may also come to identify larger ones for solution. This process may, in the long run, expand the market for large-scale tools. In order to expedite the application of tools to many different domains, it is important to define and support standards for modular and portable knowledge processors. Finally, the establishment of benchmark problems and metrics to measure solution performance needs to be done in order to guide progress in the field.

5. Conclusion

The quality of AI tools increased markedly in the 1980's. At the beginning of the decade, they started as research and development laboratory tools. Then, they passed through an adolescent stage in which they quickly became big, egocentric, clumsy, and expensive. As they have matured, they have taken the "plug and play" form of other successful, state-of-the-art commercial software. In fact, today they are leading the way in some areas of systems engineering.

The major advances in knowledge processing came from specialization of these tools along several different dimensions. One approach was to develop specialized interfaces that made it easier for non-programmer experts to perform knowledge maintenance activities. Another form of specialization was to identify generic problem-solving tasks such as diagnosis and provide standard, modular frameworks for their support. A final area of specialization was the support of co-operative problem-solving activities in distributed computing environment.

However, the field of knowledge processing as a whole is still immature. The field lacks problems of common focus that can force the development of standard frameworks, the sharing of results, and the benchmarking of progress. Currently the AI industry is not organized to promote these types of goals. Future efforts in the area of consortia development need to be put in place, in order to facilitate the growth of the field in the coming decade.

References

[1] L. D. Erman, J. S. Lark, and F. Hayes-Roth, "ABE: An environment for engineering intelligent systems", *IEEE Transactions on Software Engineering*, 14(12) (1988), 1758–1770. Reprinted in S. Andriole and S. Halpin (Ed.), Information Technology for Command and Control, IEEE, 228–239.

[2] J. Fiksel and F. Hayes-Roth, "A requirements manager for concurrent engineering in printed circuit board design and production", *Proceedings Second International Symposium on Concurrent Engineering*, Morganstown, West Virginia (1990), 373–390.

[3] F. Hayes-Roth, J. E. Davidson, L. D. Erman, and J. S. Lark, "Frameworks for developing intelligent systems: The ABE systems engineering environment", *IEEE Expert* 6(3) (1991), 30–40.

[4] F. Hayes-Roth, L. D. Erman, S. Fouse, J. S. Lark, and J. Davidson, "ABE: A cooperative operating system and development environment", *AI Tools and Techniques*, M. Richer(Ed.), Ablex Publishing, Norwood, N. J. (1989), 323–355. Reprinted in *Readings in Distributed Artificial Intelligence*, A. Bond and L. Gasser (Ed.) (1988), 457–488.

Reengineering: The AI Generation -- Billions On The Table

John Scott Minor, Jr.
International Business Machines
121 South Main Street
Akron, Ohio 44308, United States of America

ABSTRACT

Producing goods upon receipt of an order instead of a forecast has long been the dream of those concerned with management, marketing, production and inventory control. In a discrete manufacturing environment, there are many new strategies that attempt to meet the modern phenomena of shorter research and development time, shorter product life cycle, and shorter order cycle. Added to this is the impact of downsizing the organization to a minimum number of people, so the increasingly frenzied pace of meeting customer needs and competitive pressures are being handled by a smaller and smaller set of specialists. The problem then is how to handle this cumbersome issue of less people, less time, higher quality and more complexity. One approach is to address a specific subset of this problem which has provided the advent of many 1990s style manufacturing strategies. Strategy can be defined as the art of creating value. It provides the intellectual framework, conceptual model, and governing idea that allows a company's managers to identify opportunities for bringing value to a customer and for delivering that value at a profit. Some of the more prominent ones include: activity based management; continuous flow manufacturing; concurrent engineering; process reengineering; and mass customization. The new problem emerging from using these tightly focused strategies is in optimizing their cumulative impact. Each has one or more metrics and variables against which a company can compare its continuous improvement and its performance against world class manufacturing benchmarks. With this new problem as the backdrop, this chapter introduces Unit Continuous Optimization Manufacturing (UCOM): a mix of artificial intelligence software tools and advanced manufacturing strategies that provide the solution. UCOM provides the fastest customer response, lowest product/manufacturing cost, highest quality product for an order of a single unit -- and it does this in the manufacturing environmental framework of the chaos of the coming decade.

1. Unit Continuous Optimization Manufacturing - Introduction

In a discrete manufacturing environment, mass producing a single unit order is called mass customization. Producing the product at its true cost is a goal of activity based costing. Utilizing only streamlined value-added processes in the sales, development and manufacturing of the product is the goal of process reengineering. Producing the part via a just-in-time and kan ban production system is the goal of continuous flow manufacturing. Each of these strategies are provided to clients as stand alone solutions to today's manufacturing problems, but each has an underlying algorithm or software tool that can be looked at globally for the optimization of the complete variable envelope space of all of the strategy metrics or variables.

What would be the economic effect of adding an artificial intelligence based optimizing tool at the front end of today's manufacturing strategies and their underlining algorithms? The user could provide continuous improvement on a unit order by unit order scale with world class products. He could provide the lowest cost product with the highest quality and the shortest research and development lead time to market -- the industry would be his for the asking.

This is the basis for the solution that is Unit Continuous Optimization Manufacturing.

While artificial intelligence optimization algorithms are not new, their use in this very exciting capacity of global manufacturing process optimization is new. The tool has been used for years in the engineering design environment, but why stay in the micro level of an organization for optimization? Why not step back and see the overall process as the best target for optimization?

The hypothesis, based on an electronic process flow model called MINOR I, was developed after many years of seeing the impact of each new and advanced manufacturing strategy being implemented one at a time. As each new idea was put into the process flow, new focal points were established. Thus, the overall manufacturing system of processes was optimized to a new set of criteria with each new strategy. The bigger picture was never seen and no coordination of the many strategy variables was provided.

Now that the macro view is to be the target, we must decide on the variables of interest and their respective weighting factors to allow the best possible overall solution to emerge. Against what criteria do I run my business? What will differentiate my product from my competitor -- quality, price, response time? Should my model be flexible in order to allow reduced time to market to be the highest weighted factor at new product development time in order to gain maximum share -- and lowest price as the product moves into the more mature stages of its existence? MINOR I addresses these issues.

MINOR II more fully globalizes the optimization process as well as adding autonomous robotics controls on machine tools to enhance the shop floor scheduling piece of the process flow. This approach better leverages global resources and technology strong points of different nations and companies. With many singly focused and defined global research and development partnerships emerging via consortia formats, each country and each company can more fully leverage both their expertise and their financial wherewithal. The precursor of this approach is a global repository that attempts to catalog what each participant can bring to the table. Once this structure is provided, the consortia approach is very simple to communicate and implement in order to build the strongest team of players at the minimum cost.

The primary concern with MINOR II is global currency valuations and volatility. However, with real time data available via modem for all of the major exchanges, the optimization tool can simply add a new parameter to continually judge the most expedient location for each phase of a project to take place.

The real key to the whole UCOM concept of optimization is the flexibility of the underlying tool for addressing many variables with differing weights and differing output formats. The system has a built in sensitivity analysis capability so that the weighting factor can be more fully appreciated and managed.

Much more work must be done with pilot projects in order for results to be verified, but each of the strategies mentioned earlier has extensive implementation data. Just the review of the impact of the single focus strategy approaches provides some insight into how powerful a combined and optimized approach should be.

The first pilot project is expected to be a commercial aircraft engine parts manufacturer located in Pittsburgh, Pennsylvania. A following section under the MINOR I MODEL heading will detail the implementation plan, but the initial installation should allow for the removal of seven senior management positions with an annual savings to the $4 million company of approximately $550,000.

The result: a unit by unit order optimization no matter what the unit characteristics and a change from the stand alone strategies to an optimization of the overall envelope of process variables that go in to these techniques. The implementation of the model and the software tools and techniques are quite straight forward. The dynamic re-arrangement, for optimized process flow and cost minimization, of the shop floor for each single unit order is not.

New manufacturing equipment sizing strategies and the technical underpinnings of an autonomous mobile machine tool and assembly line strategy will be discussed in greater detail. Each product's individual complexity will be the driving factor for the unique set of problems a manufacturer will confront in utilizing the UCOM model. A generic implementation is presented here.

The concepts and techniques to be utilized; the MINOR I model of the complete UCOM approach and its impact on the related areas of the company; and the predictions of what may be in the next major strategy advancement, as provided in MINOR II, follow.

2. UCOM World Class Manufacturing Strategies

Today's problems stem from poor management styles of the past. The primary concern of managers today is to manage the workforce -- not the workload. They continue to be incented (usually negatively with the threat of dismissal) to meet short term objectives -- rather than think and act for the long term business goals of the overall organization. They continue to establish organizational structures that are based on functional silos rather than across the organization into project teams. They continue to get into the micro level of problems which only allows them to manage the symptoms -- rather than understanding the causes. (Please see Figure 1)

In order to meet these challenges that are etched in stone in many of our institutions and corporations, new strategies were developed that induced change. The only problem with these strategies is their singular focus on some portion of the overall problem. Again, we have gone too far into the bowels of the organization.

Traditional strategy is the art of creating value along a value-add chain, but superior companies do not just add value -- they reinvent it. They globalize their internal and external structure so that the best and the brightest are incorporated into the value-add chain. Suppliers and customers play a major role in all phases of the operation in these types of companies. They reconfigure the roles, relationships and organizational practices of everyone -- the social implications of change -- and they get to the core of their existence and optimize their value. They leverage information technology and human knowledge as their chief assets, and they enhance the utility of these assets by making sure the underlying processes are as clean and pointed and relevant as possible before they automate them. This approach provides a complete list of system constraints - some opposing - that need to be looked at as a whole and optimized as a whole. These organizations then can leverage artificial intelligence technologies for an optimized reconfiguration of this constraint set to meet the individual needs of each product and each customer.

2.1 Mass Customization

Mass Customization is the ability, in a discrete manufacturing environment, to mass produce a single unit order. This leaves behind the concept of both custom made and mass production. This is a more adaptive approach to business environment change, and it provides for the existence of contradictions in meeting customer requirements. Formed another way, the meaning relates the production and distribution of customized goods and services on a mass basis. The concept has some global parts such as time based competition, manufacture at the point of delivery, customer self-design and direct computer system

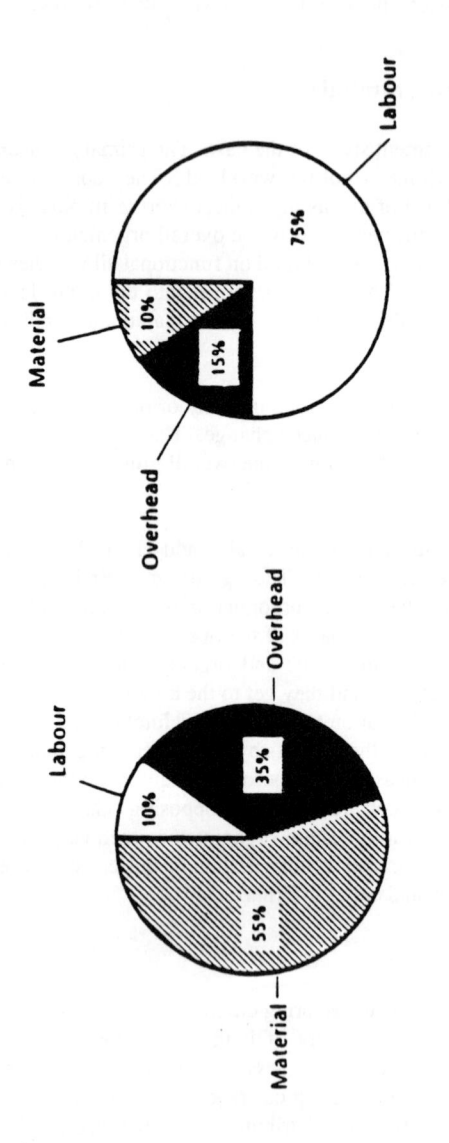

Where the costs are   Where we focus cost control

Figure 1
Management Focus

access, modularization, zero inventory, down sizing, enhanced logistics, reduced need for working capital, and light speed communications on a global basis. The problem is detail in the implementation -- how do I set up my operation to provide this type of service and structure? The devil here is in the detail, and UCOM provides the tool package to incorporate this strategy. (Please see Figure 2) (1)

2.2 Activity Based Costing

Activity Based Costing is an approach to classify and match a cost with each step in the manufacturing process of a product. Most businesses do not have a clue as to their actual cost of a specific product -- let alone what some customer mandated change will add to the base cost. Companies incent sales people on the wrong products, serve the wrong customers, design costly products, fail to see the impact of changes on the shop floor or in their engineering, manufacturing and marketing strategies, and purchase the wrong parts from outside suppliers. The cost systems of today do not meet the needs of the 90s, and business must embrace this new tool if they are to compete effectively with the truly world class competitor. Therefore, the need for this type of tool in a mass customization environment is quite obvious. The primary advantage here for UCOM is that software exists from several vendors for the development of the model for any environment. By having the ability to utilize outside software, the optimization routines can iterative run many scenarios of both cost variations and process variations in order to optimize the solution to the goals and weights of a specific customer's requirements. (Please see Figures 3, 4 and 5) (2)

2.3 Process Reengineering

Process Reengineering attempts to find only streamlined, value-added processes in the whole organization. The main idea is to move from the Adam Smith model of breaking down industrial work into simple and basic tasks to the re-unification of tasks into coherent business processes. The idea is not to fix the organization. Rather it is to start from scratch and rebuild the organization around the technology available today. The key, and the difficulty faced by management and unions, is that they must adapt their present way of thinking to more of a discontinuous model. This means identifying and abandoning the outdated rules and fundamental assumptions that underlie current business operations. Contracts and rule books alike must be torn up. The organization must start with the customer and provide him with his particular needs in the most expedient and cost effective way possible. The unfortunate part of this strategy is that a very flexible system must be designed to handle all customers or the processes must be redesigned for each individual customer. The fortunate part of this strategy is that it lends itself to rule based expert systems and the activity based costing modeling software. UCOM provides for a library of solutions and the organization can fit the end result to a specific customer's requirements. (Please see Figures 6 and 7) (3)

PRESSURES...

- CUSTOMER FOCUS
- GLOBALIZATION
- INCREASED COMPETITION
- INCREASED PRODUCT COMPLEXITY

...ARE DRIVING

DEVELOPMENT TIME	30 - 70%	LESS
ENGINEERING CHANGES	65 - 90%	FEWER
TIME TO MARKET	20 - 90%	LESS
OVERALL QUALITY	200 - 600%	HIGHER
WHITE-COLLAR PRODUCTIVITY	20 - 110%	HIGHER
DOLLAR SALES	5 - 50%	HIGHER
RETURN ON ASSETS	20 - 120%	HIGHER

Figure 2

Mass Customization

PROCESS VIEW →

| COST DRIVERS | → | ACTIVITIES | → | PERFORMANCE MEASUREMENTS |

Factor that causes a change in performance of an activity

No of new procedures

No of drawings

% defective parts received

How well is the activity carried out

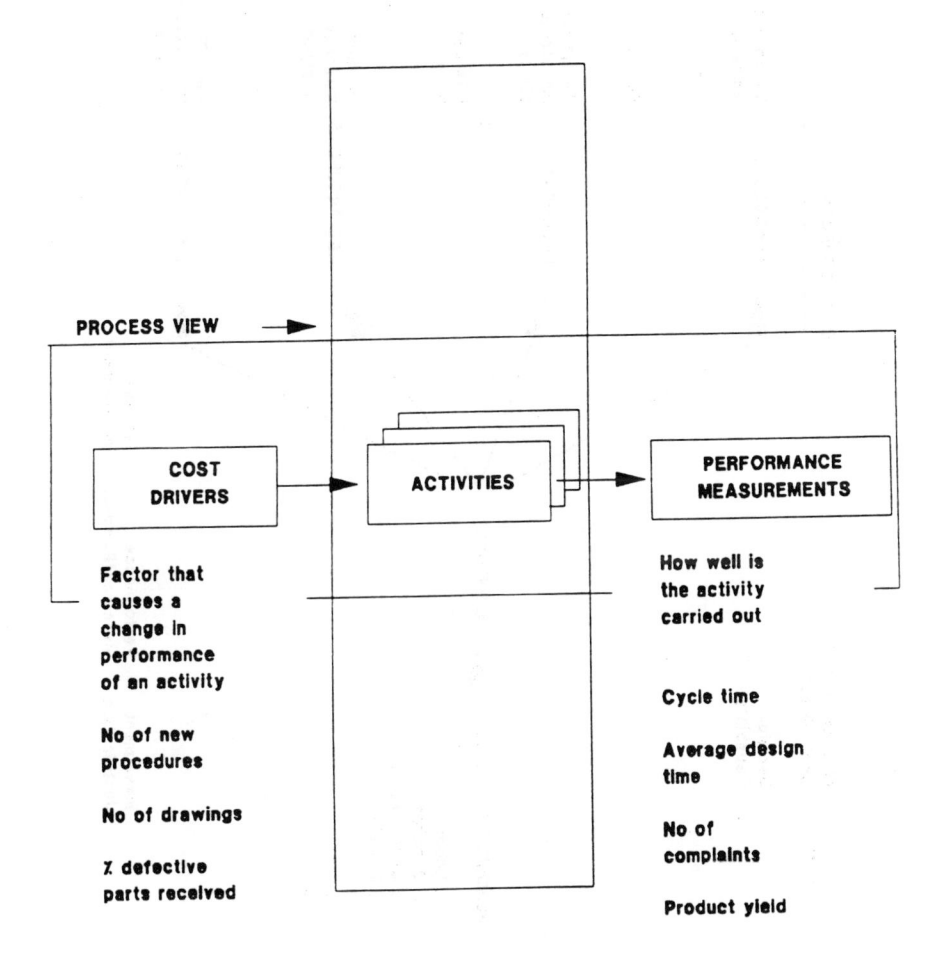

Cycle time

Average design time

No of complaints

Product yield

Figure 3

Activity Based Costing Model

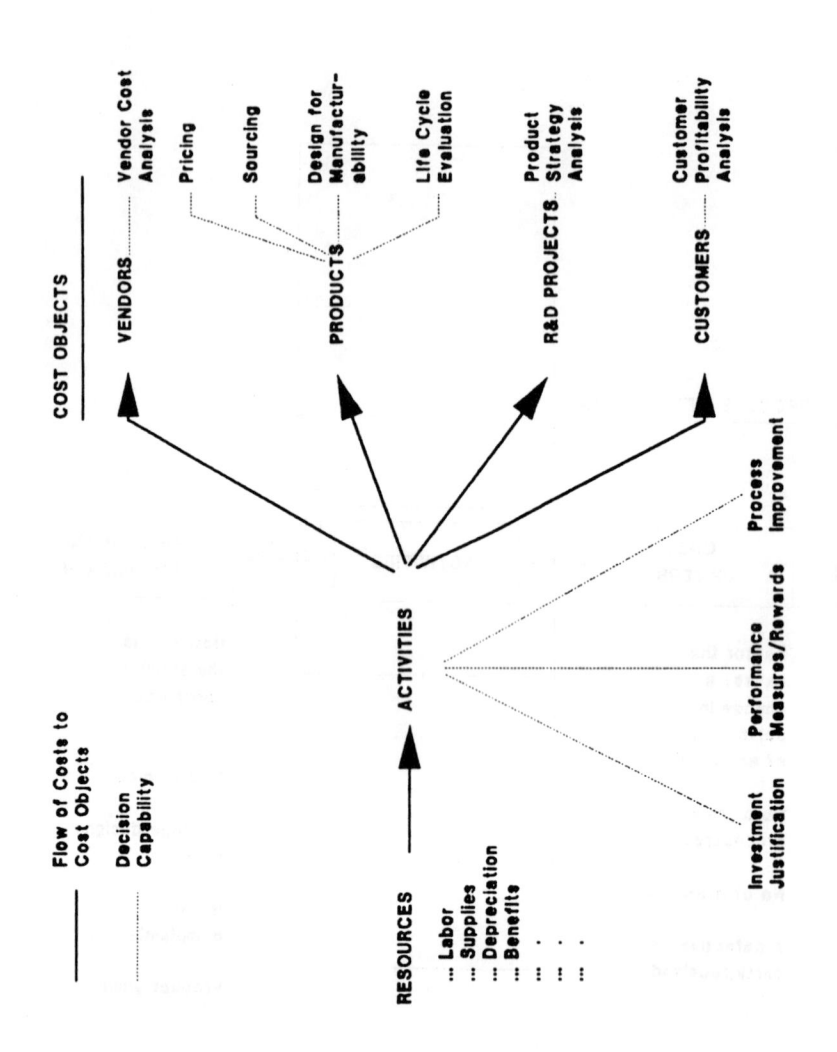

Figure 4

Activity Based Management Flow

Figure 5

Activity Based Management Detail

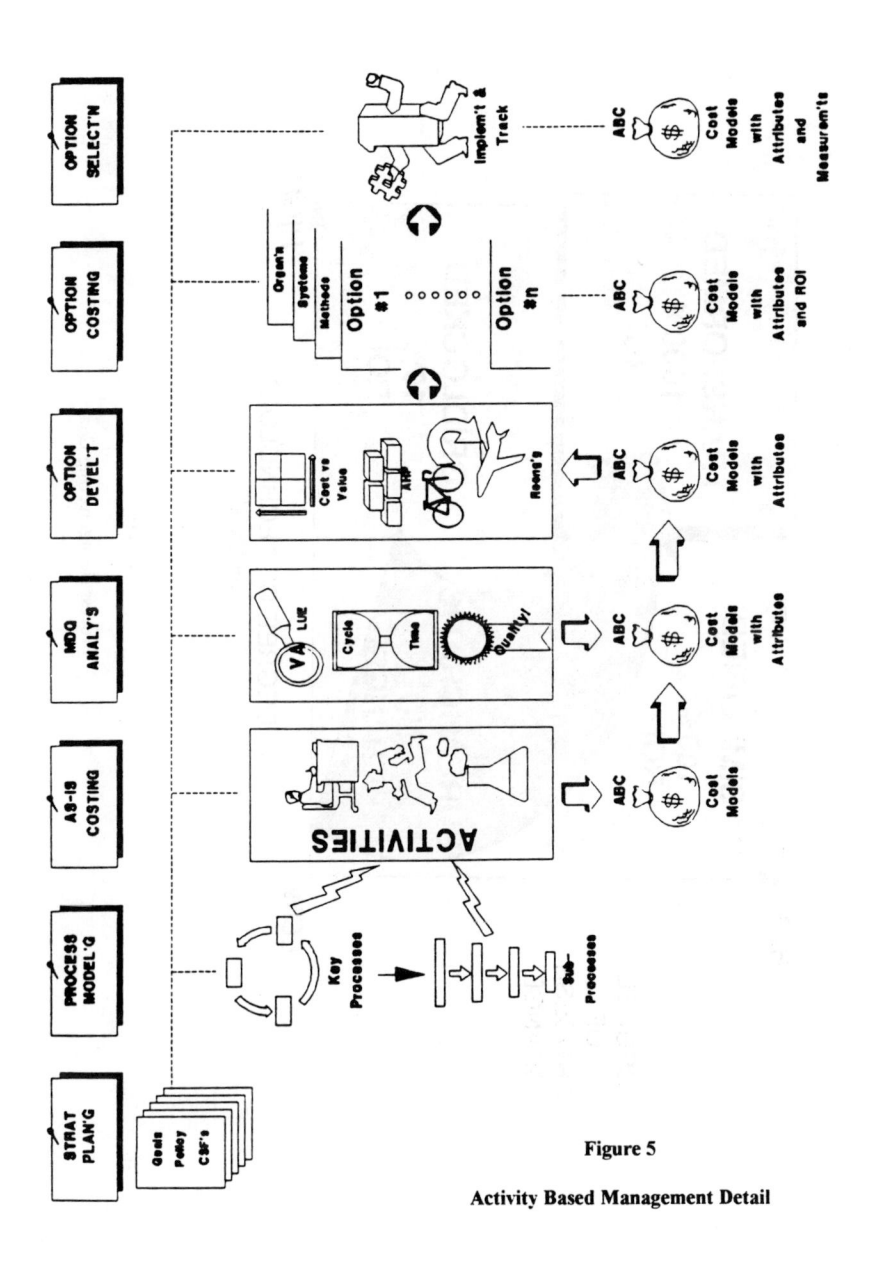

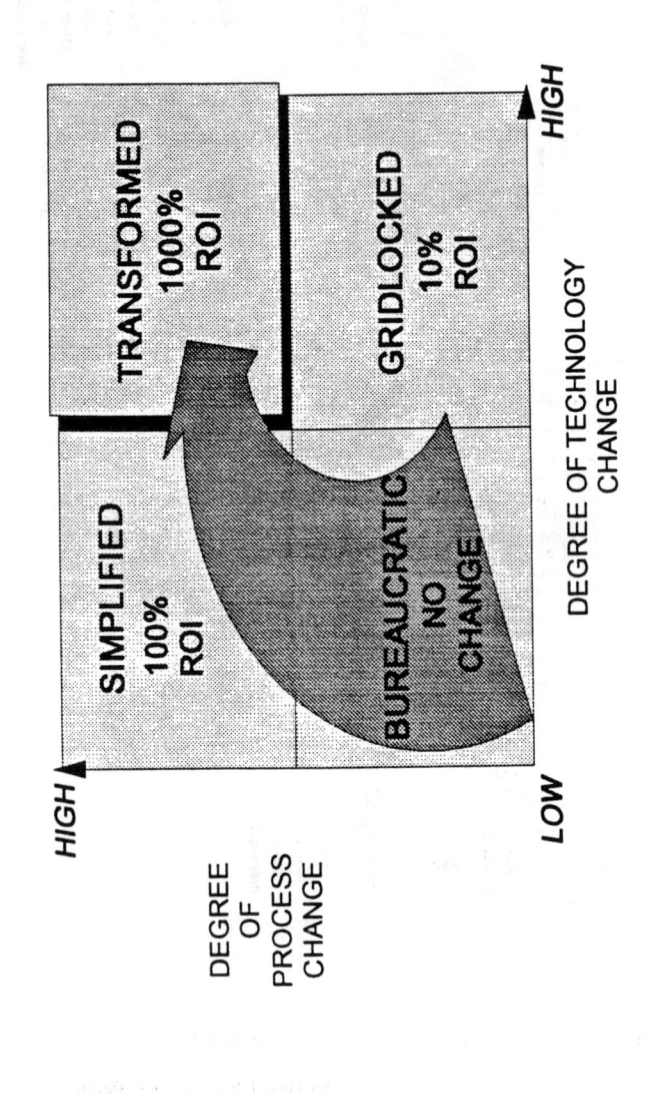

Figure 6

Process Reengineering Goal

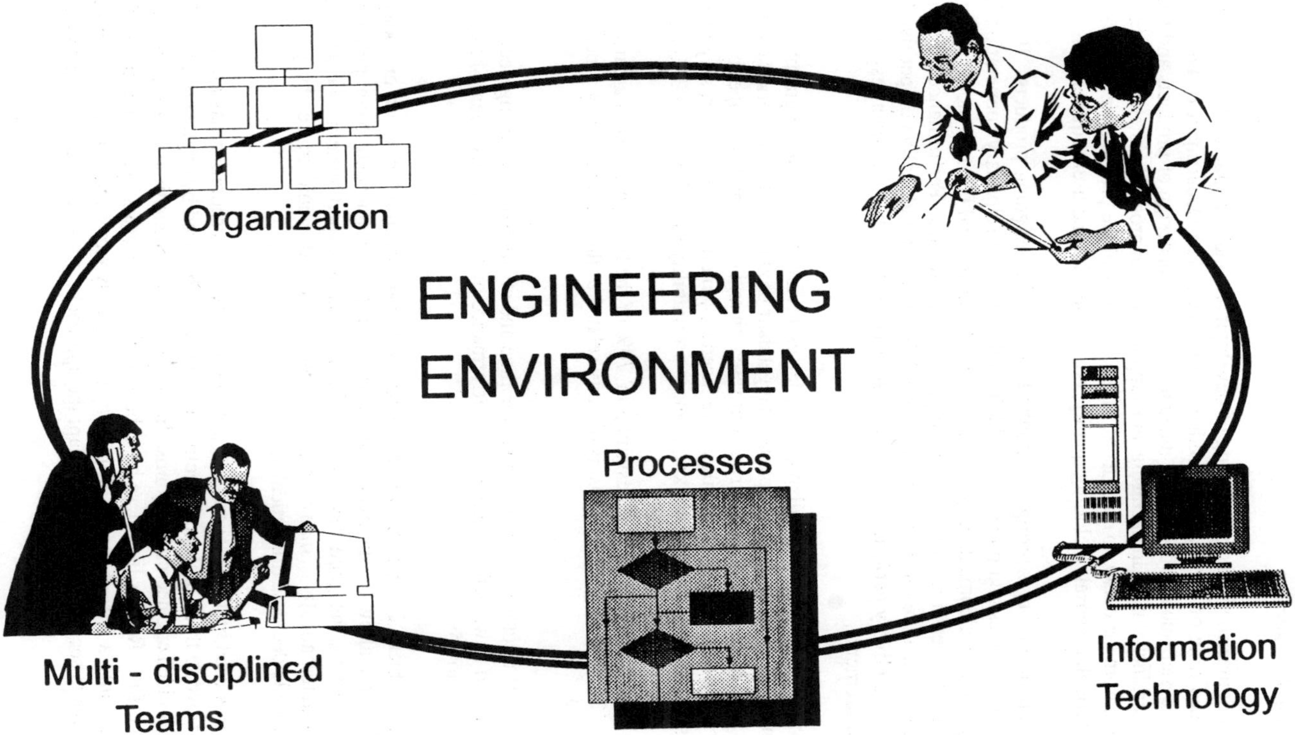

Figure 7

Process Reengineering Approach

2.4 Concurrent Engineering

Concurrent Engineering is the paralleling of development and manufacturing in order to shorten the lead time of getting new products to market. The Department of Defense's definition is "a systematic approach to the integrated, concurrent design of products and their related processes, including manufacturing and support. This approach is intended to cause the developers, from the outset, to consider all of the elements of the product life cycle from conception through disposal, including quality, cost, schedule, and user requirements."

Therefore, concurrent engineering is a method, or way of designing products. It is an attitude, a set of policies and procedures which facilitate interdisciplinary cooperation for product design and development. Engineering's role will expand and become more strategic as more pressure for newer and more advanced products accelerates. Performance, in terms of cost and time/schedule must be improved 100-200% to remain globally competitive by the late 1990s. Delivery performance must be at or better than 98% in direct response to customer requirements.

As an example of the impact available here, let us review one market segment -- the electronics industry. The first two companies that get their product out to the market capture 85% of that market. Even if you overrun your research and development budget by 50%, but be the first to market, your profitability is reduced by a mere 4%. However, staying on budget, but six months late to the market, reduces profitability by a significant 33%.

Engineering is also responsible for 'designing in' 70% of the life cycle cost of a particular product. Only 20% of the life cycle cost is directly related to the manufacturing processes.

The significance to a company of spending some time and effort cleaning up engineering is obvious. The bad news is that engineering performance over the last five years has only improved from 0 to 25%.

The primary problem is that the present organizational structure of most of today's companies does not allow for the major engineering accelerations needed to compete on a global scale. Again, the good news is that UCOM has an expert system available for addressing - automatically - many of the design issues that would make the product development cycle decrease dramatically.

One product produced by the research arm of General Electric Corporation has produced a design that is two to three times as effective as what was produced by engineers using the same computer aided engineering programs. In addition, development time was 1/10 as long as the human effort. Results like these have been available for years, but as is usual in the United States, only a few of the leading companies are dynamic enough and organized correctly enough to see the promise and adapt the approach. As a study, the vice president level of engineering of 100 companies in the greater Cleveland area was personally contacted for an invitation to discuss the GE code and its capabilities. Of that group, ONLY FOUR companies sent representatives -- and NONE even tried the product.

IBM was prepared to bring in an integration team to attach all of the computer aided engineering codes to the GE product, train the users, and provide a test period of four months in order for a good set of results to be obtained -- and still no company was willing or able to see the value.

Management of this type must change -- and change quickly if the companies they represent are to survive. (Please see Figures 8 and 9) (4)

2.5 Continuous Flow Manufacturing

Continuous Flow Manufacturing (CFM) is a strategy that produces a part via a just-in-time and kan-ban production approach. Therefore, because the production floor must continually change to meet many needs, the strategy calls for an ongoing examination/improvement effort which ultimately requires integration of all elements of the production system. The goal is an optimally balanced production line with no waste and the lowest possible cost, on-time, defect-free product. Once the company's product is set up to respond to this approach, world class leaders will move out to incorporate both their vendors and the customers. By automating much of this strategy, electronic communications between the three sets of players keeps the production loop as tightly coupled to the dynamics of the shop floor as possible. The fact that this approach has underlying algorithms as its base allows for easy integration with the UCOM methodology.

The first algorithm is "Target-Enhanced Process Capability" or Cp. Most U.S. companies are poor performers here and have a score of around 1.0. The six sigma or world class companies have a score of 2.0

The objective is to reduce the allowable variability of the finished part or product so that it is in a tolerance range that is much within the customer's allowable tolerance range. This provides for the many interfering problems of manufacturing without having a part or product fall outside the customer's specifications and into the scrap/rework area -- which wastes time and money.

It is well known that each machine tool has its own variability signature -- its own personality if you will. Therefore, one way to reduce the variability of multiple machining operations over many machine tools is to reduce the number of different tools through which the part/product must pass. This means moving away from the 'farm' approach to manufacturing, which is the congregation of like machine tools into one area, into the 'cell' approach, which congregates a subset of machine tools that are designated for the manufacture of some particular element of the final part/product. This then reduces the variations entered into the manufacturing process while also reducing set up times. The

Figure 8

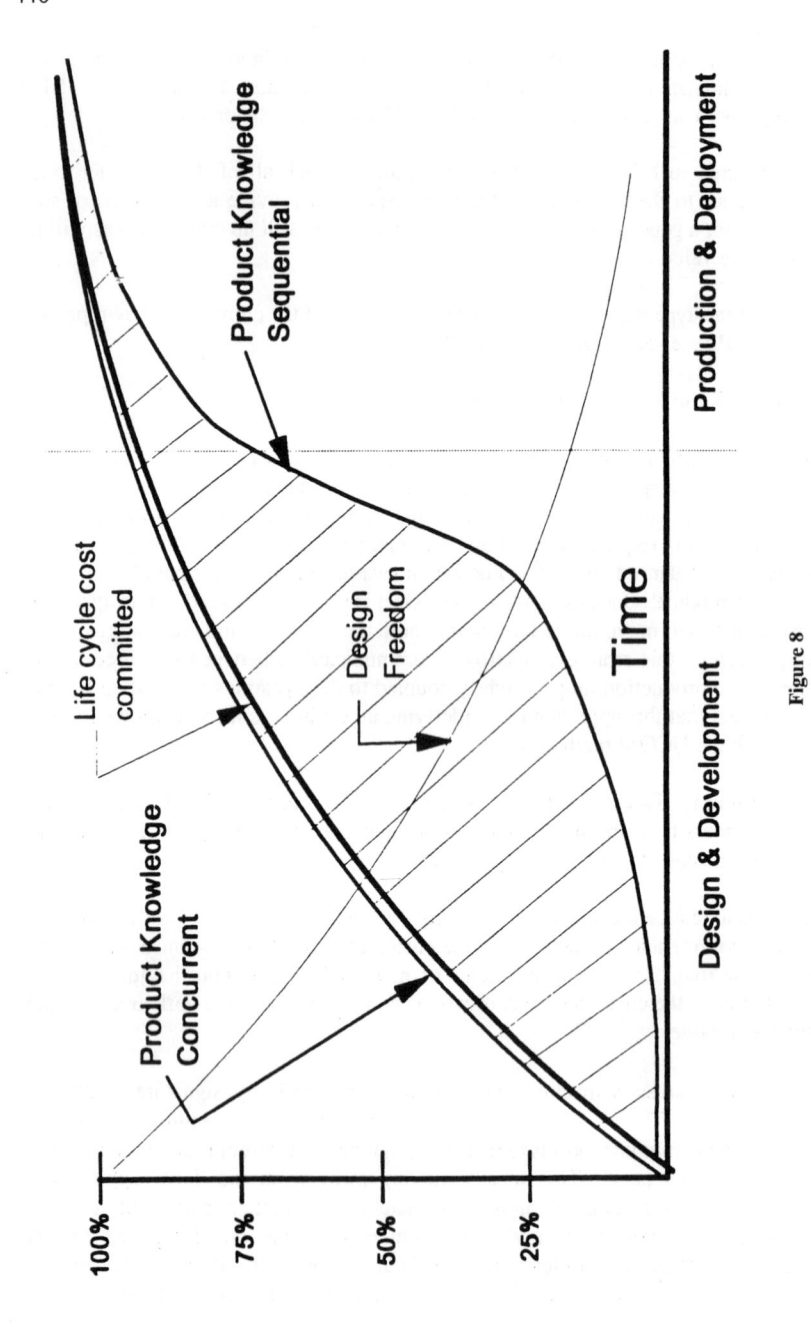

Concurrent Engineering vs Sequential

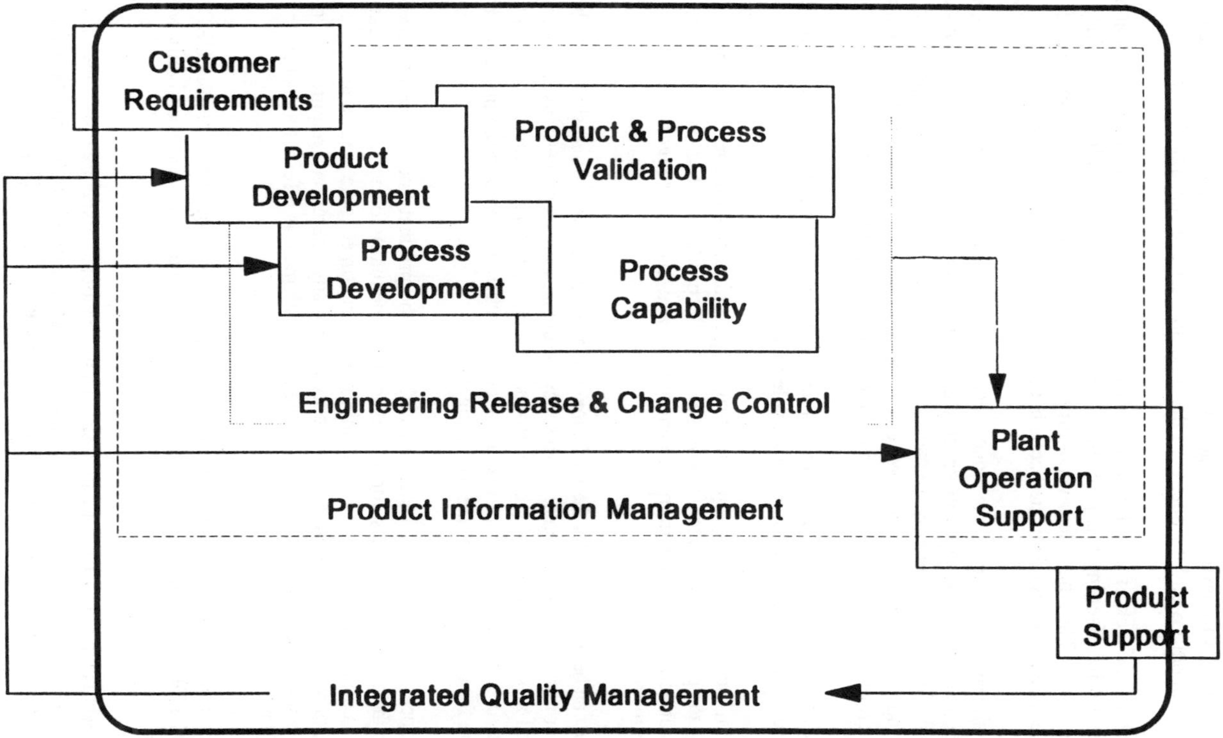

Figure 9

Concurrent Engineering Order Release Process

ultimate goal here is to never change the set up of a machine tool.: keep it dedicated to a certain set of operations and the quality of the final product goes up significantly.

The down side is the lack of use of machines that are in these cells. One approach that has been successful in offsetting this problem is the manufacture by machine tool makers of smaller machines. They have less versatility, but they are less expensive, so the cost vs quality issue is reduced and dedicated cells become much more cost effective.

The second algorithm picks up on the reduced cycle time theme mentioned above. Here, we look at cycle time efficiency. As with any efficiency rating, we look at what comes out over what goes in. Waste is the enemy, so each step in the process must be evaluated as to its need first, and its efficiency second. We cut every step that does not add value. Then we optimize each remaining step in order to cut the cycle time to a minimum.

A new piece of manufacturing equipment or a new material, etc. requires that the complete process be re-evaluated to again minimize cycle time.

This does not mean that keeping equipment running continually is the most efficient. On the contrary, the pull method, which means the end of the production line or the customer's order is what generates upstream movement of parts and material, suggests zero finished goods inventory.

If cycle time can be reduced under the lead time of the customer's requested ship date, there is no need for finished goods inventory. This may mean that work in process inventory must be generated at spot locations along the production line in order to reduce the impact of 'pinch points' or manufacturing areas that are a 'bottle neck' to the production process. This is the reason for developing the line balancing method.

Work-in-process holding areas are designated along the product line. Once an order is placed from the customer, a request for parts is made sequentially to each upstream location in the line. In order to maintain the correct minimum cycle time, some areas that have long lead times, or very costly machines that are used for many operations, require that some amount of work-in-process inventory be cached. This element is referred to as takt in the jargon of CFM.

The primary points here are: excess production equals waste and the system is ORDER driver - NOT FORECAST driven. You make only what the customer wants and only when he orders it.

The written form above may be difficult to see in a more global perspective, so I have added the following chart to give the reader a better understanding of where the numbers fall for different stage companies as they move toward world class status. (Please see Figures 10, 11 and 12) (5)

PROCESS CAPABILITY

MEASURED BY:

PROCESS CAPABILITY INDEX (Cp)

$$Cp = \frac{\text{SPECIFICATION TOLERANCE}}{\pm 3 \text{ SIGMA PROCESS CAPABILITY}} = \frac{USL - LSL}{6 \text{ SIGMA}}$$

EXAMPLES:

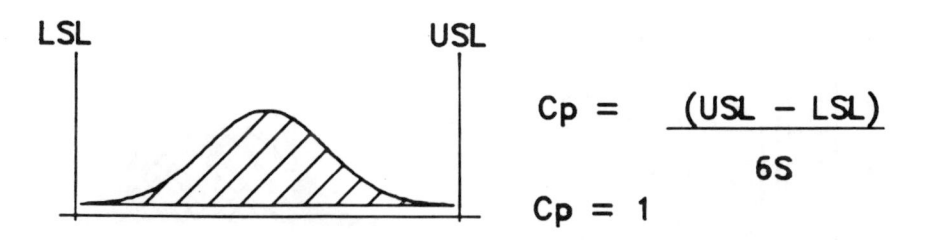

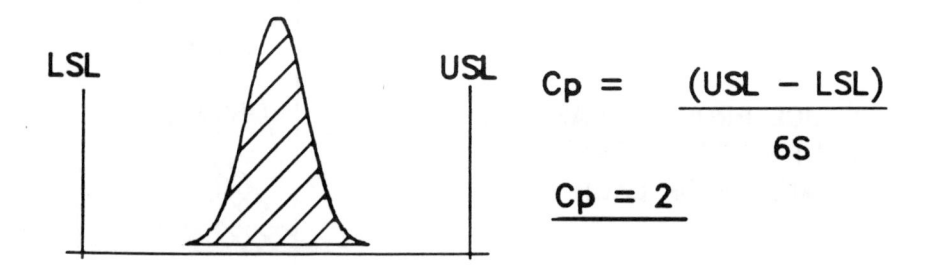

$Cp = \frac{(USL - LSL)}{6S}$

$Cp = 1$

$Cp = \frac{(USL - LSL)}{6S}$

$\underline{Cp = 2}$

Figure 10

Continuous Flow Manufacturing Metrics

TYPICAL FARM TYPE LAYOUT

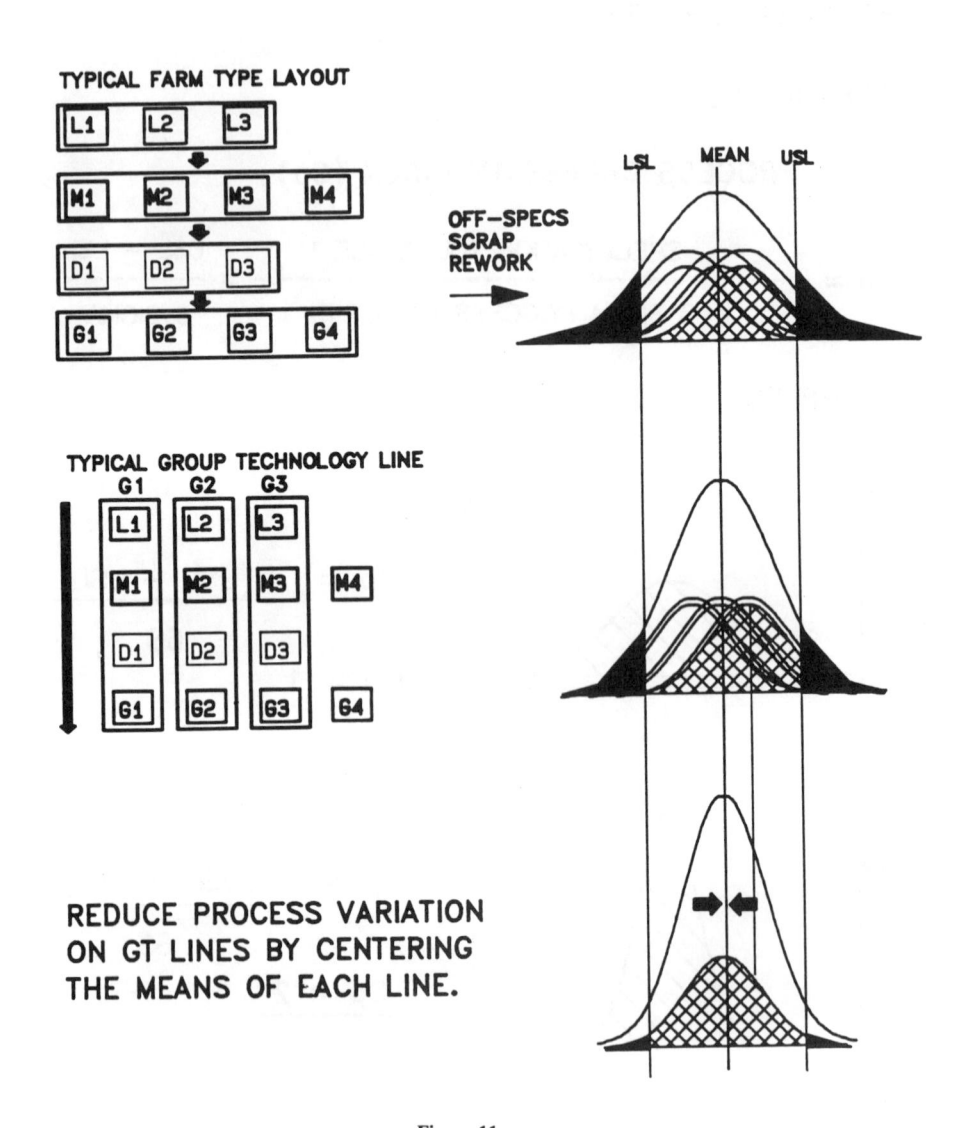

L1	L2	L3
M1	M2	M3	M4
D1	D2	D3
G1	G2	G3	G4

OFF–SPECS
SCRAP
REWORK →

TYPICAL GROUP TECHNOLOGY LINE
G1   G2   G3

L1	L2	L3
M1	M2	M3	M4
D1	D2	D3
G1	G2	G3	G4

REDUCE PROCESS VARIATION ON GT LINES BY CENTERING THE MEANS OF EACH LINE.

LSL   MEAN   USL

Figure 11

Continuous Flow Manufacturing Cell Approach

	SURVIVAL	ME TOO	COMPETITIVE	WORLD LEADER (MARKET DRIVEN)
P.L. CT RED.	------	50%	65%–85%	90%
MFG CT	>10X RPT	5X	3X	2X
Cp	<1.0	1.33	1.66	2.0
PPM	>50,000	<5000	<300	<10
SVCB	80%/MO.	95%/MO.	95%/WK	95%/Daily
PRIO	>20%	15–20%	5–10%	<5%

PIPE CT IMPROVE → TIME

Figure 12

Continuous Flow Manufacturing Stages

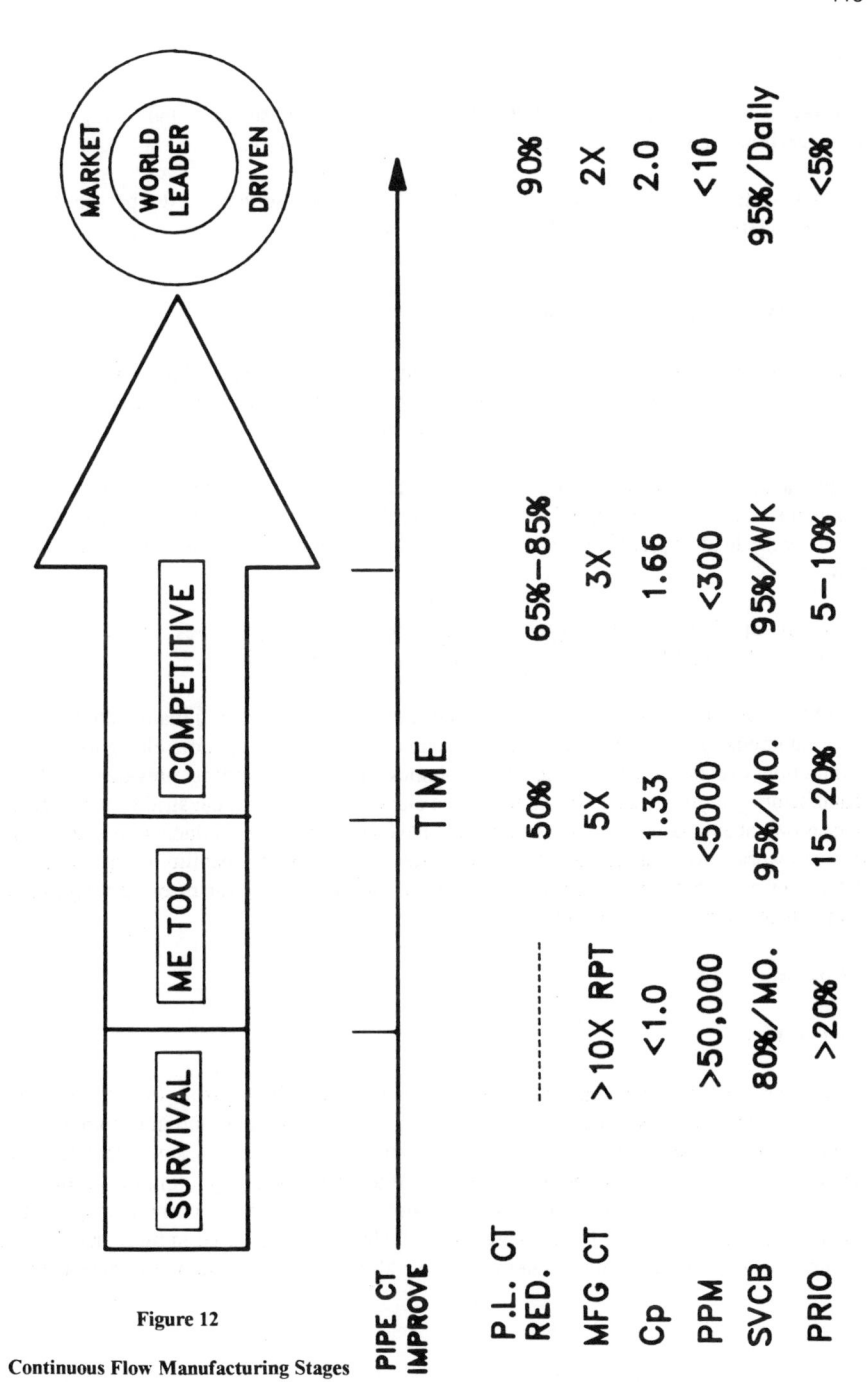

2.6 Strategies Review

This reviews the strategies of the 1990s, their particular emphasis, their underlying metrics and algorithms, and their utilization in the UCOM methodology.

The next section describes the basics of artificial intelligence so that a better perspective of the optimization tool can be explained.

3. UCOM Artificial Intelligence Review

Artificial Intelligence, as a set of products, came into being in the early 1950s. The problem at the time was excessive exaggeration over reality, and the reputation of the area suffered so badly that little was done with the field until the 1980s.

Today, again, as is the case with much that has been developed in the United States, U.S. companies have failed to capitalize on the strength of these tools. We continue to follow Japan as the lead implementer of the products that are only recently being utilized in the U.S.

Artificial intelligence technologies cover a wide array of capabilities, target uses, and flexibility. Figure 13 gives some general scope as to the technology.

All of these technologies are used in manufacturing today. The leading manufacturers may have hundreds of expert systems or neural networks addressing specific elements in their production cycles. Vision, voice and robotics are also becoming prevalent in U.S. manufacturing if you concentrate on the world class leaders. The great strides in computer horsepower at ever decreasing costs per cycle have allowed these products to permeate the fabric of world class manufacturers. The next step is to add this capability to the strategies listed above so that speed, quality and flexibility are optimized for meeting world class demands of the end of the decade.

3.1 Senses

3.1.1 Vision

Vision was one of the first areas used in the automation of quality control. Initially, lighting of the inspection area was a critical factor. New film, new cameras and new compression algorithms have reduced this problem. The basic structure is that a very small camera picks up the light reflected from some part or product at a certain critical point in the production cycle. The image is digitized in the computer and matched against a 'correct' image as the basis of the quality review. Neural networks, which will be discussed in more detail later, are used for pattern matching problems, so the addition of this technology to the camera

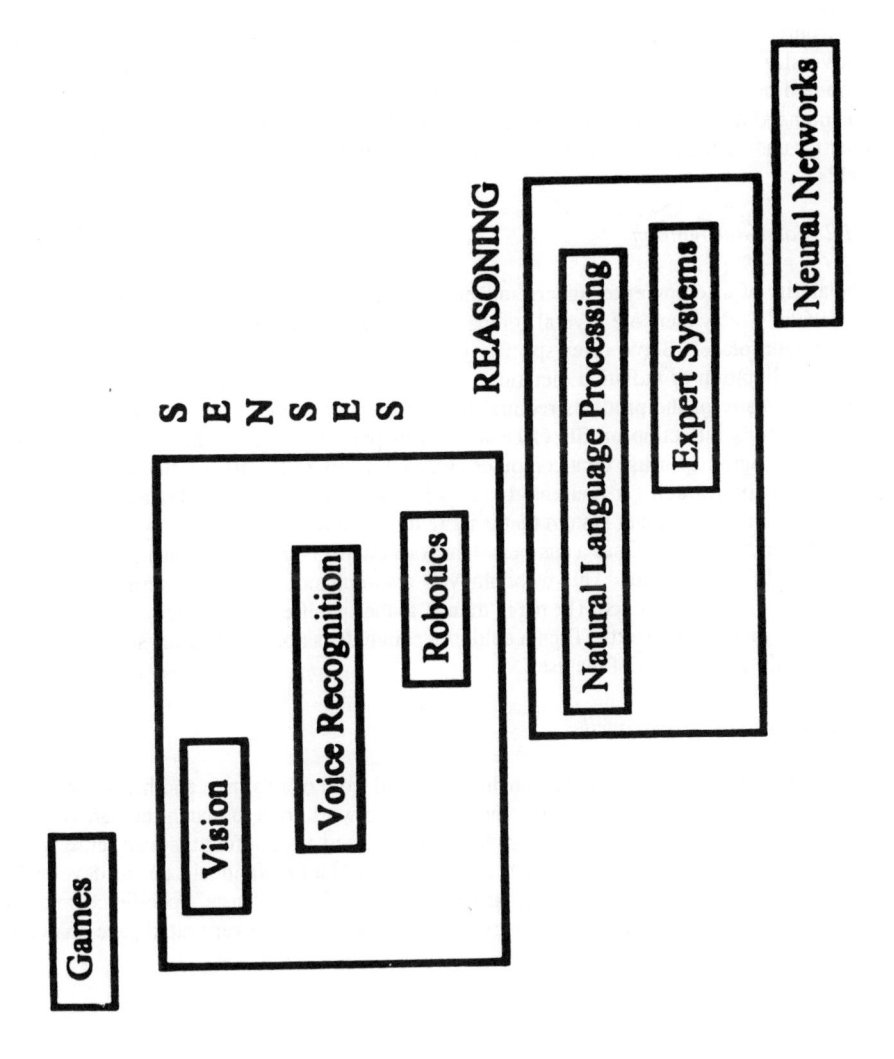

Figure 13

Artificial Intelligence Technologies

image has helped increase the speed with which a part can be reviewed. The technology has moved forward to the point where an autonomous (no human intervention and no predefined environment within which to operate) robot can drive a truck from Carnegie Mellon University in Pittsburgh to Erie, Pennsylvania and back without a road map. The two outboard cameras simply convey a stereo view of the road that the robot can interpret. Vision need not be in the visible light range either. CMU's truck has a three dimensional laser for understanding objects and the capability of infrared vision for night work without lights.

3.1.2 Voice Recognition

The ability of a computer to 'understand' human speech has been an ongoing project since the 1960s. Today there are several companies with products that address this requirement in the workplace. Some offer specific vocabularies for specific tasks such as general medical dictation or radiation dictation. Some try to be generic in their application set, however, many of the products require that the user 'train' the system to understand the specific user's inflection profile of each word or phrase. Most of the systems come as software programs for use on a computer. One company sells a device that is used in the home as a control unit for household chores like lighting, security and heating. Several companies have a computer system that is more dynamic for home use, and voice will start to appear on them soon. Once a computer (chip or system) is in the loop on a project, voice is relatively easy to add. This technology is also enhanced by the pattern recognition capability of neural networks. The net is 'trained' to the specific voice pattern of the user for each word or part of a word. Then a cold or a cough does not preclude the software from understanding the words of the user.

3.1.3 Touch, Taste and Smell

Again, transducers (something that changes a signal from one form to another) change the signal from analog into digital form and a neural network tries to match the pattern with prelearned patterns. One of the defense labs has designed a robot foot that can sense when objects are near as well as support heavy payloads. Los Alamos National Lab has developed a tiny chemical sensor that can be prefabricated to "smell" for or "taste" specific chemicals. In general, the sensor technology industry is also accelerating at a very rapid pace on every front.

3.1.4 Sensor Merging

The natural conclusion to these individual capabilities is to electronically merge them as does a human. Neural networks are being used for speech recognition also have cameras for looking at the mouth of the speaker -- reading lips while the audio sensor looks at the voice pattern. More computing power is required, but the issue is not how - it is simply when - and the answer is very soon for full commercial applications.

3.1.5 Robotics - Stationary

In the manufacturing area, robots have been hard at work for many years. The Japanese of course have the major lead in implementation, so their technology pervades the stationary robot market. Here, the robot is programmed for his environment. A specific set of tasks that are redundant are programmed into the controller, and when a controlling computer system senses the correct time frame, the task command sets are sent to the robot unit. Sensor and control algorithm advances will allow these machines to move quickly into the blue collar positions of later this decade.

3.1.6 Robotics - Autonomous

The major thrust for this technology is in the autonomous field - where robots work in and learn about their environment on a real time basis. Control programs are quickly taking on the capability to viably work in certain industries. Caterpillar, for example, has an off-road strip mining truck that works in the strip mine without the aide of humans. The main earth removing machine, called a drag line, calls the trucks as needed. In the test site, the trucks drive ten miles from the dump site to the mine, load their new cargo, and return to the tipple to dump and wait for the next signal from the drag line. Caterpillar expects to have robot controls on their whole product line within ten years.

The primary developer of much of this technology, and the lead research organization for robotics in the world is Carnegie Mellon University (CMU) in Pittsburgh, Pennsylvania. A brand new consortium with CMU as the lead player has been funded by NASA with the directive to be the primary focus for the commercialization of this technology for the United States. The U.S. has a major lead here, and the goal is to push the technology into American manufacturer's products (anything that moves) as quickly as possible. This technology, alone, has the capability of forever changing the world in which we live. The industry has been liked to the developmental stage of the automotive or aircraft industries at the turn of the 20th century. The next 20 years will see the use of these devices in everything a human now does, and one of the consortium goals is to modify the present mind-set to more fully accept and adapt to this movement.

3.1.7 Robotics - MINOR II

As mentioned above, autonomous robotics controls are now formative enough to allow some use on the shop floor. The MINOR II Model uses this technology to make the shop floor dynamic. Cost reductions will require that less machines be purchased and higher utilizations of the ones purchased realized. One easy way to accomplish this seemingly impossible task is to make the machine tools into autonomous robots. Cells would be set up in a physical fashion that would allow minimal motion of the machine tools so that they participated in more than one physical cell. A master control unit would schedule the devices in the production process. Simulation would dictate how many machine tools as a minimum would be required to move a part or product through the manufacturing process as quickly as possible. The activity based costing model software would generate a least cost approach as all the variables were run through the optimization algorithm. The inclusive artificial intelligence technology would allow optimization to take place within the constraint set defined by the customer/manufacturer/supplier alliance with each component being measured against global optimums known about other companies (benchmarking) and with each component being supplier sourced from around the world. The goal of MINOR II is to define the ultimate manufacturing enterprise by using all of the advanced technology tools currently available today -- no blue sky or vaporware.

3.2 Reasoning

3.2.1 Natural Language Processing

Voice recognition is based on some type of predefined and learned pattern -- be it audio or video or both. Natural language processing deals with 'understanding' the spoken word. This task is much more daunting, but researchers have some basis from which to develop products.

Humans learn by using many of their senses. This information is stored in the brain and recalled when another human or situation provides some stimulus. Machines can work fundamentally in the same way, however, their brain is a computer, and a computer is a poor match at present for the human brain. Recall and number crunching are unsurpassed by a computer. Real time knowledge interpretation has yet to be concurred.

IBM and Cornell have had some success in duplicating in a laboratory a portion of the human brain called the hippocampus, but the most far reaching project is underway at the University of California. Their research team and those from two supporting companies are attempting to merge silicon technology with a rat brain. The project has a 20 year time horizon, but genetic engineering and the advances of medical science and computing on a chip should allow for success in a far more near term time frame.

3.2.2 Expert Systems

Expert systems are 'intelligent' computer programs that use predefined knowledge and inference procedures to solve problems that are difficult enough to require significant human expertise for their solution. They use real world data and try to match it to the information with which they were built.

They were developed in the 1950s by using a new computer language of the time called LISP. Another product called OPS5 was developed about this time at Carnegie Mellon University. The actual field of expert systems got its start in the 1960s with the advent of knowledge bases for math equation solving (MACSYMA at Massachusetts Institute of Technology) and chemical formula analysis (DENDRAL at Carnegie Mellon University).

The 1970s brought a more complete set of solutions for pioneering the technology. DIPMETER ADVISOR was developed for the oil drilling market, PROSPECTOR was developed as a geological advisor, XCON was established to be a configurater, and PUFF and MYCIN were developed for the medical markets. Each was prebuilt for their specific task in life.

The 1980s brought out the 'shell' which is a development and run time software tool to allow user level people to build their own knowledge base. The shells contain the user interface, a knowledge base that is established by the user, and an inference 'engine' that provides the system with a way of matching the outside data with the prebuilt knowledge base.

Mrs. Fields Cookies uses an expert system for scheduling staff at their many stores, production scheduling and interviewing new employees. American Airlines uses an expert system to schedule the maintenance of their aircraft (each plane can fly only so many hours with out FAA approved maintenance and only certain airports have the required equipment to perform the work), and to schedule their flight crews (again only so many hours are allowed per flight crew member per FAA regulation, and crews are disbursed across the country). IBM is an extensive user of these tools with uses ranging from manufacturing floor scheduling to repair maintenance advisors to hardware configuraters for the field sales force. Computer systems can have so many variations in their configuration that the original 9370 model of the late 1980s took three days to configure. The expert system brought this down to a matter of minutes, greatly increased the quality of the final product for the customer, and reduced training time for the field force by orders of magnitude.

Expert systems are easy to build, easy to use, and easy to maintain because they contain none of the step by step constraints of linear programming models. They are also considered 'quick prototype' software which means they need only one rule for their knowledge base to be utilized.

And finally, the world is looking at the future for their use. The U.S. (STRATEGIC COMPUTING) has a $600 million multiyear effort for their development. Japan (ICOT) has a ten year multi-billion dollar program. Europe (ESPRIT) and Britain (ALVEY) both have projects that are smaller but multiyear. There are tremendous advantages to the winners in both dollars and time reduction for everything touched by a human that requires any thought. One university program is trying to build an expert system with all basic knowledge known to a human. Carnegie Mellon University is pursuing an 'AGENT' approach which utilizes many small specific expert systems working in unison on any given task. Products are coming to the market in prebuilt form at an ever increasing pace, and only a few years of continued development should make for major technology strides in capability while the price continues to reduce.

3.3 Neural Networks

Neural networks are quite interesting. They provide a programming attempt to replicate a portion of the capability of the human brain cell or neuron. Via programming, the artificial neural network uses data in parallel chunks that provide some pattern of interest. In most neural network forms, the pattern(s) of interest is sent to the neural network over and over until the network is 'trained' to provide the correct output for the specified pattern input. Once trained, the neural network can input random or partial data and find the specified pattern of interest.

There are many forms of networks that have specific 'training' formats and specific uses or tasks. They can be directed toward linear and nonlinear problems and the use of several 'layers' of neurons has provided a solution to the 'EXCLUSIVE OR' problem that had been a major impediment to their wide spread use before the hidden layer concept was discovered and instituted.

There are billions of neurons in the human brain, and each has approximately 150 variables. Approximately five can be reduced to a mathematical equation, and these represent the software of today. A ten year IBM/Cornell University project attempted to build the other 145 variables, which are electro-chemical in nature, into approximately 10,000 neurons of the hippocampus region of the brain. This area contains the control for human speech. Additionally, it also is the location of a strange brain wave that causes convulsions. Scientists know the location and can surgically remove the area to cure the disease, but they do not know the root cause of the wave itself. The interesting thing about the IBM project was the unexplained and non-programmed onslaught of this same brain wave ten years into their experiment. If the lack of brain wave activity is considered death -- what is the occurrence of brain wave activity by a manufactured set of neurons -- life?

Neural networks, or perceptrons as they were initially called, were developed in 1956 at Dartmouth. ADALINE, a single neuron perceptron developed by Professor Robert Widrow was one of the first commercial applications of the technology. The development of additional uses of perceptrons continued until 1969 when Marvin Minsky published a paper that criticized the technology for not being able to handle the "exclusive OR (XOR)" problem. The story got wide scientific press, and the technology lost favor until the early 1980s. This new era was based on research and development work by Dr. Hopfield in a paper presented to the National Academy of Sciences in 1982 and in work by Dr. Rumelhart on the back propagation technique in 1985. These advances allowed for the solving of non-linear problems and XOR by utilizing more than one layer of neurons in a neural network model. Once this problem was solved, the use of the tool for commercial applications has moved forward quickly.

As mentioned above, any application where data is in some pattern, but not easily manipulated with IF/THEN methodology of the expert system, is a perfect application of the neural network approach. Voice recognition for lip reading and audio signal recognition; language conversion with Swedish to Japanese as one of the first; luggage bomb detection by 'smelling' the chemical traces released by explosives; sonar mine detection in submarines where accuracy becomes a constant 92% over a human range of 88-93% after six months of training; vision inspection for quality control; and handwriting character recognition for computer input devices are all examples of problems that have patterns and the need for automation.

3.4 Hybrid

One can quickly see that the real world contains a lot of problems that provide data in both pattern forms. That is, some data can be reduced to the IF/THEN format of expert systems and other data is numeric (analog or digital) and must be handled via the neural network format.

One such problem that was faced by a major appliance manufacturer was the delivery of their products to their distribution sites in a logical manner. This situation contained multiple size delivery vehicles, various availabilities of each, and various delivery locations in the U.S. The problem then became the optimization of the available vehicles for maximum coverage of delivery sites with the least miles route. The problem was then further refined into hard constraints - number of trucks, type, and proper loading order to allow the each location to have their complete order at the back of the items on the truck, and soft constraints - the truck routes. This problem set then suggested the use of an expert system for the hard constraints and a neural network for the soft constraint. The neural network was fed the delivery locations of the orders required for the day and provided the shortest route for the deliveries. This represents the 'traveling salesman' problem, and, although there are several linear programming algorithms for its solution, they are both more time consuming and more compute intensive to solve than utilizing the neural network (Self Organizing Routing Map - Kohonen Map approach). The expert system reviewed the orders for the day, totalled the weights for the order for each location and matched those against available truck types. The solution started with the largest truck that would hold all of the deliveries. If additional trucks were required, appropriate sizes were added until the full day's orders could be delivered. The expert system then provided a 'pick list' (a pull-from-inventory order for the warehouse employees) in the reverse order from the delivery sites. The last delivery was then loaded first and the first delivery site was loaded last. Quality improved, moral improved, customer service improved, costs declined, and the time to get the scheduling job done went from hours to minutes, so productivity improved.

3.5 Fuzzy Logic

Fuzzy logic is a mathematical technique for allowing items to be in more than one mathematical set. A computer needs a digital format for interpretation -- either a one or a zero. The fuzzy logic technology allows for a more smooth transition between these end points.

One of the primary uses of this technology is in control theory. Bullet trains in Japan, elevators, and air conditioning systems all have a common thread -- the need to transition from one state to another in a smooth fashion. By adding this technique to a hardware computer chip for the fastest implementation of its capabilities, each of these systems can utilize this control approach. Trains start and stop without passengers having to hold on to overhead straps; elevators reach their destination in a more timely fashion and do not leave passenger stomachs on different floors; and air conditioning systems conserve energy by moving a room temperature between smaller upper and lower limits so that some people do not freeze while others swelter. Variations in the control cycle are minimized, so efficiency and quality increase.

Fuzzy logic was developed by Dr. Lotfi Zadeh at the University of California in the 1950s. No one saw the use for the technique until the Japanese started to focus on its applications in the control area. Now, one primary U.S. fuzzy logic chip manufacturer exists -- and almost all of his output is sent to Japan. Motorola and Intel have the capability coming on-line soon as well.

While not a true artificial intelligence product, the technology is an excellent compliment to the every increasing complexity of a global manufacturing environment. Many examples of hybrid products with neural networks and expert systems exist and provide a stronger and more diverse tool than what can be accomplished by each individually.

3.6 Genetic Algorithms

Genetic algorithms were developed in the late 1950s. They attempted to replicate the Darwinian theory of evolution via a programming structure. These first attempts did not do well because the focused on 'mutation' rather than 'mating' to generate new 'gene' combinations. Once the mating portion was properly addressed in the mid-1960s, the resulting program became the focus of attention for optimization of a solution set for many convoluted problems.

Genetic algorithms borrow from the biological models of genotype - the encoding of form, phenotype - the expression of encoded form, mutation and sexual reproduction, and (un)natural selection, for simulating evolution of form and/or function. The generic method is related to those of simulated annealing, steepest ascent, and Monte Carlo optimization. There are several methods for "steering" the essentially random search of an n dimensional space which is a bounded space defined by n independent parameters or solution space variables. A graduated evaluation of the goodness of the various solutions provided by the bounded space is used to generate an evolutionary momentum vector which determines the magnitude of progress or movement in a particular direction in the solution space. This guiding vector speeds convergence upon user defined desired traits, but simultaneously helps ensure convergent, as opposed to divergent, evolution so as not to limit the technique to stopping at any local maxima.

The solution space for these types of problems can be enormous -- and therefore too compute intensive for any realistic application with today's hardware capabilities. For example, if each move in a chess game has an average of 10 alternatives, and a typical game lasts for thirty moves on each side, then there are about 10 to the 60th power strategies for playing chess. Most of these are bad strategies. The genetic algorithm lays a surface over the three dimensional space that contains all solutions to a problem -- the solution space. Samples in many regions are compared simultaneously, and continued efforts are provided only in the areas that contain higher 'elevation' solutions -- that is, areas with higher probabilities of either a local or global optimum. Samples continue to be taken and compared until a clear global optimum emerges.

UCOM uses Engineous, a genetic algorithm based optimization tool developed by Dr. Siu Shing Tong, Corporate Research and Development Center, General Electric Company. It forms the umbrella control mechanism under which the strategy metrics, algorithms and software models are laid in order to allow a complete organization optimization to take place. This complete package comprises the computational side of the Unit Continuous Optimization Manufacturing methodology.

Because Engineous is automated, it can explore more design space options and parameter trade-offs in a given period of time. Just utilized at the engineering level, which has been GE's emphasis for the last five or so years, has produced results of a 10:1 improvement in the speed with which a new design can be created. Human error was negated, and the design was improved over a human engineer's efforts by a factor of two or three to one. The incumbent savings in manufacturing costs for GE have been valued in the millions of dollars.

The software utilizes three approaches to the optimization problem. First is an expert system that captures expert knowledge of the problem and is used to prevent Engineous from wasting time examining areas of the solution space that an expert solving the problem would know to avoid. Second is a traditional numeric technique for nonlinear optimization called gradient approximation. This approach is very good for finding a local optimum in the solution space. However, once this local 'high altitude' point has been reached, this method can proceed no further. The final phase of the optimization process then comes in to play -- the genetic algorithm. The algorithm: assigns a genetic string made up of ones and zeros to each local optimum solution provided by the gradient approximation technique, compares this population, and propagates their 'traits' (that make a local 'good' design as defined by the experts in the expert system knowledge base) into a new population. The gradient approximation method is then again utilized as a search for a new local optimum and the genetic algorithm again analyzes this new set of 'good' designs. Eventually, in a time slice defined by the user, the survival-of-the-fittest results in a superior global optimum strategy or design.

Because Engineous is generic, it can be used to connect either off the shelf or proprietary codes without the need for the source code. The user defines the target for the program which can be the global optimum or all solutions within a given range. This later format allows the user to do sensitivity analysis as one or multiple variables are allowed to change as the program iterates through possible solutions. Thus, the tool can be used in engineering, in finite scheduling, and/or in global strategy optimization -- the intent for which this chapter is dedicated.

3.7 Review

Artificial Intelligence provides a very strong solution to a wide variety of problems, and many of America's corporate leaders need to understand both their capability and potential impact in a world class global manufacturing environment. This paper was presented to an international manufacturing symposium sponsored by IBM. When the audience of manufacturing executives was asked how many of their companies were utilizing AI and how many knew the capabilities and applications of AI, one hand out of approximately 30 people was raised. This paper was also presented for an IEEE artificial intelligence audience at MIT and asked how many in the audience of approximately 50 had worked in manufacturing. Two hands were raised. This is America's problem: we make all of the discoveries and advances in a major technology, but we do not implement that technology in areas where its use will allow us world leadership. In the 1940s, 1950s and 1960s, the world looked to the United States as the top global manufacturer. That leadership has failed to continue. Now the world looks to the U.S. as a garden of good ideas, picks them freely, and takes the lead in industry after industry after implementing each one and leveraging its power to the fullest. The above unscientific but meaningful pole suggests that we are not moving to address the problem, and American manufacturing will continue to decline as both a major economic force in our economy as well as a major employment provider.

4. UCOM Simulation Review

Although the power of artificial intelligence have been demonstrated in the wide variety of projects that were reviewed above, the human must still play the master role. The knowledge that builds the artificial intelligence tools is not stagnant. Any change in the production environment must be accounted for in the knowledge base or training data set of the tools. While much of this input can be automated, the tools are not so far advanced that some human oversight is not required.

One of the best ways for this to happen is the use of graphics for the most easily understood feedback to the human portion of the manufacturing equation. One of the best dynamic graphics formats for understanding the 'big picture' of the whole manufacturing process - be it data/paper flow or physical part flow - is simulation.

There are several off-the-shelf tools available, and they run on anything from mainframes to workstations to PCs. They are usually compute intensive, but the power of today's systems have little trouble in providing a workable platform.

As in expert system shells, these tools also must have the data loaded and structured. While this step has been made simple with graphical user interface (GUI) formats, education on the total capability of the software and its extended applications must be a prerequisite to its purchase.

The present packages are primarily two dimensional in nature. This provides enough information in most cases to determine the viability of the artificial intelligence generated solution. But, there are additional problems that suggest the need for a stronger tool. And this is where the use of Virtual Reality or Cyberspace tools will prevail. These packages and their attendant hardware for 'fooling' the human senses into seeing in three dimensional space, are just starting to come to the general use market. The toy and entertainment markets are leading the way, but their volume allows the cost to be very easily justified in a major manufacturing environment.

Once implemented in the mix of tools for manufacturing excellence, additional uses such as maintenance training and final part mock-up will surface.

The Unit Continuous Optimization Manufacturing application will center on concept verification for the human aspect of the optimization process. The management team can try the solutions generated by the optimization tool before they actually move a single item. While this may seem just a waste of time to some, the general feeling is still that the human brain can add something to the optimization model. The builders are never omniscient, so a continuing review process provides the model with a quality control factor that makes the implementation of the whole UCOM strategy more palatable to the general manufacturing management team.

4.1 Review

As production, research and development, and sales/marketing sites disburse not only over the globe, but over the universe in the years to come, a quick, complete and dynamic check tool such as virtual reality will cease to be an option in a fully implemented UCOM system. The data overload and value will force the human element to adopt a simulation tool and a simulation tool that contains virtual reality elements.

5. Unit Continuous Optimization Manufacturing MINOR I Model

".....when you can measure what you are speaking about, and express it in numbers, you know something about it: but when you can not measure it, when you can not express it in numbers, your knowledge is of a meager and unsatisfactory kind"   Lord Kelvin 1883

".....any good decision is an informed decision: and the informed decision must reflect computationally optimized information in order to meet the speed, variety and quality of the global competitiveness challenge"   J. S. Minor 1993

Producing goods upon receipt of an order instead of a forecast has long been the dream of those concerned with management, marketing, production and inventory control.

In a discrete manufacturing environment, there are many new strategies that attempt to meet the modern phenomena of shorter research and development time, shorter product life cycle, and shorter order cycle. Added to this is the impact of downsizing the organization to a minimum number of people, so the increasingly frenzied pace of meeting customer needs and competitive pressures are being handled by a smaller and smaller set of specialists.

The problem then is how to handle this cumbersome issue of less people, less time, higher quality and more complexity.

One approach is to address a specific subset of this problem which has provided the advent of many 1990s style manufacturing strategies.

Strategy can be defined as the art of creating value. It provides the intellectual framework, conceptual model, and governing idea that allows a company's managers to identify opportunities for bringing value to a customer and for delivering that value at a profit. Some of the more prominent ones include: activity based management; continuous flow manufacturing; concurrent engineering; process reengineering; and mass customization.

The new problem emerging from using these tightly focused strategies is in optimizing their cumulative impact. Each has one or more metrics and variables against which a company can compare its continuous improvement and its performance against world class manufacturing benchmarks.

With this new problem as the backdrop, this chapter introduces Unit Continuous Optimization Manufacturing (UCOM): a mix of artificial intelligence software tools and advanced manufacturing strategies that provide the solution.

UCOM provides the fastest customer response, lowest product/manufacturing cost, highest quality product for an order of a single unit -- and it does this in the manufacturing environmental framework of the chaos of the coming decade.

The following provides the structure of the UCOM hypothesis development as based on an impact analysis of the UCOM methodology on each of the major subsections of the corporate entity -- suppliers through customers. UCOM is a strategy that uses only off-the-shelf hardware and software. All of the packages mentioned here are readily available, have an in-place customer base and following, and have a structured maintenance and support system that allows for their continued upgrade and development through time. Availability is not the issue -- timidity in implementation of new strategies and their support tools is.

5.1 Engineering Redesign

The real base power in any organization is the knowledge base. In the engineering area, this covers product design knowledge, customer requirements knowledge and manufacturing capability knowledge.

At the heart of this new business metaphor is the need for continuous change in organizational behavior. UCOM is responsive, modular, time-based, process-drives, and learning efficient. UCOM senses change in the environment and is able to respond rapidly in a cost effective manner. The technology infrastructure is a critical component in the transformation to the UCOM strategy.

UCOM has four basic disciplines:
Enterprise Design - Understanding the strategy, structure, design principles and tools required to create the methodology;

Human Systems - Relationship models, teaming techniques, and the total human aspects that enable adaptability;

Technology Systems - Detailed listing of hardware and software to provide the proper base for the implementation of the Enterprise Design;

Management Systems - Full process flow analysis of all management information at the initial state and the detail of the process flow map changes required to enable the transformation to the UCOM methodology.

The UCOM impact in engineering represents the first stop for a new custom designed customer order of one single unit. The UCOM system is designed to have customer requirement data, in a predefined format as required for proper utilization, electronically transmitted directly to engineering. Sales is bypassed once the initial customer relationship is established. Their role is confined to providing new customers for their product set and to redirect corporate resources should the customer encounter some problem.

Many companies are concerned with direct access to their computer system by outside parties, so several companies such as IBM have 'electronic mailboxes' available as delivery locations with global access. They can download to the engineering system on an automatic basis or the company can have an engineer sign on at various times of the day for downloading.

If there is a strong family-of-parts structure for the product set, simple parametric based rules in the more advanced computer aided engineering (CAE) tools will automatically build all needed solid three dimensional drawings, two dimensional drop off drawings, product order lists (bill of material), and routers for the shop floor. While the expert systems utilized in the CAE products are still fairly basic, they are merged seamlessly with the CAE drawing product, so the call from the genetic algorithm can be made to one software tool.

CAE tools that have this capability are becoming more prevalent in the smaller and medium size firms as their price and the corresponding price of the underlying hardware upon which they depend become less expensive.

5.2 Engineering Impact

Either the CAE tool's artificial intelligence knowledge base, or an outside vendor's expert system knowledge base must be significantly expanded to use the UCOM methodology.

Design for manufacturability, design for assembly and part number minimization rules must now be added at this step of the process. The electronic transfer of data to the parametrically defined part will automatically build all of the relevant paperwork for the customer's order. However, there is no guarantee that only one plant will build or order parts for the completed product. In addition, there is no guarantee that the customer's specification will lead to a product that can be completed within the single corporate entity. Therefore, engineering data base control tools like Product Manager systems with automated electronic management sign-off mechanisms become more important. This format keeps the human part of the equation in the sign-off loop as well as properly disbursing the workload to the facilities that have the time and capability to complete each manufacturing task.

Several companies have included vendors into their Product Manager system. This structure can provide electronic data to each as preparation for a quote, or an actual order can be automatically generated if the vendor relationship allows this to occur.

Speed enhancement is the actual target at this part of the UCOM implementation. As was reviewed earlier in the strategies section of the chapter, the first party to a market can expect the lion's share of that market. Here we simply add response speed to an order which also significantly impacts a product's market share.

If the knowledge base allows the part to be made -- with or without human intervention -- then the knowledge base of the complete organization should be properly linked to allow processes to be automatically generated. For ease of maintenance, this will mean multiple knowledge bases, but most compiled AI products are designed to work with multiple knowledge base formats as well as with other programs that provide specific access interfaces.

5.3 Sales/Marketing Impact

The direct sales force of today is slowly being replaced by the less expensive and more productive telemarketer. As technology continues to move into place to allow video images in a real time environment, salesmen will provide all of their correspondence electronically. Face to face meetings will be rare. Time constraints and overhead reductions will force companies to adopt this format in order to meet global cost constraints.

As a case in point, IBM has significantly outsourced their sales force to business partners. In a recent internal meeting for the roll out of a new product that was very technical in nature -- usually indicating the need for the internal teams -- the IBMers were told that they would be bypassed as the sales/marketing force of choice for the product rollout. The less expensive business partner people would be the only sales force educated on the workings of the product. This format provided IBM with additional revenue because they charge the business partners to get the education -- they must spend internal dollars to educate their own sales force. As this cost of doing business gets pushed further out of the organization, the internal telemarketing wing of IBM is expanding. Many of the employees in this part of the sales strategy are either contract workers with no benefits -- or they are members of another business partner that pays their overhead. IBM stays flexible on head count while cutting head count costs. You can now order any IBM computer -- PC to mainframe -- over an 800 number. When video becomes totally available, this will probably be the ONLY way you will be able to order an IBM product.

Motorola's pager division has already made great strides in automating their pager ordering system via artificial intelligence configuraters and downstream paperwork automation. A salesperson still comes to your location, but they bring a laptop computer that contains a modem with them. Before the program was started, it took many days to build the pager of your choice. When the automated configurater was developed as the front end of the order and a concurrent pull - continuous flow manufacturing - system was implemented, the pager was produced and out the door within 24 hours. Continuous improvement in this same approach has provided a manufacturing time of one hour. This type of approach is what won the Baldridge award -- and this type of approach, on a global scale - is what is going to keep a company in business in the future.

As markets become more fragmented and chaotic, the organization must try to meet the specific requests of a smaller and smaller grouping of customers. Following is an example of how the automotive industry is trying to adapt -- now that customers get more choices than the color black.

Market research has shown that supplying a customer grouping of a few hundred thousand can add significant market share because the auto maker can totally meet their needs. They will buy when this set of requirements is made available. The problem is that General Motors and Ford require over a million unit sales of a specific model to break-even on the 48 months required to design, test and build that model. Chrysler can get under a million units per model by working with a 24 to 36 month cycle. Toyota, probably the most advanced automotive manufacturer in the world, has the cycle down to 18 months - and falling. U.S. automotive manufacturers are behind again, and the Japanese are again gaining market share. The UCOM approach to meeting customer requirements is the reason, and the Japanese will continue to push U.S. makers aside until their operations are as responsive to customer needs as are the Japanese.

UCOM still relies on the sales force to initiate the sales leads, but no differentiation is made on the contact procedures as long as they provide the electronic link. Once the link is established, their only reason to again contact the customer would be to arbitrate and supervise customer complaints. Engineering, in the UCOM methodology, becomes the lead interface with the customer as their knowledge base provides all design and down stream paperwork (electronic forms) control based on the 'pull' of the customer's order.

5.4 Plant Floor Impact

UCOM takes the cell part flow program to the next step. In days past, and still unfortunately in many American manufactures, machine tools were placed in a 'farm' style layout. The farm approach puts all machines of a certain type together -- regardless of size or specialty or setup. This means all horizontal boring machines were together, all multi-axis milling machines were together, all break-presses were together, all welding machines were together, and so on. This conformed well with job classifications, which were based on the machine the operator was running, structure of the organization and with the management control theory that called for operational subsets with a certain specialty of expertise. Once the walls of the groupings were raised, real or virtual, around these machines, change has not come easily.

For the companies who have embraced the newer approach of cells, machine tools can be more specifically assigned to a smaller set of parts with which they must deal. This lessening automatically increases the Cp of the organization because the variation of each machine is more completely controlled. Reducing the machining tolerance reduces scrap and rework, so productivity and margins increase. Machine tools can be purchased with a specific subset of tasks in mind, so sizing of the tool can be reduced, saving capital equipment expenditures. Since the number of parts coming across the machine are reduced, set up for the machine can be reduced and more easily automated. The perfect cell approach would be a dedicated machine that had no setup time, but the capital costs for this approach are usually prohibitive.

UCOM accepts this constrain. The UCOM methodology calls for more automation of the location of the machine tool. This adds to the set of parts the machine will produce, but only slightly and utilization is increased. The key advances that make this possible are two. First, new foundation support structures for larger machine tools have nullified the need for major foundation work as the proper support bed for the equipment. In some cases, no specific alteration to the manufacturing floor is required at all as prep for machine tool location. This freedom to move anywhere over the shop floor for many machine tools allows the utilization of autonomous robotics to control the movement of the machines in a controlled format -- without the need for humans. Carnegie Mellon University in Pittsburgh, Pennsylvania has developed very inexpensive control algorithms that couple proximity sensors and image technology. The off-the-shelf system could be easily adapted to the machine tools for a safe but dynamic shop floor environment. Power, electricity, coolant, water, etc. would exist via quick coupled ends available at optimally designed shop floor flow locations. An artificial intelligence finite scheduling tool will decide optimum flow with the aide of the genetic algorithm. The scheduler would have the capabilities and profiles of all of the machine tools in its knowledge base, so a machine would be moved as required to reduce the time to manufacture the part while keeping quality at the maximum level. Roving 'set-up robots' would either do the work completely or assist the human.

Machines could come off-line and out of the production flow for maintenance that would cause significant down time. Machine tools of a similar nature would then be called upon to fill the manufacturing gap left by the down machine.

The just-in-time system calls for specific holding locations along the production line. With the 'peg-board' approach to machine tool placement, more access doors may have to be provided to allow flexibility in drop locations.

Engineous, the optimization tool used in the UCOM methodology has the capability to weight each factor in the optimization process. If a customer provides a certain tolerance on dimensions and a certain delivery date, the program can optimize between the conflicting elements of the production process.

New materials that take advantage of the desk top manufacturing technology, stereo lithography, will reduce the flexibility needed in different machine tools on the shop floor. CAE data is fed directly and electronically into the stereo lithography device, and some of the parts today are usable as a finished product. Materials technology is advancing rapidly, and new synthetic materials will start to replace structural steel components at an ever increasing pace.

5.5 Employee Impact

The use of computers, artificial intelligence and robotics will significantly reduce the use and need for blue collar positions in the manufacturing work place. So called steel-collar employees will slowly take over position after position. This will require very directed and very well thought out long range reeducation programs for the displaced workers.

People coming up through the basic education system must be directed to programs that will provide meaning full employment for some period of time before retraining is required.

Even today, forward thinking companies are suggesting that an employee stay flexible and add 20 to 30 days of training each year. These companies still depend on the grey matter of their human team for a great percentage of what is accomplished toward making them a viable organization. However, once automated replacements exist for a specific job or position, the human must recognize the advantage of moving on to new areas that are not yet automated. Both the education requirement and the pace of reeducation will continue to increase.

At some point in the future, genetic engineering will enhance human intelligence and ensure that humans have the mental capacity to perform in areas more advanced than their metal compatriots. World populations must decrease as employment opportunities decrease. Providing food for the masses will not be a problem, but support of many by the relative few who have jobs will not be economically feasible. As the world comes together under more and more alliances like the North American Free Trade Agreement and the European

Common Market alliance, a global strategy for addressing this problem will emerge. "Big Brother" will become more invasive as some global plan is implemented. Population sizes will be mandated, genetic engineering will be mandated -- and probably required if the offspring is to have any chance to compete -- and the variations of a life style will diminish as job flexibility moves away from the broad brush of opportunities today to the ever more complex and mentally demanding positions of tomorrow.

This future will not come without a great deal of turmoil and domestic upheaval. Change management, the capability to make change happen smoothly and for the good of all, may be a job that a human never loses. The problem will change as humans become less and less of an element in the equation.

5.6 Profitability Impact

UCOM has the promise of extending the 'rule of ten' now available in its engineering debut. This means that the engineering design cycle can be reduced by a factor of ten. What took ten months to design can now take one month -- or less. On top of this rule, the product designed can have two to three times the capability of the product designed by an engineer using the same CAE tools. Engineous simply applies more iterations and finds the global optimum in the design's solution space.

With over five years of history from which to determine effectiveness, Engineous can logically be expected to also broadly impact the impressive cost savings of each of the individual strategies reviewed in a previous section.

Just the orchestration and coordination of the individual cost saving methods will result in extending their benefit.

For example, a complete continuous flow manufacturing implementation, as a stand-alone strategy, provides: cycle time reduction of 40-60% in an old line - 70-80% in a new line; 30-60% reduction in inventory; production yield (less scrap, higher quality) increase of 5-20%; production volume capability increase of over 20%; floor space made available to allow capacity to increase by 20-50%; and the reduction of tasks that do not add value to the process of 40-70%.

This base line impact of each strategy plus the added value of optimization should make the present efforts pale by comparison.

The first complete implementation of the UCOM methodology will occur with the pilot project projected for late 1994 in a joint effort with Carnegie Mellon University in Pittsburgh, Pennsylvania.

5.7 Vendor Impact

UCOM requires the reduction of vendors for the manufacturing process. To be fully effective, UCOM also requires that the vendor set chosen all utilize the UCOM methodology. As each of their vendors also utilize the system, the complete manufacturing chain is parallelled and optimized to the full extent possible. Cost reductions are then magnified, and the chain as a whole takes the form of a multicelled entity of its own right. A complete implementation of this scope would take several years to permeate the complete chain, but no competitor could match the versatility, dexterity and agility that the full team could provide. Market share would increase many fold, and the resulting increase in business would allow further cost reduction of scale -- which would provide greater market share. The approach is a global market domination strategy that is available for the taking for the strong willed and forward thinkers.

5.8 Customer Impact

Customers must also sign up to be an electronic partner in the organism. While they have a symbiotic relationship with the lead player in the chain and his supporting set of vendors and sub-vendors, customers must nourish the feeding organism electronically. Electronic data interchange, engineering data control, electronic signatures, and automated quality assessment of the finished parts must be the price paid for the superior products and service. The UCOM organization will set a new standard for optimizing achievement and excellence in providing everything asked for by the customer. The leaders in the implementation will gather customers on an ever accelerating pace as competitors fail to see the ship coming. Strategies of this type can generate their own momentum. The exterior results are at first difficult to understand by competitors. However, the word moves quickly through the customer base community, and the result can be a surprising gain of business to the industry leaders who decide to move toward the long term goal of total automation and total customer response. Toyota is a prime example.

The handful of companies that see the vision and react will be the same handful left to work in the future. The Engineous survival-of-the-fittest approach will engulf the whole organism that supplies the customer with his every whim -- and these groupings of industry leaders will also be the ones left standing in the end.

5.9 Pilot Project

The first full pilot project should be started in late 1994 in Pittsburgh in order to take advantage of Carnegie Mellon University's prowess in manufacturing industry automation and autonomous robotics. The company has four million dollars in sales revenue and is currently break-even. They provide general purpose precision machining to the aircraft industry. Almost all machine tools are of Japanese manufacture.

The company will be moved into Pittsburgh from another state, so the shop floor and information layout will be open for a start-from-scratch approach. Every conceivable human position will be fully automated. Employment is expected to be reduced from the present 55 to approximately 35. This is without the automation of robotics on the shop floor. All equipment is computer-numerically-controlled. As a pre-step to the robotics equipment addition, manual mobile bases may be implemented, and quick couple hoses will be evenly dispersed around the production area. The robot control systems for the machine tools will require several years to complete. Employment then is expected to stabilize at some minimum number as all new work is provided by the addition of machines. (Please see Figures 14 and 15)

6. Unit Continuous Optimization Manufacturing MINOR II Model

The MINOR II Model continues the automation process and includes the globalization and dissemination of the corporate entity.

The organism that is usually confined to a country or to a set of buildings will slowly transform into more of a global organism without walls. The advances in speed and capability of electronic communications systems will allow the emergence of skills on a personal basis to be bartered with multiple companies at the same time.

In the same way that a law firm or consulting firm works for multiple clients, the knowledge worker of tomorrow will also use this approach for income. All transactions and communications will occur electronically. Companies and workers will use floating benefit plans that are funded via a pooling process. Extras will be paid for by the organization or worker that wants something different from the base package. Everything the 'employee' does will be logged against some account. A copy of each account will automatically be sent to its corresponding corporate owner each pay-period, and the funds for the work performed will automatically transfer to the 'employees' account. Allegiances on the 'employees' part will center on education opportunities and work that expands their opportunity for future work. The notion of allegiance to a certain corporation or country will slowly evaporate as the world becomes one large electronic mass of communications.

Robots will initially take over manufacturing chores and most blue collar positions. Slowly, as progress is made in merging the brain functions into silicon - or some other material - or with the physical merging of the brain with computer parts, while collar jobs will go to machines. Even in the pilot project for UCOM, one person will be able to replace all six management members: chief executive officer; chief operating officer; engineering manager; quality manager; purchasing manager; and inventory manager. Customer orders direct the UCOM software pool which in turn directs the automated efforts of the above positions.

Corporations will try to gain the expertise of the best minds on a global basis. A Russian mathematician, a Chinese physicist, an Indian programmer, an American engineer, and a

Figure 14

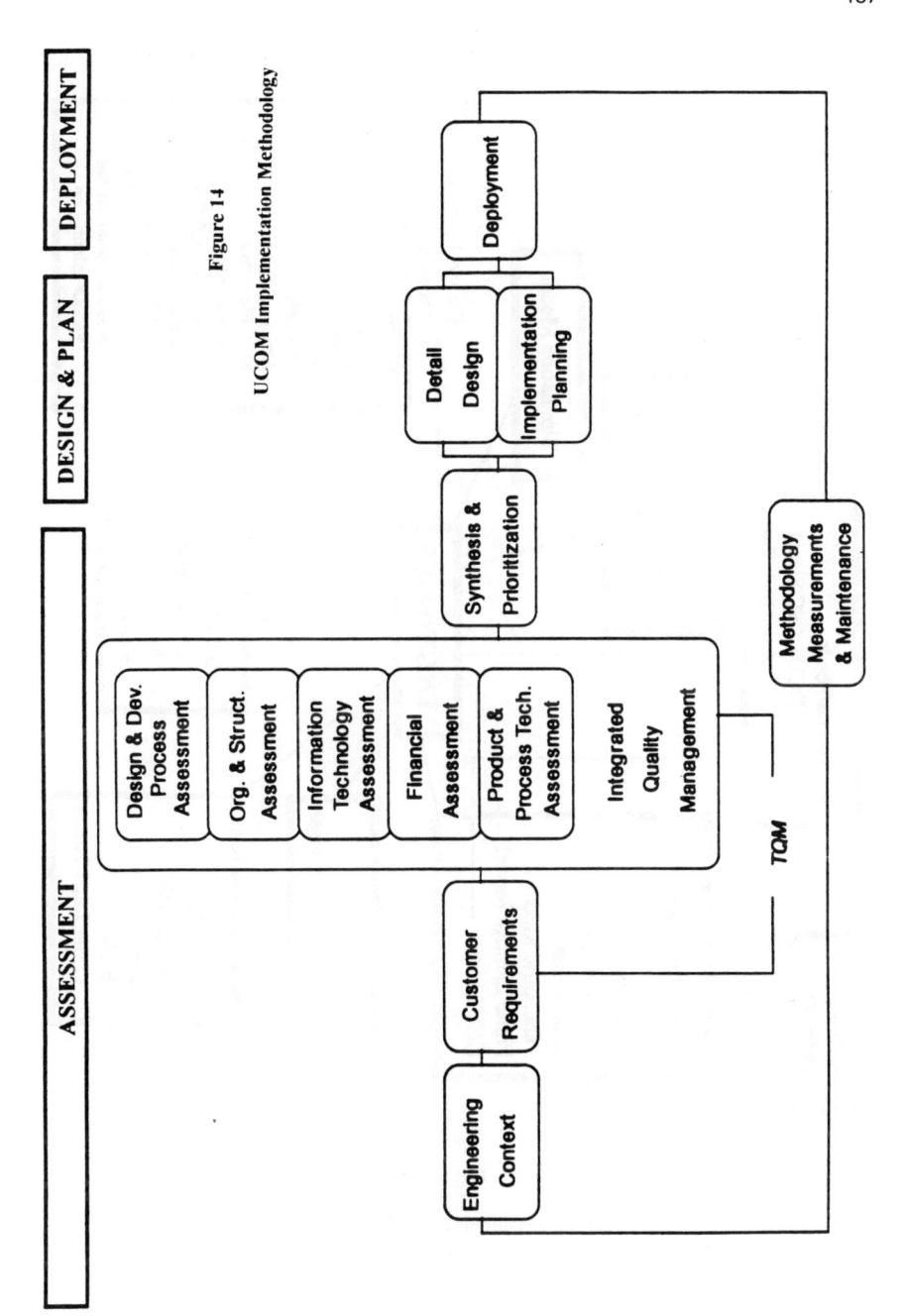

UCOM Implementation Methodology

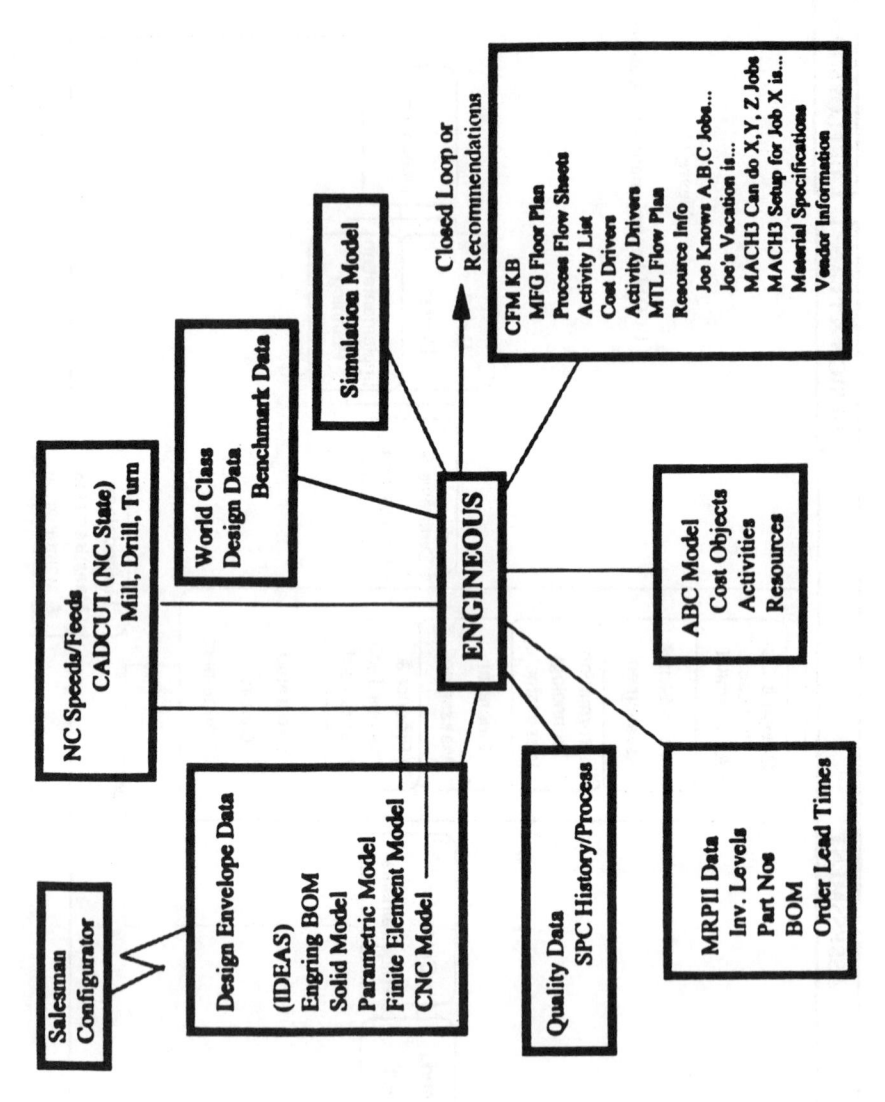

Figure 15

Unit Continuous Optimization Manufacturing Tools

Japanese designer may make up a project team and be together for only a matter of months. Physical location will be of no concern. Language will probably continue to center on English initially. Today's neural networks are already doing real-time voice conversion, so advances in this capability will allow language problems at the initial stage of the electronic worker movement to be addressed. As more and more of the basic education in developed and developing countries is provided in English, this translation requirement will diminish. Global communications networks and engineering data base control software will be the glue for this global disbursement of talent.

Country barriers will cease to exist as will trade negotiations. A global banking network will allow for the development of a single currency with no variation as to location. The U.S. dollar is again the likely candidate, but some new currency may emerge as well.

Current legislation is pending that would provide a single window into the annual $70 billion U.S. government research and development labs. As this effort is structured electronically to keep tabs on the expenditures and projects, it will be easy to add other national labs -- or corporate labs -- to the network. Once this is accomplished, the scarcity of funds and resources will mandate some jointly defined global direction for specific government funded projects. Countries will share teams of researchers for each, and the payoff in new products will be shared in some predefined fashion.

6.1 Review

In the information economy, it is imperative that executives have the correct data they need to make decisions. Understanding that ever increasing flow of data, the primary resource of tomorrow's corporations, is becoming too difficult for humans to manipulate. Humans suffer from information overload -- machines do not. Machines feed on data, and the UCOM strategy takes advantage of the second while negating the first.

The use of computers and artificial intelligence software for human augmentation will be the key to survival in the information age. Managing-by-wire is one term for handling information needs. Accessing, interpreting, structuring and managing intellectual and technology assets to enable rapid formulation and execution of any strategy is its meaning. However, just as today's fighter aircraft push the human pilot to the point of unconsciousness, information technology has surpassed the human capability to adequately address all its issues and nuances. Fighters will be 'manned' by robots in order to move the capability of the whole system forward. The weak link must be addressed -- and so it is with the information age and the corporate entity. The human brain is still the best tool for reasoning, but the volumes of data now available and required for a 'good decision' surpass the human's capability to decipher them. UCOM accomplishes this task by optimizing the data around the human defined targets in the knowledge base. The human stays in control -- the machine does the work.

UCOM provides a relational process re-engineering function. Once properly setup, UCOM can address internal or external variable changes and have all linked variables automatically adapt to the change while continuing to optimize to the corporate strategy objectives.

Warren Bennis, noted author and consultant, said: "Change is the Metaphysics of our age. Everything is in motion." The future will accelerate this process.

As all organizations continue to experience accelerated dynamic change, the need to understand the impact on our human resources has never been more critical. Organizations do not go through transitions -- people do. And the best way to quickly adapt to change WITHOUT impacting people is to start with the UCOM strategy and allow IT to adapt. Keep the problem link out of the information handling loop.

In an Information Economy, knowledge is the primary resource. Data is considered the currency in this new economy, so do we want our decision makers to be looking at the larger objectives of the organization which will be supported by the data -- or do we want them to be utilized in counting the data at the micro level? Do we want leaders or bean-counters?

Economies determine the direction of most changes. The ability to use human energy efficiently is the core of the information economy. The managers of today, if they are to survive this paradigm shift, must be visionaries -- not adapters. They must lead, and they must be dynamic enough to prepare for the future by making major changes in their information structure of today.

Data reduction must be addressed as soon as possible -- before the organization is unable to cope.

Unit Continuous Optimization Manufacturing -- a methodology for the optimization of data to provide a global optimum solution before a decision is needed -- is the strategy that provides the answer.

7 References

1. B. J. Pine, *Mass Customization*, (Harvard Press, Boston, 1993)

2. P. B. B. Turney, *Common Cents*, (Cost Technology, Hillsboro, 1991)

3. M. Hammer and J. Champy, *Reengineering The Corporation*, (HarperCollins, New York, 1993)

8 Subject Index

1. Technological innovations - Management
2. Manufacturers - Technological innovations - Management
3. New products - Management
4. Competition
5. Mass production
6. Artificial intelligence - Genetic algorithms
7. Organizational change - Corporate reorganization

I. Title

An Intelligent Tool for Discovering Data Dependencies
in Relational Databases

Prasad Gavaskar
Motorola SPS
Mesa, AZ 85202
USA

and

Forouzan Golshani
Department of Computer Science and Engineering
Arizona State University
Tempe, AZ 85287-5405
USA

Abstract

We present an automated method for inferring functional dependencies from existing relational databases. Since the identification of the candidate keys of the relation may be seen as a special case of deducing the functional dependencies, our algorithm will find the candidate keys by an intermediary step. We also develop a set of algorithms necessary for discovering multivalued and join dependencies. The technique is based on the "learning from examples" paradigm of machine learning, and uses Armstrong's axioms as heuristics to guide the discovery process. The tool is implemented in ANSI-C and SQL, and have been demonstrated to work on an Oracle database.

1. INTRODUCTION

The idea of storing data in the form of tables, proposed by Codd in the early 70's and known as the relational approach, has now gained wide acceptance in the industry. Numerous commercial Relational Database Management Systems (RDBMS) are available today on hardware ranging from Personal Computers (PCs) to mainframes. The relational approach heavily depends on the dependencies or the relationships between the data entities of the application being modeled. Defining the semantics of the data elements is not as easy as it appears and even an expert can sometimes make mistakes. Improper identification of dependencies can lead to several problems, some of which are discussed below. This research presents an automated method for inferring dependencies from existing databases. In the sections that follow we will refer to the implementation of this research as the Dependency Finding Tool (DFT).

Today, personal computers have become a corporate way of life and are available to a wide range of users including computer novices. As a result, anybody can create a database using tools which were once available only to mainframe systems personnel. While many enterprises seek professional assistance for the design of large database applications, sometimes, a lack of immediate access to such experts and the urgency of the business situation may dictate that the enterprise must do without a well-designed database. There are many databases in existence today which unknowingly violate the normalization rules -- explained later -- resulting in the storage of redundant data in relations. Our tool, DFT, can be used to analyze such databases.

Sometimes, even well designed databases may violate the rules set forth by the designer because many commercial RDBMSs do not have any means of enforcing the integrity constraints. An example, shown below, will further clarify this point. Consider the relation schema (EmpNum, EmpName, ProjectNum, Project_Hours). Here EmpNum determines the EmpName and several employees work on a given project (ProjectNum is a Foreign Key signifying a many-to-one relationship between EmpNum and ProjectNum). The number of hours that an employee works on a project is determined by the EmpNum. Suppose the above table is created with EmpNum as the key, and subsequently some tuples are inserted, as illustrated in Figure 1.

EmpNum	EmpName	ProjectNum	Project_Hours
1	Smith	100	20
2	Jones	100	40
3	Wang	200	30
4	Nguyen	200	40

Fig. 1 Employee table showing a (1:M) relationship between EmpNum and ProjectNum

Later, the engineering department decides to make project numbers unique, and let the project number determine the number of hours an employee is required to work on that project. They update the above table as shown in Figure 2. The RDBMS will allow this since there is no constraint which says that the project hours cannot be updated as above. This new table violates the third normal form, (briefly explained in section 1.2.). In this case, the DFT would point out the additional functional dependency from ProjectNum to Project_Hours and leave it up to the designer to determine the normalization violation.

EmpNum	EmpName	ProjectNum	Project_Hours
1	Smith	100	20
2	Jones	100	20
3	Wang	200	40
4	Nguyen	200	40

Fig. 2 Employee table showing a functional dependency from ProjectNum to Project_Hours

Redundant data not only consumes more space but can also lead to insert and update anomalies [16]. From a knowledge discovery standpoint, redundant information can be mistakenly discovered as knowledge, even though it is usually uninteresting to the end user [25]. Relational data model is structurally weak compared to the network and the hierarchical data models [16], because both the network and the hierarchical models allow representation of the relationships explicitly. To overcome the structural drawbacks and to minimize the data redundancy, the normalization process was proposed. Normalization converts raw tabular data to a standard form based on the constraints derived from the application domain that produces the data. Normalization requires a thorough knowledge of the dependencies between each data type that is used to model the enterprise. The definition of data dependencies is not a trivial task and may require considerable effort on the part of the database designer. Quite often, the designer is not familiar with the environment being modeled. The business experts, on the other hand, have no knowledge of the world of databases. This is a catch-22 situation and can lead to a bad model no matter who (the user or the database engineer) designs the database.

One way of resolving this problem is to model the enterprise as best as you can, add real-life data to this database, and use DFT to validate the model. This approach is sometimes referred to as rapid prototyping. The user can use the discovered dependencies either to implement constraints in the application software or to make schema changes. Several other applications where DFT may prove useful are outlined below.

There are many instances when a user would like to convert a flat file database to a relational representation in order to make use of the sophisticated data manipulation tools provided by commercial RDBMSs. The easiest way to do this is to load all the data in one table[1], design the relational schema, and then move the data into the new tables using SQL. Even though the user is familiar with his data, it would help to let DFT discover all the dependencies, since the data, or the examples, already exists in a relational format.

Although users can live with a non-normalized database there are several applications that rely on a thorough knowledge of the data dependencies. Some of these applications are:
- Coupling a Relational Database with an Object-Oriented Database (OODB), i.e. overlaying an Object Oriented (OO) schema on top of the relational schema when an OODB is used as a front end.
- Fragmentation in Distributed Databases.
- Federated databases in which a global schema is useful.
- Conversion from the Relational model to an OO or a Functional model.

1.1 Overview of the Problem

The current literature on dependency inference concentrates on database design tools and there is no reference to discovering dependencies from existing database instances. Moreover, multi-valued and join dependencies have not been thoroughly investigated. The current literature offers a definition for join dependency but does not mention when to split the relation into sub-

---
[1] Several vendors provide data loading utilities with their RDBMSs.

relations or how to split it. The goal of this research is to come up with an automated method to discover dependencies from existing databases by using traditional relational database operations. Identification of the candidate keys of the relations is considered as a special case of functional dependencies.

Data dependencies are a property of the relational schema and cannot be automatically inferred from the database instance. They must be defined by someone who knows the semantics of the data elements [16]. In line with this argument we present the discovered dependencies to the user and let the user decide how to use them. Since all the dependencies are already identified, DFT can suggest schema modifications. But this is not the intent of this research. The major benefit of this research is the investigation of algorithms to discover multivalued and join dependencies. The dependency inference technique is based on the "learning from examples" paradigm of machine learning and uses Armstrong's axioms as heuristics to guide the discovery process.

1.2 Preliminaries

Simply put, a database is a collection of related data. A Database Management System (DBMS) is a collection of programs that enable users to create and maintain a database. A DBMS should provide users with a conceptual representation of data that does not include many of the details on how the data is stored. A data model is a type of data abstraction that is used to provide this conceptual representation. In other words, a data model is a set of concepts that can be used to describe the structure of a database. The data in the database at a particular moment in time is called a database instance.

A Data definition Language (DDL) is used to define the database scheme. Once the database is populated, users must have some way of retrieving, inserting, deleting, and modifying the data. The DBMS provides a Data Manipulation Language (DML) for these purposes. A high-level DML used in a stand-alone interactive manner is called a Query Language. The DBMS stores the relation schemes, constraints, design decisions, usage standards, application program descriptions, user information, etc., in a set of tables called the **Data Dictionary**. Most commercial relational DBMSs provide a high-level **declarative language interface**. **SQL** (Structured Query Language) is one such language with capabilities for data definition, query, and update. Hence, it is both a DDL and a DML. SQL statements can also be embedded into a general purpose programming language like C.

Most databases are **centralized**, meaning that their data is stored at a single computer site. A **distributed DBMS** can have actual database distributed over many sites connected by a computer network. A **federated DBMS** is one in which the participating databases are loosely coupled and have a degree of local autonomy.

The relational data model represents a database as a collection of tables. The **network model** represents data as record types, and also represents a limited amount of 1:N relationship, called a set type. The **hierarchical model** represents data as a tree structure.

An **entity** is a "thing" in the real world with an independent existence. A **relationship type** R among n entity types $E_1, E_2,...,E_n$, is a set of associations among entities from these types. The **degree** of a relationship type is the number of participating entity types. The **cardinality ratio** constraint specifies the number of relationship instances that an entity can participate in. In section 1.0, there is a 1:N relationship between ProjectNum and EmpNum, meaning that each Project entity can be related to any number of Employee entities.

A **relation schema** R is a finite set of attribute names $\{A_1, A_2, ..., A_n\}$. Corresponding to each attribute name A_i is a set D_i, $1 \leq i \leq n$, called the **domain** of A_i. Attribute names are sometimes called attribute symbols or simply **attributes**. The domains are arbitrary, non-empty sets, finite or countably infinite. Let $D = D_1 \cup D_2 \cup ... \cup D_n$. A **relation** r on relation schema R is a finite set of mappings $\{t_1, t_2,...,t_p\}$ from R to D with the restriction that for each mapping t belonging to r, $t(A_i)$ must be in D_i, $1 \leq i \leq n$. The mappings are called **tuples**. Given a tuple t, the value of t on attribute A is called the **A-value of t**. Considering t as a mapping, the A-value of t is t(A). Interpreting t as a row in a table, the A-value of t is the entry of t in the column headed by A.

A **key** of a relation r on relation schema R is a subset $K = \{B_1, B_2,...,B_n\}$ of R with the following property. For any two distinct tuples t_1 and t_2 in r, there is an attribute B belonging to K such that $t_1(B)$ is not equal to $t_2(B)$ and no proper subset K' of K shares this property. K is a **superkey** of r if K contains a key of r. In general, a relation schema may have more than one key. In this case each of the keys is called a **candidate key**. It is common to designate one of the candidate keys as the **primary key** of the relation. A key that is composed of a single attribute is commonly referred to as a **simple key**, whereas a key that has more than one attribute is called a **composite key**. An attribute of relation schema R is called a **prime attribute** of R if it is a member of any key of R. An attribute is called **nonprime** if it is not a prime attribute.

The **SELECT** operation, denoted by

$$\sigma_{<\text{select condition}>} (<\text{relation name}>)$$

is used to select a subset of the tuples in a relation. These tuples must satisfy a selection condition which is a boolean expression specified on the attributes of the specified relation. The **PROJECT** operation, denoted by

$$\Pi_{<\text{attribute list}>} (<\text{relation name}>)$$

selects columns mentioned in the attribute-list from the relation and discards other columns.

There are three set theoretic binary operations used to merge elements from two relations which have the same number and the same type of attributes. Two attributes are of the same **type** if they belong to the same domain. The result of an **UNION** operation, denoted by R U S, is a relation that includes all tuples that are either in R or in S or in both R and S. Duplicate tuples are eliminated. The result of an **INTERSECTION** operation, denoted by R ∩ S, is a relation that includes all tuples that are in both R and S. The result of a **DIFFERENCE** operation, denoted by R - S, is a relation that includes all tuples that are in R but not in S.

The **CARTESIAN PRODUCT**, denoted by **X**, is also a binary set operation used to combine tuples from two relations so that related tuples can be identified. The result of $R(A_1,A_2,...,A_n)$ X $S(B_1,B_2,...,B_m)$ is a relation $Q(A_1,A_2,...,A_n,B_1,B_2,...,B_m)$ with n + m attributes in the order shown. Relation Q has one tuple for each combination of tuples, one from R and one from S. The **JOIN** operation, denoted by ∞, is a sequence of CARTESIAN PRODUCT followed by SELECT combined into a single operation. The result of

$$R(A_1,A_2,...,A_n) \infty_{<\text{join condition}>} S(B_1,B_2,...,B_m)$$

is a relation $Q(A_1,A_2,...,A_n,B_1,B_2,...,B_m)$ with n + m attributes in the order shown. Q has one tuple for each combination of tuples, one from R and one from S whenever the combination satisfies the join condition. The join condition is of the form <condition> AND <condition> AND ... AND <condition> where each condition is of the form $A_i \theta B_j$, A_i is an attribute of R and B_j is an attribute of S, and θ is one of the comparison operators. A join with a general join condition is called a **THETA JOIN**. A join where the only comparison operator used is =, is called an **EQUIJOIN**. A **NATURAL JOIN**, denoted by *, is an equijoin followed by the removal of the superfluous attributes.

A **Functional Dependency** (FD), denoted by X -> Y, between two sets of attributes X and Y that are subsets of relation schema R specifies a constraint on the possible tuples that can form a relation instance r of R. The constraint states that for any two tuples t_1 and t_2 in r such that $t_1(X) = t_2(X)$, we must also have $t_1(Y) = t_2(Y)$. The inference rules for FDs or the **Armstrong's Axioms** are discussed in section 4.

The **normalization process** takes a relation schema through a series of tests to "certify" whether or not it belongs to a certain normal form. Initially, Codd proposed three normal forms, which he called the first, second, and third normal form respectively. Normalization of data may be seen as a process during which unsatisfactory relation schemas are decomposed by breaking up their attributes into smaller relation schemas that possess desirable properties. **First normal form** states that the domains of attributes must include only atomic (i.e. simple and indivisible) values and that the value of any attribute in a tuple must be a single value from the domain of that attribute.

The second normal form is based on the concept of full functional dependency. A FD X -> Y is a **full functional dependency** if removal of any attribute A from X means that the dependency does not hold any more. A FD X -> Y is a **partial dependency** if there is some attribute A belonging to X that can be removed from X and the dependency will still hold. A relation schema R is in **second normal form** if every nonprime attribute A in R is fully functionally dependent on the primary key of R. The third normal form is based on the concept of a transitive dependency. A FD X -> Y in a relation schema R is a **transitive dependency** if there is a set of attributes Z that is not a subset of any key of R, and both X -> Z and Z -> Y hold. A relation schema

R is in **third normal form** if it is in second normal form and no nonprime attribute of R is transitively dependent on the primary key.

A **Multivalued Dependency** (MVD) $X \rightarrow\rightarrow Y$ specified on relation schema R, where X and Y are both subsets of R, specifies the following constraint on any relation instance r of R: If two tuples t_1 and t_2 exist in r such that $t_1(X) = t_2(X)$, then two tuples t_3 and t_4 should also exist in r with the following properties:

- $t_3(X) = t_4(X) = t_1(X) = t_2(X)$
- $t_3(Y) = t_1(Y)$, and $t_4(Y) = t_2(Y)$.
- $t_3(R - X - Y) = t_2(R - X - Y)$ and
  $t_4(R - X - Y) = t_1(R - X - Y)$.

Consider a universal relation schema $R = \{A_1, A_2, ..., A_n\}$ that includes all the attributes of the database. Using the FDs that hold on the attributes of R we can decompose R into a set of relation schemas $D = \{R_1, R_2, ..., R_m\}$ that will become the relational database schema. D is called the **decomposition of R**. **Lossless join** or nonadditive join states that no spurious tuples should be generated when a natural join operation is applied to the relations in the decomposition. It refers to the loss of information, not loss of tuples.

A **Join Dependency** (JD), denoted by $JD(R_1, R_2, ..., R_n)$ specified on relation schema R, specifies a constraint on instances r of R. The constraint states that every legal instance r of R should have a lossless join decomposition into $R_1, R_2, ..., R_n$, that is,

$* (\Pi_{<R1>}(r), \Pi_{<R2>}(r), ..., \Pi_{<Rn>}(r)) = r$

where * stands for the natural join operation and

$\Pi_{<R1>}(r)$ stands for the projection of R_1 from r.

1.3 Organization of the Chapter

This chapter is organized as follows: Section 2 contains a survey of existing literature on dependency inference. Although our algorithms use the "learning from examples" paradigm, there are other approaches for discovering knowledge from databases. These are discussed in Section 3. Section 4 contains the definition, design principles, and the algorithm for functional dependency. The definitions, design principles, and algorithms for multivalued and join dependencies are discussed in Section 5. Section 6 provides the implementation details of the prototype of DFT. The prototype is implemented in ANSI-C and SQL and uses the RDBMS from Oracle Corporation. Section 7 contains concluding remarks and some ideas for future.

2. PREVIOUS WORK

The existing literature has only a few papers on dependency inference algorithms. All of them concentrate on using examples as part of a design tool to help the database designer. Since they concentrate on the design stage when the actual database instance does not exist, the examples are typed by the designer from hand written forms related to the application. The manual typing of examples limits the number of tuples to be analyzed, and as such, most of the publications talk about examples with approximately ten tuples.

In [15] the author describes an algorithm for finding the key of a relation given a set of Functional Dependencies (FDs) for that relation. The algorithm first finds a subkey, K, which is contained in every key. K consists of the attributes which do not belong to the Right-Hand Side (RHS) of any FD. K is then extended by adding more attributes until it becomes a key.

In [7] the authors generate Armstrong relations, as examples, based on the FDs that the designer has identified. The designer is allowed to modify the example, and this modified example is then treated as a new Armstrong relation for which a set of FDs is found. The dependency inference algorithm iteratively constructs the Left Hand Side (LHS) of an FD as a set L. Each iteration compares a pair of tuples for which it considers each attribute as a possible RHS. When the tuples disagree on the current RHS attribute but agree on the attributes that are currently in L, it adds to L the attribute(s) on which the tuples disagree. If L is not empty for a particular attribute then the attributes in L form the LHS and the attribute itself forms the RHS of an FD.

In [12] the authors provide another algorithm for inferring FDs. Again, their main objective is the use of examples to help the designer. The algorithm is based on the concept of "Necessary Sets". The algorithm iteratively constructs the necessary set for each attribute in the relation. Each iteration compares a pair of tuples, s and t, to construct a set, disag(s,t), whose elements are attributes on which the two tuples disagree. If an attribute, say A, belongs to disag(s,t) then the necessary set for this attribute is set disag(s,t) without the element A or the set {disag(s,t) - {A}}. This means that the necessary set can be formed for each such element X in set disag(s,t) by subtracting that element from the set. The final necessary set for an attribute A, denoted as nec(A), is a collection of all such sets, {disag(s,t) - {A}}, for each pair of rows which disagree on attribute A. The term "necessary set" comes from the observation that if two rows s and t differ in attribute A then for any set X, if $X \rightarrow A$ holds in relation R, X must contain some attribute in disag(s,t). In order to avoid inferring redundant and transitive dependencies, the sets nec(A) are reduced before proceeding with the construction of the LHS of attribute A. Redundant dependencies are avoided by removing supersets in nec(A) and transitive dependencies are avoided by comparing nec(A) with the necessary sets of other attributes.

In [13] the authors describe algorithms for discovering dependencies based on hypothesis testing. Here too, the emphasis is on a design tool and the user is expected to provide the examples. The authors assume that a certain dependency exists. Their approach attempts to confirm or deny the dependency based on the evidence provided by the sample data. The authors deal with a limited amount of data to test the hypothesis and rely on an interactive user to confirm if the hypothesized dependency truly exists. They present an algorithm for inferring FDs in "two attribute" relations based on hypothesis testing. The algorithm for the inference of Multivalued Dependencies (MVDs) works in the same way as the FD algorithm. It first checks for the absence of an MVD and, if this is not confirmed, it then checks for the presence of certain special tuples that would support the inference of an MVD. The algorithm for Join Dependencies (JDs) first checks for the presence of a FD or a MVD among the attributes. If found, it concludes the absence of a JD. Otherwise, it checks for the presence of certain special tuples that would support the inference of a JD.

The algorithms used in DFT have some resemblance to those mentioned in [7]. The FD algorithm in [7] is driven from the Right Hand Side (RHS) of the FD, since the authors look for rows with distinct RHS and try to construct the Left Hand Side (LHS) of the FD. Since DFT also finds candidate keys, we have to drive our FD algorithm from the LHS, i.e. we look for rows with the same LHS and try to construct the RHS of the FD. DFT extracts data from the database using queries. The approach is similar to the one mentioned in [25].

3. KNOWLEDGE DISCOVERY IN DATABASES

Knowledge Discovery is the nontrivial extraction of implicit, previously unknown, and potentially useful information from data [21]. The output of a program that scans the database for some patterns of interest to the user community is called discovered knowledge. The output itself is a set of patterns, in some format, which satisfy a user specified criteria and have a degree of certainty associated with them. Most knowledge discovery problems fall into the general category of classification. Classification techniques can be categorized into Statistical Pattern Recognition, Neural Networks, and Machine Learning. Although these methods differ in the underlying models and output formats, any one of these can be applied to the same problem.

3.1 Statistical Pattern Recognition

The decision-making process of humans is somewhat related to the recognition of patterns. The goal of pattern recognition is to classify these patterns and automate the classification process using computers. Let us consider an example of wave form classification. In order to perform this type of classification, we must first measure the observable characteristics of the wave form. The most primitive way of doing this is to measure the time-sampled values, $x(t_1), x(t_2),...,x(t_n)$, of the wave form. The n measurements form a vector X. Even under normal conditions, the observed wave forms are different each time the observation is made. Therefore, $x(t_i)$ is a random variable. Likewise, X is a random vector. Each wave form is expressed by a vector in an n-dimensional space, and many wave forms form a distribution of X in the n-dimensional space.

Suppose we are given two dimensional samples of two distributions of wave forms corresponding to normal and abnormal machine conditions. Let the two of these measurements be x_1 and x_2. If we know these two distributions of X from past experience, we can create a boundary function between these two distributions, $g(x_1,x_2)=0$, which divides the two dimensional space into two regions. Once the boundary is selected, we can classify a sample depending on $g(x_1,x_2) < 0$ or $g(x_1,x_2) > 0$. We call this boundary function "g" a discriminant function. Let us say that we have a black-box which detects the sign of "g"; the black-box can be a computer program or an electronic circuit. Such a black-box will be called a categorizer or a classifier. Thus, in order to design a classifier, we must first study the characteristics of the distribution of X for each category and find a proper discriminant function. This process is called learning or training, and the samples used to design a classifier are called learning or training samples.

Pattern recognition in a broader sense can be considered as a problem of estimating density functions in a high-dimensional space and dividing the space into the regions of categories or classes. Because of this view, mathematical statistics forms the foundation of this subject. The key parameter in pattern recognition is the probability of error and hence the best classifier is one which minimizes the probability of classification error. Classifiers can be categorized as parametric and nonparametric. Parametric classifiers are based on assumed mathematical forms for either the density functions or the discriminant functions. Linear and quadratic classifiers are some of the examples of this type. When no parametric structure can be assumed for the density functions, nonparametric techniques such as k-nearest neighbor have to be used for estimating the density functions.

As the number of inputs to a classifier becomes smaller, the design of the classifier becomes simpler. As a result, feature selection or extraction of important features from observed samples plays an important role in pattern recognition. The first step to design a classifier is to collect data. The data samples are normalized and then analyzed to study the data characteristics. Data analysis includes statistical tests and feature selection. The structure of the data dictates the type of classifier to be used. In majority of cases the choice is a parametric classifier like a linear or a quadratic.

3.2 Neural Networks

Neural Networks aspire to be models of the brain's cognitive process. These models use multiple processing elements which are interconnected, and interact with each other in response to the environment. The neuron is the basic processor in neural networks. Each neuron has one output, which is generally related to the its state or activation. The output of the neuron may fan out to several other neurons. Each neuron receives several inputs over these connections, called synapses. The inputs are the activations of the incoming neurons multiplied by the weights of the synapses. The activation of the neuron is computed by applying a threshold function to this product. The threshold function is generally some form of nonlinear function like a step function.

All of the "knowledge" that a neural network possesses is stored in the "synapses" or the weights of the connections between the neurons. The network acquires the knowledge during "training". The network is trained by presenting pattern associations in sequence and adjusting the weights to capture this knowledge. Different techniques are used to calculate the weight adjustments. The chosen technique is known as the learning law of the network.

Learning methods can be categorized as supervised and unsupervised. An unsupervised learning method is one in which the weight adjustments are not made based on the comparison with some target output, i.e. there is no "teaching signal" feedback for weight adjustment. Nonparametric pattern recognition techniques like the nearest-neighbor classifier have been used for unsupervised learning. In supervised learning the difference between the network output and the target result is fed back and the weights are adjusted based on this error.

The two layer network, mentioned above, can classify only those functions which are linearly separable. In cases where the mappings between the input and output patterns are nonlinear, additional layers of neurons can be constructed between the input and the output layers. These layers are called the "hidden layers" since they do not interact directly with the outside world and hence are not visible to the outside world. A simple example of nonlinear mapping is the exclusive-OR function. Learning laws for multilayer networks are more complicated compared to the simple two layer network.

3.3 Machine Learning

Machine Learning evolved from the above two fields. Its main objective is to parallel the human learning abilities in computers. Another basic scientific objective is the exploration of alternative learning mechanisms, including the discovery of different induction algorithms, the information that must be available to the learner, the issue of coping with imperfect training data, and the creation of general techniques applicable in many task domains. Because of these objectives machine learning has become a very broad field encompassing many different methods of learning.

One way to distinguish learning strategies is by the amount of inference the learning system performs on the information provided. As the amount of inferencing capabilities of a system increase, the burden placed on the teacher or the external environment decreases. We can classify the various learning methods based on the amount of effort required of the learner and the teacher as shown below. We describe the "Learning from Examples" method in some more detail as this is the chosen method for this research. The reasons for choosing this method are given in the next section.

- **Rote learning and direct implanting of new knowledge:**

In this method no inference or other transformations of the knowledge are required on the part of the learner. Variations of this knowledge acquisition method include:

- Learning by being programmed, constructed or modified by an external entity, requiring no effort on the part of the learner (for example, the usual style of computer programming).

- Learning by memorization of given facts and data with no inferences drawn from the incoming information (for example, as performed by primitive database systems).

- **Learning from instruction:**

In this method, also known as learning by being told, the learning system acquires knowledge from an external source and transforms it into an internal representation. The system is required to perform some inference to make effective use of this new information along with the existing knowledge.

- **Learning by analogy:**

Part of the existing knowledge that is similar to the new concept, is transformed into a new form that will be useful in the new situation. New facts that will augment the existing knowledge, in the new situation, are acquired. This method might be applied to convert an existing computer program into one that performs a closely-related function for which it was not originally designed. This method requires some inferencing capabilities because a fact analogous in relevant parameters must be retrieved from the memory, transformed into the new form, applied to the new situation, and stored for future use.

- **Learning from examples:**

This is a special case of inductive learning. With this methodology, the system takes a set of examples and counterexamples of a concept, and induces a general concept that describes all of the positive examples and none of the counterexamples. The amount of inferencing capabilities of the system are much greater than "learning from instruction" and "learning by analogy". In this method positive examples force generalization whereas negative examples prevent over generalization. When only positive examples are available, over generalization might be avoided by considering only the minimal generalizations necessary or by using the domain knowledge to constrain the concept being inferred. Since our discovery algorithms fall into this category, we present a more detailed description of the relevant topics.

Since induction is a search through a description space, one must specify the goal of the search, i.e. the criteria that describes the goal description. The goal criteria depends on the domain in question, but some regularities may exist. We can distinguish between characteristic, discriminant, and taxonomic descriptions. A characteristic description is a description of a class of objects which states the facts that are true for all objects in that class. It is intended to discriminate objects in a given class from objects in all other possible classes. A discriminant description states only those properties of the objects in the given class that are necessary to distinguish them from the objects in the other classes. A taxonomic description is a description of a class of objects that subdivides the class into subclasses.

Determination of inductive assertions can be viewed as a process of consecutive application of certain "generalization rules" to initial and intermediate descriptions. A generalization rule is a transformation rule that, when applied to a classification rule, produces a more general classification rule. A generalization rule is called "selective" if the transformed rule does not include any descriptors other than those used in the initial description. If the transformed rule does include new descriptions, then the rule is called "constructive". One of the simplest generalization rules is the "dropping condition" rule. It states that to generalize a conjunction, you may drop any of its conjunctive conditions.

- **Learning from observation and discovery (also called unsupervised learning):**

This is a very general form of inductive learning that includes discovery systems, theory-formation tasks, the creation of classification criteria to form taxonomic hierarchies, and similar tasks without the benefit of an external teacher. This method requires the learning system to perform more inference than all of the approaches discussed above. This method can be subclassified, based on the degree of interaction with the external environment, as follows:

- **Passive observation:** System classifies observations from multiple aspects of the environment or observations that span several concepts.

- **Active experimentation:** In this approach, the learning system stimulates the environment and observes the response to the stimulus. Experimentation may be random and dynamically focused according to some general criteria, or strongly guided by theoretical constraints. As the system acquires knowledge and hypothesizes theories, it may be driven to confirm or disconfirm its theories and hence, explore its environment applying different observation and experimentation strategies as the need arises. This method requires the generation of examples to test hypothesized or partially acquired concepts.

3.4 Choosing a Method

The choice of a method will depend on its applicability to our dependency discovery problem, i.e. handling structured data types in databases, the ease of understanding the output results, and the theoretical basis of the framework of the method. The last criteria is important in case we have to prove or explain the derivation of a certain result, theoretically. We will consider the above three methods in the light of one or more of these criteria.

Statistical methods are data driven and cannot use the available domain knowledge. They also cannot handle the structured data types found in many databases and require the user to be literate in statistics in order to understand the output.

Neural networks are general purpose learning systems that start with random or partially random initial structure. Learning in these systems consists of incremental changes and the probabilities that threshold logic units would transmit a signal. Since the starting structure is random, it is difficult to present a formal explanation as to how the final result was obtained. We should also note that the models used in two layered networks are similar to statistical methods whose disadvantages, with respect to the problem under consideration, are discussed above. Two layered networks are too simple to handle most real world problems. Models using multiple layers and back-propagation are difficult to understand and explain.

Since Machine Learning encompasses so many different methods, it is not easy to discuss the applicability, of each one, to our dependency discovery problem. The "Learning from Examples" paradigm has been used by many researchers to discover knowledge from databases [1,2,3,4,10,13], and we will consider this method as a possible solution to our problem. With respect to our selection criteria, we find that it is based on the first order predicate logic, and hence has a formal basis. The initial structure as well as the output, which is in the form of either decision trees or production rules, is easy to understand and explain. The method is closely related to databases since the Structured Query Language (SQL) is also based on predicate calculus.

4. FUNCTIONAL DEPENDENCY ALGORITHM

This section describes an algorithm to discover Functional Dependencies that a given database instance satisfies. Identification of the candidate keys of the relations is considered as a special case of the functional dependencies. The dependency inference technique is based on the "learning from examples" paradigm of machine learning and uses Armstrong's axioms as heuristics to guide the discovery process. Before we discuss the algorithm, we formally define some key concepts, including functional dependencies.

4.1 Some Important Concepts

Definition: **(Functional Dependency)** A Functional Dependency (FD), denoted by $X \to Y$, between two sets of attributes X and Y that are subsets of relation schema R specifies a constraint on the possible tuples that can form a relation instance r of R. The constraint states that for any two tuples t_1 and t_2 in r such that $t_1(X) = t_2(X)$, we must also have $t_1(Y) = t_2(Y)$.

The definition states that the values of the X component of a tuple uniquely determines the values of the Y component. The set of attributes X is called the Left-Hand Side (LHS) of the FD, and Y is called the Right-Hand Side (RHS).

An alternative definition of FD is as follows: In a relation schema R, X functionally determines Y if and only if whenever two tuples of r agree on their X-value, they must necessarily agree on their Y-value [16]. If a constraint on R states that there cannot be more than one tuple with a given X-value in any relation instance r of R, i.e. X is a candidate key of R, this implies that $X \to Y$ holds for any subset of attributes Y of R.

Definition: **(Inference Rules for Functional Dependencies)**
Let F denote the set of FDs that are specified on relation schema R. There are usually numerous other FDs which will also hold in all legal relation instances that satisfy the dependencies in F. The set of all such FDs is called the **closure** of F and is denoted by F^*. In general, an FD $X \to Y$ is **inferred from** a set of dependencies F specified on R if $X \to Y$ holds in every relation instance r that is a legal extension of R. The closure F^* of F is a set of all FDs that can be inferred from F. The closure F^* is determined using the set of inference rules specified below. We will use the notation $F \models X \to Y$ to denote that the FD $X \to Y$ is inferred from the set of FDs F. We will denote the FD $\{X,Y\} \to Z$ by $XY \to Z$. The following six rules constitute the inference rules for FDs:

(IR1) Reflexive Rule $\quad\quad Y \subseteq X \models X \to Y$.

(IR2) Augmentation Rule $\quad\quad \{X \to Y\} \models XZ \to YZ$.

(IR3) Transitive Rule $\quad\quad \{X \to Y, Y \to Z\} \models X \to Z$.

(IR4) Decomposition Rule $\quad\quad \{X \to YZ\} \models X \to Y$.

(IR5) Union Rule $\quad\quad \{X \to Y, X \to Z\} \models X \to YZ$.

(IR6) Pseudotransitive Rule $\quad\quad \{X \to Y, WY \to Z\} \models WX \to Z$.

Inference rules IR1 to IR3 are known as **Armstrong's axioms**. They are sufficient to determine the closure F^* from F.

Definition: **(Equivalence of Sets of Functional Dependencies)** A set of FDs E is said to be **covered by** a set of FDs F, or alternatively, F is said to cover E, if every FD in E is also in F^* i.e. every dependency in E can be inferred from F. Two sets of FDs E and F are said to be **equivalent** if $E^* = F^*$. Hence, equivalence means that every FD in E can be inferred from F, and every FD in F can be inferred from E.

Definition: **(Minimal Sets of Functional Dependencies)** A set of FDs is **minimal** if it satisfies the following three conditions:
1. Every dependency in F has a single attribute for its RHS.
   2. We cannot remove any dependency from F and still have a set of dependencies that is equivalent to F.
   3. We cannot replace any dependency $X \rightarrow A$ in F with a dependency $Y \rightarrow A$, where Y is a proper subset of X, and still have a set of dependencies that is equivalent to F.

Condition 1 ensures that every dependency is in a standard or **canonical form** with a single attribute on the RHS and conditions 2 and 3 make sure that there are no redundancies in the dependencies.

A **minimal cover** of a set of FDs F is a minimal set of dependencies F_{min} that is equivalent to F. There can be several minimal covers for a set of FDs.

4.2 Inference Rules for Keys

In this section we will introduce an inference rule for keys using the theorem stated below.

> **Theorem:** Consider a relational instance r of schema R with X and M as any given attribute sets and Y as a set containing a single attribute, such that $M \cap Y = \emptyset$ and $X \cup M \neq M$. Let MY be a candidate key of R. If a FD $X \rightarrow Y$ exists in r then MX is a candidate key of R.
>
> Proof: We will give an informal proof for the above theorem.
>
> We are given that MY is a key and $X \rightarrow Y$ is a FD and we have to prove that MX is a key.
>
> Since MY is a candidate key, values of MY are distinct throughout r. Consider any two rows in r. For values of MY to be distinct in these two rows, either values of M have to be different in the two rows or values of Y have to be different or both can be different. We are not interested in the last case when values of both M and Y are distinct.
>
> Let us consider the first case when values of M are distinct in the two rows. In this case, for MX to be a candidate key, considering these two rows only, it doesn't matter whether X has the same or different values in the two rows.
>
> In the second case, where the values of M are the same in the two rows, values of X have to be distinct to make MX a candidate key. We will prove that MX is a key, for this case, by contradiction. Let us assume that the values of X are also the same in the two rows,

implying that MX is not a candidate key. For this case, values of Y are distinct in the two rows being considered. We are given that X → Y is a FD which implies that if the values of X are the same for these two rows then the values of Y have to be the same. This contradicts the fact that the values of Y are distinct for this case and forces us to infer that, in order for the FD X → Y to exist, values of X have to be distinct in the two rows thus making MX a candidate key.

Definition: **(Minimal Sets of Candidate Keys)** Let K denote the set of candidate keys, and let F denote the set of FDs that are specified on relation schema R. Based on the above theorem, we can define a **closure**, denoted by K^*, for the set K. The closure K^* of K is a set of all candidate keys that can be inferred from F and K using the above theorem.

Similarly, we can define **equivalence** for two sets of candidate keys as follows: For a given set F of FDs, two sets of candidate keys J and K are said to be equivalent if $J^* = K^*$, i.e. if every candidate key in J can be inferred from F and K and every candidate key in K can be inferred from F and J.

A set of candidate keys is **minimal** if it satisfies the following two conditions:

1. We cannot remove any key from K and still have a set of keys that is equivalent to K.

2. We cannot replace a key M in K with a key N, where N is a proper subset of M, and still have a set of keys that is equivalent to K.

A **minimal cover** of a set of keys K is a minimal set of keys K_{min} that is equivalent to K.

4.3 Design of the Functional Dependency Algorithm

Our algorithm will discover minimal FDs and keys, and will find a minimal cover for both the FDs and the keys. The discovered FDs will have a single attribute on the RHS as this is one of the conditions for a minimal FD. One of the rules that the algorithm will use to achieve its goals is: if an attribute set X is a candidate key then it determines the remaining attributes in that relation. Based on this rule, once attribute-set X is identified as a candidate key, we will not check for further dependencies with X as the LHS. We will eliminate redundant and transitive dependencies using inference rules IR1, IR3, and IR6. The other inference rules do not apply to our case of discovering minimal FDs.

As stated before, our goal is to develop an automated method to discover dependencies from existing relational databases. In line with this goal, we assume the existence of a relational database instance, and use SQL to extract the data. A database table can contain millions of rows. Processing all these rows for FDs is a monumental task. Using SQL and its selection operator, we select only those rows that fit our needs.

4.3.1 The outline of the algorithm

The first task is to formulate a query for extracting the data.

Consider a relational table T with attributes A, B, C, and D. Consider the following SQL query:

```
select * from T
    where A in (select A from T
                group by A
                having count(*) > 1) ;
```

This query projects all the attributes from table T. It selects those rows which have the same value of attribute A repeating more than once. These are the rows that fit our FD definition. Let us consider attribute A as the LHS of a possible FD. If the above query does not return any rows, then attribute A is a candidate key and, as per the rule we stated in section 4.2, we do not have to test for any further dependencies with attribute A as the LHS. If the query does return some rows, then we apply the FD definition to these rows and check if any FDs exist. The above query returns rows ordered by groups, with attribute A as the grouping column. This makes it simple to check for the FDs. The start of the next group is indicated by the change in the value of attribute A. For the same value of A in any two rows, we compare the corresponding values, of each of the remaining attributes, in these two rows. If the values of a particular attribute are different in these two rows, then this attribute cannot be the RHS of the FD with attribute A as the LHS. We mark this attribute and do not check it again while evaluating the remaining results of this query. If all the attributes are marked, then there is no need to check the remaining rows returned by this query, and so we proceed to the next query, or to the next table if all the queries for this table have been exhausted. We keep repeating this procedure, and the attributes that remain unmarked at the end represent the RHSs of the FDs with attribute A as the LHS.

For an actual implementation of the above query, we would get the table name and its column descriptions from the data dictionary of the RDBMS. We would like to check for every possible attribute set combination as the LHS. Hence, an important part of the above query is the generation of the attribute-set A. If a table has n columns, we would have (2^n - 1) different combinations for A, and hence, (2^n - 1) different queries. In the next section we shall discuss some rules which would allow us to eliminate some of the combinations. Based on what we have discussed so far in this section, let us generate an algorithm outline.

```
main() {
    while(get_next_table(T)) {
        while(get_next_attribute_set(A)) {
            formulate_the_query();
            if (rows_returned == 0)
                insert_in_key_list(A);
            else {
                for(each row returned by the query) {
                    for(each attribute not in A) {
```

160

```
                    compare_corr_values_of_curr_row_with_prev_row();
                    if (values_are_different() ) {
                        mark_the_attribute();
                        if (all_attributes_are_marked())
                            break;
                    }
                }
                if(all_attributes_are_marked)
                    break; /* get the next attribute set */
            }
            if (any_unmarked_attributes_remain()) {
                for(each unmarked attribute) {
                    form_the_FD(A, attribute);  /* A=LHS, attribute=RHS */
                    insert_in_fd_list(FD);
                }
            }
        }
    }
  }
}
```

Note that we have created two lists, viz. the fd-list and the key-list. In practice, these may be implemented as linked lists. The fd-list and the key-list will contain the minimal covers for the FDs and the Keys respectively. Another thing, we do not check any columns which are in the attribute-set itself, as possible RHSs. This is because IR1 already implies such FDs and there is no point in rechecking them. A third point reated to the above algorithm is that, when comparing the attribute values, we consider the NULL values as values that are unknown. This means, a NULL value is not considered as equal to any other value or to another NULL value. In the remaining sub-sections we will discuss how we can guarantee that the FDs and the keys, in the fd-list and the key-list respectively, are indeed minimal covers.

4.3.2 Some enhancements

We first discuss how we can eliminate some of the attribute-set combinations. It would be appropriate to add any rules we come up with, to the get_next_attribute_set(A) function, described above, so that it returns only the valid attribute-sets. Before we get into discussing the rules, let us define what we would like the get_next_attribute_set(A) function to do. We would like this function to generate combinations of attributes in a particular order. For example, if a table has three columns A, B, and C, then each time the get_next_attribute_set(A) function is called it would return an attribute-set in the following sequence: A, B, C, AB, AC, BC, ABC. We will explain the reasons in the following paragraphs. Notice that the above sequence has (2^3 - 1) combinations.

Some of these combinations may be eliminated. For example, we can make use of the following fact: if an attribute-set is a key, then concatenating any other attributes to it would make it a superkey. A key, or a superkey, being unique, will imply all other attributes in the given relation and hence, there is no point in checking for FDs with this attribute-set as the LHS. Thus, before returning an attribute-set, the function must check if one of the candidate keys, already discovered or derived is a proper subset of this attribute-set. (Candidate keys which form the minimal cover are in the key-list. Derived keys will be discussed in the next sections.) If it is, then the get_next_attribute_set() function will skip this attribute-set and try the next attribute-set in the sequence.

The above also relates to the augmentation rule given in [18]: $\{X \rightarrow Y\} \models XZ \rightarrow Y$. Based on this rule, if an attribute-set contains the LHS of an FD, that is already discovered or derived (FDs which are part of the minimal cover are in the fd-list, and the derived FDs will be discussed in one of the later sections), then it will automatically imply the RHS of this FD and hence, there is no point in checking this RHS again. This is one of the reasons for generating the attribute-sets in the above sequence.

The theorem presented below can provide further improvements.

Theorem: Consider a relational instance r of schema R with A as any given attribute-set and K as any given simple key, such that $A \cap K = \emptyset$. If a FD $A \rightarrow K$ exists in r then A is a candidate key of R.

Proof: We will give an informal proof for this theorem.

We are given that K is a simple key and $A \rightarrow K$ is a FD, and we have to prove that A is a key.

Since $A \rightarrow K$ is a FD, according to the FD definition, whenever two tuples in r have the same A-value then they must also have the same K-value. Now, since K is a key, all its values are unique throughout r. This means that for an FD $A \rightarrow K$ to exist, values of A have to be unique throughout r. This implies that A has to be a key of R, if the FD $A \rightarrow K$ has to exist.

The above theorem implies that if a column, which can be a possible RHS, is a key, then in order for an attribute-set to form the LHS of this FD, the attribute-set itself has to be a key. The fact that this attribute-set is a key or not will be revealed by the main() algorithm based on the results of the query. At this point we do not know if it is a key and hence, and the get_next_attribute_set(A) function should mark all simple keys as attributes that are ignored while comparing the rows. Let us summarize what we have stated so far, in an algorithmic manner.

```
get_next_attribute_set(A)
{
    while(TRUE) {
        if(!get_the_next_set_in_the_sequence())
            return FALSE;
        if (attribute_set_does_not_contain_key())   {
            if(attribute_set_contains_LHS_of_a_FD())
```

```
            mark_RHS_of_the_FD_as_attribute_to_be_ignored();
        for(each simple key that is not part of the attribute-set)
            mark_the_key_attribute_to_be_ignored();
        return attribute_set;
      }
   }
}
```

In the above algorithm, we mark the RHSs which can be ignored. If we mark all the RHSs, do we still need to execute the query ? The answer is YES, because at this point we do not know if the attribute-set is a key or not. We need to execute the query to check if it returns any rows, even if the attribute-set implies all other columns in the relation. This is done to make the algorithm fool-proof. If someone just duplicates a row, then even though an attribute-set implies all other columns, it will still not be unique and hence, cannot qualify as a candidate key. To incorporate this check we need to make a small change in the main algorithm given in section 4.3.1. The updated algorithm is shown below with the changes highlighted.

```
main() {
    while(get_next_table(T)) {
        while(get_next_attribute_set(A)) {
            formulate_the_query();
            if (rows_returned == 0)
                insert_in_key_list(A);
            else if (num_of_columns_to_be_ignored ==
            columns_to_be_checked) {
                ; /* do nothing */
            else {
                for(each row returned by the query) {
                    for(each attribute not in A) {
                        compare_corr_values_of_curr_row_with_prev_row();
                        if (values_are_different() ) {
                            mark_the_attribute();
                            if (all_attributes_are_marked())
                                break;
                        }
                    }
                    if(all_attributes_are_marked)
                        break; /* get the next attribute set */
                }
                if (any_unmarked_attributes_remain()) {
                    for(each unmarked attribute) {
                        form_the_FD(A, attribute);  /* A=LHS, attribute=RHS */
                        insert_in_fd_list(FD);
                    }
                }
```

                }
            }
        }
    }
}

4.3.3 Derived keys

We now discuss derived keys and then present a method for identifying derived FDs. In section 4.2, we presented a theorem which stated that, if we have a composite key MY and a FD X → Y then we can derive a composite key MX. Hence, to derive a key, we need a composite key as a starting point. If we find one, then we need to search the fd-list (i.e. the list where all the discovered FDs are stored) to see if we can find a FD whose RHS attribute is part of this composite key. If we find more than one FD satisfying this condition, then we can derive multiple keys from the same composite key. Let us store the derived keys in a list which we shall call the derived-key-list. The key-list will only contain the keys which are part of the minimal cover. All other keys are assumed to be derived and are stored in the derived-key-list. The keys in the derived-key-list are also minimal, but they just do not belong to the minimal cover, because they can be derived using the keys in the key-list.

A derived key can derive another key, if we can find a FD that satisfies the above condition. Thus, this is a recursive process and we can wind up with derived key hierarchies or, in terms of data structures, a graph. The keys in the key-list as well as the derived-key-list form the nodes or vertices of this graph. To maintain the hierarchical relationship we introduce a parent field in both the lists. The parent field contains a pointer to the parent key which, along with a FD, was involved in deriving the given key. A key can have multiple parents. Hence, instead of the parent field pointing to a single parent, it points to a parent list. Before we proceed further, let us look at an example that shows how the data structure can be represented as a directed graph.

A **graph** G is an ordered pair of sets (V, E), where V is a set of vertices, and E is a set of edges. Each edge E connects two different vertices in V. If the edges have a direction, then the graph is **directed**. Each edge in a directed graph is an ordered pair of vertices (v, w), with $v \neq w$. The tail of the edge is v, and its head is w. We sometimes say that v is a predecessor of w, and that w is a successor of v. Vertices v and w are incidents to edge (v,w). There are two principal representations for graphs, viz. the **adjacency matrix** and the **adjacency list**. The latter, which we will use, is a linked structure. To store a graph in a linked structure, we store for each vertex a list of its predecessors. Let us denote the keys in the key-list as K'_1, K'_2, and so on. Let us denote the keys in the derived-key-list as K_{d1}, K_{d2}, and so on. As will be seen later, we are interested in finding out if a path exists from a key in the derived-key-list to a key in the key-list. Due to this requirement, our edge will be directed from the child to the parent. We would also like to keep the keys in the minimal cover, i.e. the key-list, separate from others. Hence, we will keep the vertices in the key-list separate from those in the derived-key-list. Given this background information, let us consider the following sequence of events and see how they will be represented in our data structure:

- The main() function discovers a key K'_1. This is inserted in the key-list.
    - K'_1 derives two keys, K_{d1} and K_{d2}. These are inserted in the derived-key-list.
- K_{d1} derives K_{d3}. K_{d3} is also inserted in the derived-key-list.
- The main() function discovers a key K'_2.

The diagram representing the data structure is shown below.

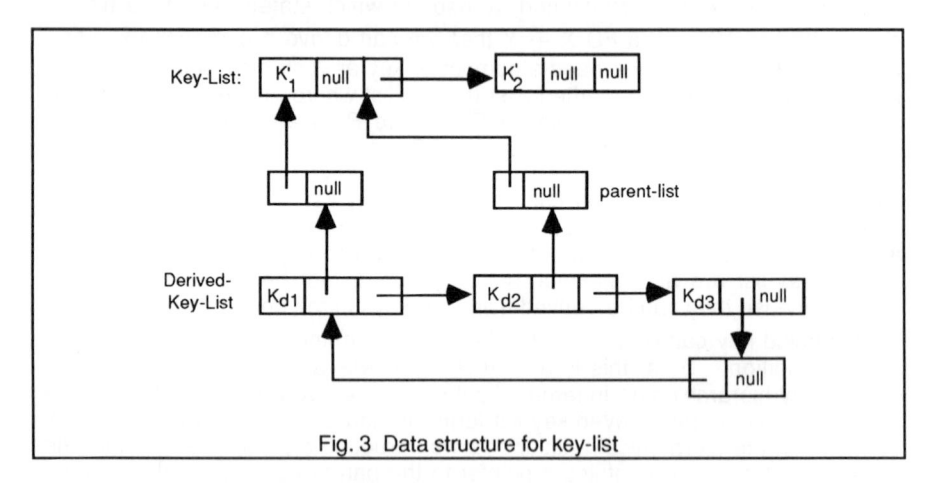

Fig. 3 Data structure for key-list

The best place to check for derived keys is the insert_key_list() function mentioned in section 4.8.1. This function gets called from the main() function when we discover a new key. It can check for derivations, before it inserts the newly discovered key, or the original key, in the key-list.

If a derived key exists in the derived-key-list, we check if the parent of the derived key is different from the one that already exists. If it is, then we store a pointer to the new parent indicating that there are two different paths to derive this key.

If a derived key exits in the key-list, then according to our definition of minimal cover for keys, it should be removed from the key-list. The minimal cover definition also states that we should be able to recreate this key from the keys in the minimal cover; in our case from the keys in the key-list. To ensure that we can recreate it, we trace the derivation hierarchy of the derived-key by following the pointers in its parent list, and make sure that it is independent of the key we are going to remove. If the key is removed from the key-list, it will however, have to be stored in the derived-key-list. This is because the get_next_attribute_set() function checks both the key and the derived key lists to ensure that it does not return an attribute-set which is a superkey. We also use the derived-key-list to check for derived FDs which we shall talk about in the next section.

If a subset of the derived-key exists in the key-list or the derived-key-list, then the derived-key becomes a superkey, and we can ignore it as we are not interested in superkeys.

The following algorithm embodies the points stated above.

```
insert_key_list(key) {
    if (is_composite(key)) {
        /* these two functions will insert any derived keys in the key_q */
        check_fd_list_if_key_derives_another_key(key);
        check_derived_fd_list_if_key_derives_another_key(key);
    }
    store_in_the_key_list(key);
    process_key_queue();
}


check_fd_list_if_key_derives_another_key(key) {
    /* Here we are given a composite key MY and we have to check if the RHS
       of any FD matches Y */
    for (every FD in the fd-list) {
        if (the RHS of the FD matches any attribute in the key) {
            build_the_new_key();
            if (new_key_exists_in_key_list()) {
                mark_the_key_in_key_list();
                if (path_to_key_list_other_than_the_marked_key_exists())
                    move_key_from_key_list_to_derived_key_list();
                if (parent_not_in_parent_list())
                    add_parent_to_parent_list();
            }
            else if (subset_of_new_key_exists_in_key_list_or_derived_key_list())
                ; /* do nothing */
            else if (exists_in_derived_key_list(new_key)) {
                if (parent_not_in_parent_list())
                    add_parent_to_parent_list();
            }
            else
                insert_in_key_q();
        }
    }
}

process_key_queue() {
    while (key=get_key_from_q()) {
        if (!exists_in_derived_key_list(key)) {
            /* the next two functions will insert any keys they find in the key-q */
            check_fd_list_if_key_derives_another_key(key);
```

```
                check_derived_fd_list_if_key_derives_another_key(key);
                store_in_the_derived_key_list(key);
            }
        }
    }
```

If you notice, in the above algorithm, we did not implement a recursive function to check for keys which are derived from derived keys. Instead we chose to use a queue to store all the derived keys and then processed the queue to check for further derivations. In other words we separated the processing of the original keys; keys discovered by the main() function when no rows are returned by the query; and the derived keys. This is because, we store the keys in two separate lists, and could not find a cleaner way to implement a recursive call without setting up some kind of flag to tell the function if any given key, that we are checking for derivations, is derived or original.

The process_key_queue() function gets the derived keys from the key-queue and checks if they further derive any key. If they do, then these derived keys are also inserted back into the key queue to be checked later for further derivations. It continues this process till the queue is empty. This is similar to using a recursive function. One thing you might notice that, after getting the key from the queue, we first check if the key exists in the derived-key-list. This is done because we do not know what is in the queue and it is quite possible that the same key as the current one could have existed in the queue, may have already been checked for derivations, and inserted in the derived-key-list. This check here saves us from going through all these steps if the key already exists in the derived-key-list.

4.3.4 Derived FDs

We can derive FDs, from the FDs we discover in the main() function, using the transitive inference rules IR1 and IR6 mentioned in section 3.1. IR6 is a generalization of IR3 and hence, we will use it for our discussion. IR6 states that if we have two FDs $X \to Y$ and $WY \to Z$, then we can derive a FD $WX \to Z$. If W is NULL, then IR6 translates to IR3. This means, we need two FDs to start with, before we can derive any new FDs.

We can apply IR6 in two different ways. We can either consider the newly discovered FD as

$WY \to Z$ and search the fd-list to see if the RHS of the FD matches any of the attributes in the LHS of the newly discovered FD, or we can consider the newly discovered FD as $X \to Y$ and search the fd-list to see if the LHS of the FD contains the RHS attribute of the newly discovered FD. The derived FDs are stored in a list called the derived-fd-list.

Once we have inserted some FDs in the derived-fd-list, we will have to search this list, in addition to the fd-list, to check for the existence of derived FDs. The two possibilities mentioned above, along with the two lists, gives us four cases to consider. A newly discovered FD can also derive a key using the theorem in section 3.2. Hence, for each newly discovered FD we will also have to search the key-list and the derived-key-list to check for the existence of derived keys. This gives us six cases overall, to check for derivations.

Just like in the case of derived keys, a derived FD can also derive another FD, provided that the IR6 conditions are satisfied. This again gives rise to the recursive process we discussed in the previous section, and we again wind up with a graph. This time, the hierarchical relationship is somewhat more complex than the previous one, because, in this case, we have two parents, or two FDs in the fd-list, that each node in the parent-list of the derived FD has to point to. The best place to check for derived FDs is the insert_in_fd_list() function that gets called from the main() function. It can check for derivations before it inserts the newly discovered FD in the fd-list.

If we derive a FD whose RHS is a subset of the LHS, then we should ignore it, because, we can always derive such FDs using IR1. If we derive an FD whose LHS contains a key, then this FD becomes a trivial FD and we are not interested in it. If we derive a FD that exists in the fd-list then, just like in the case of keys, we mark the FD in the fd-list and start tracing the paths in the parent list of both the parents that derived this FD, and make sure that both of them have at least one derivation path that does not contain the FD we just marked.

If the derived FD exists in the derived-fd-list, then, if the parents of the derived FD do not exist in the parent list of the FD in the derived-fd-list, we insert them in the parent list creating another path by which this FD can be derived. If a subset of the LHS of the derived FD exists, either in the fd-list or the derived-fd-list, we just ignore the FD, because, we can always derive this FD using the augmentation rule mentioned in the previous section.

As before, we summarize the above discussions into a routine.

```
insert_in_fd_list(fd) {
    if (!exist_in_the_derived_fd_list(fd)) {
        check_fd_list_lhs_for_derived_fd();
        check_fd_list_rhs_for_derived_fd();
        check_derived_fd_list_lhs_for_derived_fd();
        check_derived_fd_list_rhs_for_derived_fd();
        check_key_list_for_derived_key();
        check_derived_key_list_for_derived_key();
        store_fd_in_fd_list();
        process_fd_queue();
        process_key_queue();
    }
}


check_fd_list_lhs_for_derived_fd() {
    for (every FD in the fd-list) {
        if (the LHS of the FD in the fd-list contains the RHS of the new FD) {
            build_the_new_fd();
            if (RHS of new FD is not a subset of the LHS &&
```

```
            LHS of new FD does not contain a key)  {
                if (new_fd_exists_in_fd_list()) {
                    mark_the_fd_in_fd_list();
                    if (both_parents_have_at_least_one_path_that_does_not_
                    contain_the_marked_fd())
                        move_fd_from_fd_list_to_derived_fd_list();
                    if(parents_do_not_exist_in_parent_list())
                        add_parents_to_parent_list();
                }
                else if (subset_of_lhs_of_fd_exists_in_fd_list() ||
                subset_of_lhs_of_fd_exists_in_derived_fd_list())
                    ; /* do nothing */
                else if (exists_in_derived_fd_list(fd))  {
                    if(parents_do_not_exist_in_parent_list())
                        add_parents_to_parent_list();
                }
                else
                    insert_in_fd_q();
            }
        }
    }
}
```

In order to check if the FD derives a key, we have introduced two functions, namely, the check_key_list_for_derived_key(), and the check_derived_key_list_for_derived_key(). These functions are almost identical to the ones used in the insert_key_list() function mentioned in section 4.3.3.

The six functions that check for derived keys and FDs insert their findings, if any, into the key queue and the fd queue respectively. Hence, we have to process these two queues before returning from the function. We have already described the process_key_queue() function in section 4.3.3. The process_fd_queue() function is almost identical and hence, will not be described again.

5. DERIVATION OF MULTIVALUED AND JOIN DEPENDENCIES

This section describes an algorithm to discover Multivalued and Join Dependencies that a given database instance satisfies. As before, the dependency inference technique is based on the "learning from examples" paradigm of machine learning and uses a set of inference rules, similar to the Armstrong's axioms for Functional dependencies, as heuristics to guide the discovery process. Before we get into the algorithm details, we will state some preliminary definitions and rules.

5.1 Some Fundamental Concepts

We will provide formal definitions for Multivalued and Join dependencies.

Definition: **(Multivalued Dependency)** Consider a relation schema R, and X and Y as any subsets of R. A MultiValued Dependency (MVD) X ->> Y specifies a constraint on the relation instance r of R which states that if two tuples t_1 and t_2 exist in r, such that $t_1(X) = t_2(X)$, then two other tuples t_3 and t_4 should also exist in r with the following properties:
- $t_3(X) = t_4(X) = t_1(X) = t_2(X)$.
- $t_3(Y) = t_1(Y)$, and $t_4(Y) = t_2(Y)$.
- $t_3(R - X - Y) = t_2(R - X - Y)$ and
  $t_4(R - X - Y) = t_1(R - X - Y)$.

Whenever X ->> Y holds, we say that X multidetermines Y. Let us denote (R - (X U Y)) by Z. Because of the symmetry in the definition, whenever X ->> Y holds in R, so does X ->> Z. Hence, X ->> Y implies X ->> Z.

MVDs are a consequence of the first normal form, which disallowed an attribute in a tuple from having a set of values. If we have two or more multivalued independent attributes in the same relation schema, we get into a problem of having to repeat every value of one of the attributes with every value of the other attribute to keep the relation instance consistent. Informally, whenever two independent 1:N relationships are mixed in the same relation, an MVD may arise.

The formal definition, stated above, specifies that, given a particular value of X, the set of values of Y determined by this value of X is completely determined by this value of X alone and does not depend on the values of the remaining attributes Z of the relation schema R. Hence, whenever two tuples having distinct values of Y occur with the same value of X, these values of Y must be repeated with every distinct value of Z that occurs with the same value of X.

An MVD X ->> Y is called a **trivial MVD** if:
- $Y \subseteq X$ or
- X U Y = R.

A trivial MVD will hold in any relation instance r of R. It is called trivial because it does not specify any constraint on R.

Definition: **(Relation Decomposition)** Consider a universal relation schema R = {$A_1, A_2, ..., A_n$} that includes all the attributes of the database. Using the FDs that hold on the attributes of R, we can decompose R into a set of relation schemas D = {$R_1, R_2, ..., R_m$} that will become the relational database schema. D is called the decomposition of R.

Lossless join or non-additive join states that no spurious tuples should be generated when a natural join operation is applied to the relations in the decomposition. It refers to the loss of information, not to the loss of tuples.

Definition: **(Join Dependency)** Let R be a relational schema. A Join Dependency (JD), denoted by JD($R_1, R_2, ..., R_n$), specifies a constraint on instances r of R which states that, every legal instance r of R should have a lossless join decomposition into $R_1, R_2, ..., R_n$. Let us denote a natural join operation by * and a project operation by Π. The definition states that, $*(\Pi_{<R1>}(r), \Pi_{<R2>}(r), ..., \Pi_{<Rn>}(r)) = r$.

The above definition states that R can be decomposed into any number of relation schemas as long as you can recreate the original instance by joining the sub-schemas. With this definition, an MVD becomes a special case of a JD where n = 2. A join dependency $JD(R_1, R_2, ..., R_n)$, specified on relation schema R, is a **trivial JD** if one of the relation schemas R_i is equal to R. It is called trivial because it has the lossless join property for any relation instance r of R and hence does not specify any constraint on R.

5.2 Inference Rules for Multivalued Dependencies

Let M denote the set of MVDs that are specified on relation R. There are usually numerous other MVDs which will also hold in all legal relation instances that satisfy the dependencies in M. The set of all such MVDs is called the **closure** of M and is denoted by M^*. In general, an MVD X ->> Y is **inferred from** a set of dependencies M specified on R if X ->> Y holds in every relation instance r that is a legal extension of R. The closure M^* of M is a set of all MVDs that can be inferred from M. The closure M^* is determined using the set of inference rules specified below. We will use the notation M |= X ->> Y to denote that the MVD X ->> Y is inferred from the set of MVDs M. We will denote the MVD {X,Y} ->> Z by XY ->> Z. The following inference rules are well known inference rules for MVDs:

(IR1)	Augmentation	X ->> Y, Z $\subseteq$ W |= WX ->> YZ	[16]
(IR2)	Transitive	X ->> Y, Y ->> Z |= X ->> (Z - Y)	[16]
(IR3)	Replication	X $\rightarrow$ Y |= X ->> Y	[16]
(IR4)	Complementation	X ->> Y |= X ->> (R - (X U Y))	[16]
(IR6)	Augmentation	X ->> Y |= XZ ->> Y	[18]
(IR7)	Pseudo-transitive	X ->> Y, WY ->> Z |= WX ->> (Z - WY)	[17]
(IR8)	Union	X ->> Y, X ->> Z |= X ->> YZ	[17]
(IR9)	Decomposition	X ->> Y, X ->> Z |= X ->> (Y $\cap$ Z)	[17]
(IR10)	Decomposition	X ->> Y, X ->> Z |= X ->> (Y - Z)	[17]
(IR11)	Decomposition	X ->> Y, X ->> Z |= X ->> (Z - Y)	[17]

These inference rules are sufficient to determine the closure M^* from M.

<u>Definition:</u> **(Equivalence among Multivalued Dependencies)** A set of MVDs E is said to be **covered by** a set of MVDs M, or alternatively, M is said to cover E, if every MVD in E is also in M+, i.e. every dependency in E can be inferred from M. Two sets of MVDs E and M are said to be **equivalent** if $M^* = E^*$. Hence, equivalence implies that every MVD in E can be inferred from M, and every MVD in M can be inferred from E.

Definition: **(Minimal Sets of Multivalued Dependencies)** A set of MVDs is **minimal** if it satisfies the following conditions:

　　1.　We cannot remove any dependency from M and still have a set of dependencies that is equivalent to M.

2. We cannot replace any dependency X ->> A with a dependency Y ->> A, where Y is a proper subset of X, and still have a set of dependencies that is equivalent to M.

3. Every dependency, X ->> Y, in M has disjoint X and Y.

The conditions make sure that there are no redundancies in the dependencies. Condition 2 is a consequence of IR6, and condition 3 is proved as a theorem in the following sub-section.

A **minimal cover** of a set of MVDs M is a minimal set of dependencies M_{min} that is equivalent to M. There can be several minimal covers for a set of MVDs.

5.3 The Formal Framework

In this section, we will develop the formal framework that leads us to the creation of the algorithm.

Theorem (T1)

Consider a relation schema R. Let X and Y be subsets of R, such that $Y \not\subset X$, and let $Z = (R - (X \cup Y)) \neq \emptyset$. Let r be the relation instance of R, let $r_X = (\Pi_{<X>}(r))$, and let $t_i(X)$ be the value of X in tuple t_i belonging to r_X. Let $m = card(\sigma_{<X=\ t_i(X)>}(\Pi_{<XY>}(r)))$, $n = card(\sigma_{<X=\ t_i(X)>}(\Pi_{<XZ>}(r)))$, and $k = card(\sigma_{<X=\ t_i(X)>}(r))$, where card stands for cardinality, σ stands for selection, and Π stands for projection. Let

$r_Y = \Pi_{<Y>}(\sigma_{<X=t_i(X)>}(r))$, $r_Z = \Pi_{<Z>}(\sigma_{<X=t_i(X)>}(r))$, $r_{YZ} = \Pi_{<YZ>}(\sigma_{<X=t_i(X)>}(r))$, and

$r'_{YZ} = \Pi_{<YZ>}(\sigma_{<X=t_j(X)>}(r))$, for any $j \neq i$. If $k \neq mn$, for any $t_i \in r_X$, and if $r'_{YZ} \in ((r_Y \times r_Z) - r_{YZ})$, where X stands for the cartesian product, then every JD in r will be trivial.

Proof: To prove this theorem, consider a relation schema R(A,B,C). Let us create an instance r of R as follows:

X	Y	Z
x1	y1	z1
x1	y2	z2
x2	y2	z1

In this instance $r_X = \{x_1, x_2\}$, $t_1(X) = x_1$, $m = 2$, $n = 2$, $k = 2$, $r_Y = \{y_1, y_2\}$, $r_Z = \{z_1, z_2\}$,

$r_{YZ} = \{y_1\ z_1,\ y_2\ z_2\}$, and $r'_{YZ} = \{y_2\ z_1\}$. We can see that k is not equal to the product of m and n, signifying that some combinations of Y and Z are missing for this particular value of X, viz. $X = x_1$. The next part of the theorem says that if one of these missing combinations appears with another value of X then we will have a trivial JD. In this case, the combination $\{y_2\ z_1\}$ appears with x2. If we try to decompose the above relation into [XY, YZ], or [XY, XZ], or [XZ, YZ], or [XY, YZ, XZ], or into any other combination that does not contain all the attributes of the relation, we will always get extra tuples which do not exist in the original relation. This is because the combinations of Y and Z which are missing from x1 and appearing with x2, will interact with x1 during join and will not get eliminated from the final result. The only way to eliminate these extra combinations is to include the whole relation as one of the join components. If we include the whole relation in the join then we will have a trivial JD, and that is what the theorem says.

Theorem (T2)

Consider a relation schema R. Let X and Y be subsets of R, such that $Y \not\subset X$, and let $Z = (R - (X \cup Y)) \neq \emptyset$. Let r be the relation instance of R, let $r_X = (\Pi_{<X>}(r))$, and let $t_i(X)$ be the value of X in tuple t_i belonging to r_X. Let $m = \text{card}(\sigma_{<X= t_i(X)>}(\Pi_{<XY>}(r)))$, $n = \text{card}(\sigma_{<X= t_i(X)>}(\Pi_{<XZ>}(r)))$, and $k = \text{card}(\sigma_{<X= t_i(X)>}(r))$, where card stands for cardinality, σ stands for selection, and Π stands for projection. Let

$r_Y = \Pi_{<Y>}(\sigma_{<X=t_i(X)>}(r))$, $r_Z = \Pi_{<Z>}(\sigma_{<X=t_i(X)>}(r))$, $r_{YZ} = \Pi_{<YZ>}(\sigma_{<X=t_i(X)>}(r))$, and $r'_{YZ} = \Pi_{<YZ>}(\sigma_{<X=t_j(X)>}(r))$, for any $j \neq i$. If $k = mn$, for every $t_i \in r_X$, then relation r satisfies the MVD X ->> Y. Otherwise, r satisfies the JD *[XY, XZ, YZ], or a trivial JD if $r'_{YZ} \in ((r_Y \times r_Z) - r_{YZ})$, where × stands for the cartesian product.

Proof: Before we start proving this theorem, we would like to point out that the conditions, $Y \not\subset X$ and $Z \neq \emptyset$, eliminate the possibility of trivial MVDs.

To prove this theorem we will use a theorem given in [18]. The theorem in [18] uses the same definitions as T1 for X, Y, Z, and r, and states that r satisfies the MVD X ->> Y if and only if r decomposes losslessly onto relation schemas $R_1 = XY$ and $R_2 = XZ$.

We are given that r_X is a projection of X from r. To simplify the discussion, let us assume that r_X contains a single tuple, viz. $t_1(X) = a$. This means that X = a is the only distinct value of X in r. Let X = a determine m distinct values of Y and n distinct values of Z, such that the values of Z determined by X are independent from the values of Y. If we form a relation r_1, with $R_1 = XY$, and insert X=a along with the m distinct values of Y, then card(r_1) = m. Similarly, if we form a relation r_2 with $R_2 = XZ$, then card(r_2) = n. Since the relationship between X and Y is independent of the relationship between X and Z, according to the definition of MVD given in section 5.1.1, the m distinct values of Y must be repeated with every distinct value of Z that occurs with the same value of X and hence, card(r) = mn; r contains only one distinct value of X, viz. X = a. We have just shown that card(r) = card(r_1) × card(r_2), where × stands for mathematical multiplication. Let us call this result (1). Now according to the theorem in [18], stated above, in order for the MVD X ->> Y to exist, r must be equal to $r_1 * r_2$, where * stands for natural join. The common attribute between R_1 and R_2 is X. Since the relationship between X and Y is independent of the relationship between X and Z, the natural join of r_1 and r_2 will be a cartesian product between the values of Y and Z with X=a repeated in all the tuples. Thus, card(r) will be equal to mn, or equal to card(r_1) x card(r_2). This is the same as result (1) we have shown above, and hence proves the first part of our theorem, viz. if k = mn, for every $t_i \in r_x$, then relation r satisfies the MVD X ->> Y.

The last part of the theorem is already proved in T1, stated above.

Now let us prove the part for the JD *[XY, YZ, XZ]. The definition of JD states that a relation r can be decomposed losslessly into n relations. When n=2, we have an MVD, X ->> Y / Z, and r can be decomposed onto the relation schemas XY and XZ. Our theorem considers a special case of JD, when n=3, and that is what we are trying to prove here.

Let us assume that an MVD X ->> Y / Z, or a JD with n= 2, exists in r, and just like above, let us consider a single distinct value of X, viz. X=a, which determines m distinct values of Y and n distinct values of Z. We have to prove that if k ≠ mn, and if the condition in T1 is not satisfied, then we have a JD between X, Y, and Z, i.e. a JD with n=3. If card($\sigma_{<X= a>}(r)$) < mn; it cannot be greater than mn, as mn is the cartesian product of the values of Y and Z; then the decomposition of r into XY and XZ will not be lossless, i.e. it will produce spurious tuples. This is because the relationship between X and Y is no longer independent of the relationship between X and Z, and a third relationship between Y and Z now exists. In order for r to decompose losslessly, this third relationship should be taken into account by decomposing r into three relation schemas XY, XZ, and YZ. This means that a JD between X, Y, and Z will have to exist if k ≠ mn.

Theorem T3

Consider a relation schema R. Let X and Y be subsets of R such that X ∩ Y = ∅. If X is a candidate key of R, then r satisfies the MVD X ->> Y.

Proof: If X is a candidate key of R, then an FD X -> Y exists in r, and according to IR3, r satisfies the MVD X ->> Y. This is considered as a trivial MVD since it does not specify any constraint on r.

Theorem T4

Consider a relational instance r of schema R. Let X and Y be any two subsets of R such that Y ⊄ X, and let Z = (R - (X U Y)) ≠ ∅. If a simple key exists in r then any MVD that exists in r will be trivial.

Proof: Let an MVD X ->> Y / Z exist in r. We shall prove that the MVD X ->> Y/Z is trivial.

Let K be a simple key of R. Let us consider the case when X contains K or K ⊆ X. In this case X becomes a superkey of R, and according to T3 r will satisfy a trivial MVD X ->> Y.

Since K is a simple key, i.e. it has a single attribute, if X does not contain K then either Y or Z will contain it. Let us consider the case when Y contains K. In this case, Y will be unique throughout r. Let us select tuples from r where X = a. Let us say we get m tuples. Since Y is unique throughout r, there will be m distinct values of Y in these tuples.

Now, according to T2, in order for the MVD X ->> Y to exist, Z should have only one distinct value in these tuples. Let us call this value of Z as "b". This means that, in order for the MVD X ->> Y to exist, whenever X has the value "a" in the m tuples, Z should have the value "b". Now, if the values of Z correspond to the values of X, we have an FD from X to Z, or an FD X -> Z will exist. This FD will imply a trivial MVD X ->> Z/Y.

We have considered all the applicable cases and have proved that the MVD is trivial in each case, thus proving the theorem.

Theorem T5

Consider a relation instance r of R. Let X and Y be any two subsets of R such that Y $\not\subset$ X, and let Z = (R - (X U Y)) $\neq \emptyset$. If a simple key exists in r then r will always satisfy JDs of the form *[XY, YZ, XZ].

Proof: According to T4, if a simple key exists in r then every MVD existing in r will be trivial. A MVD

X ->> Y will also satisfy a JD of the form *[XY, YZ, XZ], although the third relationship between X and Z is not required. Now, if a MVD X ->> Y does not exist in r, then, using T3, we can say that X does not contain the simple key, and that either Y or Z has to contain the key. If either Y or Z contain the key, then the interaction mentioned in T1 will not exist, and r can be decomposed into XY, YZ, and XZ., satisfying our theorem.

Theorem T6

Consider a relation instance r of schema R. Let X and Y be any two subsets of R such that Y $\not\subset$ X, and let Z = (R - (X U Y)) $\neq \emptyset$. r satisfies the MVD X ->> Y/Z if all the candidate keys of R are composite and neither of X, Y, or Z contains a whole key.

Proof: We will use T4 to prove this theorem.

If we have any simple keys, then, according to T4, every MVD in r will be trivial. Using the same argument as T4, if X contains a whole composite key then X becomes a superkey making X ->> Y a trivial MVD. If either Y or Z contain a whole composite key, then one of them, the one that does not contain the whole key, will form the RHS of a FD with X as the LHS, thus making X ->> Y/Z a trivial MVD.

Lemma L1

Consider an MVD X ->> Y defined over a relation schema R, such that Y $\not\subset$ X and X U Y $\neq$ R. The following statement is true: X ->> Y, W $\subseteq$ X |= X ->> WY.

Proof: We shall prove this lemma using an example. Consider a relation schema R with attributes A, B, C, and D. Let us define an MVD X ->> Y for R, such that X = {A, B}, Y = {C}, and Z = {D}. We have to prove that we can infer X ->> WY from X ->> Y, where W $\subseteq$ X. Let us create an instance r for R such that X ->> Y holds in r.

A	B	C	D
a_1	b_1	c_1	d_1
a_1	b_1	c_2	d_1
a_1	b_1	c_1	d_2
a_1	b_1	c_2	d_2
a_1	b_2	c_3	d_1
a_1	b_2	c_3	d_3
a_2	b_3	c_4	d_4
a_2	b_3	c_5	d_4
a_2	b_3	c_6	d_4
a_3	b_1	c_4	d_4
a_3	b_1	c_5	d_4

Fig. 4 Relation instance r showing an MVD AB ->> C/D

Let W = {B}. We can see that X ->> WY holds in the above instance. In fact, the MVD X ->> WY will hold for every such instance where X ->> Y holds. This is because, according to T2, to calculate k, we select rows from r which have the same X-value, i.e. the value of every corresponding attribute in X repeats in every row that is selected. If we take any attribute from X, like B in the above example, and concatenate it to Y, it will have no effect on "m"; "m" is the number of distinct values of Y; because the value of this attribute, or a group of attributes, if we concatenate more than one, will be the same in all of the selected rows. Since "m" will not change, k = mn will still hold, thus validating our claim.

Lemma L2

Consider an MVD X ->> Y defined over a relation schema R, such that Y ⊄ X and X ∪ Y ≠ R. The following statement is true: X ->> Y |= X ->> (Y - (X ∩ Y)).

Proof: Consider a relation schema R with attributes A, B, C, and D. Let us define an MVD X ->> Y for R, such that X = {A, B}, Y = {B, C}, and Z = {D}. We have to prove that we can infer X ->> (Y - (X ∩ Y)) from X ->> Y. We will use the instance r from L1; r satisfies X ->> Y considering the X and Y as defined above.

X ∩ Y = {B}. Let Y' = Y - (X ∩ Y). Y' = {C}. We can see that X ->> Y' holds in r. In fact, the MVD X ->> Y' will hold for every such instance where X ->> Y holds. This is because, according to T2, to calculate k, we select rows from r which are grouped by the X-value. If we take out any X-attribute(s) from Y, like B in the above example, it will have no effect on "m"; "m" is the number of distinct values of Y; because the value of the removed X-attribute(s) is the same in all of the selected rows and hence, "m" is dependent on the distinct values of Y' only. Since "m" will not change, k = mn will still hold, thus validating our claim.

Theorem T7
Let M denote the set of MVDs that are specified on the relation schema R. M is not minimal if for any MVD, X ->> Y, in M, X and Y are not disjoint.

Proof: We have to prove that one of the conditions for a set of MVDs to be minimal is that, the LHS and the RHS of every MVD in the set has to be disjoint.

The goal of minimality is to avoid redundant FDs. Keeping this goal in mind, if we have an MVD X ->> Y, where $X \cap Y \neq \emptyset$, then, according to L2 there exists an MVD X ->> Y', where Y' = Y - $(X \cap Y)$. Y' is a subset of Y making the MVD X ->> Y redundant, because, given X ->> Y', we can always derive X ->> Y using L1. Hence, we have proved that one of the conditions of minimality is that X and Y be disjoint.

Our aim is to develop an algorithm that discovers minimal MVDs and also finds a minimal cover for the MVDs. The MVDs we discover will have disjoint LHSs and RHSs. This is one of the conditions for minimal MVDs. To discover the JDs, we make use T2. We eliminate redundant and transitive MVDs using inference rules IR2 through IR11, theorem T3, and lemma L2. (IR1 does not apply to our case of disjoint left and right-hand sides.)

The FD-Key algorithm, described in section 4, is a prerequisite for this algorithm. The reason is that we depend heavily on the FDs and Keys in order to eliminate redundant MVDs.

5.4 The Algorithm

The focal point of this algorithm is theorem T2. According to T2, the relation satisfies an MVD X ->> Y/Z if k = mn, where k is the number of rows with the same value of X, m is the number of distinct values of Y in the X-group, and n is the number of distinct values of Z in the X-group. X-group is the group of rows which have the same value of X repeated in all the rows. If k = 1, then m=1 and n=1 as we just have one row in the group. This means, if we have just one occurrence of a value of X, then this X-group satisfies the MVD. For the relation to satisfy the MVD, every X-group in the relation must satisfy T2. If every occurrence of X is unique, then X is a key and we will have a trivial MVD as there is no constraint specified on the relation, other than the key constraint. We have already identified all the keys in the FD algorithm. Hence, if we do not generate X whose attributes contain a key, then we will avoid checking trivial MVDs. The above discussion also reveals that any X-group which contains a single occurrence of X qualifies to be an MVD. Hence, we can save some time by skipping single-valued X-groups. Doing this, makes our SQL query exactly the same as in the FD algorithm.

As stated before, the FD algorithm is a prerequisite of this algorithm. Since the output of the FD algorithm already contains the table names, we will use its output to feed table names to our query. Along with the table name, we will also read the FDs and the keys, from the output of the FD algorithm. We will use the union rule for FDs, IR5, to store the FDs. We are not interested in minimal FDs in this algorithm. The fd-list is used to eliminate some X combinations, as we shall see in the next section, and keeping all occurrences of the FDs with the same LHS together, as a single FD, speeds up our search through the fd-list. While reading-in the keys, we will look for single-attribute keys or simple keys. If we find one, then we don't need to check for MVDs, because T4 says that every MVD will be trivial. However, T5 says that, if we have a simple key, then we will always have a JD of the form *[XY,YZ,XZ]. Hence, if we have a simple key, then, without checking the database, we can insert every [XY, YZ, XZ] combination into the jd-list.

The goal is to discover minimal MVDs. Theorem T7 states that one of the conditions for MVDs to be minimal is that, X, Y, and Z are disjoint. Obviously, we need a minimum of three columns in the table. Once we have found a table which has the right number of columns, the next step is to create the attribute set for the query. In this case, since the grouping is done by the attributes in X, the attribute-set for the query will be nothing else but the X attributes. Just like in the case of FDs, the query will return all the columns. This will enable us to check different Y combinations for the same X. We will discuss generating the XYZ combinations, in more detail, in the next section. For now, let us assume that we have a function, called get_next_xyz_comb(), which returns the next attribute set for X, and all the valid attribute set combinations for Y and Z, for this X. The Y and Z combinations can be returned in the form of a list for RHS. We will also assume that the get_next_xyz_comb() function will return an X attribute set which does not contain a key. This will ensure that the query will not return zero rows.

The query will return rows ordered by X. Since the X-values are ordered, we can easily separate the groups with the same value of X in all the rows, and for each group, we can count the number of rows in the group, and the distinct values of Y and Z. This will allow us to apply T2.

Let us capture what we have said so far in the form of an algorithm.

```
main() {
    while(!end_of_fd_file()) {
        read_next_table(T);
        read_FDs_keys();  /* key-lisy, fd-list, X -> A, X -> B |= X -> AB */
        if (get_columns_for_table(T) > 2) {
            while(get_next_xyz_comb()) {
                if(simple_key())
                    insert_comb_in_jd_list();
                else {
                    formulate_the_query();
                    /* zero rows should not occur, as X does not contain key */
                    for(each_x_group()) {
                        k = num_rows_in_group();
                        for(each_rhs_in_the_list) {
                            if (flag != NO_JD) {
                                m = distinct_y_values();
                                n = distinct_z_values();
```

```
                        if (k != m * n) {
                            flag = NO_MVD;
                            check_for_trivial_jd_with_previous_x_groups();
                            if(trivial_jd_condition_satisfied)
                                flag = NO_JD;
                            if (NO_JD_flag_set_for_all_rhs()) {
                                done = TRUE;
                                break;
                            }
                        }
                    }
                }
            }
            if (done)
                break;
        }
        insert_mvd_or_jd_list(); /* flag=NULL => MVD; flag=NO_MVD => JD */
    }
   }
  }
 }
}
```

5.5 Refinements

The above algorithm can be improved in several ways. Some refinements are presented below.

Generating the XYZ combinations

In section 5.4 we have briefly described what we would like the get_next_xyz_comb() function to do. The function has to generate combinations for the LHS as well as the RHS. Since we have to divide the relation schema R into three disjoint attribute sets, we can only use a maximum of (n - 2) attributes for the LHS (X), n is the number of attributes in R. This gives us Σ (n! / (r! * (n - r)!)) combinations for the LHS (X), where the summation is over r = 1 to r = n - 2. For each X combination, we will have Σ ((n - x)! / (r! * (n - x - r)!)) Y combinations. Here the summation is over r = 1 to r = n -x - 1, where n is the number of attributes in R and x is the number of attributes in the current X combination. Z or the third attribute set is (R - (X U Y)), which is the remaining attributes in the relation. Once we have X and Y, Z will just follow and hence, it will not be necessary to separately calculate the Z combinations.

As in the case of FDs, we will generate the X combinations in the following sequence: A, B, C, D, AB, AC, AD, BC, BD, and CD, where R = ABCD. If X contains a key then, according to T3, we can eliminate this X combination.

We will just consider Y. We will generate Y in the same sequence as X, but will eliminate any Y combination that intersects the current X. For a schema R=ABCD, if X = A and Y = B then Z = CD. If we keep generating the Y combinations in the above sequence, then at some point, for the same X, i.e. X

= A, we will have Y = CD and Z = B. This shows that the Y and Z combinations are symmetric and we can eliminate half of the Y combinations because, we would already have checked the other half, for MVDs, while checking the first half. This is because, an MVD X ->> Y/Z is the same as the MVD X ->> Z/Y.

If the current Y exists in the mvd-list; list which contains the minimal cover for MVDs; and the LHS of the mvd in the mvd-list is a subset of the current LHS, then, according to IR5, we can eliminate this Y combination. We can apply the same rule to the derived-mvd-list; list in which all the MVDs derived using the transitive and decomposition inference rules are stored.

In case a subset of the current LHS exists in the LHS of an FD in the fd-list and the RHS of this FD contains the current Y, we can use the FD inference rules IR2 and IR4, and the MVD inference rule IR3 to eliminate this Y combination.

If either Y or Z contains any key from the key-list, then, according to T6, we can eliminate this YZ combination. After eliminating all the possible RHS combinations, we will store the remaining ones into a list which we shall call the rhs-list. There will be one rhs-list for each X combination. If the rhs-list contains zero entries, then we eliminate the current X combination, and move on to the next X attribute set in the sequence.

Derived MVDs

In the main() function, we get the next valid X in the sequence and a corresponding rhs-list for this X. We check each RHS in the rhs-list to see if it satisfies T2. If it does not satisfy T2, we check it for a JD. The RHSs that remain unmarked in the end, form the RHSs of the MVDs with X as the LHS. The insert_mvd_or_jd_list() function can check if this flag is set or not, and appropriately decide which list it should insert the current XYZ combination into. Inserting an item into the jd-list is straight forward and we shall not discuss it over here. Before inserting into the mvd-list, we shall try to check for MVDs which can be derived using the transitive, union, and decomposition inference rules.

We first check for transitive derivations. IR2 and its generalization IR7 allow us to infer MVDs transitively. IR7 states that if we have X ->> Y/Z and WY ->> A/B then we can infer WX ->> (A -WY). Since we have disjoint LHSs and RHSs, A - WY will be equal to A and hence, the inferred MVD will be WX ->> A in our case. The complementation rule, IR4, will allow us to infer a second MVD, viz. WX ->> B. We can also infer another set of MVDs, viz. W'X ->> A and W'X ->> B, from X ->> Y/Z and W'Z ->> A/B. Both the sets will not be derived at the same time. The following theorem enforces this.

Theorem 8 (T8):

Consider a relation instance r of R. Let X ->> Y be any minimal MVD defined on R, such that Z = (R - (X U Y)) ≠ ∅. A minimal MVD, which, along with X ->> Y, transitively derives another minimal MVD, cannot have a left-hand side which contains both Y and Z.

Proof: To prove this theorem, we will assume that we are given an MVD X ->> Y/Z and another MVD VYZ ->> A/B; LHS contains both Y and Z; and we will prove that we cannot transitively derive another minimal MVD from these two.

Using the transitive rule IR7, and taking into account that the LHS and the RHS are disjoint for minimal MVDs, the derived MVDs would be: VZX ->> A, VZX ->> B, VYX ->> A, VYX ->> B. Now, by the definition of minimal MVD, (X U Y U Z) = R. Since VYZ, A, and B, in the MVD VYZ ->> A/B, are all disjoint, A and B have to be proper subsets of X. This means, the LHSs and the RHSs of the MVDs, VZX ->> A, VZX ->> B, VYX ->> A, and VYX ->> B, cannot be disjoint without A and B being NULL. If A and B are NULL, we have a trivial MVD. This proves that we cannot transitively derive minimal MVDs if the LHS of an MVD contains the complete RHS of another MVD.

If the subset of either of the RHS attribute sets, Y or Z, exists in the LHS of another MVD, either in the mvd-list or the derived-mvd-list, we can derive two new MVDs transitively using IR7.

In the above discussion of transitive derivation, we have considered the current MVD as X ->> Y/Z. Similar to the case of FDs, we can consider the current MVD as WY ->> A/B or W'Z ->> A/B, and search the RHSs of the MVDs in the two lists to check if either Y or Z, belonging to the RHS of the MVD in the list, is contained in the LHS of the current MVD. Again, according to T8, the LHS will not contain both Y and Z. For each match, we can derive two new MVDs.

For each MVD we derive, we need to check if its LHS contains a key. If it does, then it becomes a trivial MVD and we can ignore it. While forming the new MVDs, we will use L2 to make the LHS and RHS disjoint. We will also make sure that, after forming the new MVD, say L ->> M/N, the condition N = (R - (L U M)) ≠ ∅ is satisfied. After this, we will check if the derived MVD already exists in the mvd-list or the derived-mvd-list. Similar cases were discussed for FDs in section 4.

If a subset of the LHS of the derived MVD exists in either of the two lists, then we can always derive this MVD using IR6. Hence, we will just ignore the derived MVD. If a subset of the derived MVD exists in the fd-list, then we will check if the RHS of that FD contains the RHS of the MVD. If it does, then the MVD becomes a trivial MVD and hence, we will ignore it. The derived MVD can recursively derive other MVDs. Hence, if it does not exist in the mvd-list, the derived-mvd-list, or the fd-list, we shall insert it in a queue called the mvd-queue. This queue will be processed later to check for recursive derivations. Before considering the derivations due to the union and the decomposition rules, we would like to state another theorem which will help us save a few steps.

Theorem 9 (T9):

Consider a relation instance r of schema R. Let X and Y be any two disjoint subsets of R and let X ->> Y/Z be an MVD defined on R such that Z = (R - (X U Y)) ≠ ∅. Let n be the total attributes in R. If X contains n - 2 attributes, then the MVD X ->> Y cannot participate in deriving other MVDs using the Union and Decomposition inference rules.

Proof: This theorem has a simple proof. If you look closely at the Union and the Decomposition inference rules, you will see that they just infer

MVDs with different RHS combinations and the same LHS. If the LHS of the parent MVD has (n - 2) attributes then we are left with just two attributes to form the two RHS attribute sets. If we put one attribute in each attribute set, then we come up with just one RHS combination. Since no other RHS combinations remain, no more MVDs with the same LHS can be derived.

We have so far discussed only the transitive derivations. We now discuss the derivations due to the union and decomposition inference rules, IR8 - IR11. If the MVD we want to insert is X ->> Y/Z, all four of these inference rules expect an MVD X ->> A/B to exist either in the mvd-list or the derived-mvd-list. This means, they expect to find an MVD whose LHS matches the LHS of the MVD we want to insert. Let us say that the main() function has discovered the MVD X ->> Y/Z, and before inserting it in the mvd-list, we would like to check if it derives other MVDs. Let us further say that while checking the mvd-list we found X ->> A/B. Now by the Union rule, IR8, we can derive: X ->> YA, X ->> YB, X ->> ZA, and X ->> ZB. Since X, Y, and Z are disjoint, and so are X, A, and B, none of the four RHSs, viz. YA, YB, ZA, or ZB will intersect X. There will however, be common attributes between Y and A, Z and B, etc., and we may very well wind up MVDs which are equal to the parent MVDs, or MVDs which have zero attributes in one of the RHS attribute sets. We will check for such MVDs and ignore them. Just like in the case of transitive derivations, we will check if the RHSs contain a key. In this case, we have copied the LHS (X) as-is from the parents, and hence, it will not contain any keys. If the RHSs contain a key, then the MVD becomes trivial, and we will ignore it. After this, we will check if the MVDs exist in the mvd-list, the derived-mvd-list, and the fd-list. These cases are handled in the same way as before. Finally we will insert the MVDs, which survive all these tests, into the mvd-queue. Before we go further, we will state another theorem, which will help us eliminate derivations resulting from the use of the three decomposition rules.

Theorem 10 (T10)

Consider a relation instance r of schema R. Consider X and Y as any two disjoint subsets of R. Let a minimal MVD X ->> Y/Z exist in r such that Z = R - (X U Y) ≠ ∅. Let M be another subset of R such that X and M are disjoint and M U Y ≠ Y. Let another minimal MVD X ->> M/N exist in r, such that N = R - (X U M) ≠ ∅. The three Decomposition inference rules applied to these MVDs will **each** infer the same set of MVDs as the Union inference rule applied to these MVDs.

Proof sketch: We are given two minimal MVDs and have to prove that IR8 to IR11 applied to these MVDs will each derive the same set of MVDs.

Let us assume the MVDs are: X ->> Y/Z and X ->> M/N.

IR8 will derive the following MVDs: X ->> YM, X ->> YN, X ->> ZM, and X ->> ZN. If we just consider the RHS, the output set will be {YM, YN, ZM, ZN}.

Similarly, the output set for IR9 will be: { Y ∩ M, Y ∩ N, Z ∩ M, Z ∩ N}, that for IR10 will be: { Y - M, Y - N, Z - M, Z - N}, and that for IR11 will be: {M - Y, N - Y, M - Z, N - Z}.

Let R = ABCDE.
Case 1: Let X = A, Y = BC, Z = DE, M = CD, and N = BE.
  YM = BCD/E, YN = BCE/D, ZM = CDE/B, ZN = BDE/C.
  Y ∩ M = C/BDE, Y ∩ N = B/CDE, Z ∩ M = D/BCE, Z ∩ N = E/BCD.
  Y - M = B/CDE, Y - N = C/BDE, Z - M = E/BCD, Z - N = D/BCE.
  M - Y = D/BCE, N - Y = E/BCD, M - Z = C/BDE, N - Z = B/CDE.
Case 2: Let X = AB, Y = CE, Z = D, M = E, and N = CD.
  YM = CE/D, YN = CDE/∅, ZM = DE/C, ZN = CD/E.
  Y ∩ M = E/CD, Y ∩ N = C/DE, Z ∩ M = ∅/CDE, Z ∩ N = D/CE.
  Y - M = C/DE, Y - N = E/CD, Z - M = D/CE, Z - N = ∅/CDE.
  M - Y = ∅/CDE, N - Y = D/CE, M - Z = E/CD, N - Z = C/DE.

Looking at case 1, we find that the same set of derived MVDs are produced in each of the four sub-cases. The same result is observed in case 2. In fact, if we take any such case, we will find that the Union and Decomposition inference rules will derive the same MVDs.

Using T10, we can eliminate checking for derivations using the decomposition rules IR9 to IR11. We have already considered the derivations using the Union rule in the above paragraph. These are all the MVDs that we would like to derive.

6. REALIZATION OF THE TOOL

This section describes the implementation of the algorithms discussed in sections 4 and 5. The prototype is implemented in the C language on a Data General (DG) MV/20000 minicomputer (which is rated to be 12 times faster than a DEC VAX 780) and uses the Oracle RDBMS. As mentioned in section 4, the FD algorithm uses the data dictionary to get the table names, the columns names and their data types. Since this information is unknown at compile time, the SQL query as well as the data-type binding is done at runtime. Most RDBMSs refer to this runtime binding method as "Dynamic SQL". The steps involved in embedding dynamic SQL in "C" are specific to the particular RDBMS. Oracle uses a eighteen step process for Dynamic SQL. Since this process is specific to Oracle and is well described in [24], we will not describe the details here.

The FD-Key algorithm has 5200 lines of code and the MVD-JD algorithm has 6300 lines, with a total of 11,500 lines, without the user interface.

6.1 The User Interface

We have mentioned at the beginning of this chapter that, one of the goals of this tool was to minimize user intervention. We wanted to automate the discovery process as much as possible. In line with this goal, we did not concentrate much on the user interface. The username and password are normally entered from the command line when the program is executed. The status line displays messages like "Connecting to the database", "Selecting table names from the data dictionary", "Searching for FDs in table AAA", "FD: X -> Y found in table AAA", and so on. The error line displays any errors that are encountered. It displays the error number as well as a short description.

6.2 Example

In this section we will show a sample table, run it through our prototype, and analyze the results. Let us consider the following table:

PROPERTYID	COUNTY	LOTNUMBER	AREA	PRICE	TAXRATE
1	Maricopa	800	2000	80000	5
2	Maricopa	602	2000	80000	5
3	Maricopa	803	2200	85000	5
4	Maricopa	804	2200	85000	5
201	Yavapai	601	3000	88000	4
202	Yavapai	602	3000	88000	4
203	Yavapai	603	3500	95000	4
204	Yavapai	604	3500	95000	4

Fig. 5 Property table

This table was designed with the following dependencies in mind:

Candidate Keys: PropertyId
(County, Lotnumber)
FDs: County -> TaxRate
Area -> Price
Area -> County

Let us use this table as input to our FD-Key algorithm. The algorithm will find the table name "PROPERTY" in the data dictionary. It will then get the column names and their data types from the data dictionary. The algorithm will select the column name "PropertyId" as the LHS and then formulate the following query:

select * from Property where (PropertyId) in
　　(select PropertyId from Property
　　　　group by PropertyId having count(*) > 1);

This query will not return any rows, signifying that PropertyId is a key. PropertyId will be inserted in the key-list. As this is the first key and there are no FDs discovered so far, the check for derived keys will be unsuccessful. The algorithm will then select "County" as the next LHS and will run the same query as above with the attribute "County" substituted for "PropertyId". This query will return the following results:

PROPERTYID	COUNTY	LOTNUMBER	AREA	PRICE	TAXRATE
1	Maricopa	800	2000	80000	5
2	Maricopa	602	2000	80000	5
3	Maricopa	803	2200	85000	5
4	Maricopa	804	2200	85000	5
201	Yavapai	601	3000	88000	4
202	Yavapai	602	3000	88000	4
203	Yavapai	603	3500	95000	4
204	Yavapai	604	3500	95000	4

Fig. 6 Result of query with County as the LHS

The FD algorithm will find the FD County -> TaxRate, and will insert it in the fd-list. It will then select "Lotnumber" as the next LHS. The query using Lotnumber will return the following results:

PROPERTYID	COUNTY	LOTNUMBER	AREA	PRICE	TAXRATE
2	Maricopa	602	2000	80000	5
202	Yavapai	602	3000	88000	4

Fig. 7 Result of query with LOTNUMBER as the LHS

As you can see, there are no FDs or Keys in the above subset. The next LHS will be "Area". The query will select the following subset:

PROPERTYID	COUNTY	LOTNUMBER	AREA	PRICE	TAXRATE
1	Maricopa	800	2000	80000	5
2	Maricopa	602	2000	80000	5
3	Maricopa	803	2200	85000	5
4	Maricopa	804	2200	85000	5
201	Yavapai	601	3000	88000	4
202	Yavapai	602	3000	88000	4
203	Yavapai	603	3500	95000	4
204	Yavapai	604	3500	95000	4

Fig. 8 Result of query with AREA as the LHS

The algorithm will find the FD Area -> County. It will derive the FD Area -> TaxRate from Area -> County and, County -> TaxRate which is already in the fd-list. The algorithm will insert Area -> TaxRate in the FD-queue and then insert Area -> County in the fd-list. It will then pull Area -> TaxRate from the FD-queue. Since Area -> TaxRate does not derive any further FDs, it will be inserted into the derived-fd-list. The algorithm will then find the FD Area -> Price from the returned subset; remember that the algorithm is in a loop checking every RHS for a possible FD, and Price is the next RHS it finds after County. Since Area -> Price does not derive any other FDs, it will be inserted in the fd-list. The algorithm will then find the FD Area -> TaxRate from the subset. Since this FD already exists in the derived-fd-list, the algorithm will try to insert the second set of parents in the parent-list of Area -> TaxRate which is in the derived-fd-list; in this case the discovered FD has parents set to NULL and hence, there is no second set of parents for this case. Before looking at the next LHS, let us see what the fd-list and the key-list contain so far.

FD-List:	County -> TaxRate
	Area -> County
	Area -> Price
Key-List:	PropertyId
Dervd-FD-List:	Area -> TaxRate

The next LHS will be Price. The query using Price as the LHS will return the rows shown in fig. 9. The first FD the algorithm will find will be Price -> County. This will derive the FD Price -> TaxRate. The algorithm will put Price -> TaxRate in the FD-Queue and check if Price -> County derives any other FDs. In this case, it derives a second FD Area -> County. Area -> County exists in the fd-list. The algorithm has to see if an alternate path to derive Area -> County exists and so it invokes the AND-OR graph search algorithm we described in the previous section. The algorithm will push (Area -> Price, NULL) as an AND node on the stack. It will then call search_graph with Price -> County as the argument. Price -> County was just discovered from the subset. It has no parents and is waiting to be inserted in the fd-list. It is also not marked.

PROPERTYID	COUNTY	LOTNUMBER	AREA	PRICE	TAXRATE
1	Maricopa	800	2000	80000	5
2	Maricopa	602	2000	80000	5
3	Maricopa	803	2200	85000	5
4	Maricopa	804	2200	85000	5
201	Yavapai	601	3000	88000	4
202	Yavapai	602	3000	88000	4
203	Yavapai	603	3500	95000	4
204	Yavapai	604	3500	95000	4

Fig. 9 Result of query with Price as the LHS

This means we will hit the success step in the search_graph algorithm and will pop the next AND node, which in this case happens to be Area -> Price. Area -> Price is in the fd-list and is not marked. It also happens to be the last node on the stack. The search_graph algorithm will thus return a success and our FD-Key algorithm will now remove Area -> County from the fd-list and insert it into the derived-fd-list. It will then insert Price -> County into the fd-list. The next thing it will do is pull Price -> TaxRate from the FD-queue and check if it derives any other FDs. Price -> TaxRate derives Area -> TaxRate which happens to exist in the derived-fd-list. The algorithm will insert a second set of parents for Area ->TaxRate and then insert Price -> TaxRate in the derived-fd-list. The algorithm keeps going for this query and then for other LHSs and so on. We will stop our analysis at this point and see what the three lists contain.

FD-List:	County -> TaxRate	Dervd-Fd-List:	Area -> TaxRate
	Area -> Price		Area -> County
	Price -> County		Price -> TaxRate
		Key-List:	PropertyId

Since we have not found any composite keys as yet, the derived-key-list is still empty. The algorithm will finally produce the following output:

Candidate Keys:	PropertyId
	(County, Lotnumber)
Derived Keys:	**(Lotnumber, Price)**
	(Lotnumber, TaxRate)
	(Lotnumber, Area)
FDs:	County -> TaxRate
	Area -> Price
	Price -> County
	Price -> Area
	TaxRate -> County
Derived FDs:	Area -> TaxRate
	Area -> County
	Price -> TaxRate
MVDs:	none

The candidate keys discovered by the algorithm are the same as what we had intended, but in addition to them the algorithm found three other derived keys. Now, the data implies these keys and we will leave it up to the application domain expert to decide whether these keys will really hold after more data is added to the table.

The algorithm produces interesting results for the FDs in this case. It tells us that only the first two FDs from our original design are part of the minimal cover. It also finds three other FDs to be part of the minimal cover and says that the third FD in our design, viz. Area -> County, can be derived from the minimal cover.

The algorithm does not find any MVDs as this table contains a simple key. In the next paragraph we will look at the performance of the algorithm.

6.3 Performance Analysis

The example shown in section 6.2 took 8 seconds to run. There were over 100 other processes running on the DG computer. Fig. 10 shows the results of a series of experiments we ran to analyze the performance of this algorithm.

# of Rows	# of Columns	Time spent	Comment
3413	4	3:41	Non-indexed
3413	4	3:05	Indexed
4413	4	3:42	Mixed
5413	4	3:46	Mixed
5413	5	10:43	Mixed
5413	6	23:22	Mixed

Note: Mixed means some columns were indexed.
Fig. 10. Performance Data

The first experiment, shown in the first two rows, was done keeping the rows and columns constant and running the algorithm with the columns indexed, and then after dropping the indexes. We did not see a drastic change in performance.

In the next set of experiments, shown in rows three through five, we changed the number of rows, and again did not see a drastic change in performance.

In the last set of experiments, shown in rows four through six of the above table, the number of columns in the table were increased. This produced a drastic change in the performance.

From the above experiments we can conclude that the performance depends on the following factors (the factors are shown in the order of their effect on the algorithm performance):

- Number of columns in the table.
- Number of rows in the table.
- If the column combination for the query is indexed or not.

We can provide an estimation of the complexity of the algorithms. Since disk access is much more expensive than computations in memory, we will ignore the performance cost of those parts of the algorithm that execute in RAM (Random Access Memory). With this assumption and by considering the number of attribute combinations for the algorithms (as discussed in sections 4 and 5), we come up with the following <u>worst-case</u> results.

For a table with n columns:

- FD-Key algorithm: $2^n - 1$ queries
- MVD algorithm: $\sum (n! / (r! * (n - r)!))$
  where the summation is over $r = 1$ to $r = n - 2$.

If the table has m rows and there are no indexes, then for each query we have to do a sequential search through the m rows. Actually a query plan for the query shown in section 6.2 looks as follows:

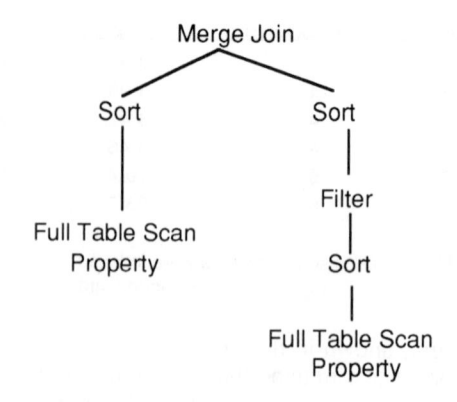

```
                        Merge Join
                       /          \
                   Sort            Sort
                    |               |
                    |              Filter
                    |               |
              Full Table Scan      Sort
                 Property           |
                                    |
                              Full Table Scan
                                 Property
```

Fig. 11 Query Plan

In the above diagram, "Filter" is used for restricting the number of rows returned from a table. A *merge join* is performed by merging two sorted sets of operands. If a query does not use an index, which is the worst case scenario we are considering, the DBMS must perform a full table scan to execute the query. A full table scan involves reading all rows of a table sequentially and then examining each row to determine whether it meets the criteria of the query's WHERE clause. To perform a full table scan a DBMS must read every disk block that belongs to this table. Some of the newer DBMS versions do a multi-block read. For our analysis, we consider single-block reads, which means the DBMS has to access the disk for reading each block. Since we are interested in the disk accesses, we ignore the cost of sorts, filter, and the merge-join operations, since they are normally done in the main memory. If a table has m rows and the size of the disk block is "s", we will have (2m / s) disk accesses for each query. Multiplying this factor by the number of queries, we still get an exponential worst-case performance cost.

By using the refinements described in sections 4 and 5 we eliminate up to sixty percent of the query combinations. The elimination of combinations depends on how quickly we can find a key or an FD. If a 10 column table has all keys with more than 6 attributes and the LHS of every FD for this table has more than 5 attributes, then we will waste time checking for all one, two, three, and four attribute combinations. But such heuristics are not known to us at the beginning of the algorithm.

By performing experiments, similar to those described above, we saw a severe performance degradation when the number of columns increased to more than 10 and the number of rows (non-indexed columns) was above 50,000. Hence, these can be considered as the upper bounds for this tool from a performance standpoint.

7. CONCLUSION

Relational database theory proclaims that the key to a good relational database design is the thorough identification of the underlying data dependencies. However, defining the semantics of the data elements is not a trivial task. In addition, with the exception of some newer versions, commercial relational database management systems do not make any attempt to uphold the prescribed dependencies.

Most database designers define the primary key but overlook other candidate keys which might exist. One of the reasons for this, is that, most RDBMSs use a unique index to represent keys and have no means to differentiate the primary key from other candidate keys. This tool highlights all such candidate keys which might have been neglected by the designer and leaves it upto the application domain exert to make the final decision.

The tool discovers FDs, Keys, MVDs, and JDs from large databases. While our algorithms are simple, they make good use of the methodology of rule induction from examples. Previous work was limited to small data-sets in main memory. For example, in [16] the authors state that there is no simple algorithm to find a minimal cover for a set of FDs. Our tool not only does this but also finds a minimal cover for a set of candidate keys. Unlike the previous methods presented in section 2, this tool does not depend on any user input, and once started, do not need any user intervention. In addition, many authors have admitted that it is very difficult to discover MVDs and JDs from existing databases. From the discussions in chapters four through six, and the complexity of the algorithms, we can attest to this statement.

Thus far, we have not considered noise, and as such we reject a dependency even if we find only one exception in the entire database. Confidence factors or weight measures based on number of positive examples can be attached to the dependencies as a way of handling noise. Once a suitable way of handling noise is developed, one can look at incremental learning to take into account the dynamic nature of databases.

Once the dependencies between the various data elements are identified, one can look at discovering other knowledge about the enterprise model. One such interesting problem would be checking for referential integrity.

REFERENCES

1. Y. Cai, N. Cercone, and J. Han, "Learning Characteristics Rules fromRelational Databases", In Proceedings of the International Symposium on Computational Intelligence, Milan, Italy, Sept., 1989, 187-196.
2. Y. Cai, N. Cercone, and J. Han, "An Attribute-Oriented Approach for Learning Classification Rules from Relational Databases", In Proceedings of the 6th International Conference on Data Engineering, Los Angeles, California, Feb., 1990, 281-288.
3. Y. Cai, N. Cercone, and J. Han, "Attribute-Oriented Induction in Relational Databases", In Knowledge Discovery in Databases, ed. by G. Piatetsky-Shapiro and W. Frawley, AAAI Press/MIT Press, 1991, 213-228.
4. J. Han, Y. Cai, and N. Cercone, "Discovery of Quantitative Rules from Large Databases", In Proceedings of the 5th International Symposium on

Methodologies for Intelligent Systems, Knoxville, Tennessee, Oct., 1990, 157-165.
5. G. Piatetsky-Shapiro, "Discovery, Analysis, and Presentation of Strong Rules", In Knowledge Discovery in Databases, ed. by G. Piatetsky-Shapiro and W. Frawley, AAAI Press/MIT Press, 1991, 229-248.
6. M. Kantola, H. Mannila, K. Raiha, and H. Siirtola, "Discovering Functional and Inclusion Dependencies in Relational Databases", In International Journal of Intelligent Systems, Vol. 7, 1992, 591-607.
7. H. Mannila and K. Raiha, "Design by Example: An Application of Armstrong Relations", In Journal of Computer and System Sciences, Vol. 33, No. 2, 1986, 126-141.
8. D. Bitton, J. Millman, S. Torgersen, "A Feasibility and Performance Study of Dependency Inference", In Proceedings of the 5th International Conference on Data Engineering, Los Angeles, California, Feb., 1989, 635-641.
9. C. Date and R. Fagin, "Simple Conditions for Guaranteeing Higher Normal Forms in Relational Databases", In ACM Transactions on Database Systems, Vol. 17, No. 3, Sept., 1992, 465-476.
10. R. Yasdi, "Learning Classification Rules from Database in the Context of Knowledge Acquisition and Representation", IEEE Transactions on Knowledge and Data Engineering, Vol. 3, No. 3, Sept., 1991, 293-306.
11. H. Mannila and K. Raiha, "Practical Algorithms for Finding Prime Attributes and Testing Normal Forms", In Proceedings of the 8th ACM SIGACT-SIGMOD-SIGART Symposium on Principles of Database Systems, (PODS'89), 1989, 128-133.
12. H. Mannila and K. Raiha, "Dependency Inference", In Proceedings of the 13th International Conference on Very Large Databases, Brighton, Sept., 1987, 155-158.
13. T. Bagchi, V. Rao Baratam, and S. Saha, "Dependency Inference Algorithms for Relational Database Design", In Computers in Industry, Vol. 14, Elsevier Science Publishers B.V., 1990, 319-350.
14. H. Mannila and K. Raiha, "Small Armstrong Relations for Database Design", In Proceedings of the 4th ACM SIGACT-SIGMOD-SIGART Symposium on Principles of Database Systems (PODS'85), Portland, Oregon, Mar., 1985, 245-250.
15. S. Kundu, "An Improved Algorithm for Finding a Key of a Relation", In Proceedings of the 4th ACM SIGACT-SIGMOD-SIGART Symposium on Principles of Database Systems (PODS'85), Portland, Oregon, Mar., 1985, 189-192.
16. R. Elmasri and S. Navathe, Fundamentals of Database Systems, The Benjamin/Cummings Publishing Co, Inc., 1989.
17. J. Ullman, Principles of Database and Knowledge-Base Systems, Vol. I, Computer Science Press, Inc., 1988.
18. D. Maier, The Theory of Relational Databases, Computer Science Press, Inc., 1983.
19. R. Michalski, J. Carbonell, and T. Mitchell, Machine Learning An Artificial Intelligence Approach, Vol. I, Morgan Kaufmann Publishers, Inc., 1983.

20. R. Michalski, J. Carbonell, T. Mitchell, Machine Learning - An Artificial Intelligence Approach, Vol. II, Morgan Kaufmann Publish
21. G. Piatetsky-Shapiro and W. J. Frawley, Knowledge Discovery in Databases, AAAI Press, 1991.
22. K. Fukunaga, Introduction to Statistical Pattern Recognition, Academic Press, 1990.
23. V. Vemuri, Artificial Neural Networks: Theoretical Concepts, IEEE Computer Society Press, 1988.
24. Oracle Corp. - Pro*C Supplement to the Oracle Precompilers Guide, Ver. 1.3, 1989.
25. C. J. Matheus, P. K. Chan, and G. Piatetsky-Shapiro, "Systems for Knowledge Discovery in Databases", In IEEE Transactions on Knowledge and Data Engineering, Vol. 5, No. 6, Dec., 1993, 903 - 913.

A CASE-BASED REASONING (CBR) TOOL TO ASSIST AIR TRAFFIC FLOW

Dr. Bikas Das
The MITRE Corporation
7525 Colshire Drive
McLean, Virginia 22102-3481

and

Scott Bayles
The MITRE Corporation
7525 Colshire Drive
McLean, Virginia 22102-3481

Abstract

This paper documents the results of a study performed under a research project entitled "Using Artificial Intelligence to Support Traffic Flow Management (TFM) Problem Resolution." As part of this research, case-based reasoning technologies where exploited to capture and analyze previous experiences of TFM specialists at the Air Traffic Control System Command Center (ATCSCC). These experiences were then used to help guide decisions on how to solve current TFM problems. As specialists ponder a problem that is predicted to occur in the next few hours, they can access various levels of data recorded from experiences relevant to the current problem. The specialists can then see how the previous problems are similar to the current (or predicted) situation and how those situations were resolved.

This paper describes how we determined that artificial intelligence, and case-based reasoning in particular, can be applied in a useful manner to the TFM domain. A prototype was developed, and the knowledge contained therein was successfully reviewed by many TFM specialists. Our scenario concentrated on severe weather in, and en route to, the New York Air Route Traffic Control Center area during the summer and its effect on the entire National Airspace System (NAS).

1. Introduction

The Federal Aviation Administration (FAA) is in the process of defining and documenting the operations concept and architecture for the future Traffic Flow Management (TFM) system. The challenge of future TFM is to organize complex air traffic flows through busy areas in the National Airspace System (NAS), manage the volume of traffic into and out of congested airport areas, and minimize delay related problems in the advent of continued growth of air traffic and its complexity. The complex and adaptive behavior of the NAS is the result of many, often simultaneous decisions that are made by controllers located at local Air Traffic Control (ATC) facilities, traffic flow managers, both at local facilities and at a national facility dedicated to system wide concern, and NAS users (e.g., airlines).

In this paper we will focus on a particular class of decision making that relies on past experiences. We have adopted a Case-Based Reasoning (CBR) approach [1] to recognize "similar" problems and to guide our decision making by reasoning about past situations. The paper will begin with a brief description on the background of the problem. Next, the suitability of the CBR paradigm will be argued. The problem scenario will be described in some detail along with an overview of our CBR prototype. Our initial results will be summarized as lessons learned. The paper will conclude after proposing several research topics for possible future work.

2. Background

Many TFM problems that occur in the NAS repeat themselves in a similar manner on a fairly regular basis. Much of the information that describes these problems has been recorded. By using historic data, problem patterns (or cases) can be recognized and stored in a case database along with the actions that were taken to resolve the problems. TFM problems are generally created whenever the capacity of the system components (airports, runways, navigational equipment, etc.) is reduced and can not meet existing demand (such as, flight schedules). External events, for example weather or equipment failures, often cause this imbalance of demand and capacity (i.e., closing of runways).

The purpose of this project is to determine whether CBR can be useful in supporting the decision making associated with TFM problem resolution. Recently the TFM Architecture and Requirements Team (TFM-ART), a panel of representatives from various branches of the FAA, emphasized the need for developing an historic database of TFM situations and matching of their respective patterns. Our work will test some of the concepts discussed by TFM-ART.

The FAA currently has 20 Air Route Traffic Control Centers (ARTCCs) in the contiguous United States, responsible for varying volumes of airspace. Figure 1 is an illustration of the various ARTCCs that cover the United States and highlights the New York ARTCC (ZNY), which will be focused on throughout this paper. Any TFM problem that can not be handled internally by an ARTCC, or that is of interest to other parties nationally (including the airlines), is coordinated through the Air Traffic Control System Command Center (ATCSCC).[1] FAA personnel typically rotate on 2-3 year tours of duty in various facilities, with only a few individuals staying permanent in one location. Through these changes in duty and normal retirement cycles, valuable job experience is lost because currently there is no way of formally capturing it. This experience is critical when dealing with problems that often occur on a day-to-day basis.

Obviously, when dealing with the dynamics of weather and air traffic operations, no two days are ever exactly the same. Our goal is to generalize some events into similar "type" problems. Given the fact that these problems are regularly occurring, we can use previous experiences and solution methods to help resolve the current problem. Through the course of this project, we focused on problems with "national" traffic implications, as

---

[1] ATCSCC was known previously as the Central Flow Control Facility.

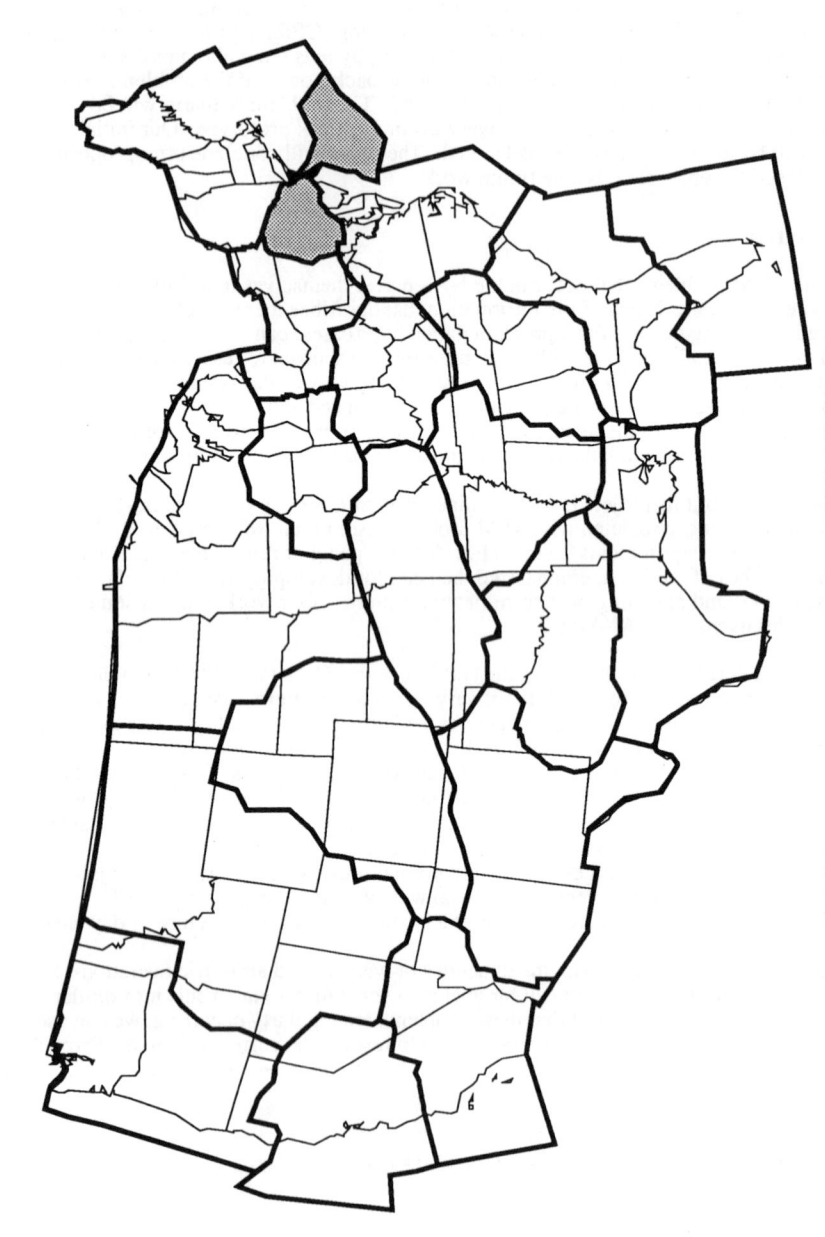

Figure 1. ARTCCs in the Conterminous United States (ZNY is highlighted)

opposed to "local" implications. Problems with national implications are most suitably handled at the ATCSCC, while local problems are dealt with by the local ARTCCs. The example we use in our scenario is severe summer thunderstorms in the New York ARTCC which impacts traffic nationwide due to the volume of airplanes and the constraints of the airspace in this area. The New York ARTCC (ZNY) is highlighted in Figure 1. These storms happen often in the months of June–September, and are similar in nature, although never exactly the same. The CBR approach is proposed to capture previous experiences in dealing with these TFM problems and to use those experiences to support the current decision-making process.

There are a number of groups in the FAA working on several decision-support tools (i.e., decision generation, optimization, visualization and simulation of the decision process) that address various aspects of the TFM problem. Our efforts were combined with other efforts in order to complement them, prevent duplication, and provide mutual benefits. This exposed us to another set of experts, and different approaches to the information management of the TFM problem.

3. Problem Analysis : The CBR Approach

CBR is a field of AI that has gained tremendous popularity in recent years. In CBR, a reasoner remembers previous situations similar to the current one and uses them to help solve the new problem. The number of applications using CBR has increased rapidly in recent years and has proved to be effective both for experimental and applied systems [2]. CBR is applicable to a decision domain if it has the following characteristics:

- Historical data is available in the domain
- The domain can not be formalized with rules or formalization requires too many rules
- The domain is already precedent based
- The application area evolves or changes over time requiring frequent updates to handle new problem types

After extensive analysis of the TFM decision domain, we found that all of the above criteria were met. In TFM, a traffic situation is usually triggered by seasonal weather patterns that are repetitive and require precedent based problem solving. The experience-base is continually updated and refined as new cases occur. Although the weather patterns may not change much, the complexity of traffic patterns will increase in the future with the ever increasing demand for more flights. Thus this application will evolve and our case-base will provide the framework for a learning paradigm. A detailed description of the domain is provided in Section 4.0.

Rule-Based Reasoning (RBR) is the traditional reasoning model used in expert systems. This model embodies an expert's knowledge in the form of domain specific rules about problem solving. The TFM domain is very dynamic and hard to formalize.

Even for an experienced specialist, rules to solve the problem appear to be complicated and hard to explicate. Often, rules are too numerous. RBR systems are suitable for those domains where rules are readily available and are relatively stable with time. Within the FAA there are efforts to research RBR approaches in the TFM domain. While it is possible that rules can be explicated and then used to model certain aspects of the domain, RBR alone may not be appropriate for the entire domain. A mixed paradigm approach, that is, a combination of CBR and RBR methods has been found effective and complementary to each other for several applications [3, 4].

4. The CBR Prototype Overview

To demonstrate the feasibility of applying CBR techniques to TFM, we chose a scenario representative of the difficulty involved in resolving TFM problems. We started with the premise that we could only deal with one type of problem in this initial prototype due to the dynamics of the domain. This problem also had to be somewhat repeatable, allowing us to capture many cases. Finally, this problem had to have nationwide repercussions, or system-wide effects. It was decided that severe weather (e.g. summer thunderstorms) in the New York ARTCC (ZNY) area would fit these constraints. Hence the scenario consists of severe weather (summer thunderstorms) moving across the north eastern part of the country, (Figure 2) thus impacting ZNY and consequently the entire NAS.[2] To expedite the development of the proof-of-concept prototype, we have chosen a commercial-of-the-self (COTS) tool called Remind [5, 6] (Remind™ is the concentrated registered trademark of Cognitive Systems, Inc.). Some of the tool's capabilities will be apparent as we describe case representation and retrieval mechanisms.

4.1 Knowledge Engineering

Over the last 12 months, we interviewed and observed TFM experts in the field. The experts in this domain are the TFM specialists who work in the ATCSCC, and the Traffic Flow Management Coordinators who work in the Traffic Management Units (TMUs) of the ARTCCs[3]. Over 40 TFM specialists including both supervisors and staff, with varying levels of experience and backgrounds, participated in our knowledge engineering activities. While getting the general idea of the details of their job, we solicited their ideas on useful capabilities in a CBR tool.

Once we were confident that we understood the TFM operations, we began to focus on specifics that needed to be included in our prototype. During this period we concentrated our knowledge engineering sessions on a few TFM specialists based on their experiences with the New York airspace and dealing with thunderstorm related problems.

---

[2] Because of the linkage of the traffic flow on the north east of the country.

[3] For simplicity TFM specialists at the ATCSCC, and Traffic Flow Managment Coordinators at the ARTCC are both referred to as TFM specialists in this paper.

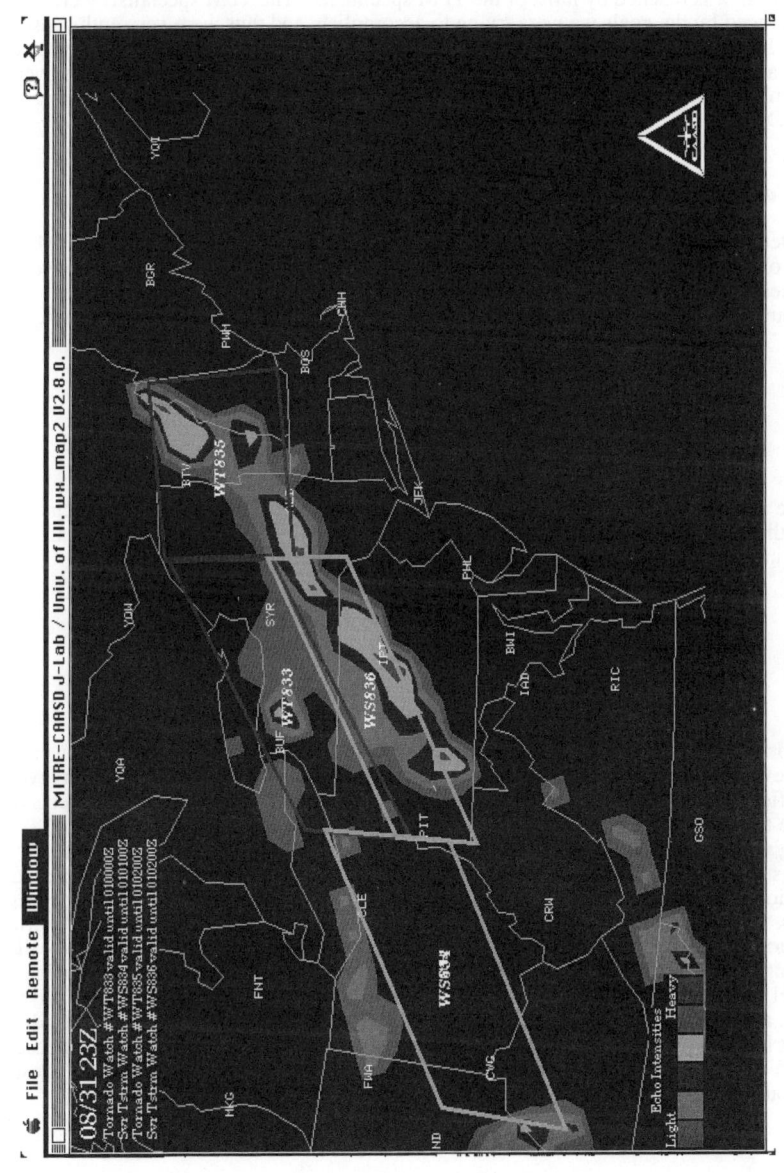

Figure 2. A Line of Thunderstorms Moving Through ZNY and Neighboring Centers

The process of refining the attributes that defined these problems was iterative until a consensus was reached by most of the TFM specialists. The TFM specialists were very encouraged by the goals we were trying to accomplish, and thus were very enthusiastic about helping us. Both the TFM specialists in the ATCSCC, and the ZNY have been an integral part of the design of this prototype, and have truly been included since the very beginning. We spent approximately 200 hours at the ATCSCC, and 50 hours in the ARTCCs.

4.2 Case Representation

Determining appropriate case granularities for the problem at hand, that is, at what level of detail the case must be represented consumed a great deal of effort. Knowledge engineering sessions were of considerable help in this matter. The following five attributes are the primary discriminators and best represent the developing problem.

- Density of thunderstorm
- Jet routes affected
- Major airports involved
- Type of problem
- Day of the week

These attributes relate to probable effects due to predicted weather. A TFM specialist uses these key features to retrieve situations from the past. Secondary attributes are filled in as information becomes available and include the following:

- Duration
- Push
- Demand
- Arrival rates
- Runway configuration

The primary and secondary attributes are used in indexing cases. The implications of this two level representation of the episode are discussed in the Section 5.0. The first level signifies effects due to the weather event, while the second represents mostly effects due to the traffic in addition to weather. A typical example of a case is shown in Table 1 along with the data type (i.e., symbol, integer, text, etc.) for each attribute. A case consists of other attributes besides those in the two categories. These attributes provide additional information to the user about a case and are not necessarily used in indexing cases. For example, the attributes resolution-arrivals and resolution-departures describe solutions or resolution methods. The case outcome is represented by the attribute called resolution performance. However, current understanding of the system is not sufficient to model this attribute appropriately. Therefore, we have used qualitative indicators like good, average, and bad as values for the resolution-performance attribute. A great deal of effort has been spent in other projects trying to determine the best performance metrics. When conclusive results are obtained from these projects, we can easily insert the results into our case representation.

Table 1. Typical Example of a Case

	Field Name	Field Value	Field Type
S	arrival rate for program	48	Integer
S	arrival rates during program	60, 49, 37, 28	List of Integer
	ARTCCs included in program	ALL	List of Symbol
	date	9/3/93 19:00:00	Date
S	day of the week	Friday	Symbol
S	demand during program	60, 52, 24, 14	List of Integer
P	density of thunderstorms	solid line	Symbol
	departure rates during program	24, 28, 28, 37	List of Integer
S	duration of problem	4	Integer
P	jet routes affected	64, 60, 6, 75, 80	List of Integer
P	major airport involved	JFK	Symbol
	max depart. delay during pgm	210	Integer
	other pgms affecting this area	SWAP, ground stops	List of Symbol
	resolution performance	average	Symbol
	resolution-arrivals	1500–implement pgm with 48 rate...	Text
	resolution-departures	2000–West gates closed, reroute ...	Text
S	runway configuration	arr-22L, depart-22L	Text
	speed of thunderstorms	n/a	Integer
	time of day (Z)	1900	Integer
P	type of problem	enroute	Symbol
	weather description	line of tstms from Jamestown	Text
S	where is push coming from?	n/a	Symbol
	winds at airport	14	Integer

P = primary attributes
S = possible secondary attributes

4.3 Case Retrieval

We have explored various retrieval options that Remind provides. Currently there are thirty cases in the case-base. At this time, the nearest-neighbor method of retrieval appears to be the most suitable. In this method, attributes are appropriately weighted according to their relative importance and a scoring scheme is used to index cases. A linear search is employed to find top scoring cases. The retrieved case can be compared against the new case in the case-comparison window (see Table 2). The similarity number is an estimate of how close the case is to the new one and falls in the range of 1 to 100.

Table 2. Comparison of a Retrieved Case with the Input Case

Field Name	Input Case Field Value	Retrieved Case Field Value
arrival rate for program	43	48
arrival rates during program	n/a	60, 49, 37, 28
ARTCCs included in program	n/a	ALL
date	11/11/1993 22:00:00	9/3/93 19:00:00
day of the week	Thursday	Friday
demand during pgm	n/a	60, 52, 24, 14
density of thunderstorms	solid line	solid line
departure rates during program	n/a	24, 28, 28, 37
duration of problem	3	4
jet routes affected	80, 75, 70	64, 60, 6, 75, 80
major airport involved	JFK	JFK
max depart. delay during pgm	n/a	210
other pgms affecting this area	n/a	SWAP, Ground Stops
resolution performance	n/a	average
resolution-arrivals	n/a	1500- implement pgm with 48 rat
resolution-departures	n/a	lack of departure flexibility causing
runway configuration	n/a	arr.-22L, depart-22L
speed of thunderstorms	n/a	n/a
time of day (Z)	2200	1900
type of problem	enroute	enroute
weather description	line of tstms from Jamestown, NY	line of tstms from Jamestown, NY
where is push coming from?	n/a	n/a
winds at airport	n/a	14

Input Case, id: = 27   Retrieved Case, id: = 23

SIMILARITY: 79.17

In our prototype, a case is described by a set of features or attributes leading to an outcome called resolution performance. When this outcome is well defined, inductive indexing is the best method to use to index cases. It is desirable to have many examples of each type of outcome value. In Remind, the inductive algorithm examines all the features to determine which ones are most responsible for the difference in outcome values. Then it partitions the case population successively into two homogeneous groups called clusters, for each of those discriminating features. The clusters are examples of cases with similar outcome values. Results of this clustering process is a binary decision tree containing "splits" (e.g., decision points in the tree). The discriminating features according to their relative importance are the driving force for splitting cases. The main strength of induction is that it can objectively and rigorously analyze cases to determine the best features for distinguishing between outcome values, and use those case features to build an index that will be used for case retrievals. Inductive retrievals are efficient because of the hierarchical structure of the tree, and the retrieval time only increases by the log of the number of cases, rather than linearly. Such an approach requires a well defined outcome and a large number of representative cases.

There is yet another indexing mechanism called knowledge-guided induction, which is a combination of pure induction and knowledge-based indexing. Pure induction derives rules for determining discriminating attributes based on the past experiences represented by cases themselves, but does not take advantage of the domain specific knowledge stored inside an expert's head. Domain specific knowledge is codified in the Remind tool through the use of a qualitative model.

5. Lessons Learned

Knowledge engineering is traditionally a very painstaking task. However we feel that the knowledge engineering process in CBR reduces the time and cost as compared to that of RBR systems. The "experts" are more at ease describing experiences and situations as opposed to exact rules. CBR also captures negative or contradictory cases with no additional effort. It is relatively easy to use the same case-base to arrive at different outcomes just by changing the index structure. Currently we have a concise and simple set of attributes. Most of them can be captured using existing information. However, some attributes are difficult to capture (i.e., what crew was on duty, what significant interactions did various aircraft have with controllers before they reached their destinations) and are currently not represented in the prototype.

Retrieval of cases using "surface features" [7] can be misleading and often requires validation. The surface features are those that are readily available in the existing information about the operation and often are the obvious attributes of the problem at hand. However there are certain features that are more discriminating than the surface features and could be important in indexing cases. As an illustration, use of surface features may retrieve a plastic hammer from the case-base of hammers unless a more discriminating feature such as "hammer's ability to drive a nail" is used in indexing cases. Unfortunately those features are not readily available and need to be derived. This requires a complete model about the operation, which is often not present and almost impossible to have. However one can make some progress by building an approximate model using surface features which usually takes either the form of a qualitative model (i.e., causal relationship among features and their qualitative contribution to the outcome) [8, 9] or the form of rules. In the TFM situation a human is in the loop, and retrieved cases are presented to the human who validates them. Using a casual model for the effects of weather and traffic leading to the resolution, we have developed a framework to further explore knowledge-based indexing of cases (Figure 3) by using Remind's knowledge-guided induction mechanism.

Modeling the resolution performance itself is the hardest task. Identification and characterization of all the measures of resolution performance will be challenging work. Currently resolution performance is represented fairly subjectively by three values: good, bad, and average. As more cases are added to the case-base, these resolution performance values can be tested against the actual performance of new cases. Thus it will open up a more objective analyses of this attribute. Then an appropriate clustering mechanism may provide information about which attributes are most discriminating for this outcome.

Our prototype will be useful for a TFM specialist who is inexperienced in this particular problem domain, but has overall exposure to TFM operations. As described earlier, initial retrievals of cases are made using the primary set of attributes (i.e., weather related attributes). These cases may cover a variety of traffic patterns based on arrival/departure push, and capacity-demand profile. Examination of the presented cases opens up a range of opportunities that a specialist may take advantage of in manipulating traffic (e.g., traffic may be heavy in the west corridor but light in the south, where re-routing traffic to the south is a good idea). However the specialists can always perform a secondary retrieval when they obtain traffic pattern information for this new case, or whenever information about all other attributes becomes available. It is our observation that the specialists do not like canned solutions but prefer textual and animated replays of similar problems.

Finally, we have learned that the specialists have different philosophies on moving traffic, mostly determined by level of experience. However the predominant opinion is to push traffic until it cannot go anymore. For example, if thunderstorms are just forming, the specialist can push the traffic without implementing a ground delay program, until the pilots start complaining that the weather is too severe.

203

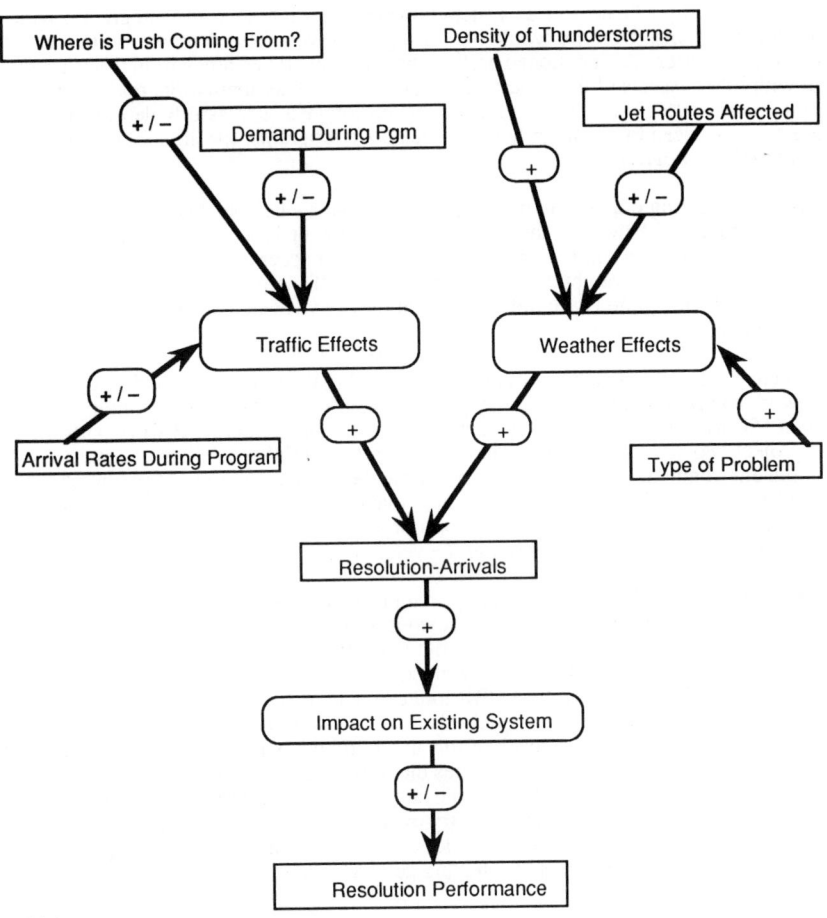

Links
- +   Directly Proportional Relationship
- −   Inversely Proportional Relationship (not shown in Figure 3-1)
- +/−   Unspecific Relationship

Figure 3. A Model of Casual Relationships of Attributes to be Used in Indexing

6. Future Work

The construction of the current prototype has accomplished most of our initial goals. However, it is still a "bare bones" system lacking a user interface and integration with other relevant tools. Here we would like to suggest several research possibilities that this prototype can offer. As mentioned earlier, there are several efforts within the FAA that deal with other aspects of the general TFM problem such as simulation, optimization, and evaluation of strategies. In order to make this prototype a successfully deployed system, these concepts need to be integrated with the above programs, specifically the rule-based portions of the programs.

After evaluating our current prototype, the specialists believe that CBR concepts can be helpful to them. However, the specialists are adamant about having all decision-support tools integrated into one package that they can access from their individual workstations. They have noticed, as we have, that CBR helps solve a piece of the puzzle but needs to be integrated with other tools that they have evaluated and that support their functionality.

There has been quite a bit of interest in improving the training methodology in TFM. Our work has captured the attention of the intelligent-training-system community both here at MITRE and the FAA. A system that builds on the initial CBR work performed in this project, would have immediate payoffs to training TFM specialists. Also, in a training environment, the dynamics of real-time data and uncertainty of predicted data could be controlled, making the CBR system even more effective. This type of tool can contribute to some extent by including appropriate explanation and presentation methodology, as well as experiences of the veteran specialists.

Knowledge-based indexing is an important research activity [9]. Humans are pretty good at indexing information based on the context and can dynamically change the indexing when given more knowledge about that particular information. To achieve a better indexing scheme in CBR, we rely mostly on the analysis of attributes and available heuristics of TFM specialists. The CBR tool can help us build a qualitative model based on this analysis and can use this model to index our cases. As the case library grows, we should be able to study the feedback of this model on indexing cases. The clustering of cases using induction reveals the attributes that discriminate one case from another. We can perform experiments that allow us to examine rules for determining those distinguishing features. As the case library evolves over time, the rules will be refined or fine-tuned. Inductive learning, as we understand it, is the process of deriving those rules, or in other words, is the process of inferring general principles from specific examples or cases.

This project has concentrated on the potential operational benefits of a CBR system. In this section we have also addressed the benefits of using CBR for training. Finally, we would like to suggest using this CBR system for post-analysis of how problems were resolved. In a non-time-critical environment, previous cases could be reviewed with an analysis of the actions that were taken to determine how the system can be improved.

7. Conclusion

We have described a CBR approach to help resolve a particular class of TFM problems. A proof-of-concept CBR prototype was developed to demonstrate the effectiveness of this approach, as well as the validity of our problem description. The attributes that we have settled on for describing TFM problems provide a robust definition. It is our opinion that only minor modifications would be needed to use this representation to describe all TFM problems.

Through the support of the eventual users (the specialists) of this prototype, we have determined that CBR can be helpful in a decision-support role for TFM problem resolution, both at the national ATCSCC and the local ARTCCs. This project did not perform formal analyses of whether using this tool provides *better* decisions; it did illustrate what features of CBR the specialists found most useful.

The specialists are eager to investigate tools that will assist them, but are very selective in their support of the numerous tools proposed to them. The specialists appreciated that this CBR tool provides suggestions of how things were handled in the past, as opposed to direct solutions. They liked the ability to be able to access a common "experience-base" of veteran's procedures, as well as innovative ideas from the less experienced specialists. Finally, they appreciated the idea of being able to support their own decisions by using portions of previous solutions that worked well, or improving portions that did not work well. It was important to the specialists to be able to view the previous cases that performed well, and just as important, the cases that did not perform well.

Another goal of this project has been achieved by creating user interest in this prototype and in CBR technology. While the CBR system is in the early stage of development, a number of interesting research possibilities exist. Using previous experiences is a straightforward concept that the users can easily understand. This, combined with their involvement from the very early stages of this project, has led to strong support in helping this work proceed.

Acknowledgments

The authors would like to thank TFM specialists at the FAA for their operational inputs. We also want to thank George Swetnam, Steve Christey, and Alice Mulvehill for their reviews, and Elizabeth May for her patience and meticulousness in preparing this paper.

References

1. Kolodner, J., Summer 1991, "Improving Human Decision Making Through Case-Based Decision Aiding," *AI Magazine*, Vol. 12, No. 2., pp. 52–68.

2. Kolodner, J., 1993, *Case-Based Reasoning*, San Mateo, CA: Morgan Kaufmann Publishers, Inc.

3. Golding, A., and P. Rosenbloom, "Improving Rule-Based Systems through Case-Based Reasoning," Proceedings of Ninth National Conference on Artificial Intelligence (AAAI-91) (MIT Press, Cambridge, MA,1991).

4. Skalak, D., and E. Rissland, 1990, "Inductive Learning in a Mixed Paradigm Setting," Proceedings of Eighth National Conference on Artificial Intelligence (AAAI-90), (MIT Press, Cambridge, MA).

5. Cognitive Systems, 1993, *The ReMind Users Manual*, (Boston, MA).

6. Pierce, M., et al., 1992, "Case-Based Design Support: A Case Study in Architectural Design," *IEEE Expert,* Vol. 7, No. 5, pp.14–20.

7. Simoudis, E., 1992, "Using Case-Based Retrieval for Customer Technical Support," *IEEE Expert*, Vol. 7, No. 5, pp. 7–11.

8. Goel, A., 1991, "A Model-Based Approach to Case Adaptation," Proceedings of 13th Annual Conference of Cognitive Science Society (Laurence Erlbaum, Hillsdale, NJ).

9. Barletta, R., and W. Mark, 1988, "Explanation-Based Indexing of Cases," Proceedings of Seventh National Conference on Artificial Intelligence (AAAI-88), (MIT Press, Cambridge, MA).

A Study of Financial Expert System Based on FLOPS

Takaomi Kaneko
Professor of Business Management,
Department of Engineering, Kyushu Tokai University,
9-1-1 Toroku, Kumamoto-shi, JAPAN

Kumiko Takenaka
HOYA Service Co. Ltd.
Tomiou Dai-2 build. 4F, 2-1-4 Shibazakimachi,
Tachikawa-shi, JAPAN

ABSTRACT

The purpose of this study is to build a financial expert system based on fuzzy theory and Fuzzy LOgic Production System (FLOPS), which is an expert tool for processing the ambiguity.

The study consists of four parts. For the first part, the basic features of expert systems are presented. For the second part, fuzzy concepts and the evaluation of classical expert systems to fuzzy expert systems will be presented. For the third part, the expert system shell (FLOPS) used in this study will be described. For the last part, it will be presented the financial diagnosis system, developed by using the Wall's seven ratios, traditional seven ratios and also 34 ratios selected by a financial expert.

After analyzing and investigating these three kinds of methods, financial diagnosis system will be developed as a fuzzy expert system which used a membership function based on averages and standard deviation. At the last step, the new approach will be tried by increasing the fuzzy sets for five membership functions.

Some practical examples will be given. Throughout the paper, the way of building a financial diagnosis system based on fuzzy expert system is stressed.

INTRODUCTION

It is the hope of managers to process lots of information and makes their decisions. Hence, it can be expected that his expertise and experience will be used by the computer, that is, a creation of expert system. Artificial Intelligence and Expert System are new challenge for computer technology experts and management science specialists.

The purpose of this study is to build a financial expert system based on fuzzy theory and **F**uzzy **LO**gic **P**roduction **S**ystem (FLOPS), which is an expert tool for processing the ambiguity. Some practical examples will be given. Throughout the paper, the way of building a financial diagnosis system based on fuzzy expert system is stressed.

This study consists of four parts. For the first part, the basic features of expert systems are presented. Expert systems are advanced computer systems that can solve specialized problems at the level of a human expert. An expert is a person who has knowledge or special skills in a specific area. The development of expert systems requires a new kind of person called knowledge engineer, who must extracts human expertise and develops reasoning methods that match the expert's knowledge. Strictly speaking, an expert system consists of main two modules: knowledge base and inference engine. The knowledge base is a collection of information about a certain domain for use in solving problem. And the inference engine is a mechanism for manipulating the expertise from the knowledge base and to form inferences and draws conclusions.

For the second part, fuzzy concepts and the evolution of classical expert systems to fuzzy expert systems will be presented. Fuzzy theory was first introduced by Lofti A. Zadeh as an alternative to ordinary sets theory. The central concept of fuzzy theory is the membership function which represents the degree to which an element belongs to a set. Fuzzy expert system is an expert system which incorporates fuzzy logic and/or fuzzy sets into its reasoning process and /or knowledge representation.

For the third part, the expert system shell (FLOPS) used in this study will be described. FLOPS is a fuzzy rule-based production expert system tool developed by William Siler, James Buckley and Douglas Tucker at the Kemp-Carraway Heart Institute in 1985. Its basic syntax is derived on the non-fuzzy expert system shell OPS5 (Official Production System) developed by Charles L. Forgy at Carnegie-Mellon University in 1984. This new inference engine is characterized by the use of fuzzy reasoning techniques to deal with ambiguities, contradictions, and uncertainties.

For the last part, it will be presented the financial diagnosis system, developed by using the former theories. At a first step of the financial analysis, the Wall's index method will be tried. It is a total ratio method with some weights. Actually, he has weighted the following sevens: Current ratio, Debt ratio, Fixed assets, Inventories turnover, Receivables sales, Fixed assets turnover and Net worth turnover. It is a part of this study that the Wall's index weights will be used in fuzzy rule. Using his weights, more clear answer will be obtained by including the important degree. Outlines of the system are:

1. Data: four companies (NEC, FUJITSU, OKI, Japan NCR) from 1986 to 1991 in the Nikkei financial analysis books.
2. The average and standard deviation among four companies. And then the membership value of the average is 0.5, that of the average plus standard deviation is 1.0. Now, the fuzzy numbers have been calculated as the membership functions.
3. Rule: all ratios are inferred from this fuzzy rule:
   IF the ratio is more than the average **THEN** the financial condition is better.
4. By the above fuzzy rule, seven ratio's membership values will be determined. Next, the total value for changing seven degrees to one degree will be calculated by :

$$\frac{\sum x_i \times w_i}{\sum w_i}$$

x_i : the calculated membership values
w_i : the weights of Wall's index method

At the next step, traditional ratios are used: Total assets turnover, Receivables sales, Inventories turnover, Payables sales, Quick ratio, Current ratio, and Fixed assets invested capital. In the latter case, they don't have their weights. So they have been counted as the same weights. Someone would make a complaint about the seven financial ratios. Certainly, they may be a very few against many financial ratios. At the third step, it will be tried 34 tools in the Nikkei financial analysis books: six profitability ratios, four productivity ratios, five efficiency ratios, five stability ratios, five liquidity ratios, five scale ratios, four growth ratios. These ratios, however, don't have their weights, too. So, they will be weighted by a financial expert. After analyzing and investigating these three kinds methods, financial diagnosis system will be developed as a fuzzy expert system which used a membership function based on averages and standard deviation. However, it is needed a method of more efficient evaluation system for the companies. At the last step, five fuzzy sets will be used: very bad, bad, so so, good, and very good.

EXPERT SYSTEM

2.1 Overview

Expert System is an advanced computer system that can solve specialized problems at the level of a human expert. An expert is a person who has knowledge or special skills in a specific area. Expert System, like human expert, is designed to be experts in one specific problem. It provides a way to store human knowledge, expertise, and experience in computers. Prof. Edward Feigenbaum at Standford University has defined an expert system as ".. an intelligent computer program that uses knowledge and inference procedures to solve problems that are difficult enough to require significant human expertise for their solution." [1].

Expert system is a subfield of Artificial Intelligence (AI) (see Fig. 2-1), which is the branch of computer science concerned with developing programs that exhibit intelligent behavior. Expert system is the area that has developed more practical applications and it has invoked much interest in business and industry. More than 25% of the AI market are for expert systems [2].

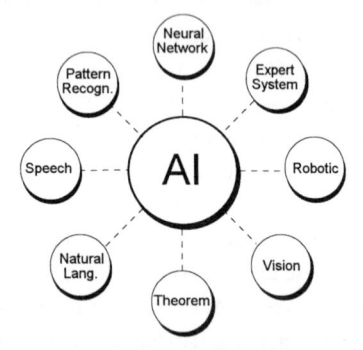

Fig. 2-1 Subfields of Artificial Intelligence

The development of expert systems requires a new kind of person, who must extracts human expertise and develops reasoning procedures that match the expert's knowledge. **Knowledge Engineer** is the computer scientist who creates expert systems. A knowledge engineer develops the knowledge representation and the reasoning strategies.

The user is the person who supplies information to the expert system and receives expert advice in response.

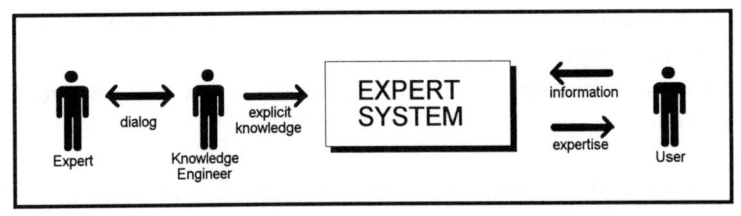

Fig. 2-2 Basic Expert System Architecture [3]

2.2 Characteristics of an Expert System:

An expert system is usually designed to have the following characteristics [4].
- The quality of the advice given by the system must be equal or better than by a human expert.
- The system should be able to explain, in the same way that human expert can explain, why a certain conclusion was reached. This feature is useful to check the human's reasoning and it is important to confirm that the knowledge has been correctly acquired and has been correctly used by the system.
- The expert system has separate problem-specific knowledge and problem-solving methodology.
- The expert system must have an efficient mechanism to add, change and delete knowledge.
- The system must perform in an equal or faster time required by a human expert. This feature is especially important in real-time systems.
- The system must be able to reason under uncertainty's conditions.

Advantages of expert systems:
The expert system has a number of features:
- *Availability*: in an expert system the expertise is more available, specially when there are a limited number of human experts in a field.
- *Permanence*: unlike human expert, who may retire or die, the expert system's knowledge is permanent.
- *Multiple Expertise*: the knowledge of multiple experts can be made available in one expert system. The combination of the expertise of several experts exceeds the single human expert.
- *Explicit Knowledge*: the knowledge of the expert has not been known implicit in the expert's mind, but has been known explicitly. This automated expertise is the basis for automating other related tasks.

- *Fast Response*: depending on the computer used, an expert system may respond faster than a human expert.
- *Explanation*: the expert system can always explain in details the reasoning of the conclusion.

Some of the capabilities and features that distinguish an expert system from conventional computer programs are shown as follows: [5)]

Table 2-1 Expert System and Conventional Programs

Expert System	Conventional Program
• Makes decisions	• Calculates results
• Based on heuristics	• Based on algorithms
• More flexibility	• Less flexibility
• Expert system tool/shell	• Programming language
• Knowledge engineering	• Software engineering / programmer analysts
• Can handle uncertainty	• Cannot handle uncertainty
• Can work with partial information, inconsistencies, partial beliefs	• Require complete information
• Can provide explanations of results	• Gives results without explanation
• Symbolic reasoning	• Numeric calculations
• Primarily declarative	• Primarily procedural
• Control and knowledge separated	• Control and knowledge interlaced
• Uses inferencing to reason toward a hypothesis	• Uses processing in the search for or calculation of answers
• Qualitative data	• Quantitative data

2.3 Components

The essential components of an expert system (Fig. 2-3) are:
1) the **knowledge acquisition** module to assist with the development of the knowledge base.
2) the **knowledge base** containing knowledge (facts, information, rules) about a problem domain.
3) the **inference engine** for manipulating the stored knowledge to produce solutions to problems.
4) the **explanatory interface** to handle communication with the user in natural language.

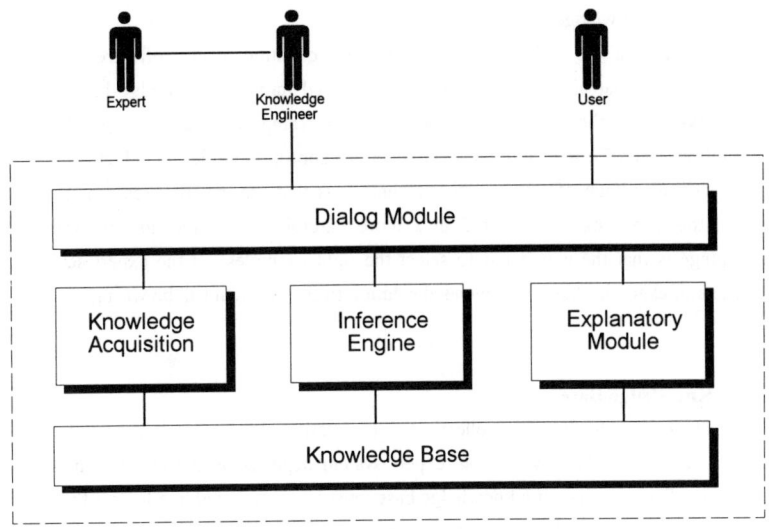

Fig. 2-3 Structure of a Expert System [6]

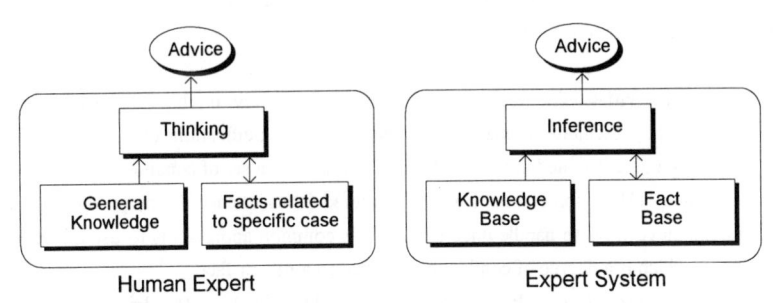

Fig. 2-4 An analogy between human experts and expert systems [7]

2.3.1 Knowledge Acquisition Module

Prof. Edward Feigenbaum, AI expert and system developer, said that knowledge acquisition is the bottleneck in the development of expert systems [8]. It will be defined as the input of the human expert knowledge into the system along with their coding into internal forms and the development of appropriate system outputs.

Knowledge engineers try to determine the manner in which experts solve the specific problem that the computer is attempting to solve. The most prevalent method of knowledge acquisition has been the interviewing. The knowledge engineer obtains knowledge from the human expert asking questions about how to solve the domain task. This method is time consuming. Here the knowledge engineer plays a central role. The quality of the system depends on the knowledge engineer's ability to extract the correct expert's domain knowledge.

Because of the difficulties in knowledge acquisition, many software tools have become available that allow information to be entered into the knowledge base in more natural ways. With these tools, the expert supplies examples of particular problems and their solutions, then the shell automatically generates appropriate rules by induction.

The advantage of this inductive technique is that the expert takes a more active role and the knowledge is acquired in a more natural way to the human expert. A disadvantage is that the expert has to select the right examples. If the expert selects the wrong sets of examples, the recommended solution to the problem will be wrong.

2.3.2 Knowledge Base

The knowledge base is a collection of information about a certain domain for use in solving a problem. The power of an expert system depends on the effective use of the knowledge. The expertise of a knowledge base must be represented in a format that allows the inference engine to perform deductions with it.

The techniques used most frequently to represent the expert knowledge are production rules, semantics nets, frames and predicate calculus [9]~[11].

- **Production rules**: The technique used most frequently is the rule-based system, because they facilitate human understanding and modification of systems. The knowledge base is coded in form of IF-THEN rules. IF a set of conditions (antecedent) is satisfied, THEN some actions will be executed (consequent). Many production rule systems have ways to handle uncertainties. A common approach is to use a certainty factor, which is a degree of confidence attached to a rule or fact.
  FLOPS (**F**uzzy **L**ogic **P**roduction **S**ystem), DENDRAL (**Dendr**itic **Al**gorithm), MYCIN, OPS5 (**O**fficial **P**roduction **S**ystem), CLIPS are examples of rule-based expert system.

```
IF antecedent THEN consequent
```

Fig. 2-5 Production System

- **Semantics nets**: The structure of a semantic net is represented graphically in terms of **nodes** and the **links** connecting them. The nodes represent objects, concepts or situations in the domain. The links describe the relation between these nodes. In semantic net, relationships are of primary importance, because they provide the basic structure for organizing knowledge.
  Semantic net can inherit properties across the net, so one node may inherit all its

properties to another node. Another advantage is their flexibility to add new nodes and relations as required.

Examples of expert systems using this technique are Smalltalk, CASNET (**C**ausal **AS**sociational **Net**work).

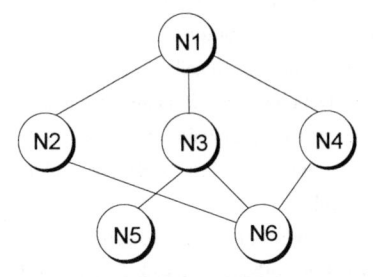

Fig. 2-6 Semantic Net Structure

- **Frames**: Frames were originally proposed by Minsky (1975) as a basis for understanding visual perception, natural language dialogs, and other complex behaviors[12]. A frame is made up of a set of slots. This slot contains attributes, descriptions, procedure, data or pointer to another frame. Like semantic nets, frames are usually organized in hierarchical structure that enables frames to inherit attributes from other frames located above them in the hierarchy.

The main difference between frame and rule is that frame can represent default values, references to other frames, rules or procedures for which values can be specified. An example of this method is KEE (**K**nowledge **E**ngineering **E**nvironment).

Slot$_1$	Value$_1$
Slot$_2$	Value$_2$
...	...
Slot$_n$	Value$_n$

Fig. 2-7 Frame Structure

- **Predicate calculus**: the facts are represented as a combination of predicate and argument. There are expressions in the form of predicate logic. Predicate calculus is used to represent statements about specific objects or *individuals*. A predicate may have one or more arguments.

Prolog is an example of this method.

Predicate_name (term$_1$, term$_2$, ..., term$_n$)

Fig. 2-8 Predicate Calculus

2.3.3 Inference Engine

The inference engine uses the domain knowledge and the acquired information in order to arrive at an expert solution or offer expert recommendation or diagnosis. It contains the general problem solving knowledge. The inference engine is responsible to determine what knowledge to use next, when to ask a question to the user, or when to search the knowledge base for any information.

Two general methods of inferencing are typically used:

- **Forward-chaining** (or Data Driven): it begins from some data or symptoms and moves down until it arrives to a conclusion, final node or frame. Forward chaining is easy to computerize and is available when the data is produced automatically. Also it is much easier to comprehend.

  Planning, control and monitoring problems are better solved with forward chaining. It is useful identifying which hypothesis to check first when there are a large number of possible faults.

  XCON, FLOPS, OPS5, CLIPS use the forward chaining method.

- **Backward-chaining** (or Goal Driven): begin from some hypothesis, conclusion, final node or frame and go backward until it finds some evidence to support or refute the initial hypothesis. Each subgoal becomes a hypothesis during the reasoning process. It is normally programmed in a recursive way.

  Backward chaining is particularly useful in diagnosis problems. It is used to prove a particular fault by proving that certain other facts are true.

  MYCIN is a backward chaining expert system.

  Others expert systems, like SPERILL II, ART, KEE, uses a combination of forward and backward chaining.

2.3.4 Explanatory Interface

This feature enables the expert system to handle communication with the user in natural language, explaining the reasoning of the system. The user can ask why the system requires some specific information or how it justifies that conclusion.

This facility is very important to identify and correct errors, omissions, and inconsistencies. This is a critical element of expert systems because it determines how well the system will be accepted by the end users. The consultation will be more acceptable to the end user depending of the reasoning's explanation it does.

2.4 Applications

In the 1960s NASA (National Aeronautics and Space Administration) was planning to send a vehicle to Mars and needed a program that would go on board the craft to perform chemical analysis of the soil on Mars[13]. The resulting program was DENDRAL, developed in 1971 by Lindsay, Buchanan, Feigenbaum, and Lederberg at Standford University. Its immediate successor, MYCIN (Shortliffe, 1976) was also developed at Standford University. MYCIN is a computer system that diagnoses bacterial infections of the blood and prescribes suitable drug therapy.

The first commercial expert system was XCON, originally called R1, which was started in 1978 as a collaborates venture between Digital Equipment Corporation (DEC) and John McDermott at Carnegie-Mellon University. XCON is an expert system configuring system for DEC computer systems.

The number of expert systems reported has grown tremendously. They are classified into a number of different categories [14]:

Table 2-2 Categories of Expert Systems

Category	Description	Application
• Control	Regulate the whole operation of a system or machine.	Air traffic Control, business management systems.
• Debugging	Prescribe remedies for malfunctions.	Electronic circuit board testing.
• Design	Configuration system, building design, configuring departmental structures, microchips.	
• Diagnosis	Identify the cause of a given set of symptoms.	Electronic, engineering, medical diagnosis.
• Instruction / Training	The student can ask questions just as human teachers.	Educational systems.
• Interpretation	take an information and infer results from the input.	Signal interpretation systems, identifying geological structures.
• Monitoring	Compare the actual state of a system with expected data.	Nuclear power, chemical plants, computer systems.
• Planning	Devise actions to yield a desired result.	Robotics, salesman problem, job scheduling.
• Prediction	Infer the probable consequences of a real of hypothetical situation.	Stock market trends, electricity demand, Waterhouse stock level.

Some examples of expert systems:

Name	Chemistry
CRYSALIS	Interpret a protein's 3-D structure
DENDRAL	Interpret molecular structure
TQMSTUNE	Fine-tune a Triple Quadruple Mass Spectrometer
CLONER	Design new biological molecules
MOLGEN	Design gene-cloning experiments
SECS	Design complex organic molecules
SPEX	Plan molecular biology experiments

Name	Geology
DIPMETER	Interpret dipmeter logs
LITHO	Interpret oil well log data
MUD	Diagnosis/remedy drilling problems
PROSPECTOR	Interpret geologic data for minerals

Name	Engineering
REACTOR	Diagnosis/remedy nuclear reactor accidents
DELTA	Diagnosis/remedy GE locomotives
STEAMER	Instruct operation - stream powerplant

Name	Electronics
ACE	Diagnosis telephone network faults
IN-ATE	Diagnosis oscilloscope faults
NDS	Diagnose national communication net
EURISKO	Design 3-D microelectronics
PALLADIO	Design and test new VLSI circuits
REDESIGN	Redesign digital circuits to new
CADHELP	Instruct for computer aided design
SOPHIE	Instruct circuit fault diagnosis

Name	Medicine
PUFF	Diagnosis lung disease
VM	Monitors intensive-care patients
ABEL	Diagnosis acid-base/electrolytes
AI/COAG	Diagnosis blood disease
AI/RHEUM	Diagnosis rheumatoid disease
CADUCEUS	Diagnosis internal medicine disease
ANNA	Monitor digitalis therapy
BLUE BOX	Diagnosis/remedy depression
MYCIN	Diagnosis/remedy bacterial infections
ONCOCIN	Remedy/manage chemotherapy patients
ATTENDING	Instruct in anesthetic management
GUIDON	Instruct in bacterial infections

Name	Computer Systems
PTRANS	Prognosis for managing DEC computers
BDS	Diagnosis bad parts in switching net
XCON	Configure DEC computer systems
XSEL	Configure DEC computer sales order
XSITE	Configure customer site for DEC computers
YES/MVS	Monitor/control IBM MVS operating system
TIMM	Diagnosis DEC computers

FUZZY EXPERT SYSTEM

3.1 Uncertainty in Expert Systems

There are several sources of uncertainties and imprecision in most area of an expert system. The solution to problems may be uncertain. The process of acquiring knowledge is uncertain, because the knowledge acquired can not exactly represent the human expert reasoning process. The way in which the expert uses the knowledge to make inferences and evaluate the problem solution is an uncertain process. The information that will be used may be incomplete. The data cannot be expected to be 100% precise. Therefore the expert system should operate with real-world situations and reason with this uncertain information.

A number of theories have been devised to deal with uncertainty. These theories include classical probability, Bayesian probability, Hartley theory based on classical sets, Shannon theory based on probability, Dempster-Shafer theory, and Zadeh's Fuzzy theory [15]. The designer of the expert system has to determine which method will be used, according to the nature of the uncertainty being handled.

In this thesis, the *fuzzy theory* will be used. Fuzzy theory has been applied in many areas including mathematics, business, engineering, and psychology. It will offer a way of dealing with linguistically defined objects, so the rules generated using this method have the ability to represent vagueness in a very natural way.

3.2 Fuzzy Sets [16] [17]

In a classical set or crisp set theory, a set is defined as a finite collection of objects grouped together. A given object belongs to the set or either not belongs to the set, is represented in terms of a *characteristic function* or *discriminatory function*. If an object x is an element of the set, then its characteristic function is 1. If an object doesn't belong to the set, then its characteristic function is 0.

$$\mu_A(x) = \begin{cases} 1 & \text{if } x \text{ is an element of the set A} \\ 0 & \text{if } x \text{ is not an element of the set A} \end{cases}$$

where x is an element of the universe X. The characteristic function can be defined in terms of the following functional mapping:

$$\mu_A(x) : x \to \{0, 1\}, \quad x \in X$$

which means that the characteristic function maps a universal set X consisting of 0 and 1.

For example, if X consists of months of the year then the set
$A = \{ x \mid x$ has 28 days $\}$ is a subset of X, with membership 1.
$B = \{ x \mid x$ has 32 days $\}$ is not a subset of X, with membership 0.

However, this classical set theory is not always adequate to describe real world systems, where we think and make decisions with uncertainty information. In real life many things have generally more than only true/false possibilities.

Several methods have been proposed for dealing with uncertain information in expert systems, such as Probabilities, Certainty Theory, and Fuzzy Logic. The last method will be used in this research.

The Fuzzy Set Theory was first introduced by Lofti Zadeh in 1962. It provides a strict mathematical framework in which vague conceptual phenomena can be precisely and rigorously studied. A fuzzy set can be regarded as the label applied to a linguistically expressed concept that has not precise boundary and such concept, with all their associate vagueness, is one of the important method by which we exchange ideas, information and understanding.

The fuzzy set theory allows each element of the set to belong to the set in a various degree of membership, indicated by a value in a range of 0 and 1. The degree of membership in a fuzzy set is measured by the *membership function* or *compatibility function* $\mu_{\tilde{A}}(x)$. A particular value of the membership function is called *grade of membership*. The value 0 represents absolute falsity and 1 represents absolute true.

In other words, conventional set theory with membership 0 or 1 is a limited case of the more general fuzzy set theory, which has membership between 0 and 1. So, when the grade of membership of the fuzzy set is 0 or 1, the fuzzy set is reduced to ordinary set.

The fuzzy set $\tilde{A}$ will be defined as [18]:
$$\tilde{A} = \{ (x, \mu_{\tilde{A}}(x)) \mid x \in X \}$$
where $\tilde{A}$ is the fuzzy set A, X is a collection of objects, x is one object of the collection X, and $\mu_{\tilde{A}}(x)$ is the membership function or grade of membership of x in $\tilde{A}$ which maps each object in X onto the interval of real numbers between 0 and 1, written as [0, 1].

There are some other ways to denote fuzzy sets:
- A fuzzy set is represented by its membership function:

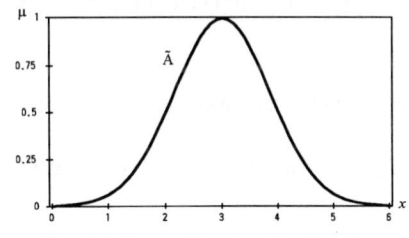

Fig. 3-1 Fuzzy Membership Function.

- $\tilde{A} = \mu_{\tilde{A}}(x_1)/x_1 + \mu_{\tilde{A}}(x_2)/x_2 \ldots = \sum_{i=1}^{n} \mu_{\tilde{A}}(x_i)/x_i$

$$\text{or } \int_x \mu_{\tilde{A}}(x)/x$$

3.2.1 Set Theoretic Operations

The basic set theoretic operations and relations of complement, subset, union and intersection will be defined as:

Set complement:
The membership function of the complement of a fuzzy set $\tilde{A}$, $\mu_{\tilde{A}'}(x)$ is defined by

$$\mu_{\tilde{A}'}(x) = 1 - \mu_{\tilde{A}}(x), \quad x \in X$$

Set subset:
The membership function of the fuzzy set $\tilde{A}$ subset of fuzzy set $\tilde{B}$ is defined by

$$\mu_{\tilde{B}}(x) \geq= \mu_{\tilde{A}}(x), \quad x \in X$$

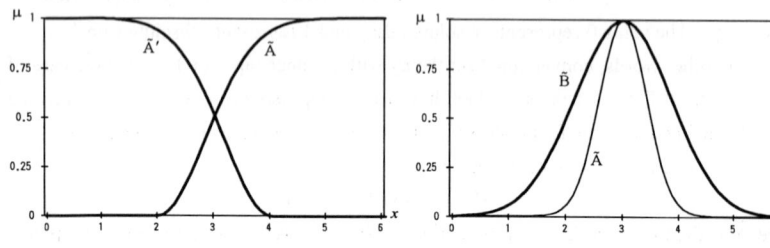

Fig. 3-2 Fuzzy Complement.

Fig. 3-3 Subset of fuzzy sets.

Set Intersection:
The membership function $\mu_{\tilde{A} \cap \tilde{B}}(x)$ of the intersection $\tilde{A} \cap \tilde{B}$ is defined by

$$\mu_{\tilde{A} \cap \tilde{B}}(x) = \min\{\mu_{\tilde{A}}(x), \mu_{\tilde{B}}(x)\}, \quad x \in X$$

Set Union:
The membership function $\mu_{\tilde{A} \cup \tilde{B}}(x)$ of the union $\tilde{A} \cup \tilde{B}$ is defined by

$$\mu_{\tilde{A} \cup \tilde{B}}(x) = \max\{\mu_{\tilde{A}}(x), \mu_{\tilde{B}}(x)\}, \quad x \in X$$

The range of the membership function is a subset of the non-negative real numbers whose maximum is finite.

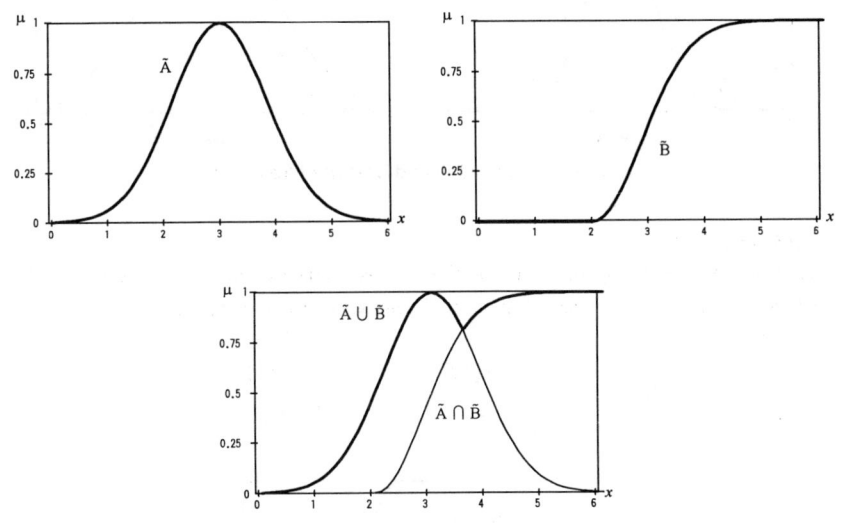

Fig. 3-4 Union and intersection of fuzzy set.

The above definitions are the most commonly used operators in fuzzy sets. However there are not the only definitions available. Others include:

Set product: $\quad \mu_{\tilde{A} \times \tilde{B}}(x) = \mu_{\tilde{A}}(x) \times \mu_{\tilde{B}}(x)$

Probabilistic Sum: $\quad \mu_{\tilde{A} \hat{+} \tilde{B}}(x) = \mu_{\tilde{A}}(x) + \mu_{\tilde{B}}(x) - \mu_{\tilde{A}}(x) \cdot \mu_{\tilde{B}}(x)$
$$= 1 - (1 - \mu_{\tilde{A}}(x))(1 - \mu_{\tilde{B}}(x))$$

Bounded Sum or Bold Union: $\quad \mu_{\tilde{A} \oplus \tilde{B}}(x) = \min(1, (\mu_{\tilde{A}}(x) + \mu_{\tilde{B}}(x)))$

Bounded Product or Bold Intersection: $\quad \mu_{\tilde{A} \otimes \tilde{B}}(x) = \max(0, (\mu_{\tilde{A}}(x) + \mu_{\tilde{B}}(x) - 1))$

Bounded Difference: $\quad \mu_{\tilde{A}|-|\tilde{B}}(x) = \max(0, (\mu_{\tilde{A}}(x) - \mu_{\tilde{B}}(x)))$

Concentration:
This operation reduces the grade of membership in elements that have smaller grades of membership. The CON operator is approximate to the linguistic modifier *very*.

$$\mu_{CON(\tilde{A})}(x) = (\mu_{\tilde{A}}(x))^2$$

Dilation:
This operation dilates fuzzy elements by increasing the grade of membership in elements with smaller grade of membership. It is approximate to the linguistic modifier *more or less*.

$$\mu_{DIL(\tilde{A})}(x) = (\mu_{\tilde{A}}(x))^{0.5}$$

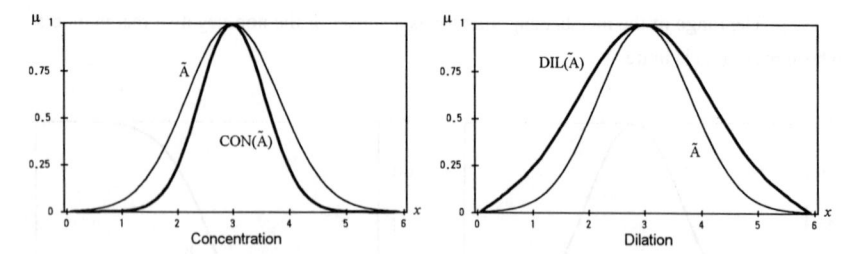

Fig. 3-5 Concentration and Dilation of fuzzy sets.

Intensification:

This function raises the grade of membership in those elements within the crossover points and reduces the grade of membership of those outside the crossover points.

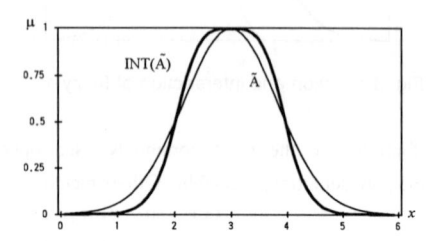

$$\mu_{INT(\tilde{A})}(x) = \begin{cases} 2(\mu_{\tilde{A}}(x))^2 & 0 \leq \mu_{\tilde{A}}(x) \leq 0.5 \\ 1 - 2(1 - \mu_{\tilde{A}}(x))^2 & 0.5 \leq \mu_{\tilde{A}}(x) \leq 1 \end{cases}$$

Fig. 3-6 Intensification of fuzzy sets.

Normalization:

Max is the maximum grade of membership for all element x. If the maximum grade of membership is < 1, then all grades of membership will be increased. If max = 1, then the grades of membership are unchanged.

$$\mu_{NORM(\tilde{A})}(x) = \mu_{\tilde{A}}(x) / \max(\mu_{\tilde{A}}(x))$$

3.3   Fuzzy Numbers [19) 20)]

The members of a fuzzy number are the universe of all the real numbers from $-\infty$ to $+\infty$. The fuzzy number has a sign ~ under the number to indicate that this is a fuzzy number. The function that produces the grade of membership of a fuzzy number is called a

fuzzifier. It expresses the quantitative measure of the uncertainty involved in the fuzzy number.

The standard membership functions of a fuzzy number are detailed as follows:

3.3.1 Triangular Fuzzy Number:

This fuzzifier has three parameters: the value at which the grade of membership starts to increase from 0 to 1 (c_0); the value at which the grade of membership is 1 (c_1); and the value at which the grade of membership is declining from 1 to zero (c_2).

$$\mu_{\tilde{C}}(x) = 0, \qquad x <= c_0$$
$$= (x - c_0) / (c_1 - c_0), \qquad c_0 < x <= c_1$$
$$= (c_2 - x) / (c_2 - c_1), \qquad c_1 < x <= c_2$$
$$= 0, \qquad x > c_2$$

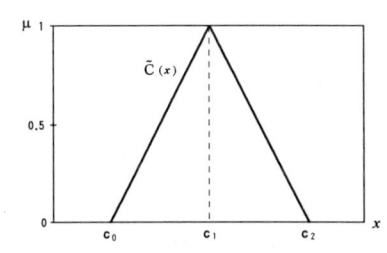

Fig. 3-7 Triangular Fuzzy Number

3.2.2 Bell Shaped Fuzzy Number:

This fuzzifier has two parameters: the central value **c** and the standard deviation **s**. The c is the value of x that has grade of membership 1, and s is deviation from the central value at which x has grade of membership 0.5. The s is symmetrical about the central value.

$$\mu_{\tilde{C}}(x) = \exp\left\{ \ln 0.5 \times \frac{(x-c)^2}{s^2} \right\}$$

$$s = \sqrt{abs^2 + (rel \times c)^2}$$

abs is the absolute uncertainty and **rel** is the relative uncertainty. abs is independent of the central value, and rel is the portion of its central value that contributes to the net uncertainty.

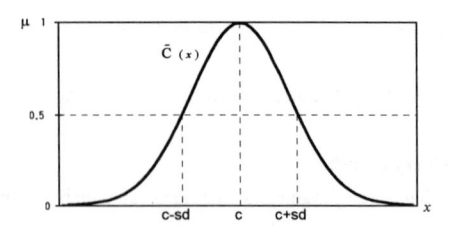

Fig. 3-8 Bell Shaped Fuzzy Number

3.3.3 Trapezoidal Fuzzy Number:

This fuzzifier is used in more complex fuzzy numbers. It has a grade of membership of 0 at $-\infty$ to c_0, increases the value from 0 (at c_0) to 1 (at c_1) in a straight line, has a constant value of 1 at c_1 to c_2, declines in a straight line to zero at c_3, and continue with zero until $+\infty$.

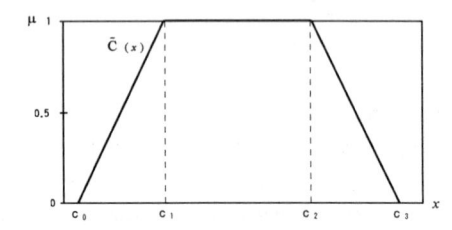

$$\begin{aligned}
\mu_{\bar{C}}(x) &= 0, & x &\leq c_0 \\
&= (x - c_0) / (c_1 - c_0), & c_0 &< x \leq c_1 \\
&= 1, & c_1 &< x \leq c_2 \\
&= (c_3 - x) / (c_3 - c_2), & c_2 &< x \leq c_3 \\
&= 0, & x &> c_3
\end{aligned}$$

Fig. 3-9 Trapezoidal Fuzzy Number

3.4 Fuzzy Inference [21] [22]

Just as classical logic is the basis of conventional expert systems, fuzzy logic forms the basis of fuzzy expert systems. The main advantage of the classical logic is that systems based on true/false valued logic is easy to model deductively and so the inferences can be exact. The main disadvantage is that very little in the real word is really two-valued.

In traditional logic, one of the most used inference rule is the *modus ponens*. It is defined as follows:

Premise	p
Implication	p → q
Conclusion	∴ q

where p and q are crisp propositions.

If $\tilde{a}, \tilde{a}', \tilde{b}$, and $\tilde{b}'$ are fuzzy sets, using the modus ponens, the fuzzy propositions will be defined as follows:

Premise	x is $\tilde{a}'$	current ratio is very high
Implication	if x is $\tilde{a}$ → y is $\tilde{b}$	if current ratio is high → good
Conclusion	∴ y is $\tilde{b}'$	∴ very good

Fuzzy inference is based on two concepts: fuzzy implication and a compositional rule of inference.

- Fuzzy implication is represented as

$$\tilde{a} \to \tilde{b}$$

where $\tilde{a}$ and $\tilde{b}$ are fuzzy sets. Fuzzy implication is defined by a fuzzy *relation*. The fuzzy relation R associated with the implication $\tilde{a} \to \tilde{b}$ is a fuzzy set of the Cartesian product $I \times J$, where $\tilde{a} \in I$ and $\tilde{b} \in J$. The *min* operator is one of the most frequently used in fuzzy implications. R is defined as

$$\mu_R(i,j) = \min(\mu_{\tilde{a}}(i), \mu_{\tilde{b}}(j)) \quad i \in I, j \in J$$

- A composition of relations will be defined when applying one relation after another. If R is a fuzzy relation from $I \times J$, and $\tilde{x}$ is a fuzzy subset of I, then the fuzzy subset $\tilde{y}$ of J is given by the composition of R and $\tilde{x}$, as follows:

$$\tilde{y} = \tilde{x} \circ R$$

where ∘ is the composition operator. The most commonly used method is the *max-min* composition.

$$\mu_{\tilde{b}}(j) = \max_{i \in I} \min(\mu_{\tilde{a}}(i), \mu_R(i,j))$$

3.5 Fuzzy Expert System [23) 24)]

Fuzzy Expert System is an expert system which incorporates fuzzy logic and/or fuzzy sets into its reasoning process and/or knowledge representation.

Reflecting human expertise, much of the information in the knowledge base of a typical expert system is imprecise, incomplete, or not totally reliable. Different methods have been used to handle it (see 3.1), specially the Certainty Factor (MYCIN) and

Dempster-Shafer theory, and most of them are probability based. These methods may be effective in specific cases, but, the experts often do not thing in probability value. Fuzzy theory offers a more natural way to handle it. Although some people claim the fuzzy treatment is unreliable, the elasticity of fuzzy sets gives them a number of advantages over conventional sets. Some of them includes

- They avoid the rigidity of conventional mathematical reasoning and computer programming.
- Fuzzy sets simplify the task of translation between human reasoning, which is inherently elastic, and the rigid operation of digital computers.
- Fuzzy sets allow computers to use the type of human knowledge called common sense. Common-sense knowledge exists in the form of statements that are usually, but not always, true.
- A feature of fuzzy logic, which is of particular importance to the management of uncertainty in expert systems is that it provides a systematic framework of dealing with fuzzy quantifiers, such as "most", "very", "frequently", ... In this way fuzzy logic subsumes both predicate logic and probability theory and makes it possible to deal with different types of uncertainty with a single conceptual framework.

3.6 Applications [25] [26]

Many fuzzy expert systems have been developed since 1984. Some of these are listed as follows:

- **SPERILL II**: It is and expert system for damage assessment of existing structures, using fuzzy sets to represent imprecise data.
- **FLOPS** (Fuzzy Logic Production System): It is a fuzzy rule-based production expert system. It makes uses of fuzzy numbers and provides a complete set of fuzzy inequality operators. FLOPS has been successfully used to model several different expert domains.
- **SYSTEM Z-II**, It was developed by Chinese University of Hong Kong. It is a fuzzy expert system shell, which effectively deals with both uncertainty and imprecision. The domains of Z-II have been medical diagnosis, psychoanalysis, and risk analysis. It allows knowledge to be expressed in fuzzy linguistical terms.
- **FRIL** (Fuzzy Relational Inference Language): It was developed by J. Baldwin. This expert system shell permits inference under uncertainty and incompleteness.
- **COFESS** (Cooperative Fuzzy Expert Systems): It is an expert system which utilizes fuzzy theory to recognize patterns.

FLOPS

4.1 Overview of FLOPS

Fuzzy LOgic Production System (FLOPS) is a fuzzy rule-based production expert system, developed by William Siler, James Buckley, and Douglas Tucker at the Kemp-Carraway Heart Institute in 1985 [27]~[29].

The principal objective for the creation of this new expert system shell was the need to solve a "problem", and the inadequate tools to solve it. "The problem was to process echocardiogram images of the heart by computer, unsupervised by humans, to yield chamber volumes and measure of regional heart muscle function. FLOPS was developed so that we could classify the region clusters by having the computer look at the images in the same way as a cardiologist." [30].

FLOPS's basic syntax is derived on the non-fuzzy expert system shell OPS5 (Official Production System) developed by Charles L. Forgy at Carnegie-Mellon University in 1984. This new inference engine is increased for the use of fuzzy reasoning techniques to handle ambiguities, contradictions, and uncertainties. FLOPS uses fuzzy number's and fuzzy set's data type, and it provides a complete set of fuzzy inequality operators to handle this imprecise information.

Another feature of FLOPS is the ability to remember what it already learned during the search process, so it gives considerable power and the minimum time to run a program.

A difference between FLOPS and other expert systems is that it is designed to be programmed by the end-user, not necessary by a highly trained knowledge engineer. All these special features (like data-driven and fuzzy concepts) make it a difficult language based on concepts little known to the most programmers.

Another features of FLOPS include as follows:
- data type: int, flt, atm, fzset, fznum, TT (Time Tag), xx.cf
- fuzzy operators: ~, ~<=, ~=, ~>=, ~> and ~<>.
- two file types: type I (external FLOPS file), type II (ASCII data type).
- two call types: open and transfer command.
- an editor to create/modify type II data file: FLEDIT (FLOPS editor)
- rules to generate new rules.
- two versions of FLOPS: sequential FLOPS (deductive reasoning) and parallel FLOPS (inductive reasoning).

FLOPS runs on IBM PC/XT/AT compatible computers, Nihon IBM PS/55 Series, the DEC Microvax II computer, Toshiba J-3100 Series, Sanyo AX Computer and NEC PC-9901 Series and PS/55 Series [31][32].

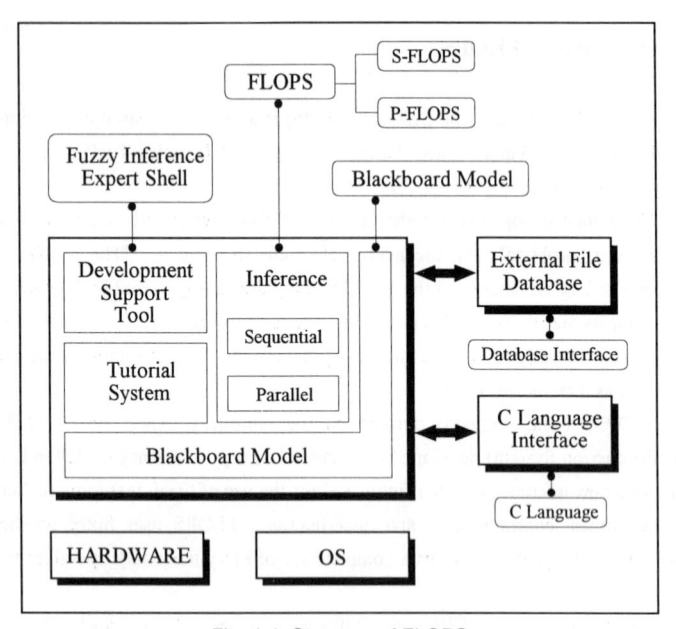

Fig. 4-1 Structure of FLOPS

4.2 Production System

A **production system** is an inference engine based on IF-THEN rules. Each IF pattern will be referred as an **antecedent** and each THEN pattern as a **consequent**. The antecedent of a rule is a set of conditions to be met to make the rule fireable. The consequent is a set of commands to be executed if the rule is fired.

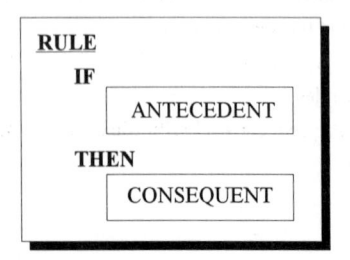

Fig. 4-2 Rule-based Expert System

In FLOPS, a typical rule is:
```
rule 1000  4  ( index  ^liability.cf 0 )
   -->
   reset ,
   write 'Enter ratio of liability: ' ,
   read 1  ^liability ;
```

Production system is a non-procedural language, so the order in which rules are executed is totally independent of the order in which they appear in the program. FLOPS is a data-driven language, because it only depends of the data available in the working memory. When more than one rule is fireable, a rule-conflict algorithm selects one rule for firing, and stacked the others not-selected rules for later backtracking. When there are not more fireable rules, FLOPS will be backtracked, and fired by one fireable but not-selected rule. The process will end when all rules have been fired or when FLOPS encountered the stop command.

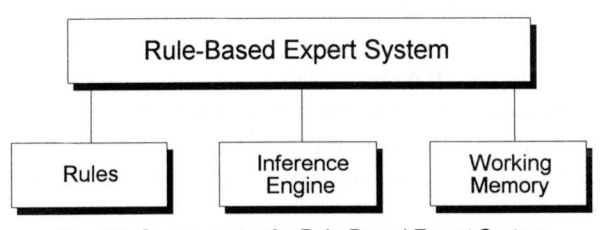

Fig. 4-3 Components of a Rule-Based Expert System

4.2.1 Fireability

A rule is fireable when any data in the memory element matches all the conditions specified in the rule's antecedent or LHS (Left Hand Side). Let's make an example here:

```
:r3
rule 1000  4  ( index  ^current.cf 0 )
   -->
   reset ,
   write 'Enter Current ratio: ' ,
   read 1  ^current ;
:r4
rule 1000  4  ( index  ^liability.cf 0 )
   -->
   reset ,
   write 'Enter ratio of liability: ' ,
   read 1  ^liability ;
```

One memory element's data will match many rules at the same time. In this example, if we create a memory element index
> make index ^year 1992 ^company "ABC" ;

the element index is fireable in rules r3 and r4.

All the program's rules will be examined simultaneously by all the data in the memory element. If more than one data instance match the LHS's conditions, all of these will be fireable for the same rule. In the above example, lets make three instances of memory element index:

> make index ^year 1992 ^company "ABC" ;
> make index ^year 1992 ^company "XYZ" ;
> make index ^year 1992 ^company "IJK" ^current 116.58 ;

In this example, all the rules in the program are examined 3 times by fireability. In case of rule r3, two instances of memory element index are fireable for it. The program has two instances of the rule r3, one for ABC, and one for XYZ.

When more than one rule are equally fireable, FLOPS fires the rule with highest confidence level or uses the **MEA** (Means Ends Analysis) rule-conflict algorithm like OPS5. This rule conflict algorithm chooses the rule to be fired by the following sequence:
- The rules which have been already fired are discarded.
- It chooses the rule which has the most recent time tags of the data in the first pattern of the LHS.
- If there are more than one dominant rule, then it checks the remaining patterns.
- If more than one rule remains, then it selects the most complex LHS.
- Eventually, if there is no domain rule, the algorithm chooses one arbitrary rule.

4.2.2 Backtracking

Backtracking is the process to pop up one fireable rule from the unfired rule stack when the last fireable rule has no long new fireable rule. When more than one rule are fireable at the same time, at first only one of them will be fired and the other non-fired rules are placed in an unfired rule stack for later execution.

The unfired rule stack is LIFO (Last In First Out), so the last rule put in the stack is executed when the program is backtracking to another rule for firing.

4.2.3 Decision Tree

Non-procedural languages (like OPS5, PROLOG, FLOPS) describe the program structure in the form of decision tree. Decision tree is the rules represented by tree structure. Each node contains an instance of a rule. Let's explain it with an example:

233

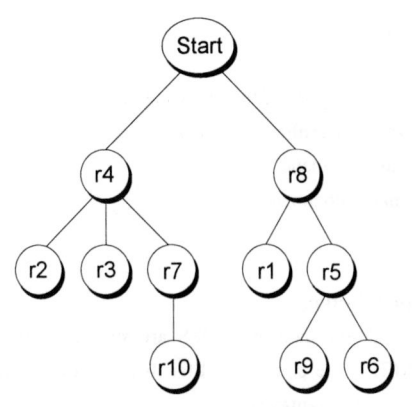

Fig. 4-4 Decision Tree

Initially, the data in memory element makes rules r4, and r8 fireable. In this example, the rule-conflict algorithm selects rule r4 for firing and places rule r8 in the unfired rule stack. r4 is fired and it will make either rules r2, r3, and r7 fireable. One rule is selected for firing, say r3, and the others are stacked. In this point, the unfired stack contains rules r8, r2, and r7. The rule r3 is executed and makes no new rules fireable because r3 is a terminal node. Now FLOPS backtracks, and r7 (the last rule in the stack) is popped up from the stack. The last rule is fired because the unfired rule stack is LIFO.

The rule r7 makes only the r10 fireable. Rule r10 makes no new rules fireable, so, the program backtracks again and pops r2 from the unfired stack, which contains in this moment only the rule r8. It backtracks again, firing r8. The stack is empty. r8 makes r1 and r5 fireable, and so on. This process continues until one rule contains the halt or stop command, or until no more new fireable rules exist and the unfired stack is empty.

The next table shows the new fireable rules, the rule fired in this moment, and the contents of the unfired rule stack, created by following the above example rule tree.

Table 4-1 Example of Decision Tree

	fireable stack	fired	unfired rule stack
1.	r4, r8	r4	r8
2.	r2, r3, r7	r3	r8, r2, r7
3.	backtracking r7	r7	r8, r2
4.	r10	r10	r8, r2
5.	backtracking r2	r2	r8
6.	backtracking r8	r8	-
7.	r1, r5	r5	r1
8.	r9, r6	r9	r1, r6
9.	backtracking r6	r6	r1
10.	backtracking r1	r1	-
11.	stop		

4.3 FLOPS Anatomy

A FLOPS program consists of the following sections:
- The Declaration section: **literalize** command.
- The Rule section: **rule** command.
- The Input section: **make** command.

4.3.1 The Declaration Section

Like in OPS5, all the data in FLOPS are stored in structures called Working Memory Element (WME). The command "**literalize**" creates a memory template of the working memory element. The syntax is:

```
literalize memory_element_name
        attribute_name  data_type
      [ attribute_name  data_type ...] ;
```

Example:
```
literalize index
    year       int
    company    atm
    current    flt
    liability  flt
    ratio      flt ;
```

In the above example, **index** is the name of the memory element. One WME may contain more than one attribute, and each attribute has a specific data type. In this example, the WME has five attributes: "year", "company", "current", "liability", and "ratio", which are "int", "atm", "flt", "flt", and "flt" respectively. FLOPS includes by default the Time Tag (TT is unique, assigned in the sequence in which element was created) of the memory element and the confidence level in all attributes, 0 by default.

The data type defined by literalize are:
- **atm**   character string (atom)
- **int**   integer
- **flt**   floating point number
- **fznum**   fuzzy number
- **fzset**   fuzzy set

The literalize command just create a **template** of the WME, it does not put any data in the WME.

4.3.2 The Rules Section

Remember that FLOPS is based on IF-THEN type rules. The syntax of the rule command is:

```
rule {rule_name} {rule_priority {block_number}}  LHS (Left hand Side)
-->
RHS (Right Hand Side) ;
```

Example:
```
rule 1000 4 ( index ^liability.cf 0 )
-->
reset ,
write 'Enter ratio of liability: ' ,
read 1 ^liability ;
```

The rule's components are:
1) the word "**rule**" is a FLOPS command.
2) the rule **Name** is a unique rule identifier. It must begin with an alphanumeric character. FLOPS assigns by default r0, r1, r2, ...
3) the rule **Priority** must be an integer among 0 and 1000, by default 1000.
4) the **Block** number must be an unsigned integer, by default 0. This value is given only when the rule priority was assigned.
5) the **Antecedent or LHS** (Left Hand Side) is conditions to be met to make the rule fireable. It consists of a set of **patterns** enclosed in parenthesis "( )". Each pattern must be referred to only one memory element, but may have more than one subpattern. Each subpattern must be referred to only one attribute of this memory element. The antecedent's syntax is:

```
( memory_element_name ^attribute_name {operator} (value/variable)
                      [ ^attribute_name {operator} (value/variable) ... ] )
```

Example:
```
rule 900 ( index ^year <YR> ^liability <RT> ^ratio.cf 0 )
         ( ratio_name ^name "liability" ^weight <WT> )
         ( std ^year <YR> ^ratio <SR> )
-->
modify 1 ^ratio ( <WT> * <RT> / <SR> ) ;
```

In the above example there are three memory elements: index, ratio_name and std. So the rule has three patterns, and the pattern number one (element index) has three subpatterns: ^year <YR>, ^liability <RT>, and ^ratio.cf 0.

In the Rule and Input sections, the **attribute name** must be preceded by a caret sign (^). When the **operator** is omitted, FLOPS assumes the equal sign (=). Variable is an alpha string enclosed in angle backers (< >), which may match any data.
6) the symbol "-->", which separates the antecedent from the consequent.
7) the **Consequent or RHS** (Right Hand Side) contains instructions to be executed when the rule is fired. The syntax is:

> command [, command] ;

The commands permitted in FLOPS are as follow:
- File: call, close, open, transfer
- Debug: debug, pconf, prdes, prmem, prule, prwme, prstack
- Memory: clear, coreleft, delete, literalize, make, modify
- User/FLOPS interaction: cls, halt, move, read, run, stop, write
- Rules: fire, reset, rule

When the RHS of a rule has many commands, each one is separate with a comma (,), and the last command terminates with a semicolon (;).

The conditions of the **antecedent** determine if a rule is **fireable** or not, and the rule **priority** may determine the **order of execution**, when more than one rule is fireable. The rule with highest priority will be fired first.

4.3.3 The Input Section

The "**make**" command creates a new instance of the working memory element. If the program does not create any instance of memory element by the make rule, no rules will be fireable, and so that the program will be stopped. Each command creates only one instance of a specific WME, but more than one instance of this element can be created. The limit for the number of WME is the memory space.

The syntax of the make command is:

> make memory_element_name [attribute_name attribute_value ...] ;

Example:
    make index ^year 1607 ^company "ABC" ;

The memory element is "index", the attribute "year" is 1992 and the attribute "company" is "ABC". This element index will make any rule fireable.

4.4 Confidence Level

In FLOPS, there is the confidence level assigned to rule, data, and comparisons between data. Also it exists other confidences (antecedent and posterior confidences) which are calculated.

The confidence level is a range among **0** (not confidence) and **1000** (full confidence). In attribute of type atm, int, and flt, FLOPS creates, by default, attribute confidences by adding ".cf" to the attribute name.

- **Attribute Values Confidence**:
Each attribute in the WME has a confidence level assigned to it. FLOPS does it automatically by adding an additional attribute, when the data types are flt, int, and atm. You can access this attribute confidence level by appending ".cf" to the attribute name. For example, if the attribute name is ^year, then the confidence level is ^year.cf.
When the user not assigns any attribute confidence level, FLOPS takes a default value of 1000.

- **Comparison Operator Confidence**:
In the FLOPS program, all the operations executed in the antecedent involve comparisons between attribute's values. These comparisons are made with Boolean equality, Boolean inequalities, fuzzy equality, and fuzzy inequalities.
Boolean comparisons always return a confidence level of 0 (false) or 1000 (true). Instead of this, the fuzzy comparisons may return a confidence among 0 and 1000.

- **Antecedent Confidence**:
The antecedent confidence level is the minimum (fuzzy AND) between the confidence level of all the attribute values of the rule and the confidence returned by all the comparison operators.

> antecedent cf = min (attribute values cf, operator cf)

Fig. 4-5 Antecedent Confidence Calculation

A rule is fireable, if the antecedent's conditions are met with a confidence level greater or equal than a threshold value. The default value of the rule firing threshold is 500. If you want to change this value, append "-t" and the new value / 1000 when you execute FLOPS.

   FLOPS -t 0.25

The new value of the threshold is 250.

- **Rule's Confidence**:
The prior confidence level or rule priority is the confidence assigned at the rule when it was created, and appear immediately after the word "rule". If there is not confidence assigned to the rule, a default value of 1000 is taken. This rule priority controls the order of execution of the rules, when many rules are fireable at one time.

- **Posterior Confidence**:
The posterior confidence level is the minimum (fuzzy AND) between the rule's confidence and the antecedent confidence level.

> posterior cf = min (rule cf, antecedent cf)

Fig. 4-6 Posterior Confidence Calculation

FLOPS assigns this posterior confidence level to the confidence level of all attributes which data has been made, modified, or read by this rule.

4.5 Representation of Fuzziness in FLOPS

As mentioned above, FLOPS is an expert system that permits the use of fuzzy concepts to handle uncertainties. It is done by different ways.
- Fuzzy Values
- Fuzzy numbers
- Fuzzy sets
- Fuzzy Operators

4.5.1 Fuzzy Values:

All the attribute values (int, flt, atm, fznum types) have a confidence level associated with them. When an attribute is created or modified, FLOPS automatically creates or modifies the confidence level associated with it. For example:

```
rule  700  ( index  ^liability.cf 0  ^year >= 1988  ^year <= 1992 )
    -->
    reset ,
    write 'Enter ratio of liability: ' ,
    read  1  ^liability ;
```

In this example the attribute "liability" was read with a confidence level assigned to ^liability.cf. This confidence level (^liability.cf) reflects the measure of confidence that we place in the attribute value "liability". If we are sure that the liability confidence level is zero (1000) and the year is among 1988 and 1992 (1000), then the ^liability.cf is 700 (the confidence of the rule). However, if we are not sure that the year is among 1988 and 1992 (550), the ^liability.cf is 550. This confidence level of 700 or 550 means that we are not very sure about the attribute liability's value read.

You can see with this example that the posterior confidence level of the rule is assigned to the attribute confidence level, ^liability.cf in this case.

4.5.2 Fuzzy Numbers:

FLOPS uses the bell-shaped membership function to handle fuzzy numbers. The input of fuzzy number consists of three parameters:
- a central value
- an absolute uncertainty
- a relative uncertainty, a proportion of the central value.

For example, when we want to create a new instance of memory element with a fuzzy number, we do like that:

> make area ^x 450, 10, 0.1 ;

In this case, the **central value** (c) is 450, the **absolute uncertainty** (a) is 10 and the **relative uncertainty** (r) is 0.1. FLOPS will calculate authomatically the variance and standard deviation using these three values. The variance is the absolute value squared plus the relative value * the central value squared (2125). The standard deviation is the root squared of the variance (46.0977). So the net uncertainty will be calculated by

$$standard\ deviation = \sqrt{absolute^2 + (central \times relative)^2}$$

This 46.0977 means that the number 496.0977 has a confidence level of **500**. Intermediate values are normally distributed.

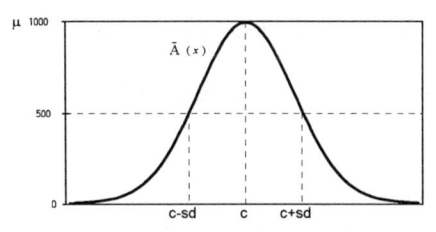

Fig. 4-7 Normal Distribution

4.5.3 Fuzzy Sets:

Fuzzy set is a collection of descriptor that applies to some attribute in various degrees. The grade of membership is the measure of the confidence level that we place in the validity of this object. FLOPS uses the attribute of fzset data type to implement the fuzzy sets.

Example:
```
literalize season
    name        atm
    weather     fzset (
                    hot
                    warm
                    cold
                ) ;
```

The fuzzy set members (hot, warm, ...) are character strings, and the grades of membership are numbers from 0 (not confidence) to 1000 (complete confidence). For example, in spring the weather will be:

```
name "spring"
weather fzset (
    hot  0
    warm 400
    cold 800
    ) ;
```

All the members of fuzzy sets must be declared when the program is written. So, FLOPS only can change the value assigned to them, with the "modify" command, it can not add new members in the run time. We can access a specific member of the attribute, appending to the name of the attribute (^weather) a dot (.) plus member's name (.cold).

Example:
```
rule 1000 ( season ^name <NM> ^weather.cold <COLD> )
    -->
    write 1 'The confidence of cold in <NM> is <COLD>' ;
```

In this case the attribute is ^weather and a member of this attribute is ^weather.cold.

4.5.4 Fuzzy Operators:

FLOPS uses the following fuzzy operators: ~<, ~<=, ~=, ~>=, ~>, ~<>. These operators return a confidence level in a range anywhere from 0 to 1000.

Fuzzy operators may be used to compare:
- two scalar numbers: In this type of test, the operators require information about the uncertainty associated with the numbers, an absolute uncertainty and a relative uncertainty.

Example:
    literalize area
        a   int
        b   flt
        x   fznum
        y   fznum ;

    rule r1 (area ^a <A> ^b <B> ^b ~> <A> (3, 0, 1))
    -->
        write 1 '<B> is fuzzily greater than <A> ' ;

If the two values compared, ^a and ^b, were displaced by their mean by exactly the standard deviation (net uncertainty), the confidence returned by the operation is 500 (the threshold value). If the values are displaced by less than the net uncertainty, the confidence is more than 500 (see fig. 4-7). We must have a confidence of at least 500 (default value) for rule r1 to be fireable.

- two fuzzy numbers or one fuzzy number and one scalar number: In this type of operation you don't need to specify the absolute and relative uncertainty, because they are stored with the fuzzy numbers. So, they look like an ordinary operation.

    Example:
        rule r2 (area ^a <A> ^x <X> ^a ~< <X>)
        -->
            write 1 '<A> fuzzily less than <X>' ;

        rule r3 (area ^x <X> ^y <Y> ^x ~>= <Y>)
        -->
            write 1 '<X> fuzzily greater or equal than <Y>' ;

A confidence level greater than or equal to 500 is required to rule r2 or r3 be fireable.

4.6   Parallel FLOPS

There are two versions of FLOPS: sequential and parallel. In sequential FLOPS, when more than one rule are fireable, only one is fired and the other rules are placed on a stack to be processed later. In Parallel FLOPS (PFLOPS), when many rules are fireables, all rules are fired simultaneously. So, there is no unfired stack and the PFLOPS does not backtrack. Also no rule conflict algorithm is involved, because all rules are fired.

PFLOPS makes a substantial reduction in system overhead, therefore one program in PFLOPS usually runs much faster than the same program in sequential FLOPS.

When more than one rule is fireable on the same memory instance and tries to modify this instance, a memory conflict is presented. FLOPS uses weak monotonicity to solve the memory conflicts problems. In weak monotonicity, the value of the attribute may be replaced by another, if the new value has a confidence level greater than or equal to the old value.

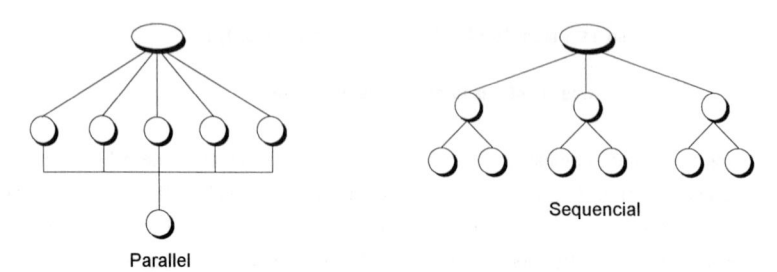

Fig. 4-8 Sequential and Parallel FLOPS

BUILDING A FINANCIAL DIAGNOSIS SYSTEM

5.1 Overview

There are two important factors in the diagnosis of a system, *element* to handle it and *method* to get a conclusion. In this research, it will be used the financial ratios as the elements and fuzzy expert system's concepts, especially Fuzzy Logic Production System, as the method to manipulate expert's uncertain rules. Specifically, throughout this research, the way of getting a total index value for financial ratios analysis system will be tried. This total index getting from companies every year will be displayed in a 2-dimensional graph. A next subject will be tried the way of building for a prediction system based on this total index.

The process of this study includes the following steps:
1) Wall's index method will be used to get an index value. It is based on financial ratios with some weight.
2) A total value will be calculated by fuzzy IF-THEN rules using the Wall's weight.
3) Five fuzzy sets, tuning the membership function, and 3-dimensional graph will be synthetically used to diagnose the companies.

5.2 Wall's Index Method [33]

There are lots of financial ratios to decide the profitability, safety, and growth, etc. of the companies. They are enough to evaluate the companies partially but not totally. Wall's index method is called "total ratio method with some weight" and is able to diagnose the companies synthetically taking into account the effects of each ratios. Actually, he has weighted the seven ratios [34]~[39] for the manufacturing companies as follows:

Table 5-1 Wall's weights

Current ratio	25%
Debt ratio	25%
Fixed assets	15%
Inventories turnover	10%
Receivable sales	10%
Fixed assets turnover	10%
Net worth turnover	5%

1) **Current Ratio**: The current ratio is a measure of liquidity that relates total current assets to total current liabilities. It indicates that the current assets will be used in a short period of time to bring cash into the company. If the total current assets exceed the current liabilities, it means that the company will be able to pay off its current liabilities. It is desirable that the current assets will be at least twice for the current liabilities.

$$\text{Current ratio (\%)} = \frac{\text{Current assets}}{\text{Current liabilities}} \times 100$$

2) **Debt ratio**: Debt ratio is a coverage ratio that shows the percentage of total capital provided by the creditors of a business. Debt includes current and long-term obligations. A ratio of 100% means that the creditors and owners are furnishing same amounts of funds for business activity.

$$\text{Debt ratio (\%)} = \frac{\text{Total liabilities}}{\text{Stockholders' equity}} \times 100$$

3) **Fixed assets to net worth ratio**: This ratio is calculated by dividing the total fixed assets by the stockholder's equity. The higher ratio is the better.

$$\text{Fixed assets ratio (\%)} = \frac{\text{Total Fixed assets}}{\text{Stockholders' equity}} \times 100$$

4) **Inventory turnover**: This ratio is an expression of the number of times required to convert an inventory into cash or receivable. A high inventory turnover is an indicative of good sign, denoting that the inventory is readily marketable.

$$\text{Inventory turnover (turn)} = \frac{\text{Sales}}{\text{Average Inventories}}$$

5) **Receivable sales**: It is a measure of how many days will be taken to collect receivable from sales on account. A collection period significantly in excess of 60 days indicates a problem with either the granting of credit, collection policies, or both.

$$\text{Receivables sales (days)} = \frac{\text{Notes receivable} + \text{Accounts receivable} + \text{Notes receivable discounted}}{\text{Sales}} \times 365$$

6) **Fixed assets turnover**: This radio focuses only on the property, plant, and equipment item. It is calculated by dividing the sales by the average fixed assets.

$$\text{Fixed assets turnover (turn)} = \frac{\text{Sales}}{\text{Average Fixed assets}}$$

7) **Net worth turnover**: This ratio measures the profits generated on funds provided by common stockholders. It is computed by dividing sales by stockholders. The more turnover is better.

$$\text{Net worth turnover (turn)} = \frac{\text{Sales}}{\text{Stockholders}}$$

Concretely, each index will be calculated as follow,

standard ratio = arithmetic mean + median + mode
index = real ratio ÷ standard ratio × weight

then the total index value is obtained by aggregating each index or totality. The evaluation will be done by comparing the value with the sum of the weight 100.

For the first step of this financial analysis system, it will be examined with this Wall's Index Method. As an example, the following table shows the obtained result of Fujitsu Co. Ltd. in 1986.

Table 5-2 Wall's Index Value of Fujitsu Co. Ltd. in 1986

Financial Ratio	Weight	Real ratio	Standard ratio	Index
1 Current Ratio	25	144.27	146.92	24.55
2 Debt ratio	25	52.69	42.49	31.00
3 Fixed assets	15	77.25	80.23	14.44
4 Inventory turnover	10	5.03	5.57	9.03
5 Receivable sales	10	3.77	3.84	9.82
6 Fixed assets turnover	10	2.12	2.68	7.91
7 Net worth turnover	5	5.04	5.77	4.37
Total	100			101.12

In this example, the total index became 101.12. This value is the measure of the system's total evaluation. The diagnosis of companies will be done by comparing this value with the total weight 100.

The purpose of this paper is to get such total index by using fuzzy IF-THEN rules.

5.3 Rules of Diagnosis

Because of the weight assigned to each ratio, Wall's index method is generally admitted the total evaluation method for the companies. These weights are not calculated with scientific foundation but they are based on the experience of an expert, Wall himself. So, these are like to the weights assigned by the other experts when they make a total evaluation. These characteristics make possible the total evaluation of a company. On the contrary, it was rather difficult for the traditional method.

As a part of this study, the Wall's weights will be used with the fuzzy rules to get a more clear-cut solution. To obtain a total index value, the following membership function will be used:

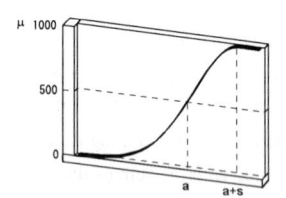

Fig. 5-1 Membership Function

In the Fig. 5-1, "a" is the average of each ratio among the companies and "s" is the standard deviation. In this membership function, the membership value of the average "a" applies to 500, and average + standard deviation applies to 1000.

This research will compute the total value using FLOPS commands to deal with membership functions. An example of FLOPS program is shown in Fig. 5-2. The "pconf" command displays the membership value. The total value will be calculated as shown in Fig. 5-2. For example, if there are 7 ratios, like in the Wall's method, then 7 membership values "x_i" will be determined. Therefore, the total value will be calculated as follows: at first, each membership value x_i will be multiplied by its weight, then it will be divided by the average of the weight.

```
: Declaration
    literalize input
        x    flt
        mem  fznum ;
```

```
: Data Input
    make input ^mem a+s, s, 0;

: Example of Rules
    rule 1000 ( input ^x ~≥ ^mem )
        -->
        pconf
```

Fig. 5-2 FLOPS program

247

1. Current Ratio

2. Debt Ratio

⋮

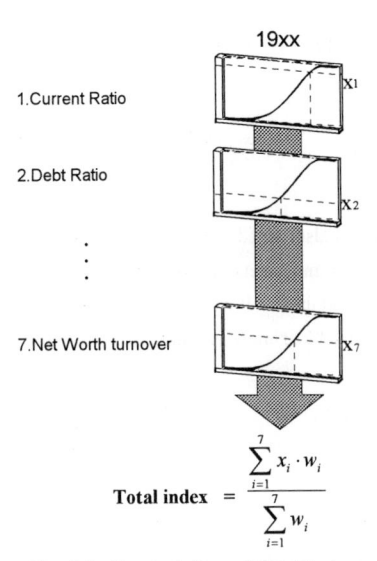

7. Net Worth turnover

$$\text{Total index} = \frac{\sum_{i=1}^{7} x_i \cdot w_i}{\sum_{i=1}^{7} w_i}$$

Fig. 5-3 Computation of Total Index

5.4 Financial Data

The financial data was taken from the Nikkei Financial Analysis book [40] at 1986-1991. It was chosen four companies from Tokyo Stock Exchange Market, NEC, Fujitsu, OKI, and Japan NCR. Some ratios, for example Net worth turnover, etc., that were not included in the book, were calculated in advance.

As a first step, the total index of the 4 companies was calculated using the data of the 7 ratios of the Wall's method. The result was shown in the Fig. 5-4.

For the comparison, traditional ratios were used in parallel with the Wall's ratios with weights. These traditional ratios were calculated as follows:

1) **Total Assets turnover**: It is a measure of how well management utilized all of the company's assets to gain income. It is computed as follows:

$$\text{Total assets turnover (turn)} = \frac{\text{Sales}}{\text{Average total assets}}$$

2) **Receivable sales** (see 5.2)

3) **Inventory turnover** (see 5.2)

4) **Payables turnover**: This ratio is calculated by dividing the operating payables with the sales.

$$\text{Payables turnover (days)} = \frac{\text{Notes payable - trade} + \text{Accounts payable - trade}}{\text{Sales}} \times 365$$

5) **Quick ratio**: This ratio is also called the *acid test or liquidity ratio*. It is calculated by dividing the liquid current assets (cash, marketable investment, and accounts receivable) with current liabilities. It excludes the inventory and prepaid items from the asset. A quick ratio of 100% or more is regarded as desirability.

$$\text{Quick ratio (\%)} = \frac{\text{Liquid assets}}{\text{Current liabilities}} \times 100$$

6) **Current Ratio** (see 5.2)

7) **Fixed Assets to invested capital**: The purpose of this analysis is to reveal the profitability of an investment in stock. This ratio is calculated by dividing the fixed assets with the stockholders' equity plus the long-term liabilities, as shown below.

$$\text{Fixed assets to invested capital (\%)} = \frac{\text{Fixed assets}}{\text{Stockholders' equity} + \text{Long - term liabilities}} \times 100$$

However, it is assumed that the traditional method of company's diagnosis is based on the same weight to all the ratios. The result of this method is shown in Fig. 5-5.

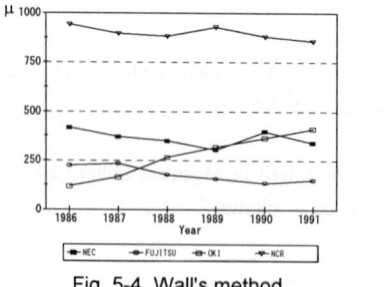

Fig. 5-4 Wall's method

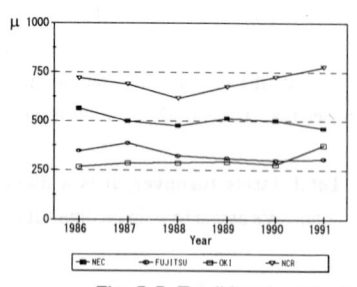

Fig. 5-5 Traditional method

In the former two methods, it was used only 7 ratios. However, as a matter of fact, there are so many real ratios, perhaps more than 150.

So, at a next step, an expert chooses more relevant ratios from the Nikkei Financial Analysis book. The expert selected 34 ratios from more than 150 ratios, and assigned a weight to each ratio. These ratios are grouped as follows:

1) **Profitability ratio**:
   Ordinary income/sales, return on sales, selling, general and administrative expenses/sales, return on total assets, return on stockholders' equity, and ordinary income/total assets.
2) **Productivity ratio**:
   Sales per employee, current income/employee, value added/employee, and labor expenses/value added.
3) **Efficiency ratio**:
   Total assets turnover, inventory turnover, fixed assets turnover, receivables/sales, and payables/sales.
4) **Stability ratio**:
   Fixed assets/invested capital, stockholders' equity ratio, debt ratio (A), fixed assets/ stockholders' equity, and debt ratio (B).
5) **Liquidity ratio**:
   Ordinary flow ratio, current ratio, current assets turnover, receivables/payables, and quick ratio.
6) **Scale ratio**:
   Sales, return on sales, total assets, stockholders' equity/capital stock, and number of employees.
7) **Growth ratio**:
   Sales, ordinary income, current income, and operating income before depreciation.

It was applied these ratios with their respective weights to the previous system. The following graph was the result:

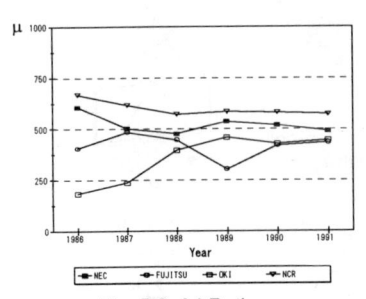

Fig. 5-6 34 Ratios

5.5 Process of Diagnosis

In addition to the mentioned system, it is tried to develop the more effective method for sequential diagnosis of the companies. As mentioned, only one membership function was used for all the ratios. However, in the diagnosis system made by a human expert, this membership is different from the company data and/or the ratios used. Therefore, the new approach will be tried by increasing the fuzzy sets for five membership functions: very bad, bad, so so, good, and very good, as shown in Fig. 5-7.

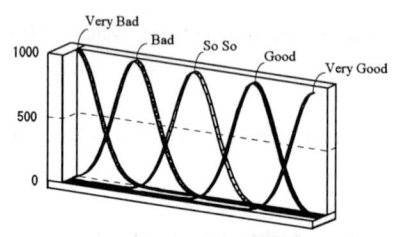

Fig. 5-7 Five Fuzzy Memberships

Each membership function will be constructed with each professional subjective judgment. By tuning specific membership function for each ratio, the ideal function will be built. The ratios and the weights previously assigned will be used here.

Using these 5 fuzzy membership functions shown above, a total value for each fuzzy set will be counted and calculated as follows:

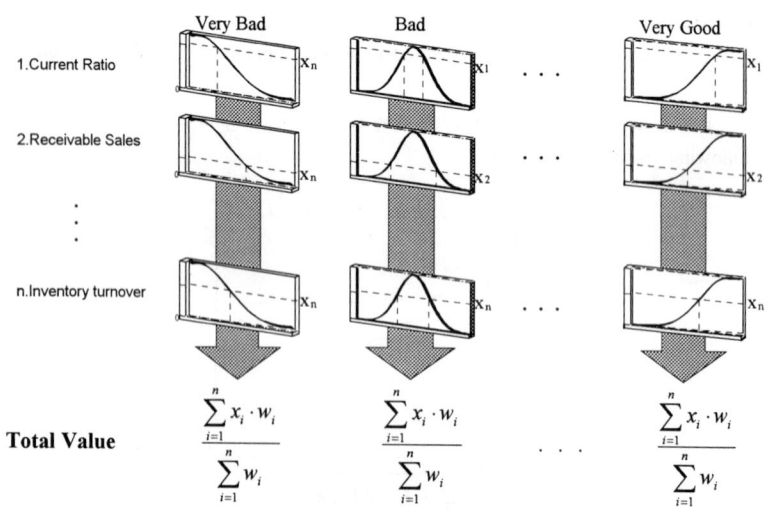

Total Value $\quad \dfrac{\sum_{i=1}^{n} x_i \cdot w_i}{\sum_{i=1}^{n} w_i}$

Fig. 5-8 Calculation of each total value

For this time, the membership value for each fuzzy set will be shown by tridimensional diagrams (Fig. 5-9). This diagram is a kind of explanatory module that makes clear the degree of each fuzzy set and shows the whole situation of the company in each year. Also, it will be useful in the tuning of the membership function.

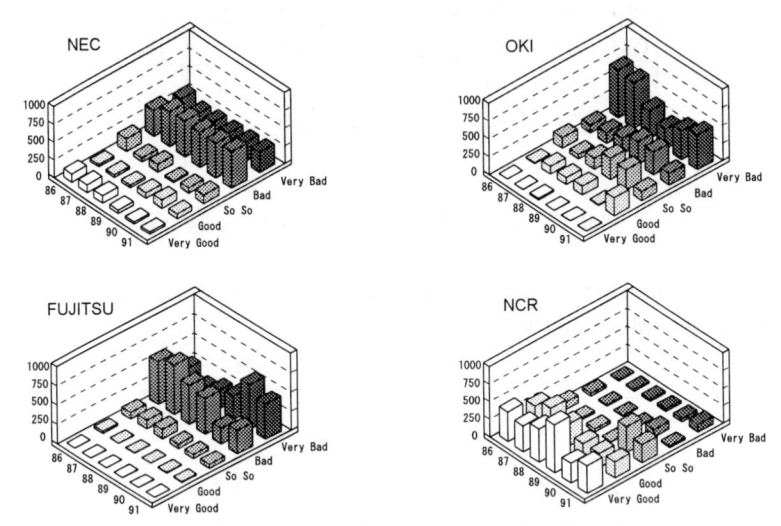

Fig. 5-9 Tridimensional Graph of the membership values

Finally, these five values are converted into a single degree with the following method: each fuzzy set is labeled like as VB (very bad), B (bad), S (so so), G (Good), and VG (very good). The total evaluation value is T.

$$T = 500 - (VB/2 + B/4) + (G/4 + VG/2)$$

This formula means that the total value is the consequence of the calculation as follows: medium membership 500 minus the weight of each membership very bad, bad, good, and very good. In an extreme case, in which each membership value is 1000, the total value T looks like:

Table 5-3 Example of Total value T

	VB	B	S	G	VG	T
Example 1	0	0	0	0	1000	1000
Example 2	0	0	0	1000	0	750
Example 3	0	0	1000	0	0	500
Example 4	0	1000	0	0	0	250
Example 5	1000	0	0	0	0	0

Table 5-4 Total value of OKI Electric Co. Ltd.

	VB	B	S	G	VG	T
OKI 1986	663	112	149	12	1	144
OKI 1987	633	148	61	105	2	174
OKI 1988	384	217	153	97	22	289
OKI 1989	278	316	301	125	2	314
OKI 1990	396	318	284	14	1	227
OKI 1991	487	163	128	250	3	280

The total value T in 1986 is:

$$T = 500 - (663/2 + 112/4) + (1/2 + 12/4) = 144$$

Using the previous method, the results of the calculation of the total value for the 4 companies are shown as follows:

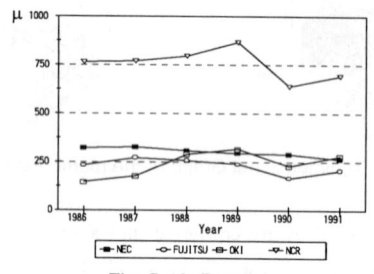

Fig. 5-10 Results

5.6 Evaluation of the System

The purpose of this research is to build the total evaluation system for the companies. The Wall's index is one of the most frequently used method for evaluation, and it is applied these characteristics in this study.

At a first step in this study, a simple membership function created by using average and standard deviation was used to calculate three types of financial data, such as Wall's ratios, traditional 7 ratios, and the 34 ratios selected by an expert. The aim of the evaluation system was to judge the tendency of each company's financial conditions from 1986 to 1991, especially between 1989 and 1990, when the market was bad and the times were hard.

By using the Wall's index method, it is shown in Fig. 5-4 that the data of Fujitsu and Japan NCR were descended between 1989 and 1990. In fact, the market was as bad as the company's conditions. NEC also reflect the fact in the year before and it was recovered in the 1990. However, using the traditional ratios, the Fig. 5-5 shows that Japan NCR and OKI don't reflect the fact. Also, the other companies, NEC and Fujitsu, showed mild movement. This is because the traditional ratios have the same weight applied to all the ratios, and they don't include human subjective ambiguity, like the Wall's ratios. In the case of the 34 ratios selected by an expert, it didn't give a correct answer about Fujitsu, in spite of the weight assigned by the expert (see Fig. 5-6).

At the next step, five fuzzy sets were used for more detailed evaluation. The results applying this method has more reliability than the former systems. We can find in Fig. 5-10 that the four companies reflect the fact. It shows the fluctuating markets between 1989 and 1990. From the overall viewpoint, we may say that the last proposed method is best for the total evaluation system.

CONCLUSION

Most of expert systems have been programmed by expert system tools. This is because there are plenty of flexibility for high-level reasoning strategies in these tools. Among many commercial base tools, FLOPS has been selected for the purpose of dealing with the ambiguity.

In this research, the Wall's seven ratios, traditional seven ratios, and 34 ratios were selected as financial ratios data. These ratios were chosen by experts' experience. However, it is necessary for us that we should originate a scientifical method for the selection of the appropriate ratios. Actually, a research by Numeric Analysis to select the ratios is being started.

In this study, the input data was restricted only to four companies of computer market (NEC, FUJITSU, OKI, Japan NCR). To prove the generality and flexibility of this system, it will be interesting to apply this theory not only in these computer business companies but also in different type of industries.

It was made use of the NEC PC-9801 FA to run the proposed diagnosis program written in parallel FLOPS. The use of FLOPS makes it very convenient to set up and modify fuzzy operations, but it takes more time to run the system than programs coded in a procedural computer language, such as C language. If higher speed is desired, the expert system must be recoded in a compiled language, such as Pascal or C/C++. It is desirable that a study will come into view, in which FLOPS's features (for example, fuzzy numbers...) will have been built by C++ language.

BIBLIOGRAPHY

[1] Giarratano, J., Riley, G., Expert Systems, principles and programming, PWS-Kent Publishing Company, Boston, p1, 1989.
[2] Hu, D., Expert Systems for Software Engineering and Managers, Chapman and Hall, New York, p3, 1987.
[3] Prerau, D., Developing and Managing Expert Systems, Addison-Wesley Publishing Co., p18, 1990.
[4] Giarratano, J., Riley, G., op. cit., pp8~10.
[5] Prerau, D., op. cit., p8.
[6] Zimmerman, H. J., Fuzzy Set Theory and its Applications, Kluwer Academic Publishers, Massachusetts, p174, 1991.
[7] Parsaye, K., Chignell, M. Expert Systems for Experts, John Wiley & Sons, Inc., Canada, p32, 1988.
[8] Giarratano, J., Riley, G., op. cit., p8.
[9] Zimmerman, H. J., op. cit., pp175~177.
[10] Barr, A., Feigenbaum, E., The Handbook of Artificial Intelligence, Addison-Wesley Publishing Co., pp153~222, Vol I, 1981.
[11] Forsyth, R., Expert Systems, Principles and case studies, Chapman and Hall Computing, pp10~12, pp~142160, 1989.
[12] Barr, A., Feigenbaum, E., op. cit., p216.
[13] Parsaye, K., op. cit., pp15~16.
[14] Giarratano, J., Riley, G., op. cit., pp18~20.
[15] Giarratano, J., Riley, G., op. cit., p185.
[16] Zimmerman, H. J., op. cit., pp11~20.
[17] Kandel, A., Schneider, M., Fuzzy Sets and Their Applications to Artificial Intelligence, in "Advances in Computers", Yovits, M., Academic Press, Inc., pp69~105, Vol 28, 1989.
[18] Zimmerman, H. J., op. cit., p12.
[19] Siler, W., Theory and Practice of Fuzzy Expert System, Proc. IEEE, pp20~29, 1989.
[20] Hirota, K., Introduction of Fuzzy Expert System, Omu Co., Japan, pp103~105, 1993.
[21] Zadeh, L., Kacprzyk, J., Fuzzy Logic for the Management of Uncertainty, John Wiley & Sons, Inc, USA, pp215~218, 1992.

22) Negoita, C. V., Expert Systems and Fuzzy Systems, The Benjamin/Cummings Publishing Company, Menlo Park, CA, pp95~98, 1985.
23) Kandel, A., Fuzzy Expert Systems, CRC Press, Inc., Florida, pp17~19, 1992.
24) Kandel, A., Schneider, M., op. cit., pp69~105.
25) Nihon Fuzzy Gakkai, Fuzzy Expert System, Nikkan Kogyou Shimbunsha, pp8~9, pp192~215, 1993.
26) Kandel, A., op. cit., pp18~19.
27) Siler, W., op. cit., pp40~79.
28) Siler, W., Tucker, D., FLOPS: A Fuzzy Logic Production System User's Manual, Kemp-Carraway Heart Institute, Birmingham, AL, pp1~297, 1986.
29) Buckley, J., Elementary learning in a Fuzzy Expert System, in Fuzzy Logic for the Management of Uncertainty, Zadeh, L., Kacprzyk, J., John Wiley & Sons, Inc. USA, pp447~464, 1992.
30) Siler, W., Buckley, J., Tucker, D., Functional Requirements for a Fuzzy Expert System Shell, in Approximate Reasoning in Intelligent Systems, Decision and Control, ed. E. Sanchez and L. A. Zadeh, Pergamon Press, New York, pp21~31, 1987.
31) Ying, H., Siler, W., Tucker, D., A New Type of Fuzzy Controller based upon a Fuzzy Expert System Shell FLOPS, in International Workshop on Artificial Inteligence for Industrial Applications, pp382~386. 1988.
32) Siler, W., Hirota, K., Theory and Practice of Fuzzy Expert System, Denki Shoin, Co., Japan, pp201~202, 1990.
33) Bamba, K., Kaikeigaku Daijiten, Chuo Keizaisha, Tokyo, 1992.
34) Cohen, J., Robbins, S., The Financial Manager, Basic Aspects for Financial Administration, Harper & Row, pp192~203, 1968.
35) Bedford, N., Introduction to Modern Accounting, The Ronald Press, Co., New York, pp686~704, 1968.
36) Li, D., Accounting for Management Analysis, Charles Merill Books, Inc., Columbus, Ohio, pp454~472, 1964.
37) Anthony, R., Reece, J., Accounting, Text and Cases, Richard Irwin, Inc., Illinois, pp411~425, 1983.
38) Solomon, L., Vargo, R., Walther, L., Financial Accounting, Harper & Row, Publishers, New York, pp656~673, 1985.
39) Imdieke, L., Smith, R., Financial Accounting, John Wiley & Sons, Inc., pp689~701, 1987.
40) Maeda, T., Nikkei Keiei Shihyo, Nihon Keizai Shinbunsha, Tokyo, 1986-1991.

AN ASSOCIATIVE DATA PARALLEL COMPILATION MODEL FOR TIGHT INTEGRATION OF HIGH PERFORMANCE KNOWLEDGE RETRIEVAL AND COMPUTING

ARVIND K. BANSAL
Department of Mathematics and Computer Science
Kent State University, Kent, OH 44242 - 0001, USA
E-mail: arvind@mcs.kent.edu

Abstract

Associative Computation is characterized by intertwining of search by content and data parallel computation. An algebra for associative computation is described. A compilation based model and a novel abstract machine for associative logic programming are presented. The model uses loose coupling of left hand side of the program, treated as data, and right hand side of the program, treated as low level code. This representation achieves efficiency by associative computation and data alignment during goal reduction and during execution of low level abstract instructions. Data alignment reduces the overhead of data movement. Novel schemes for associative manipulation of aliased uninstantiated variables, data parallel goal reduction in the presence multiple occurrences of the same variables in a goal. The architecture, behavior, and performance evaluation of the model are presented.

Keywords: Artificial intelligence, Associative computing, Data parallel Computing, High performance, Knowledge bases, Knowledge retrieval, Logic programming

1 Introduction

Many problems such as modeling of aero-elasticity in aircrafts, image understanding systems, natural language understanding systems, geographic information systems, financial accounting systems, robot navigation, and genome sequencing require efficient integration of high performance intelligent reasoning, efficient information retrieval from large knowledge bases, and massive parallel scientific computing. In recent years, the logic programming paradigm [1, 2] has become a popular tool for AI systems and knowledge representation due to its natural capability to store databases, the declarative nature of programming, and the power of nondeterministic computation which facilitates alternate solutions to a problem. Previous approaches for high performance intelligent processing under logic programming paradigm are as follows:

- Spawning and mapping concurrent processes either on MIMD architectures or pipelined vector computers [3, 4]: these models suffer from combinatorial explosion of processes. In process based models overhead of spawning processes is linearly proportional to the number of data elements. Such schemes are not suitable for data intensive problems which need same abstract computation on a large amount of data.

- Simulation of conventional abstract machines [5] on massive parallel machines: these models simulate pointer based data representation, and indexing based upon predicate name to identify procedures used for goal reduction. The use of pointer based data representation causes inherent sequentiality, and indexing scheme restricts the query power.

- Exploitation of associative search and data parallel computation [6, 7, 8]: these models do not exploit intertwining of search by content and data parallel computation.

- Loose coupling of intelligent reasoning on conventional architectures using conventional data representation and data parallel scientific computing on massively parallel supercomputers: these schemes suffer from data transfer overhead between architectures and run time overhead to transform data between conventional data representation schemes and data representation schemes on massively parallel supercomputers.

The implementations based upon conventional models, which use indexing on predicate-name to select a clause and pointer-based data representation, are incapable of answering a large class of queries needed in real world knowledge retrieval and data parallel computation. Some examples are

(a) queries which derive the set of objects based upon incomplete information about their attributes,

(b) queries which relate one or more objects without apriori knowledge of the relationship,

(c) queries which reason about meta-relations - relations relating relations - about the objects, and

(d) queries integrating inequalities, ranges and data parallel computations.

The following two examples illustrate the above deficiencies: Example 1 illustrates the power of associative search to derive unspecified relations for the given attributes and handling of queries with meta-relations; Example 2 illustrates intertwining of search by content, inequality tests, and data parallel computation.

Example 1:

Conventional models answer queries such as "Who is a sister of Tom?"; "Who is a brother of Mary"?; "Who is a parent of Mary?"; "Who is a parent of John" etc. These models can not answer queries such as "How are Tom and Mary related?"; "How are Mary and John related?"; "Specify the relatives and their corresponding relationships to Tom?"; and "Who are relatives of Tom?". The corresponding query formulations are given in Table 1: the first column illustrates the class of queries, the second column illustrates formulation, the third column shows the query processing scheme, and the fourth column shows the processing capability of conventional systems.

% **Meta relations**

$relative(sister).\ relative(brother).\ relative(parent).$

% **Concrete relations**

$brother(X,\ Y)\ :\text{-}\ parent(Z,\ X),\ parent(Z,\ Y),\ male(X),\ not\_same(X,\ Y).$

$sister(X,\ Y)\ :\text{-}\ parent(Z,\ X),\ parent(Z,\ Y),\ female(X),\ not\_same(X,\ Y).$

$parent(jane,\ mary).\quad parent(ram,\ mary).$
$parent(jane,\ tom).\quad parent(ram,\ john).$
$male(tom).\quad male(john)\quad male(ram).$
$female(mary)\quad female(jane).$

Table 1: Query power extension in associative model
GR: Goal reduction, FL: fact lookup, MR: meta relation

Class	Goal	Type	Con.
brother of mary?	$brother(mary,\ X)?$	GR	Yes
relationship of tom with mary?	$P(tom,\ mary)?$	FL	No
relationship of mary with tom ?	$P(mary,\ tom)?$	GR	No
relationships and relatives of tom?	$P(tom,\ X)?$	GR	No
relatives of tom?	$relative(P)\ \bigwedge$	MR	No
	$P(tom,\ X)?$	GR	

Example 2:
Consider a large data base of employees in a company. We have to identify the set of employees within the salary range $25,000\ <\ salary\ <\ 35,000$ in a specific department "D". The salaries of these employees have to be raised by 5 %. First the subset of employees is selected using associative search with inequality tests $salary\ >\ 25,000$ and $salary\ <\ 35,000$, and equality test $department\ =\ $"D". After the selection of the subset, data parallel computation is used to compute the raise. This intertwining of associative search and data parallel computation is not possible in conventional systems.

In this paper, we present an algebra for associative computing, and use this algebra to develop a novel computation model of associative logic programming which achieves tight integration of associative knowledge retrieval, data parallel computation, and rule based programming without data transformation and data transfer overhead. The model introduces many novel implementation techniques as follows:

- Abstract data is represented to exploit seamless integration of associative search and data parallel computation [9, 10].

- Data parallel matches and data parallel computations are used to release bindings to reduce overhead of finding alternate solutions,

- A data parallel environment is used to facilitate associative computation on bags of bindings, vectors, sets, and sequences.

- Different components of the abstract machine are aligned to reduce the overhead of data transfer and data transformation.

The compiler and emulator for the proposed model have been implemented. The compiler has been written in C^{++}. The emulator has been written using ANSI C. The model is portable to any architecture which supports a data parallel version of C. The benchmark results show:

- Knowledge retrieval is independent of the number of ground facts.

- Knowledge retrieval is possible by incomplete information which makes knowledge discovery possible.

- Relations with a large number of arguments are handled efficiently with little overhead.

- Associative lookup speed is quite comparable to data parallel computations, which allows the tight integration of high performance knowledge retrieval and data parallel computation without any overhead of data movement and data transformation.

- The model is efficient for both scalar and data parallel computations on various abstract data such as sequences, matrices, bags, and sets.

Implications of these results are that the model can be successfully applied to data intensive problems such as geographical information systems, image understanding systems, statistical knowledge bases, and genome sequencing. For example, in geographical information systems, spatial data structures such as quad-trees and oct-trees are represented associatively; different regions having the same values can be identified using associative search on values [10], and integrated with intelligent rule based reasoning. Integration of data parallel scientific computing, knowledge base retrieval, and rule based reasoning provides necessary tool for image understanding systems. Statistical queries can directly benefit from associative search by content, associative representation of structures, data parallel arithmetic computations, and data parallel aggregate functions. Genome sequencing requires integration of knowledge retrieval, efficient insertion and deletion of data elements, and efficient manipulation of matrices for heuristic matching of sequences.

Section 2 describes the associative computing paradigm and an algebra for associative computation. Section 3 discusses definitions of logic programming, Warren Abstract Machine (WAM) - the compilation model on conventional machines, and advantages of incorporating associative computing over conventional schemes. Section 4 describes an associative abstract machine. Section 5 describes the advantages achieved by seamless integration of associative search, data parallel computation, and data alignment in the abstract machine. Section 6 describes the model behavior. Section 7 discusses the performance evaluation and compares the work with other related works. The last section concludes the paper.

2 Background

In this section, the definitions and mathematical notations, associative architecture [10, 11], associative computing [10], and associative representation of abstract data such as matrix, tree, and quad-tree are presented.

2.1 Preliminary Definitions and Notations

A bag is a collection of items such that there can be multiple occurrences of an element. A *D-bag*, denoted by $\mathcal{D}$, is defined as an ordered bag which also contains null elements $\perp$. For example, { 2, $\perp$, 3 } is a D-bag. However, { 2, $\perp$, 3 } $\neq$ { $\perp$, 2, 3 } since D-bags are ordered. $\perp \preceq$ every element in the D-bag. A D-bag $\mathcal{D}_1 =$ { d_{11}, ..., d_{1N} } is D-included in another D-bag $\mathcal{D}_2 = \{d_{21}, ..., d_{2N}\}$ if $\forall I_{(1 \leq I \leq N)}$ $d_{1I} \preceq d_{2I}$. For example, { 4, $\perp$, 5, 6 } is a *D-subbag* of { 4, 3, 5, 6 } since $\perp \preceq 3$. D-union of two D-subbags of a D-bag derives a new D-bag { d_{31}, ..., d_{3N} } such that $\forall I_{(1 \leq I \leq N)}$ $d_{3I} = d_{1I}$ if $d_{2I} \preceq d_{1I}$, $d_{3I} = d_{2I}$ if $d_{1I} \preceq d_{2I}$, or $d_{3I} = d_{1I}$ if $d_{1I} = d_{2I}$. For example, D-union of the D-subbags { $\perp$, b, c } and { a, b, $\perp$ } derives the D-bag { a, b, c }. D-intersection of two D-subbags of a D-bag derives a new D-bag such that $\forall I_{(1 \leq I \leq N)}$ $d_{3I} = d_{1I}$ if $d_{1I} \preceq d_{2I}$, $d_{3I} = d_{2I}$ if $d_{2I} \preceq d_{1I}$, or $d_{3I} = d_{1I}$ if $d_{1I} = d_{2I}$. For example, D-intersection of D-subbags { 2, 3,, $\perp$ } and { $\perp$, 3, 4 } derives the D-subbag { $\perp$, 3, $\perp$ }. Note that D-inclusion, D-union, and D-intersection of two D-subbags are different than usual definitions of union and intersection since D-inclusion, D-union, and D-intersection are based on pairwise comparison of elements in two D-bags. The truth values *true* and *false* are treated synonymously with the values "1" and "0" respectively. An F-bag is a D-bag which has either *true* and *false*. We treat *false* (or "0") $\preceq$ *true* (or "1"). A F-bag of *1s* is denoted by $\mathcal{F}^1$, a F-bag of *0s* is denoted by $\mathcal{F}^0$, and a F-bag containing both *1s* and *0s* is denoted by $\mathcal{F}$. F-bags are realized by logical bit-vectors. Under the assumption *false* $\preceq$ *true*, D-union of F-bags and logical-OR of the corresponding logical bit-vectors are equivalent, and D-intersection of F-bags and logical-AND of the corresponding logical bit-vectors are equivalent. Cartesian product is denoted by $\times$, equivalence is denoted by $\equiv$, isomorphism is denoted by $\cong$, inclusion is denoted by $\subseteq$, D-inclusion is denoted by $\sqsubseteq$, intersection of two bags is denoted by $\cap$, D-intersection of two D-bags is denoted by $\sqcap$, union of two bags is denoted by $\cup$, D-union of two D-bags is denoted by $\sqcup$, logical-AND is denoted by $\wedge$, logical-OR is denoted by $\vee$, underived value is denoted by $\top$, data parallel computation on D-bags is denoted by $\odot^D$, and computation on a pair of data element or a singleton data element is denoted by $\odot$, associative update is denoted by $\uplus$, associative deletion is denoted

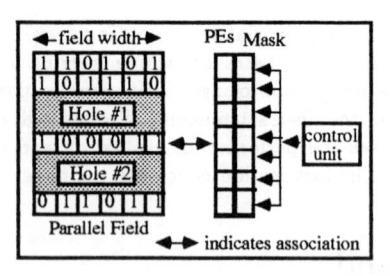

Figure 1: SIMD Architecture and Parallel Field

by $\ominus$, and communication cost between two data elements d_I and d_J is denoted by $e(d_I, d_J)$. Both $\odot$ and $\odot$ are generic representation of an operator. A subscript is used to denote a specific instance of an abstract data, and a superscript is used to denote a subclass.

2.2 Architecture for Associative Computing

An architecture is a sequence of *processing cells*. Each cell is a quadruple $< C_i, R_i, S_i, M_i >$ where C_i denotes a processing element (PE), R_i denotes a set of local registers, S_i denotes local storage, and M_i denotes a mask-bit. The D-bag of mask bits is set selectively to filter instructions. An instruction is broadcast to each cell simultaneously. The flow of control is effected by generating, saving, and restoring the mask bits based on the results of tests on local data [10, 11]. An associative search of a field for a specific value sets up the corresponding mask bit which is stored and manipulated during computations. SIMD architecture with content addressable memory [11] (see Figure 1) satisfies this criteria. However, we do not limit the scope of the associative computation to SIMD architectures.

2.3 Associative Computing Paradigm

Associative computing [10] is characterized by a seamless integration of *data-parallel computation* [12], *association of data elements* [13], and *search by content* [11]. *Data-parallelism* is characterized by performing the same abstract computations concurrently on a bag of data [12].

Data parallel computations are independent of the number of elements in a D-bag since each data element can be mapped on a single PE in a unit time. For SIMD computers, the data parallel computation is performed in a lock-step fashion. However, the notion of data parallel computation is more general: same abstract computations on individual elements are performed concurrently without any control dependency. A computation is *data sequential* if the same abstract computations are performed individually on the elements, one element at a time,

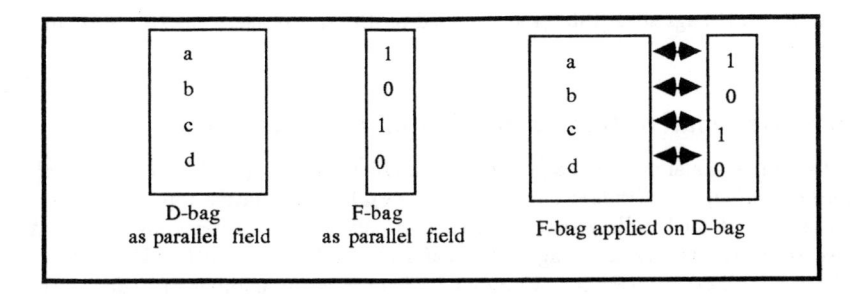

Figure 2: Associative representation of data

of the same bag. Partial data parallelism is exploited if M elements (M ≤ number of elements in the bag) are processed at a time. If the number of elements in a field is larger than the maximum number of processing elements, the elements are folded, and the processing time is multiplied by O(number of elements/number of processing elements).

2.4 Representing Data on Associative Architecture

Data items with the same data type are stored in a *parallel field* - a two dimensional memory organization of columns and rows with each cell attached to each row. Each row stores a data item as illustrated in Figure 1. A D-bag is realized using a parallel field. ⊥ is realized using a hole which means absence of an element. A F-bag is realized using *logical bit-vectors* as illustrated in Figure 2. D-union of two F-bags is realized by logical-OR of the corresponding logical bit-vectors; D-intersection of two F-bags is realized by logical-AND of the corresponding logical bit-vectors. For example, { $0, 1, 0$ } $\bigsqcup$ { $1, 0, 1$ } ≡ { $0 \vee 1, 1 \vee 0, 0 \vee 1$ } ≡ { $1, 1, 1$ }. Similarly, { $1, 1, 0$ } $\sqcap$ { $0, 1, 0$ } ≡ { $1 \wedge 0, 1 \wedge 1, 0 \wedge 0$ } ≡ { $0, 1, 0$ }.

2.4.1 Notations related to associative data

Number of D-bags are denoted by M (where $M \geq 1$), and number of elements in a D-bag are denoted by N (where $N \geq 1$). Using association of parallel fields, a D-bag of M-tuples of the form { $< d_{11}, ..., d_{M1} >, ..., < d_{1N}, ..., d_{MN} >$ } is stored as M aligned D-bags. M D-bags are aligned if the communication distance $\forall$ $(I, J)_{1 \leq (I,J) \leq M}$ $e(d_{IK} \in \mathcal{D}_I, d_{JK} \in \mathcal{D}_J)$ is same for all tuples.

An association of M D-bags is denoted by $\mathcal{D}_1 \bigoplus \mathcal{D}_2 \bigoplus ... \bigoplus \mathcal{D}_M$ where $\bigoplus$ represents the association of two D-bags. A projection of a D-bag from an association of D-bags is denoted by by Π_I, and selection of a specific element from a tuple is denoted by π_I. For example, $\Pi_{I(1 \leq I \leq M)}$ ($\mathcal{D}_1 \bigoplus \mathcal{D}_2 \bigoplus ... \bigoplus \mathcal{D}_M$) = $\mathcal{D}_I$, and $\pi_{I(1 \leq I \leq N)}$ ($< d_{11}, ..., d_{1N} >$) = d_{1I}. An association of a D-bag with an F-bag is denoted by $\bigotimes$. Association of a D-bag and an F-bag is different from an association of two D-bags since F-bags are applied on D-bags to select the corresponding elements.

2.5 Associative Representation of Abstract Data

A two-dimensional matrix is represented as a D-bag of triples of the form $\{(index_1^1, index_1^2, value_1), ..., (index_N^1, index_M^2, value_{NM})\}$. Associative representation is quite efficient for handling sparse matrices as only non-zero entries are stored. A tree is represented as a D-bag of the pairs of the form $\{$ $(Path_1, Value_1)$, ..., $(Path_I, Value_I)$, ..., $(Path_N, Value_N)$ $\}$ where N is the number of nodes in the tree; $Path_I$ is the path encoding of the node from the root node, and $Value_I$ is information stored at the node. For details of tree representation and manipulation, the reader may refer to [9, 14]. A quad-tree, an example of spatial data structure, is represented as a D-bag of 5-tuples of the form $\{$ $(X\_lb_1, X\_ub_1, Y\_lb_1, Y\_ub_1, Value_1)$, ..., $\{$ $(X\_lb_I, X\_ub_I, Y\_lb_I, Y\_ub_I, Value_I)$, ..., $\{$ $(X\_lb_N, X\_ub_N, Y\_lb_N, Y\_ub_N, Value_N)$ $\}$ where N is the number of nodes in a quad-tree. Where "lb" denotes lower bound, and "ub" denotes upper bound. Similarly, an oct-tree is represented a D-bag of 7-tuples of the form $\{$ $(X\_lb_1, X\_ub_1, Y\_lb_1, Y\_ub_1, Z\_lb_1, Z\_ub_1, Value_1)$, ..., $\{$ $< X\_lb_I, X\_ub_I, Y\_lb_I, Y\_ub_I, Z\_lb_I, Z\_ub_I, Value_I >$, ..., $\{$ $<X\_lb_N, X\_ub_N, Y\_lb_N, Y\_ub_N, Z\_lb_N, Z\_ub_N, Value_N>$ $\}$. Any partial data structure satisfying a specific property is represented as a D-subbag $\mathcal{D} \bigotimes \mathcal{F}$ where $\mathcal{D}$ is the original D-bag and $\mathcal{F}$ is derived by searching for a specific attribute.

Example 3:

Consider the various abstract representation of data. For example, the sparse matrix in Figure 3A is represented as $\{$ $< 1, 1, 10 >, < 100, 50, 30 >, < 100, 90, 50 >$ $\}$. The binary tree in Figure 3B is represented as $\{$ $< 100, a >, < 110, b >, < 120, c >, < 111, b >, < 112, d >$. The quad tree in Figure 3C is represented as $\{$ $< 0, 50, 0, 50,$ "corn" $>, < 50, 75, 50, 75,$ "wheat" $>, < 75, 100, 50, 75,$ "corn" $>, < 75, 100, 75, 100,$ "water" $>, < 50, 100, 0, 50,$ "water" $>$ $\}$.

3 Overview of Logic Programming Concepts

This section reviews definitions of logic programming [1, 2], describes WAM - the conventional implementation schemes of logic programs [5], and presents advantages of associative computation [6, 9, 14] over conventional computation schemes.

A *logic program* is a set of *facts* and *rules*. The set of facts and rules having the same name and number of arguments form a *procedure*. Each rule (Horn clause) is of the form $A := B_1, ..., B_N$ ($N \geq 0$). For a fact, N is 0. The left hand side of a rule is a *clause-head*, and each literal B_i on the right-hand side is a *subgoal*. While facts form the data part of a program, rules are used to reduce a query to simpler subqueries. The execution model is based upon reduction of a query to a conjunction of subqueries (subgoals) in a repeated manner. The heart of the reduction process is *unification*: a pattern matching process, which is used for matching the values of the same type of data objects, associating a value with a variable, and passing the values between procedures. The process of unbinding caused by the failure

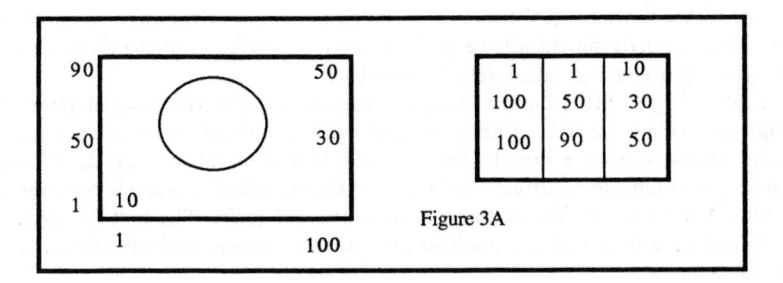

Figure 3A

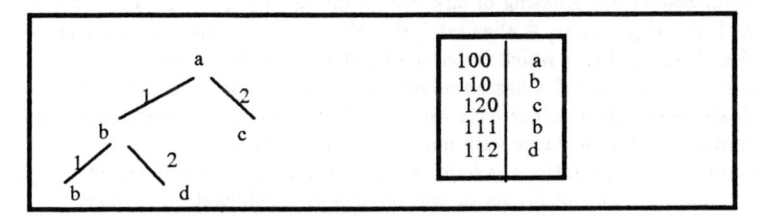

Figure 3B

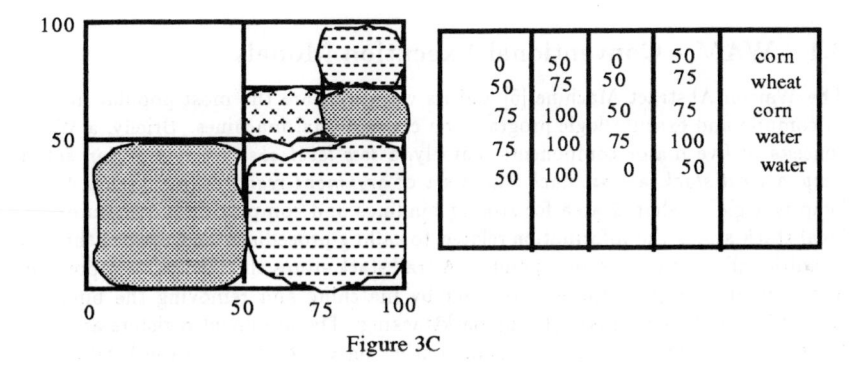

Figure 3C

Figure 3: Associative representation of abstract data

of the unification between a goal term and a clause-head, and the selection of an alternate clause during a goal reduction is called *shallow backtracking*. The process of restoring the environment when a goal can not unify with any of the clause-heads in the given procedure is called *deep backtracking*.

Variables begin with a capital letter, and constants begin with a small letter. A *multiple-occurrence variable* occurs more than once in a literal. A *single-occurrence variable* occurs once in a literal. For example, in the literal $a(X, X, b, Y)$, the variable X is a multiple-occurrence variable, and the variable Y is a single-occurrence variable. A shared variable occurs in at least two subgoals. The first occurrence of a shared variable is called a *producer* and other occurrences of the variable are called *consumers*.

Two uninstantiated variables are aliased if they share the same value. Aliasing is an equivalence relationship: any variable V_1 is aliased to itself; if V_1 is aliased to V_2 and V_2 is aliased to V_3 then V_1 is alias to V_3; if the variable V_1 is aliased to V_2 then V_2 is aliased to V_1. Aliasing of any two variables $V_{I(1 \leq I \leq N)} \in$ alias set $\{ V_1, V_2, ..., V_N \}$ and $W_{J(1 \leq J \leq M)} \in$ alias set $\{ W_1, W_2, ..., W_M \}$ gives a new alias set $\{ V_1, ..., V_N, W_1, ..., W_M \}$ which is derived by the union of two sets.

A fact is a *simple fact* if it has no aliased variables. A fact is a *complex fact* if it has at least one multiple-occurrence variable. A clause is a *complex clause* if it is either a complex fact or a clause with non-empty clause-body.

Meta relations treat relations as objects. In Example 1, the predicate *relative/1* is a meta relation. The positions of the objects is significant in a query. For example, "Who is Tom related to under what relationship?" and "Who are related to Tom under what relationship"? are expressed as "$P(tom, X)$?" and "$P(X, tom)$?" respectively. The uninstantiated variable P indicates unspecified relation names, and the variable X indicates unspecified objects.

3.1 WAM - Conventional Execution Models

The Warren Abstract Machine [5] and its variations are the most popular models to compile and execute logic programs on conventional machines. Briefly, a WAM consists of five major components, namely, a *low level abstract instruction set*, a *heap*, a *local stack*, a *trail stack*, and a set of *argument registers* (see Figure 4). A heap is a global shared area for storing bindings and complex data structures. A local stack stores the information related to calling procedures and control threads (traditionally known as choice-points). A trail stack stores the bindings temporarily and is used to restore the environment by matching and removing the bindings caused by the failed clauses, during backtracking. The argument registers are used to store the pointers to the arguments (either on the heap or on the local stack) of the procedure being executed. The use of argument registers reduces the data transfer time significantly during a procedure call, and has the same application as the use of registers in assemblers. Storage of previous environment consists of argument registers to be altered and multiple pointers such as the pointer in the heap, pointer in the trail stack, pointer in the local stack. Previous environment is restored after successful execution of the called procedure or during backtracking. Complex data structures are stored and manipulated in the heap using chain of

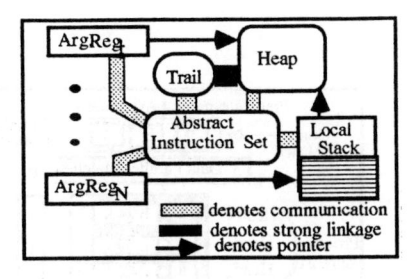

Figure 4: WAM architecture

pointers. Allocation of these pointer-based structures cause run time overhead of memory allocation, and manipulation of data items is sequential due to traversal of chain of pointers.

3.2 Advantages of Associative Computation

The advantages of the associative computation model, exploited in the previous approaches [6, 9, 14, 15] are as follows:

(1) A goal position can be searched in one unit operation. An association of D-bags is used to hold all the elements of an argument in the set of clauses of a program. This process matches the ground values to identify non-unifiable clause-heads, and derives the bindings for the goal variables. The data parallel match reduces the shallow backtracking significantly. Figure 5 illustrates the concept. The goal $P(H, k)$ matches the whole set of clause-heads { $a(b, k)$, $a(c, k)$, $a(d, c)$, ..., $c(d, g)$, $c(c, k$ }in three unit data parallel matches.

(2) Bindings are associatively released in a constant number of operations during backtracking. Figure 6 illustrates the concept. The bindings related to time-stamp *20* are released in a unit data parallel match.

(3) Association of D-bags is used instead of conventional stacks to unify two logical terms. Use of associative search significantly reduces the overhead of searching for the bindings and the restoration of control thread.

(4) Data parallel computation is perfomed on all the elements of a D-bag simultaneously. In conventional schemes, such computations are dependent on the size of the input data, use tail recursion, and are at best close to linear iteration time.

4 An Algebra of Associative Computation

This section describes an algebra for associative computation. This algebra is the basis of associative logic programming model. There are five types of laws: *laws for data association, laws for associative search, laws for associative selection, laws*

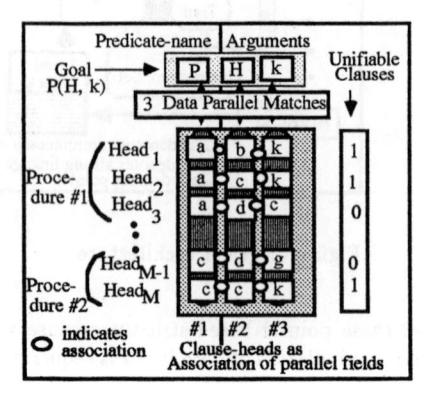

Figure 5: Data parallel match in goal reduction

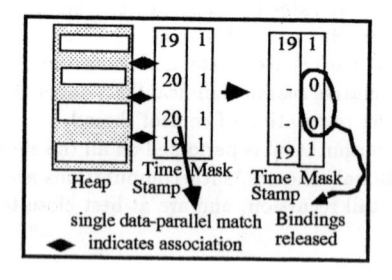

Figure 6: Data parallel release during backtracking

for data parallel computations and *laws for associative update*. These laws have been used extensively in the development of the model. Each rule is denoted by $R^A_{I(1 \leq I \leq 17)}$.

4.1 Laws of Data Association

The first rule states that a D-bag of M-tuples is given by the association of M D-bags such that corresponding elements in every tuple are aligned. For example, { a, 2, 3, 4 } $\bigoplus$ { b, 5, 6, 7 } is equivalent to { $< a, b >, < 2, 5 >, < 3, 6 >, < 4, 7 >$ }.

$$\mathcal{D}_1 \bigoplus ... \bigoplus \mathcal{D}_M \equiv$$
$$\{ < d_{11}, ..., d_{M1} >, ..., < d_{1N}, ..., d_{MN} > \} \qquad \text{-Construction (4.1)}$$

The second rule states that given a D-bag of tuples, a D-bag of elements can be projected in a unit operation.

$$\Pi_I(\mathcal{D}_1 \bigoplus ... \bigoplus \mathcal{D}_M) \Rightarrow \mathcal{D}_I. \qquad \text{- Projection (4.2)}$$

The third rule states that resulting association is independent of level of sub-associations formed by individual D-bags. For example, { 1, 2 } $\bigoplus$ ({ 3, 4 } $\bigoplus$ { 5, 6 }) is equivalent to ({ 1, 2 } $\bigoplus$ { 3, 4 }) $\bigoplus$ { 5, 6 }, and both are equivalent to { $< 1, 3, 5 >, < 2, 4, 6 >$ }.

$$\mathcal{D}_1 \bigoplus (\mathcal{D}_2 \bigoplus \mathcal{D}_3) \equiv (\mathcal{D}_1 \bigoplus \mathcal{D}_2) \bigoplus \mathcal{D}_3 \qquad \text{- Associativity (4.3)}$$

The fourth rule states that associations of D-bags are not symmetric. However, the associations are isomorphic: a pair (x, y) $\in \mathcal{S}_1 \bigoplus \mathcal{D}_2$ (such that $x \in \mathcal{D}_1$ and $y \in \mathcal{D}_2$) has a bijective mapping to (y, x) $\in \mathcal{D}_2 \bigoplus \mathcal{D}_1$. The implication of this rule is that the same information can be represented equivalently by permuting the order of association, and the communication distance between elements of a tuple is independent of the order of association.

$$(\mathcal{D}_1 \bigoplus \mathcal{D}_2 \not\equiv \mathcal{D}_2 \bigoplus \mathcal{D}_1) \bigwedge (\mathcal{D}_1 \bigoplus \mathcal{D}_2 \cong \mathcal{D}_2 \bigoplus \mathcal{D}_1) \qquad \text{-Antisymmetry (4.4)}$$

4.2 Laws of Associative Search

The fifth rule states that associative search of a data element d in a D-bag $\mathcal{D}$ derives an F-bag $\mathcal{F}$ such that for every $d_j = d$, the corresponding element in F-bag is "1" otherwise the corresponding element in F-bag is "0". For example, associative search of an element 4 in the D-bag { 3, 5, 4, 7, 4, 9 } gives an F-bag $\{0, 0, 1, 0, 1, 0\}$ and { 3, 5, 4, 7, 4, 9 } $\bigotimes \{0, 0, 1, 0, 1, 0\}$ derives $\{0, 0, 3, 0, 3, 0\}$.

$$d \in \mathcal{D} \Rightarrow \mathcal{D} \bigotimes \mathcal{F} \text{ such that}$$
$$\forall I_{(1 \leq I \leq N)}(d = d_I) \Rightarrow (\pi_I(\mathcal{F}) = 1)$$
$$\forall I_{(1 \leq I \leq N)}(d \neq d_I) \Rightarrow (\pi_I(\mathcal{F}) = 0) \qquad \text{Membership (4.5)}$$

The sixth rule states that by associatively searching in one field, the associated data elements in the other field can be extracted. For example, associative search for a tuple $\{4, \top, \top\}$ in a D-bag { $< 4, 5, 6 >, < 3, 7, 9 >, ..., < 4, 9, 10 >$ }

derives the F-bag { 1, 0, ..., 1 }. The F-bag when applied on the D-bag selects the D-subbag { < 4, 5, 6>, ⊥, ..., < 4, 9, 10 > }.

$$(\mathcal{D}_1 \oplus \mathcal{D}_2 \oplus, ..., \mathcal{D}_M) \wedge d \in \mathcal{D}_{J(1 \leq J \leq M)} \equiv$$
$$(\mathcal{D}_1 \oplus \mathcal{D}_2 \oplus, ..., \mathcal{D}_M) \otimes \mathcal{F} \text{ such that}$$
$$\forall I_{(1 \leq I \leq N)} ((\pi_J(< d_{1I}, d_{2I}, ..., d_{MI} >) = d) \Rightarrow \pi_I(\mathcal{F}) = 1)$$
$$\forall I_{(1 \leq I \leq N)} ((\pi_J(< d_{1I}, d_{2I}, ..., d_{MI} >) \neq d) \Rightarrow \pi_I(\mathcal{F}) = 0). \text{ Search (4.6)}$$

4.3 Laws of Associative Selection

The seventh rule states that association of an F-bag with a D-bag selects the data elements whenever the corresponding element in F-bag is 1. For example, { 3, 5, 6 } ⊗ {0, 1, 0} ≡ { ⊥, 5, ⊥ }.

$$\mathcal{D}_1 \otimes \mathcal{F} \Rightarrow \mathcal{D}_2 \text{ such that } (\mathcal{D}_2 \sqsubseteq \mathcal{D}_1) \text{ and}$$
$$\forall I_{(1 \leq I \leq N)} (\pi_I(\mathcal{F} = 1)) \Rightarrow (d_{2I} = d_{1I})$$
$$\forall I_{(1 \leq I \leq N)} (\pi_I(\mathcal{F} = 0)) \Rightarrow (d_{2I} = \bot). \qquad \text{- Selection (4.7)}$$

The eighth rule states that selecting data elements from two associated D-bags is equivalent to selecting data elements from individual D-bags and then associating the D-subbags. For example, ({ 4, 5, 6 } ⊗ { 1, 0, 1 }) ⊕ ({ a, b, c } ⊗ { 1, 0, 1 }) ≡ { < 4, a >, < 5, b >, < 6, c > } ⊗ { 1, 0, 1 } ⇒ { < 4, a >, ⊥, < 6, c > }.

$$(\mathcal{D}_1 \oplus \mathcal{D}_2) \otimes \mathcal{F} \equiv (\mathcal{D}_1 \otimes \mathcal{F}) \oplus (\mathcal{D}_2 \otimes \mathcal{F}). \qquad \text{- Distributivity (4.8)}$$

The ninth rule states that data elements of a D-bag $\mathcal{D}$ selected by applying a F-bag $\mathcal{F}_1$ includes the data elements selected by applying another F-bag $\mathcal{F}_2$ if $\mathcal{F}_1 \sqsubseteq \mathcal{F}_2$. For example, ({ 5, 6, 7 } ⊗ { 1, 0, 0 }) ≡ { 5, ⊥, ⊥ } ⊑ ({ 5, 6, 7 } ⊗ { 1, 0, 1 }) ⇒ { 5, ⊥, 6 }

$$\mathcal{F}_1 \sqsubseteq \mathcal{F}_2 \Rightarrow (\mathcal{D} \otimes \mathcal{F}_1) \sqsubseteq (\mathcal{D} \otimes \mathcal{F}_2). \qquad \text{- Monotonicity (4.9)}$$

The tenth rule states that data elements of a D-bag selected by applying D-union of two F-bags is same as D-union of two D-subbags derived by by applying individual F-bags on the D-bag. For example, ({2, 3, 4} ⊗ {1, 0, 0}) ⊔ ({2, 3, 4} ⊗ {0, 1, 0}) ≡ {2, 3, 4} ⊗ ({1, 0, 0} ⊔ {0, 1, 0}) ⇒ ({2, 3, 4} ⊗ {1, 1, 0}) ⇒ ({2, ⊥, ⊥} ⊔ {⊥, 3, ⊥}) ⇒ < 2, 3, ⊥ >.

$$(\mathcal{D} \otimes \mathcal{F}_1) \sqcup (\mathcal{D} \otimes \mathcal{F}_2) \equiv \mathcal{D} \otimes (\mathcal{F}_1 \sqcup \mathcal{F}_2) \qquad \text{- D-union (4.10)}$$

The eleventh rule states that data elements of a D-bag selected by applying D-intersection of two F-bags is the same as D-intersection of two D-subbags derived by applying individual F-bag on the D-bag. For example, ({2, 3, 4} ⊗ {1, 1, 0}) ⊓ ({2, 3, 4} ⊗ {0, 1, 1}) ≡ {2, 3, 4} ⊗ ({1, 1, 0} ⊓ {0, 1, 1 }) ⇒ ({2, 3, 4} ⊗ {0, 1, 0}) ≡ ({2, 3, ⊥} ⊓ {⊥, 3, 4}) ⇒ { ⊥, 3, ⊥ }.

$$(\mathcal{D} \otimes \mathcal{F}_1) \sqcap (\mathcal{D} \otimes \mathcal{F}_2) \equiv \mathcal{D} \otimes (\mathcal{F}_1 \sqcap \mathcal{F}_2) \qquad \text{- D-intersection (4.11)}$$

The twelfth rule states that Cartesian product of a bag $\mathcal{D}$ with *true* is equivalent to applying $\mathcal{F}^1$ on $\mathcal{D}$, and is equivalent to $\mathcal{D}$ itself. For example, {2, 3, 4} × { *true*

$\} \equiv \{2, 3, 4\} \bigotimes \{1, 1, 1\} \equiv \{2, 3, 4\}$.

$\mathcal{D}_1 \times \{ \text{ true } \} \equiv \mathcal{D}_1 \bigotimes \mathcal{F}^1 \equiv \mathcal{D}$.   - Identity (4.12)

The thirteenth rule states Cartesian product of a D-bag $\mathcal{D}$ with *false* is equivalent to applying $\mathcal{F}^0$ on $\mathcal{D}$, and derives a null set ϕ. A null set is also represented as a D-bag with every element as $\perp$.

$\mathcal{D} \times \{ \text{ false } \} \equiv \mathcal{D} \bigotimes \mathcal{F} \equiv \phi$.   - Zero (4.13)

4.4 Laws of Data Parallel Computation

The fourteenth rule states that data parallel computation on two associated D-bags is equivalent to same computation on every pair of corresponding elements of the two D-bags. Any computation involving $\perp$ maps onto $\perp$. For example, $\{2, 3, \perp\}$ $*^D \{3, 4, \perp\} \equiv \{ 2 * 3, 3 * 4, \perp * \perp \} \Rightarrow \{6, 12, \perp\}$.

$(\mathcal{D}_1 \bigoplus \mathcal{D}_2) \bigwedge (\mathcal{D}_1 \bigodot^D \mathcal{D}_2) \Rightarrow$
$\{ (d_{11} \odot d_{21}), ..., (d_{1N} \odot d_{2N}) \}$ - Vector-vector $\Rightarrow$ vector (4.14)

The fifteenth rule states that a data parallel computation involving a scalar value d and a D-bag $\mathcal{D}$ is equivalent to taking Cartesian product of the singleton set $\{ d \}$ with $\mathcal{F}^1$, and performing data parallel computation on the association ($\{ d \} \times \mathcal{F}^1) \bigoplus \mathcal{D}_1$. For example, $4 *^D \{2, 3, 4\} \equiv (\{ 4 \} \times \{ 1, 1, 1 \}) *^D \{ 2, 3, 4 \} \equiv \{4, 4, 4\} *^D \{2, 3, 4\} \Rightarrow \{8, 12, 16\}$.

$d \bigodot^D \mathcal{D} \equiv ((\{ d \} \times \mathcal{F}^1) \bigoplus \mathcal{D}) \bigwedge ((\{ d \} \times \mathcal{F}^1) \bigodot^D \mathcal{D}) \Rightarrow$
$\{ (d \odot d_{11}), ..., (d \odot d_{1N}) \}$.   - Scalar-vector $\Rightarrow$ vector (4.15)

4.5 Laws of Associative Update

The sixteenth rule concerns *associative update*. The rule states that insertion of a tuple $< d_1, ..., d_M >$ in an association $\mathcal{D}_1 \bigoplus ... \bigoplus \mathcal{D}_M$, derives $(\mathcal{D}_1^U \bigoplus ... \bigoplus \mathcal{D}_M^U)$ where $\mathcal{D}_I^U$ denotes the updated D-bag. Each $\mathcal{D}_I^U$ is equal to $\mathcal{D}_I \bigcup \{d_I\}$. During insertion alignment is preserved. However, the position of insertion need not be fixed. For example, $< a, b, c > \uplus \{ < 2, 3 > < 4, 5 >, < 6, 7 > \}$ derives $\{ < 2, 3, a > < 4, 5, b >, < 6, 7, c > \}$.

$(< d_1, ..., d_M > \uplus (\mathcal{D}_1 \bigoplus ... \bigoplus \mathcal{D}_M)) \equiv$
$\quad (\mathcal{D}_1 \bigcup \{d_1\}) \bigoplus ... \bigoplus (\mathcal{D}_M \bigcup \{d_M\}) \bigwedge$
$\quad \exists K_{(1 \leq K \leq N+1)}(\pi_K((\mathcal{D}_1^U \bigoplus ... \bigoplus \mathcal{D}_M^U) = < d_1, ..., d_M >) \bigwedge$
$\quad \pi_K((\mathcal{D}_1^U \bigoplus ... \bigoplus \mathcal{D}_M^U) \notin (\mathcal{D}_1 \bigoplus ... \bigoplus \mathcal{D}_M)$ - Aligned update (4.16)

The seventeenth rule concerns *associative release*. The rule states that by associatively searching in one field, the associated data elements in the other field can be released by applying the complement of the F-bag derived during associative search (R_5^A). For example, associative release of a tuple $\{4, \top, \top \}$ from a D-bag $\mathcal{D} = \{ < 4, 5, 6 >, < 3, 7, 9 >, ..., < 4, 9, 10 > \}$ is derived by a sequence of computations: membership of the tuple in $\mathcal{D}$ derives a F-bag $\mathcal{F} = \{ 1, 0, ..., 1 \}$.

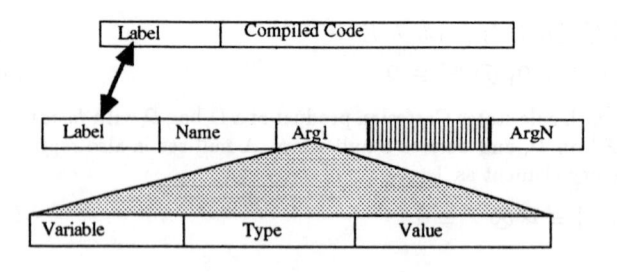

Figure 7: Associative program representation

Complement of $\mathcal{F}$ derives $\neg \mathcal{F} = \{\ 0,\ 1,\ ...,\ 0\}$. $\mathcal{D} \otimes \neg \mathcal{F}$ derives the D-subbag $\{\ \perp, < 3,\ 7,\ 9 >, ..., \perp\ \}$.

$(< d_1, ..., d_M > \ominus (\mathcal{D}_1 \oplus ... \oplus \mathcal{D}_M)) \equiv$
$(\mathcal{D}_1 \oplus \mathcal{D}_2 \oplus, ..., \mathcal{D}_M) \otimes \neg \mathcal{F}$ such that
$\forall I_{(1 \leq I \leq N)} ((\ \pi_J(< d_{1I}, d_{2I}, ..., d_{MI} >) = d) \Rightarrow \pi_I(\mathcal{F}) = 1)$
$\forall I_{(1 \leq I \leq N)} ((\ \pi_J(< d_{1I}, d_{2I}, ..., d_{MI} >) \neq d) \Rightarrow \pi_I(\mathcal{F}) = 0)$. Deletion (4.17)

5 Structure of the Model

The compilation model maps a program, denoted by $\mathcal{P}$, as a pair of associations of the form $< \mathcal{L} \oplus \mathcal{P}^N \oplus \mathcal{A}_1 \oplus ... \oplus \mathcal{A}_N, \mathcal{L} \oplus \mathcal{P}^C >$. The first element of the pair, denoted by $\mathcal{P}^H$, represents a D-bag of clause-head tuples, and second element of the pair, denoted by $\mathcal{P}^B$, represents a D-bag of clause-body tuples. $\mathcal{L}$ is a D-bag of labels connecting clause-heads to low level code of the corresponding clause-body, $\mathcal{P}^N$ is a D-bag of procedure-names, $\mathcal{A}_I$ is a D-bag of I_{th} argument in set of the clause-heads in a program, $\mathcal{P}^C$ is a D-bag of a sequence of compiled instructions corresponding to the set of clause-bodies in the program such that each element $c_i \in \mathcal{P}^C$ is a sequence of instructions corresponding to one clause-body. A schematic of program representation is given in Figure 7.

The abstract machine architecture is a 7-tuple of the form $< \mathcal{I},\ \mathcal{E}^D,\ \mathcal{E}^G,\ \mathcal{E}^V,\ \mathcal{E}^H,\ \mathcal{E}^S,\ \mathcal{R} >$. $\mathcal{I}$ is a set of abstract instructions; $\mathcal{E}^D$ is a data parallel environment which holds the D-bag of bindings for a non-scalar variable; $\mathcal{E}^H$ is a heap which holds the scalar bindings for variables and references of bag of bindings for variables with multiple values; $\mathcal{E}^H$ and $\mathcal{E}^V$ are the associations used to handle alising of uninstantiated variables; $\mathcal{R}$ is a bag of global registers to store temporary values; and $\mathcal{E}^S$ is a control stack which is used to store the control thread during program execution. The abstract instruction set consists of instructions for associa-

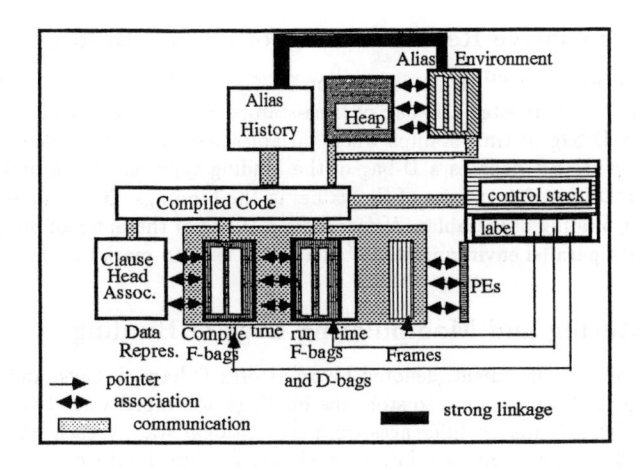

Figure 8: Associative abstract machine with data alignment

tive matching of the goal arguments to the corresponding clause heads, binding a goal variable to a D-bag of values, extracting a scalar value from a D-bag binding, D-intersection and D-union of F-bags, testing and backtracking for null F-bag representing unifiable clauses, shallow backtracking, data movement between the heap and the global registers, and procedure calls. The details of individual instructions and their operational semantics are outside the scope of this paper, and are given in [16, 17]. The individual components of the architecture are aligned to make use of the associative computing as detailed in Section 4. The overall scheme to execute a logic program is a tuple $< \mathcal{P}^H \bigoplus E^D, \mathcal{P}^B, \mathcal{E}^G \bigoplus \mathcal{A}^V \bigoplus \mathcal{R}, \mathcal{E}^S, \mathcal{E}^H >$ as demonstrated in Figure 8.

5.1 Associative Representation of Clause-heads

The association of clause-heads is represented as $\mathcal{L} \bigoplus \mathcal{P}^N \bigoplus \mathcal{A}_1 \bigoplus ... \bigoplus \mathcal{A}_M$. Each argument itself is an association of the form $\mathcal{V} \bigoplus \mathcal{B}^D \bigoplus \mathcal{B}^V$ where $\mathcal{V}$ is a D-bag of variables, $\mathcal{B}^D$ is a D-bag of binding type (integer, real, atom etc.), and $\mathcal{B}^V$ is a D-bag of actual values of the variables. A clause-head is treated as (M + 2) tuple of the form $< \pi_I(\mathcal{L}), \pi_I(\mathcal{P}^N), \pi_I(\mathcal{A}_1), ..., \pi_I(\mathcal{A}_M) >$ where $\pi_I(\mathcal{L})$ is connects clause-head and the clause-body, $\pi_I(\mathcal{P}^N)$ is the name of the literal, and $\pi_I(\mathcal{A}_{J(1 \leq J \leq M)})$ is an argument in the clause-head.

5.2 Associative Representation of Global Bindings

A heap, denoted by $\mathcal{E}^G$, is an association $\mathcal{T}^1 \bigoplus \mathcal{T}^2 \bigoplus \mathcal{V} \bigoplus \mathcal{B}^B \bigoplus \mathcal{B}^D \bigoplus \mathcal{B}^V \bigoplus \mathcal{N}^D \bigoplus \mathcal{F}^R$. $\mathcal{T}^1$ denotes a D-bag of time-stamps when variables were created. $\mathcal{T}^2$ denotes a D-bag of time-stamps when variables were bound. $\mathcal{V}$ denotes a D-bag of variable-id. $\mathcal{B}^B$ denotes a D-bag of the binding type (singleton or D-bag). $\mathcal{B}^D$ denotes a D-bag of data type of the values in the bindings. $\mathcal{B}^V$ denotes a D-bag of the scalar binding of variables. $\mathcal{N}^D$ denotes a D-bag of the index of bindings stored in the data parallel environment. $\mathcal{F}^R$ is a F-bag used to mark the shared variables.

5.3 Storing and Manipulating Bags of Bindings

Data parallel environment, denoted by $\mathcal{E}^D$, stores D-bags, F-bags, and association of D-bags. D-bags are used to store the bindings of a multi-valued variable. $\mathcal{E}^D$ is a stack with two major differences from conventional stacks: each data element is a D-bag, F-bag, or their association. Each bag is aligned with processing elements to reduce overhead of data movement during data parallel computation. It is due to this data parallel environment that *vector-vector* → *vector* and *vector-scalar* → *vector* computations are independent of number of data elements.

5.4 Holding Temporary Values - A Data Parallel Version

Global registers, denoted by $\mathcal{E}^R$, are exact replica of heap except that global registers use the large set of registers present in PE's: each register keeps a tuple. Once all the registers in PEs are used then an image of these registers is stored in the data parallel binding environment. In order to reduce data movement overhead, previous frames of registers are aligned to PEs. The information about each frame of registers is stored in the control stack, and is restored during backtracking. Usage of large number of registers in a frame provides *persistence* which reduces the overhead of storage and retrieval of temporary values during procedure calls and backtracking.

5.5 Handling Control Flow

Control stack, denoted by $\mathcal{E}^S$, is an association $\mathcal{T} \bigoplus \mathcal{W} \bigoplus \mathcal{N}^C \bigoplus \mathcal{N}^D \bigoplus \mathcal{N}^A \bigoplus \mathcal{L}^B \bigoplus \mathcal{L}^R \bigoplus \mathcal{L}^C$. $\mathcal{T}$ is a D-bag of time-stamps needed to restore previous environment. During backtracking, the current time-stamp is decremented, the corresponding tuple in $\mathcal{E}^S$ is searched using the new value of time-stamp, and the the environment is restored. $\mathcal{W}$ is a D-bag of index of frames of the global registers. $\mathcal{N}^C$ is a D-bag of index of the base-element of the bags generated at compile time. $\mathcal{N}^D$ is a D-bag of index of the base-element of the data parallel bindings, generated at run time, for a goal. $\mathcal{L}^B$ is a D-bag of labels of the code where the control has to jump during backtracking. $\mathcal{L}^R$ is a D-bag of the labels of the code where the control has to jump to fetch next scalar value from a producer. $\mathcal{L}^C$ is a D-bag of labels of the code where control will jump after the successful execution.

5.6 Handling Aliasing

Aliased variables are handled associatively using two D-bags, namely, *alias set environment*, denoted by $\mathcal{E}^V$, and *alias history*, denoted by $\mathcal{E}^H$. $\mathcal{E}^V$ is a sequence of F-bags which are used to select different sets of aliased variables. $\mathcal{E}^V$ is aligned with $\mathcal{E}^G$ to derive the attributes of aliased variables. $\mathcal{E}^H$ is a D-bag of the form $\mathcal{T} \oplus \mathcal{N}^O \oplus \mathcal{N}^U$, and is used to restore the previous aliasing information during backtracking. $\mathcal{T}$ denotes a D-bag of time-stamps when new alias set was created. $\mathcal{N}^O$ denotes the D-bag of *F-bag index* (in $\mathcal{E}^V$) representing previous alias set of a variable, and $\mathcal{N}^U$ denotes the D-bag of *F-bag index* (in $\mathcal{E}^V$) representing current alias set of a variable.

6 Integrating Retrieval and Computation

This section describes the advantages achieved by alignment of D-bags and F-bags to reduce the overhead during goal reduction, retrieval of information from aligned fields, and data movement.

6.1 Alignment and Data Parallel Computation

Alignment of the arguments in the same clause facilitates

(i) data parallel pruning of those clauses which do not share the same values for multiple occurrence goal variables: arguments of the same clauses are pairwise equated in a data parallel manner. For example, data parallel equality on first two arguments for a goal $a(X, X, Y)$ and a set of clauses { $a(5, 6, 4)$, ..., $a(4, 4, 2)$, ..., $a(6, 7, 9)$}, derives a F-bag {0, ..., 1, ..., 0 } "(4.14)" indicating non-unifiability of clause-heads $a(5, 6, 4)$ and $a(6, 7, 9)$ "(4.7)". Figure 9 illustrates the implementation. Data parallel equality on $\mathcal{A}_I$ = { a, b, c } and $\mathcal{A}_J$ = { b, b, c } gives a F-bag { $0, 1, 1$ }.

(ii) data parallel computation on two arguments of the same clause. For example, arithmetic comparison to identify whether second argument is greater than the third argument "(4.14)" for a goal $a(X, X, Z) \wedge X > Z$ and a set of clauses { $a(5, 6, 4)$, ..., $a(4, 4, 2)$, ..., $a(6, 7, 9)$}, a unit data parallel inequality test derives the F-bag {1, ..., 1, .., 0 } which selects the clause $a(6, 7, 9)$ "(4.7)".

(iii) data parallel derivation of clauses satisfying complex conditions. For example, finding out the set of clauses which satisfy the goal $a(X, X, Z) \wedge X > Z$ will need D-intersection of two F-bags {0, ..., 1, ..., 0 } and { 1, ..., 1, ..., 0 }. Using "(4.10)" F-bag { 0, ..., 1, ..., 0 } is derived, and using "(4.7)" the unifiable clause $a(4, 4, 2)$ is selected.

6.2 Alignment and Data Movement

In conjunction with alignment of the program association with data parallel environment, the alignment of the arguments in a clause facilitates

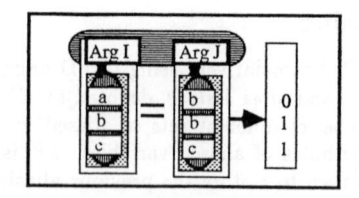

Figure 9: Handling multiple occurrence goal variables

(iv) selection of a subset of data in $\mathcal{P}^H$ without actual movement of data "(4.7)" and "(4.8)". For example, given a set of clauses { $a(5, 6, 4)$, ..., $a(4, 4, 2)$, ..., $a(6, 7, 9)$}, and an F-bag { $1, ..., 1, ..., 0$ }, the arguments of clauses $a(5, 6, 4)$ and $a(4, 4, 2)$ are selected for data parallel computation without data movement.

(v) movement of selected data from $\mathcal{P}^H$ to $\mathcal{E}^D$ using a constant number of data parallel computations. For example, given a set of clauses $\mathcal{D} = \{\ a(5, 6, 4),$ ..., $a(4, 4, 2), ..., a(6, 7, 9)\ \}$ and a F-bag $\mathcal{B} = \{1, ..., 1, ..., 0\}$, $\Pi_3(S \otimes B)$ derives { $4, 2, \perp$ } in a unit time.

6.3 Alignment and Aliasing

One of the concerns in logic programs is to efficiently access and manipulate the aliased variables such that binding of one variable is also seen by the other aliased variable. Conventional systems use chain of references to access the bindings. In contrast, this model benefits by associative search (see Eq. (4.5)) to derive for the bindings of aliased variables; $\mathcal{V} \otimes \mathcal{F}$ is used to select an alias set and derive corresponding attributes in $\mathcal{E}^G$ using "(4.7)" and "(4.8)"; D-union of two F-bags is used to derive a new alias set "(4.10)"; bindings of all variables in an alias set are updated in constant number of data parallel computations using "(4.16)"; and bindings are accessed using "(4.5)" and "(4.6)".

6.4 Alignment and Associative Goal Reduction

There are four rules for associative goal reduction, namely, *matching constants, pairwise data parallel equality, conjunctive pruning,* and *unification*. we use $\mathcal{F}_T^U$ as F-bag of unifiable clauses at time stamp T.

The first rule of matching constants states that matching the constant goal argument Arg_I^G with the corresponding D-bag $\mathcal{A}_I$ derives an F-bag $\pi_I(\mathcal{F}_T^U)$ (see Eq. (4.5)) which is used to prune those clauses which do not match with Arg_I^G using "(4.7)".

$((type(Arg_I^G) = \text{"constant"}) \bigwedge (Arg_I^G \in \mathcal{A}_I)) \Rightarrow (\mathcal{P}^H \otimes \Pi_I(\mathcal{F}_T^U))$ such that
$type(\ \pi_{I(1 \leq I \leq N)}(\mathcal{A}_I)) \in \{\ \text{"constant"}, \text{"variable"}\}) \Rightarrow$
$\pi_{I(1 \leq I\ N)}\ (\pi_{J(1 \leq J \leq M)}(\mathcal{F}_T^U)) = \text{"1"}$

type($\pi_{I(1 \leq I \leq N)}(\mathcal{A}_I)) \notin \{$ "constant", "variable"$\}) \Rightarrow$
$\pi_{I(1 \leq I \leq N)} (\pi_{J(1 \leq J \leq M)}(\mathcal{F}_T^U)) =$ "0". Matching constants (5.1)

The second rule of *data parallel equality* states that if two goal arguments Arg_I^G and Arg_J^G share the same multiple-occurrence variable, then non-unifiable clauses are pruned by equating the D-bags $\mathcal{A}_I$ and $\mathcal{A}_J$ (rule R_{13}^A).

$(type(Arg_I^G) =$ "variable") $\bigwedge (type(Arg_J^G) =$ "variable") $\bigwedge$
$(Arg_I^G = Arg_J^G) \equiv \mathcal{P}^H \bigotimes \Pi_{(I,J)(1 \leq (I,J) \leq M)}(\mathcal{F}_T^U)$ such that
$\Pi_{(I,J)(1 \leq (I,J) \leq M)}(\mathcal{F}_T^U) = (\mathcal{A}_I =^\mathbf{D} \mathcal{A}_J)$ - Data parallel equality (5.2)

The third rule of *conjunctive pruning* combines rules R_{10}^A and R_1^L and R_2^L to prune all the clauses which do not unify with the given goal. The rule takes intersection of every F-bag generated by using R_1^L and R_2^L and uses rule R_{10}^A to derive D-intersection.

$\mathcal{F}_T^U = \Pi_1(\mathcal{F}_T^U) \sqcap ... \sqcap \Pi_M(\mathcal{F}_T^U)$ -Pruning non-unifiable clauses (5.3)

The fourth rule of unification is used for matching remaining goal arguments, and is generally used to handle aliased variables.

$\exists K_{(1 \leq K \leq N)} (\pi_K(\mathcal{F}_T^U) =$ 'true') $\Rightarrow$
$\forall J_{(1 \leq J \leq M)} ((type (\pi_J (\pi_K(\mathcal{A}_1 \bigoplus ... \bigoplus \mathcal{A}_M))) =$ "variable") $\Rightarrow$
unify $(\pi_J (\pi_K(\mathcal{A}_1 \bigoplus ... \bigoplus \mathcal{A}_M)), Arg_J^G))$ -Partial unification "(5.4)"

6.5 Deriving Unspecified Relations for Objects

Given a query of the form $P(\mathcal{A}_1^G, ..., \mathcal{A}_N^G)$, any of the $N + 1$ positions may be uninstantiated. Only those queries where predicate variable P occurs as one of the arguments can not be answered. The query where all $(N + 1)$ elements are variables is meaningless. The query about unspecified predicate-names is processed in the same manner as to the queries with the instantiated predicate names. However, there is one major difference: the presence of an uninstantiated variable in the position of predicate names may instantiate the variable to multiple predicate name. Since different predicate names represent different procedures, multiple procedures (with the same number of arguments) have to be handled in contrast to the case when the predicate names are instantiated.

7 The Model Behavior

The model exploits run time execution efficiency both at the data level during data parallel goal reduction by treating the clause-heads as association of D-bags for efficient pattern-matching, and control level during the execution of the code of the corresponding subgoals in the selected clause. The forward control flow is divided into three parts: *pre-call processing*, *pre-clause processing*, and *clause processing*.

Pre-call processing is used to perform data parallel goal reduction, and setting up the bindings for goal variables. Pre-call processing uses "(5.1)", "(5.2)", and "(5.3)" to to derive unifiable clauses.

Pre-clause processing is used to test the presence of unifiable clauses in $\mathcal{F}_T^U$, and pass control to the right clause. Pre-clause processing is used to derive the potential bindings for a goal variable using R_4^L, and to jump to the code area associated with unifiable clause. If $\mathcal{F}_T^U$ is non-empty then next tuple $\pi_K(\mathcal{L} \oplus \mathcal{P}^N \oplus \mathcal{A}_1 \oplus ... \oplus \mathcal{A}_M)$ is selected in a non-deterministic manner, the corresponding clause-label $\pi_K(\mathcal{L})$ is picked, and control jumps to the corresponding instructions sharing the label (see Figure 10). The order of selection of procedures after data parallel goal reduction is non-deterministic. Within a procedure, the order of selection of clauses within a procedure is non-deterministic.

Clause processing is used to handle aliasing of variables, setting up the global registers, handling alternate bindings of shared variables during backtracking, and storing the current state into control stack, before starting next cycle.

During forward control flow, the global registers corresponding to the subgoal arguments are set, the time-stamp T is incremented by one, the old environment is stored in $\mathcal{E}^S$, and the pre-call processing is invoked. If at any time $\mathcal{F}_T^U = \phi$, deep backtracking occurs, information is retrived from the control stack, and the previous environment is restored. Upon backtracking, the old time-stamp T^O is picked up from the global register; the previous environments from the control stack are released using "(4.17)", and the data parallel environment is also restored back by setting the corresponding pointers; $\mathcal{E}^H$ and $\mathcal{E}^V$ are also restored. After deep backtracking, the control may jump back either to pre-clause processing area where alternate clause is selected for processing, or to clause area to pick up the next scalar value for a producer. The exact nature of backtracking is dependent upon the entry in $\mathcal{E}^S$.

Example 4

For example, in Figure 10, after the data parallel goal reduction of a goal with an uninstantiated variable in the place of the predicate-name, three clauses with label 1$ in the procedure with predicate name *p1*, and labels 4$, and 6$ in the procedure with the predicate-name *p2* are selected. One of the procedures *p2* is selected, the control jumps to the code area starting with one of the labels 4$, and the corresponding instructions are executed. Upon backtracking, another label 6$ (in the same procedure) is selected, and control jumps to the code area starting with the label 6$. Upon the failure of the code with label 6$, the control jumps to the code area in the second procedure, and the instructions starting with the label 1$ are executed.

7.1 Handling Aliased Variables

The process of aliasing two variables V_1 and V_2 involves four steps as follows:

(i) The current F-bags $\mathcal{F}_1$ and $\mathcal{F}_2$ corresponding to alias sets of V_1 and V_2 are identified. This information is derived using associative search in $\mathcal{E}^G$, and the F-bag indices are picked from the D-bag $\mathcal{N}$ in $\mathcal{E}^G$ using "(4.6)".

(ii) D-union of alias sets represented by F-bags $\mathcal{F}_1$ and $\mathcal{F}_2$ (see Eq. (4.10)) is used to generate a new alias set selected by F-bag $\mathcal{F}_3$ which is stored in $\mathcal{E}^V$.

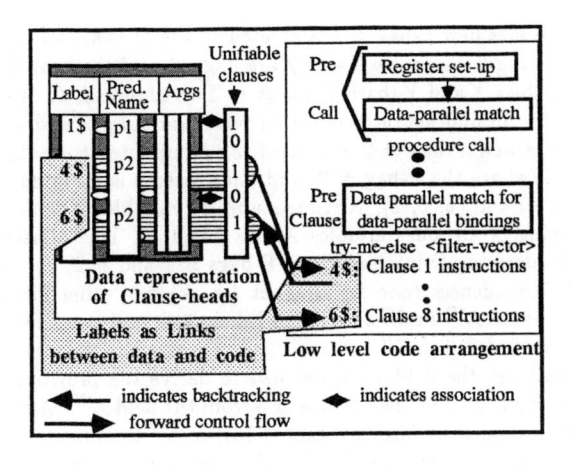

Figure 10: Model behavior

(iii) Two new entries are inserted in $\mathcal{E}^H$ to store the triple (time-stamp of creation of new alias set, index of previous alias set, index of current alias set) using "(4.16)". This history is needed to restore the environment during backtracking.

(iv) The index of the current alias set for each variable is updated in $\mathcal{E}^G$ using "(4.16)".

Instantiating any of the variables, automatically instantiates all the aliased variables by using an associate search (see Eq. (4.5)) and associative update (see Eq. (4.16)) in constant number of associative computations. Accessing the binding for aliased variables needs no extra access cost under this scheme "(4.5)" and "(4.6)". During backtracking, previous alias sets are recovered as follows:

(i) The indices for the current alias sets are identified by searching $\mathcal{T}$ (in $\mathcal{E}^H$) for the current value of time-stamp using "(4.6)".

(ii) The indices of old alias sets are selected from $\mathcal{N}^0$ in $\mathcal{E}^H$ using associative search on $\mathcal{N}^U$ using "(4.6)".

(iii) Data parallel update (see Eq. (4.16)) is used to replace the alias set index by the indices of old alias sets for each variable in the alias set.

(iv) The entries from $\mathcal{E}^H$ are released using "(4.17)".

Example 5

In Figure 11, the current alias set reference in the heap is initialized to 0. At time stamp 4, variables X and Y are aliased. Since the alias set references for variables X, Y are θ, a new F-bag # 1 is created in the alias set environment, and the bits corresponding to variables X and Y are flagged, and the heap is updated to indicate that variables X and Y share F-bag # 1. Two tuples $(4, 0, 1)$ and $(4, 0, 1)$ are inserted in $\mathcal{E}^H$. Similar action is repeated when variables Z and W are aliased at time stamp 6: a new F-bag # 2 is created; $\mathcal{E}^G$ is updated to indicate that the variables Z and W share the F-bag # 2, and alias history is updated to store two new tuples $(6, 0, 2)$ and $(6, 0, 2)$. At time stamp 8, variables X and Z are aliased resulting in aliasing of all four variables X, Y, Z and W. This aliasing results in the F-bag # 3 by deriving D-intersection of F-bags # 1 and # 2; two data parallel search-and-update sequences (one for alias set $\{X, Y\}$ and other for alias set $\{Z, W\}$ are used to update $\mathcal{N}^D$ in $\mathcal{E}^G$ to refer to F-bag # 3; and two new tuples $(8, 1, 3)$ and $(8, 2, 3)$ are entered in $\mathcal{E}^H$.

Upon backtracking, the field $\mathcal{T}$ is searched to derive the previous F-bags # 1 and # 2 from the field $\mathcal{N}^O$, and new F-bag # 3 from the field $\mathcal{N}^U$. Alignment of $\mathcal{E}^G$ and the $\mathcal{E}^V$ allows aligned update (see Eq. (4.16)) of $\mathcal{E}^G$ to indicate that variables X and Y share F-bag # 1 and variables Z and W share F-bag # 2.

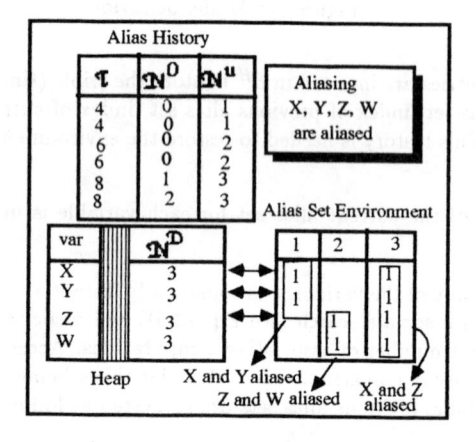

Figure 11: Aliasing variables associatively

7.1.1 Handling shared variables

In a conjunctive goal (or subgoals), a producer may be bound to a D-bag. In such cases, the index of the corresponding D-bag of bindings is stored in $\mathcal{E}^G$ and the D-bag is stored in $\mathcal{E}^D$. If the consumer uses one scalar value at a time, then one value is picked from the D-binding at a time, and the selected binding is released from the D-bag. This process is continued using an iterative loop until the D-bag is ϕ. For a goal having two or more producers, tuple of bindings is derived from association of D-bags. The abstract instruction for repetitive selection of values is as follows:

>   repeat-else-backtrack <label>
>   <label>: next-value <binding-vector>

The first instruction stores the <label> into the control stack along with the other information, and goes to the next instruction to pick up a value from the D-bag. Upon backtracking, the control comes back to the instruction "next-value", and the next iteration takes place.

7.2 Handling Multiple Procedures

While deriving the queries for the goals with uninstantiated predicate names, there may be multiple procedures which are unifiable to the same goal, since they share the same object. However, each of these procedure must be solved separately to derive different relations and the related argument values. This is achieved by picking up one clause, identifying the corresponding procedure, and flagging all the unifiable clauses having the same procedure name in a different F-bag. During backtracking, the next clause is picked up in two different steps: if the current procedure has an unprocessed clause, then the next clause of the current procedure is selected. After all the clauses in a procedure are over, a F-bag corresponding to unifiable clauses in next unifiable procedure is picked.

Meta-relations are treated as any other relation with the exception that uninstantiated variables also occur in the place of predicate names.

8 Performance Evaluation

The prototype emulator, which also includes the data parallel SIMD operations, has been implemented using ANSI C. The data demonstrates that the number of operations needed for associative lookup is independent of the number of ground facts. Thirty operations are needed to match a ground fact with two arguments. The number of operations is linearly dependent upon the number of arguments in a query as shown in Table 2. For each extra argument, twelve extra operations are needed to load the value in registers, perform data parallel match, and perform logical ANDing of the previous bit-vector with the new bit-vector obtained during data parallel match.

For a 20 ns clock supported by current technology, and three clock cycles (load-execute-store cycle), the associative look up speed is eight hundred thousands × number of facts for a set of facts with two arguments. In the presence of data parallel scientific computations intertwined with associative lookup, the peak execution speed is limited by the associative look-up speed which is eighty MCPS (million computations per second) for thousand facts.

The worst case of execution occurs when consumer of a shared variable uses one scalar value at a time. In such cases, the execution speed reduces to two hundred thousand logical inferences per second (LIPS). The slow down is caused primarily due to the overhead of storing the control thread during forward control flow, register set up, and retrieving the control thread during backtracking. Our results show that the overhead of data parallel matching is less than the overhead of storing

Table 2: Number of non-shared arguments vs. execution speed

Number of arguments	1	2	3	4	5	6	7	8	9	10
Parallel operations	18	30	42	54	66	78	90	102	114	126
Scalar operations	2	3	4	5	6	7	8	9	10	11
Total operations	20	33	46	59	72	85	98	111	124	137

the control thread during forward control flow which makes the model suitable for handling flat programs with relations having a large number of arguments.

9 Other Related Works

Earlier work on applying search by content for Prolog was described elsewhere [6]. Recently, concurrent work on DAP (Distributed Array Processor) Prolog [8] exploits three paradigms, namely, set based programming, data parallel scientific computation, and associative search to a limited extent. The scheme tries to simulate WAM which causes some limitations since WAM was not designed to exploit associative computation. In comparison to our recent work [15], the limitations of this scheme are as follows:

(i) Intertwining of associative search and data parallel computation is negligible, and is not based upon any formal model.

(ii) Left hand side of the set of clauses is not aligned with data parallel environment which causes the overhead of data movement. The concept of data parallel environment is very limited. Only arrays are used for data parallel computation.

(iii) Derivation and reasoning using unspecified relations is not supported.

(iv) Lack of run time allocation in parallel field does not allow indefinite size of sequences.

Associative computation is not exploited for handling aliased variables.

Other schemes [7] simulate pointer based representations using associative memories. The scheme does not exploit the power of association to represent data associatively. The scheme also does not intertwine associative search with data parallel computation.

The major differences between the proposed model and WAM based models are as follows:

(i) Unlike WAM, alternative clauses are not traversed sequentially. In our model, clause-body and clause-head are loosely connected through a label. Associative non-deterministic search is used to pick up the next clause. In WAM there is no concept of association of D-bags and associative search on D-bags. At best, WAM makes use of indexing on limited number of arguments.

(ii) Unlike WAM based models, our model is based upon alignment of association D-bags. The alignment of data reduces the data movement during selection of the data elements from program association, and provides tighter integration between logic programming and data parallel scientific computing. In our model, data parallelism is exploited during data retrieval, computation, and data movement.

(iii) Use of data parallel environment in our model allows *vector-vector* → *vector* and *vector-scalar* → *vector* computations in unit time.

(iv) Unlike WAM based models, our model uses associative computation for aliasing, which reduces the cost of update and access of bindings for aliased variables.

In contrast to the interpretation based conventional data-parallel model proposed in [9, 14], this model does not suffer from data sequentiality caused by the presence of multiple-occurrence variables in the goals. The model also handles variable aliasing in the clauses efficiently using the associative data-parallel search and data-parallel assignment property. The major advantage is the synergy of exploiting associative computation and execution of low level code. This work extends our previously reported work [15] by introducing an algebra for associative computation, and reports the benchmark results.

10 Conclusion

In this paper, we described an associative compilation model which integrates high performance intelligent processing, data parallel computing, and associative information retrieval under the framework of logic programming. The model is capable of answering queries with unspecified relations about the given objects. This model benefits from the synergy resulting from associative search, data-parallelism during goal reduction, the use of low level code, data alignment, and data parallel computation to reduce data-transfer and data transformation overhead. Use of persistent global registers reduces the overhead of storing and restoring the environment. The model has been implemented using C^{++} and ANSI C [16, 17]. The benchmarks of the emulator are are very encouraging for data intensive computations.

Acknowledgements

I thank Prasad Lokam for compiling the model, Madhavi Ghandikota for implementing the emulator for the abstract instruction sets, and Jerry Potter for some useful discussions related to SIMD computations. I also thank Leon Sterling for proof reading the paper.

References

[1] R. Kowalski, *Logic for Problem Solving*, Elsevier-North Holland (1979).

[2] L. Sterling and E. Shapiro, *The Art of Prolog*, MIT Press, (1986).

[3] Kanada Y., Sugaya, M., A Vectorization Technique for Prolog without Explosion, Proceedings of the International Joint Conference of Artificial Intelligence, Detroit, Michigan, USA (1989) pp. 151 - 156.

[4] A. Takeuchi and K. Furukawa, *Parallel Logic Programming Languages*, Lecture Notes In Computer Science, Vol. 225, Springer Verlag, Newyork (1986) pp. 242 - 254.

[5] D. H. D. Warren, *An Abstract Prolog Instruction Set*, Technical Report 309 , SRI International (1983).

[6] P. Kacsuk, and A. Bale, *DAP Prolog: A Set Oriented Approach to Prolog*, The Computer Journal, Vol. 30, No. 5 (1987) pp. 393-403.

[7] Stormon, C. D., Brule, M. R., Riberio, J. C. D. F., *An Architecture Based on Content Addressable Memory for the Rapid Execution of Prolog*, Proc. Fifth International Conference of Logic Programming, Seattle, USA (1988) pp. 1448 - 1473.

[8] P. Kacksuk, *DAP Prolog*, in Execution Models of Prolog for Parallel Computers , Research Monograph, MIT Press (1990).

[9] A. K. Bansal and J. L. Potter, *An Associative Model to Minimize Matching and Backtracking Overhead in Logic Programs with Large Knowledge Bases*, The International Journal of Engineering Applications of Artificial Intelligence , Permagon Press, Volume 5, Number 3 (1992) pp. 247 - 262.

[10] J. L. Potter, *Associative Computing*, Plenum Publishers, Newyork (1992).

[11] C. C. Foster, *Content Addressable Parallel Processors*, Van Nostrand Reinhold Co., New York, (1976).

[12] W. D. Hillis and G. L. Steele Jr., *Data Parallel Algorithms*, Communications of the ACM 29 (1986) pp. 1170 - 1183.

[13] J. A. Feldman and D. Rovner, *An Algol Based Associative Language*, Communications of the ACM , Volume 12, No. 8 (1969) pp. 439 - 449.

[14] A. K. Bansal and J. Potter, *Exploiting Data Parallelism for Efficient Execution of Logic Programs with Large Knowledge Bases*, Proceedings of the Tools for Artificial Intelligence 1990 , Herndon, USA (1990) pp. 674 - 681.

[15] A. K. Bansal, J. L. Potter, and L. V. Prasad, *Data Parallel Compilation and Extending Query Power of Large Knowledge Bases*, In the Proceedings of the International Conference of Tools for Artificial Intelligence (1992) pp. 276 - 283.

[16] M. Ghandikota, *Implementing Abstract Instruction Set for Logic Programs on Associative Supercomputers*, MS Thesis, Department of Mathematics and Computer Science, Kent State University , Kent, OH, USA (1993).

[17] L. V. Prasad, *Compiling Logic Programs to Incorporate Data-parallelism on Associative Supercomputers*, MS Thesis, Department of Mathematics and Computer Science, Kent State University, Kent, OH, USA (1993).

SECTION : 3
AUTOMATION

Software Automation: from "SILLY" to "INTELLIGENT"

Xu Jiafu, Chen Daoxu, Wang zhijian, Dong Lijun

Institute of Computer Software, Nanjing University,

Nanjing 210093, P. R. China

ABSTRACT

In this paper the adaptation of some AI techniques to software automation is proposed. The inductive reasoning is used as the basis of inductive program synthesis, while explanation-based learning is employed to learn algorithm design strategies. Methods for elementary algorithm learning, optimizing method learning and analogical program derivation are also described. Future work on this aspect is outlined in the end.

1. Introduction

By software automation we mean to rely on computer systems as much as possible in software development[1], in other words, to generate programs from informal requirements automatically. AI is playing an increasingly important role in this field. In the State Key Laboratory of Novel Software Technology of China, the Software Development Automation project is conducted, the aim of which is to exploring the methods of applying various AI techniques such as knowledge representation, automatic deduction, analogical reasoning and explanation-based learning, in software development to automate the whole development process.

It will be a long time before it is possible to fully achieve this ambitious goal. We consider software automation at three different abstraction levels:

(1) Requirement level: from an informal, probably vague, problem description to a formal and precise software functional specification;

(2) Functional level: from the functional specification to its design specification;

(3) Design level: from the design specification to the corresponding executable program on a particular target machine.

Working from design level to requirement level, which is the strategy we selected to build experimental systems, the Software Development Automation project has been pursuing the ultimate goal for nearly eight years. The early emphasis of our work was placed on automatic transformation techniques of both design and functional levels and verification techniques of software specification and transformation rules. The main results include:

(1) A hierarchical decomposition model has been implemented. FGSPEC is its formal representation. The language is wide-spectrum and graphically presented. Its corresponding design specification language is GSPEC[2].

(2) An experimental software automation system NDAUTO has been operational since August, 1987. By combining the transformational and procedural approaches, the system can transform software specification written in a graphical specification language GSPEC to executable programs

automatically[2].

(3) An experimental algorithm automation system NDADAS was completed in September, 1988[3]. The system can automatically or semi-automatically generate GSPEC design specification from the functional specifications written in FGSPEC.

(4) Some work on the requirement level has been done with the emphasis on the transformation from informal specifications written in Chinese to formal specification and a theory of Combinatory Type Grammar is presented[4].

During this period some other software automation systems such as CYPRESS/KIDS[5], have been reported in the literature. But up to now, almost all the software automation systems are experimental. In fact, no general-purpose automation systems have been put into practical use. One of the key limitations, and hence a direction for further researches relates to "the silliness of the systems".

Following are among what make a system look silly:

(1) Man-system interface are purely formal.

Almost all software systems accept only purely formal functional specifications. That means software specifications must be expressed in some precisely defined way if any success is expected in automating software development. On the other hand, the specification should be understandable for user to validate its correctness in describing their requirements. From end-users' point of view, the ideal man-system interface language is a natural language or a multimedia language. Specification by examples may be natural and easier for users to formulate. As a complement for purely formal methods, it may be alleviated the problem to some extent.

(2) System's capabilities are brittle

Almost all software automation systems are limited in their ability. Some of them apply only to very narrow domains, and others can only solve relatively simple problems. One of the important reasons for the limitation is lacking the ability of learning. The system cannot acquire knowledge from problem solving practices and user-system interaction. That means the system capability will be solidified when it is put to use. Since a great deal of knowledge is needed for programming, it is very difficult to develop a strong system at one stroke. It is believed that the complexity of software automation system may be such that direct programming becomes infeasible as a method of construction. So we must enable systems to learn for themselves to achieve desired level of performance.

Overcoming the limitations mentioned above sets agenda for our current research tasks. The research on "intelligent" software automation system, of which the heart is the learning ability such as inductive reasoning, analogical reasoning and explanation-based learning, etc., has resulted in a formalism called PD-Cal[6] and two experimental systems, NDIPS[7]and NDSAIL[8].

2. Incorporating machine learning into software automation

2.1 Inductive program synthesis

Users prefer providing the requirements by the concepts or terms in the application domain they are familiar with. The derivation of programs from incomplete specifications, such as input/output pairs or computation traces, is one of alluring forms of requirements, and has long been an attractive research field in software automation. Less formal specifications improve the readability, so less training is needed for users to write and understand them.

The development of the inductive program synthesis system NDIPS is aimed at:

(1) The intended program can be synthesized from a less number of examples.

(2) The burden on users to provide suitable examples can be partially transferred to the system.

(3) The target programs can be verified automatically to some extent.

(4) The system can support generating programs hierarchically.

AI techniques have been found their greenhouse in the field of inductive program synthesis, which is based on the common point of view that concepts can be treated as programs in the framework of logic programming language. Here we only focus on the AI techniques adopted in NDIPS.

AI techniques involve knowledge acquisition, knowledge representation and knowledge processing. Reasonable induction relies on the domain theory or background knowledge the system has possessed. The background knowledge plays its roles in decreasing the potential search space of hypothesis and alleviating the burden of the users to provide examples.

The facts and rules presented by users in their requirements contain some kind of knowledge in the application domain. The knowledge should be incorporated into the system in a suitable form for later uses. By knowledge-based systems we mean those which not only use their built-in knowledge base to solve the problem, but also learn more and more knowledge during their working runs.

In NDIPS, the process of program synthesis is treated as a problem solving task. Primitive constructors, input/output pairs and computation traces correspond naturally to the operator set, initial/goal state and solution path respectively. This makes it easier to utilize AI techniques. For example, while generating the computation traces from input output pairs, the heuristic rules are learned through a failure driven mechanism. While searching for the solution, every time the dead path or cyclic path is discovered, the reasons are recorded to avoid the same situations occurring. The trials and errors will be decreased until the suitable conditions for each operator is determined.

The hypothesis space is constructed by the generalized subsumption relation, which combines the redundant hypotheses into the equivalents under the background knowledge, to improve the searching efficiency. The construction algorithm and the underlying theory have been thoroughly studied.

The key to successful induction is choosing the suitable examples, the number of input/output pairs being decided and the critical ones being selected. The critical input/output pairs are mainly used to control the searching direction and the correctness of the target program, which is partially done by mode inference and type inference mechanisms in NDIPS. The system can find critical examples and ask users for answers actively, instead of being fed by users blindly.

The synthesized programs are treated as the higher-level constructors and can be reused in other synthesis tasks.

2.2 Machine learning in reductive problem solving

Using problem reduction as the fundamental approach in problem solving, the system should have the following capabilities:

(1) Knowing how to decompose those problems which the system cannot solve directly into simpler ones;

(2) Being able to solve a group of relatively simple problems directly;

So among the target concepts of the learning for the automatic problem solver are:

(1) The algorithm scheme, it is the formal representation of the method for decomposing problems;

(2) The elementary algorithm, elementary algorithms are the basic components to be used to synthesize suitable algorithms as problem descriptions require.

From the viewpoint of machine learning, the software automation system has the following peculiarities:

(1) The targets of learning, especially the algorithm schemes, are highly complicated. Users will have difficulties in preparing many examples; but on the other hand, the programming knowledge formalization is very difficult, so less demand on background knowledge will make the implementation easier.

(2) The learning environment is relatively simple. Training examples used in learning are provided mainly by user purposely.

(3) The correctness of learned results must be guaranteed so that the system can solve problems correctly.

Based on the discussion above, two paradigms of learning, i.e. inductive and analytic, are adopted in the software automation system to improve its ability.

Explanation-based learning [8] is used in algorithm scheme learning, which learns algorithm design strategies from user provided training examples.

Inductive techniques are used in elementary algorithm learning, which learns the functional descriptions from input/output pairs.

2.2.1 Algorithm scheme learning

The algorithm scheme learning mechanism accepts training examples provided by the user and automatically generates algorithm schemes which can be used as templates to deal with the problems of the similar type during problem-solving sessions. In the framework of EBG, the learning procedure ML can be viewed as:

$$ML:<TC,TE,DT,OC>-->SD,$$

where

.TC is the target concept, which is the algorithm scheme,

.TE is the training example provided by the user,

.DT is the domain theory, which is a group of rules describing general programming knowledge such as methods of constructing functions, and properties of data structures and programming knowledge associated with some specific fields,

.OC is the operationality criterion, which is a group of predicates on the set of algorithm scheme. These predicates are used to decide whether or not a scheme is in the form that can be instantiated successfully.

SD, the result of the procedure ML, is
$$<f=FC:in\text{-}type\text{-->}out\text{-}type,SC>,$$
where
.FC is the instantiation of TC,
.FC is the generalization of the algorithm description in TE,
.SC is the predicate formula set in which there are some constraints satisfying OC for every function variable in FC.

The form of FC can be defined recursively from some primitive functions and function variables with some functions constructing patterns. In fact, the FC is the algorithm description with some of its constituent functions parameterized. The details of FC may be found in [9].

The algorithm scheme learning procedure itself can be divided into two phases, the explanation phase(analysing the training examples)and generalization phase(generating algorithm schemes).

In the explanation phase, the following are established:
.Data relation diagram;
.The generalizing paths, which assure the forms of function pattern;
.Data type and the associated operators analysis.

The generalization in the second phase is strictly based on the results of the first phase. Two kinds of generalization are put into effect:

.Operator generalization: selecting suitable operators and generalizing them to operator variables with a group of predicate logic formula as the constraints to limit the generalization scope.

.Data type generalization: broadening the definition of the data type appearing in the training examples, so that the learned algorithm schemes can be used in solving problems with different but similar data type.

In generalizing the data type, an algebraic model is introduced. A data type is an axiomatic algebra. By abstracting procedure, two different data types can be showed to have some structure. So the training examples can be generalized with abstract data typing. When the result algorithm scheme is used as a template in solving a new problem, the abstract type can be instantiated as different types. The difference can be that of integer and list or set.

We can call it a type variable, and a group of axioms function similarly as the semantic constraints do for the operator generalization.

The operationality criterion is a vague concept in the EBG framework and one of the key issues in the design of an explanation-based learning system. As mentioned above, for each generalized operator variable are introduced a group of semantic constraints, which delimit the variable's instantiating scope explicitly. However, the system's inferring procedure is influenced by the knowledge it possesses. With its knowledge increasing, the inferred constraints may be changed. Therefore, the meaning of "operationality" will vary with the system capability.

2.2.2 Elementary algorithm learning

In the context of algorithm synthesis, there should be a reasonable quantity of elementary algorithms which are used as basic units to configure more complicated algorithms. It is often the case that no functional descriptions of these elementary algorithms make them hard to be used in the automated synthesis process. Therefore, a mechanism that can construct the specification of an algorithm is designed. In NDSAIL, the algorithm is in the form of tree structure, so its functional description can be constructed hierarchically from self-specified leaf nodes when the tree contains no recursive nodes.

However, the specification of subtree rooted at the recursive definition node cannot be deductively constructed because the specification of the leaf node that calls recursively cannot be obtained. Two facilities, an interpretative executor and an inductive learner, are introduced to deal with this problem. The interpretative execution facility produces the simulated input data by analysing the algorithm tree and gets the input/output pairs after the suggestive specification of the tree with recursive nodes. The inductive algorithm is based on Fringe algorithm[10], which is an improvement of ID3, and ours is more adaptable to the requirements of inductive learning.

The reduction has been performed on the results of decision-tree to simplify the logical expression produced. Furthermore, a verification unit has been designed to decide whether or not the inductive results are correct. It is designed by taking the consideration of the specific background of elementary algorithm learning. Thus, even if the inductive mechanism can not guarantee the correctness by itself, the system results are always correct.

2.2.3 Optimizing method learning

Inefficient results are often generated by the mechanisms described in the previous sections due to the difficulties to use the method specific enough to deal with different problems. An explanation-based transformation mechanism is designed in an attempt to improve the efficiency of the synthesized algorithms. The training examples are generated automatically by the system. An interpretative executor records the trail of the example running and tries to find optimizable structures, especially the repeated recursions. The behavior on the specific running is then generalized to be applicable for the original algorithm. The eureka function of fold-unfold transformation [11] is defined to eliminate the inefficient parts.

The three kinds of learning have been implemented as three parts of the Software Automation System NDSAIL on SUN workstation. The scheme learned from one example algorithm has been successfully used to solve other problems which can not be solved before learning, and the running time decreases obviously when optimizing mechanism is applied. It is in this sense we say that the ability of the software automation system has been improved through self learning.

2.3 Analogical program derivation

Analogy is an important method in human learning behavior, and has found applications in machine learning. The utilization of analogical reasoning in software automation is a challenging research area.

We have proposed a program derivation calculus, PD-Cal, which can uniformly express specifications, programs and the derivation from the former to the latter. Based on the notion of generalization, a kind of analogical correspondence is defined.

On the basis of the work stated above, we have defined an analogical program derivation process [6].

3. Future Work

As we have mentioned above, software development automation is very important in the software development, because in some sense it is the radical approach to improve productivity, but based on the current status of the art in this area, until now, its progress has not been satisfied with the users, software developers, etc. It is still a long-term and very difficult task to put experimental software automation systems into practical use. According to the status of our laboratory, we are planning to lay emphasis on the following three points in the near future.

3.1 Putting experimental systems on the design level into practice

As we know, the ultimate goal of developing software automation systems is to ease human effort in software development. Although in the past eight years we have developed six systems, but all of these are experimental. In order to put them into practice as soon as possible, we have to do more engineering work to them, especially to make the man-system interface more user-friendly, system documents more readable and complete. In this context, on the one hand, we are now engaging in the formalization of the syntax and semantics of a restricted Chinese, and on the other hand, we rewrite our systems in C++, and extend the original design with the object-oriented approach.

3.2 Designing a new system

In the experiences and lessons of the previous experimental systems, we are planning to design a new system primarily by the object-oriented approach with a textual and graphical specification language. In the mean time, we look for more appropriate mechanisms for software automation systems.

3.3 Looking for analogical approach

First of all, we are going to see and clarify some technical issues possibly facing in order to design software automation system based on analogical approach. Afterwards, we are entering into the design proper.

References
1. Xu Jiafu, *Computer Research and Development*, **25(11)**: 7-13, 1988 (in Chinese)
2. Xu Jiafu etal, *J. of Computer Science and Technology*, **12(2)**: 92-97, 1989
3. Xu Jiafu, *Computer Research and Development*, **27(2)**:1-5,1990 (in Chinese)
4. Zhai Chenxiang, *Ph.D. Dissertation, Nanjing University*, 1990 (in Chinese)
5. Smith, D. R. KIDS, *IEEE Trans. on software Engineering* SE-**16(9)**:1024-43,1990
6. Lu Jiangguo, *Ph.D. Dissertation, Nanjing University*, 1991 (in Chinese)
7. Xu Jiafu etal., *University Computing*, **12(3)**, 92-97,.1990
8. Mitchell, T. etal., *Machine Learning*, 147-80,1986
9. Fei Zongming, *Ph.D. Dissertation, Nanjing University*, 1991(in Chinese)
10. Pagallo, G, *Proceedings of the 11th IJCAI* , 639-1989
11. Burstall, R. M. etal., *JACM* **24(1)**:44-67,1977

SOFTWARE ENGINEERING USING ARTIFICIAL INTELLIGENCE:
THE KNOWLEDGE-BASED SOFTWARE ASSISTANT

DOUGLAS A. WHITE
Rome Laboratory/C3CA
525 Brooks Road
Griffiss AFB, NY 13441-4505

ABSTRACT

The impact of software on everyday life has reached significant proportions. Its widespread use is the source of both great benefit and great frustration. The increasing capabilities and decreasing cost of computers have encouraged their widespread use and the development of increasingly complex systems. Cost and reliability, once primarily the concern of government and large businesses, are now the concern of everyone. The development of software consists of many knowledge intensive, intellectual activities related to understanding the problem to be solved and the design of the solution to that problem. These activities are largely informal, subjective, and undocumented. This condition leads to software that frequently is costly, incorrect, unreliable, and difficult to support. Knowledge-Based Software Engineering (KBSE) is a new technology that has the potential for providing orders-of-magnitude improvement in productivity and, at the same time, much higher quality. It applies Artificial Intelligence (AI) technology to enable computers to assist and automate the software development processes. The U.S. Air Force's Rome Laboratory has been pursuing one approach to KBSE, the Knowledge-Based Software Assistant (KBSA), since 1983. The KBSA will provide intelligent assistance to developers by formalizing the processes and products of software development. This will enable the creation of a knowledge base used by the computer to remember and reason about the activities leading to the creation of a software product. Software developed by the KBSA will be more affordable, capable, and reliable.

Keywords: artificial intelligence, automatic programming, deductive synthesis, knowledge-based system, software engineering, transformational synthesis

1. Background

1.1 Opportunity

Software has been expensive and a source of problems and frustration since computers were first introduced. Numerous solutions adopted over the decades have not eliminated the software problems that continue to plague those who seek the flexibility it promises to provide. The late 1970's and early 1980's were a time of significant change in the computer industry. The DoD, well advanced in the efforts to develop and standardize the common programming language Ada®, recognized that a language alone was not the solution to its increasingly difficult software problems. Affordable, personal workstations were becoming more available as the cost of computers decreased and their power increased. Applications of ever greater variety and complexity were being undertaken throughout both industry and government. Demand for software was projected to grow at a rapid pace. Much greater than potential increases in productivity and manpower to be found within the existing software development paradigm [1]. It was during this time that the relatively low cost machines became available that could support the development and use of large programs written in the favorite language for Artificial Intelligence research, Lisp. With these machines, AI emerged as a viable approach for attacking certain types of problems. Notable successes in constructing expert systems that attacked problems such as locomotive maintenance and computer system configuration helped to lift the reputation of AI. It was in this environment that the USAF Rome Laboratory (then the Rome Air Development Center) embarked upon the bold program of research and development that promises to revolutionize the process by which software is developed and supported. The goal of this research is to develop a computerized, intelligent assistant that carries out the instructions of the human expert. While doing this it will learn the tasks sufficiently well to be able to provide increasingly autonomous support. The research for this integrated set of tools draws from the technology belonging to the field of Artificial Intelligence.

1.2 The software problem

Complaints of current software technology's failings are familiar [2]. The demand for software is increasing more rapidly than the ability to produce it. Costs are growing to where needed systems are not affordable. The complexity of systems is outstripping our ability to produce and maintain them. The illusory promise of software is that it will be flexible and low in cost because it is not implemented as a physical object. It has not lived up to this promise. Software is expensive to develop and impossible to maintain. A 1979 analysis of nine DoD software projects costing approximately seven million dollars indicated that most of the money was wasted [3]. For these nine projects forty-seven percent of the expenditures were for software that was never used. Of the remaining costs twenty-nine percent was for software that was never delivered. Nineteen percent was spent for software that required extensive rework after delivery. Of the final five percent of expenditures, only two percent were for software that was used as delivered.

Decreasing cost and the growing availability and power of computing resources have enabled a computer invasion of our lives. Activities once requiring humans are now automated. Appliances and facilities that formerly were controlled through simple mechanical or electromechanical mechanisms are now controlled by embedded computers. This widespread use of computers combined with an explosion of information has created a demand for software that can not be satisfied. In the early 1980's, the projected rate of growth for software was estimated to be 12 percent per year. Based upon this estimate, a 20 percent improvement in productivity was projected to be worth $45 billion to the United States in the year 1995 [4]. It was clear that to afford to build modern systems an increase in productivity was needed. The STARS Program was the DoD's answer to this challenge. With Ada® as a cornerstone, an improvement of the state of practice was sought through an evolutionary improvement of the existing software development paradigm [5].

1.3 Expert systems and automatic programming

The evolution of traditional software development technology was not the only approach for solving the software problem. Expert systems, although useful for maintaining hardware were not suitable for maintaining software systems. Efforts undertaken to capture the "expertise" required for maintaining software systems reached the conclusion that it is impossible to build such an expert system [6]. The knowledge used in creating software is not retained in any of the artifacts of the development such as the documentation or code. Conventional software development is a manual, intellectual process. The only formal record of a system development are the products such as documents, specifications and code. The processes, reflecting the knowledge used in the development, that would enable the understanding necessary for expert system supported maintenance are informal and unrecorded. With the resurgence of research in the field of AI also came a natural growth in automatic programming efforts. The goals of AI and automatic programming are very similar. AI addresses problems by reasoning about potential plans for achieving the desired goal. Automatic programming attempts to understand problem specifications and reason about plans to achieve the desired outcome. The purpose of applying AI to software development is to enable users to specify their requirements and have the computer automatically produce a program that exhibits the correct behavior. For fully automatic programming the ultimate specification language for Americans would be English. The search for this capability has existed from the very early days of computing and early compilers were thought to be automatic programming systems. Three of the approaches to automatic programming pursued at that time included deductive synthesis, program transformation, and intelligent assistants. Representative work in these and other areas is described in Barr and Feigenbaun [7] and Rich and Waters [8]. Each of these approaches has its own shortcomings leading to the belief that fully automatic programming, a perennial hope for solving the software problem, is unlikely to be achieved soon.

Deductive synthesis approaches are based upon specifying input and output conditions for a program in a predicate calculus formalism. Theorem proving techniques are then applied and the program is synthesized as consequence of the sequence of the steps necessary to construct the proof. This approach suffers from the unmanageable growth of the search process as the size of the program to be produced increases. Also, the writing of specifications and application of theorem proving techniques is not a simple process, requiring skills and effort at least comparable to the task of writing programs.

The transformational approach is a top-down approach in which a specification of a program is gradually rewritten in a more efficient and definite form. Portions of the program specification are selectively replaced with new constructs that preserve the behavior or meaning of the specification while adding some benefit such as efficiency or executability. The transformational approach generally includes the use of a wide spectrum language that accommodates both abstract specifications and low level executable specifications or code. The transformational approach introduces its own problems of limiting or controlling the transformations to be applied. Each transformation is a design decision and in a large system there are potentially many such decisions that must be made. In a given situation, many different transformations may be applicable with significantly different results depending on the selection. The problem becomes similar to that of deductive synthesis. The space of potential solutions that must be considered grows very large and limits the size of problem that may be addressed.

A third approach to automatic programming is that of intelligent assistants. Unlike the previous approaches which seek total automation of the programming tasks, this approach provides assistance to developers rather than eliminating them. An intelligent assistant is a computerized agent that emphasizes the strong points of both the computer and the human. The computer may perform analysis, assist in managing details and assist in routine and tedious details. Intelligent assistants are typically knowledge-based systems in which the knowledge of the activity to be performed is formally codified to enable the computer to participate. The benefit of the intelligent assistant approach is that they do not have to completely solve all the problems before they can be useful. Although the Rome Laboratory Knowledge-Based Software Assistant draws from all of AI, it is most closely associated with this category of automatic programming.

2. An AI Approach

2.1 Failings of conventional approaches

It is useful to consider why conventional approaches have failed to significantly reduce software's problems to better understand what must be done in a knowledge-based assistant approach. Attempts to improve productivity and quality have traditionally focused on improving languages and programming styles, requiring better documentation and enforcing elaborate management procedures that promote quality control. From very early in the history of computing new programming languages were developed that promised to enable computers to understand more human-like languages and ease the job of writing programs. Standards and methods for structuring code have been pursued to make software more reliable and maintainable. Management procedures have been instituted to insure validity of design, correctness of implementation and adequacy of documentation. Software engineering environments have been developed which manage the configuration of development products and support management procedures. More recently Computer Aided Software Engineering (CASE) tools have provided graphic languages for expressing design, and Object-Oriented paradigms have been proposed as holding great promise for improving productivity and quality. The improvements achieved to date have at best been modest with recently reported estimates ranging from fifteen percent [9] to one hundred percent [10].

The reason that these advances have not resulted in the needed productivity and quality improvements is that they only deal with the products of development and not the intellectual processes. Current development environments rely on manual methods for insuring that requirements are traceable to code. Requirements and specifications are only informally and imprecisely stated. Specifications rapidly lose their correspondence to the implementation as the system evolves. The relationship between domain processes and procedures and system representations and support for those processes and procedures is not systematically recorded and maintained. Maintenance continues to be performed at the code level, for in the words of one manager, "the code is the only accurate documentation of the product." CASE tools tell you what products to produce but give little assistance regarding how to produce them. Support for reuse is not provided, and current attempts are being focused on code. None of these approaches get to the root problem of understanding the system in all of its forms and complexity.

Programming languages, from FORTRAN to the graphic notations of CASE tools, have only slightly eased the process of communicating an algorithmic process to the computer by making the code more readable. The current popularity of C/C++ can not be explained if language readability has a major impact on productivity and quality. Developers still must provide the intellectual transformation of design concepts into efficient algorithms. Structured programming and other programming styles have made code more readable, but even when enforced by a modern language, do not provide any knowledge about the algorithm that is encoded or what considerations occurred in transforming the design to the particular implementation. We must still rely on the authors comments to gain this understanding.

Software development standards (such as MIL STD 2167A) and process capability maturity levels have resulted in more rigid control of the movement of artifacts through the life cycle, but do not manage or capture the intellectual activities that are used in transforming the artifacts of one stage to the next. The "Waterfall Model" [11] shown in Figure 1 is commonly used to represent the life cycle processes of software development and support as they exist today and are specified in MIL STD-2167A. Although there is movement toward an evolutionary "spiral model" many of the basic problems remain [12]. The traditional software life cycle is composed of discrete stages each of which yield some form of documentation as the sole product. These products may only be validated informally and manually with the previous stages product.

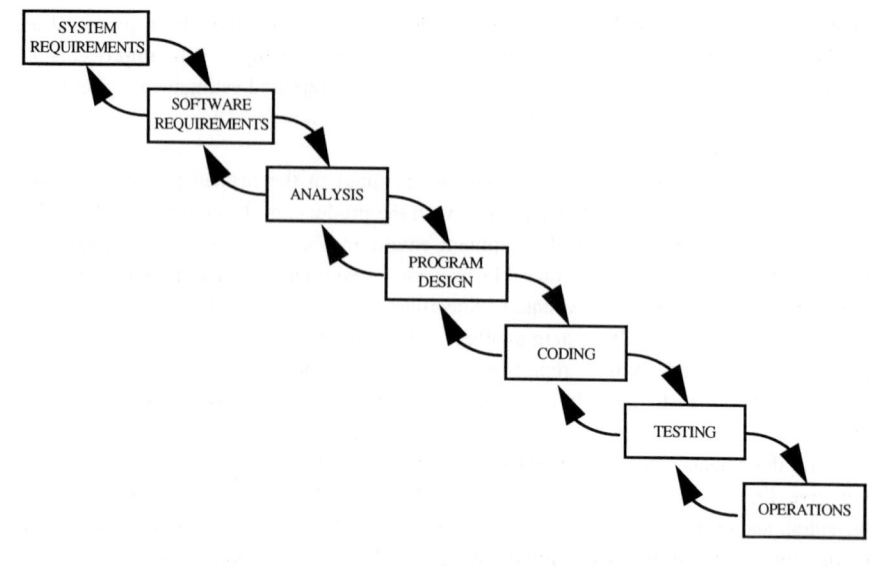

Figure 1

Following the MIL STD-2167A life cycle model for a typical development, requirements are defined in a Software Requirements Analysis phase and documented in a Software Requirements Specification (SRS). The SRS is validated by manual review and serves as the defining document throughout the remainder of the life cycle. Although called the "requirements" specification, this document contains some design detail since Computer Software Configuration Items (CSCI's) are the basic structural element. Authors of the SRS must think in terms of software functions and this can lead to the most costly type of error, requirements errors. Problems with the current paradigm at this stage include requirements that are: incomplete and inconsistent; finalized with minimal user involvement and before design implications are known; formed around programming solutions rather than domain needs; and validated by a manual review of the SRS. Specifications are then created by software designers as they translate the requirements of the SRS into descriptions of software structure and function. Prototypes are rare and the design specification is manually validated with the SRS. Problems that can occur at this stage include misunderstanding the requirements, misunderstanding the specification, inappropriate individual design preferences, and failure to maintain coordinated documentation. Coding or implementation is a complex and highly intellectual activity in which individuals perform the simultaneous acts of translation (of specifications from English to a formal programming language), design, and optimization. Decisions and rationale related to implementation activities are unrecorded and lost. The completed code becomes the final and probably the only accurate specification of the developed system. Once delivered, support frequently consists of interpreting and directly modifying optimized code, using documentation only as a rough guide since it usually does not accurately reflect the software. This results in systems that are high in cost, poor quality and fail to meet user needs. The source of these problems is not a lack of maturity in the administrative processes, but is lack of formal understanding of the products and the intellectual processes that are performed at each stage. Even fully automatic programming as represented by deductive synthesis or transformational programming will not achieve the necessary improvement in productivity. This will require more than a system that can generate programs from specifications. It requires an environment that supports the whole life cycle of a system from collection and organization of needs and desires to the explanation of an implementations behavior and identification of the requirement for that behavior. This is a situation that will benefit greatly from the assistance and automated reasoning that can be provided by artificial intelligence.

2.2 Objectives

The goal of knowledge-based software engineering technology is to shift from informal manual development to formalized computer-assisted development, addressing the failings of conventional software technology and dramatically improving both productivity and quality. Automation, intelligent assistance and increased involvement of expert application engineers and end-users will reduce the time to develop and update software, and will ensure that the systems being developed meet the user's requirements. A formal implementation methodology, deriving code from specifications will insure that the implementations are correct. Decision processes as well as products will be represented and recorded to form a knowledge base that is the "corporate memory" of the system development. This knowledge base will provide the basis for various types of analysis and automation and will also be available for life time system support and evolution. Systems will no longer be maintained by modifying code. Modifications will be introduced at the requirements level and new versions will be derived reusing the previously captured development process. The time required to produce new versions will be reduced, making systems much more responsive and less costly. Systems of much greater complexity will become feasible because of the significant ability of computers to efficiently organize and manipulate large quantities of knowledge in cooperation with the creative and common sense abilities of humans [13].

2.3 Concept

In 1982, the Rome Laboratory (formerly the Rome Air Development Center) initiated the program to develop a knowledge-based system addressing the entire software system life cycle. The Knowledge-Based Software Assistant is based upon the belief that by retaining the human in the process many of the unsolved problems encountered in automatic programming may be avoided. It proposes a new programming paradigm in which software activities are machine mediated and supported throughout the life cycle. The underlying concept of the KBSA, described in the 1983 report [14], is that the processes in addition to the products of software development will be formalized and automated. This enables a knowledge base to evolve during development that will capture the history of the life cycle processes and support automated reasoning about the software being constructed. The impact of this formalization of the processes is that software will be derived from requirements and specifications through a series of formal transformations. Maintenance, consisting of enhancement and change will be introduced as changes to requirements and specifications. It will be possible to "replay" the process of implementation using the development history recorded in the knowledge base. KBSA will provide a corporate memory of how objects are related, the reasoning that took place during design, the rationale behind decisions, the relationships among requirements, specifications, and code, and an explanation of the development process. This assistance and design capture will be accomplished through a collection of integrated life cycle facets, each tailored to its particular role, and an underlying common environment.

The KBSA provides an environment where design takes place at a higher level of abstraction than is current practice. KBSA mediates all activities and provides process coordination and guidance to users. It assists in translating informal application domain representations into formal executable specifications. The majority of software development activities focus on the specification. Early validation is provided through prototyping, symbolic evaluation, and simulation. Implementations are formally derived from specifications through a series of meaning preserving transformations, insuring that the implementation correctly represents the specification. Post deployment support of the developed application system is also focused at the requirements/specification level with subsequent implementations being efficiently generated through a largely automated "replay" process. This capability provides the additional benefit of reuse of designs as families of systems can spawn from the original application. Management policies are formally stated enabling machine assisted enforcement and structuring of the software life cycle processes.

The techniques for achieving these goals are:
(I) formal representation and automatic recording of all the processes and objects associated with the software life cycle,
(ii) extensible knowledge-based representation and inferencing to represent and utilize knowledge in the software development and application domains,
(iii) a wide-spectrum specification language in which high-level constructs are freely mixed with implementation-level constructs,
(iv) correctness/meaning preserving transformations that enable the iterative refinement of high-level constructs into implementation-level constructs as the KBSA carries out the design decisions of the developer.

2.4 Approach

KBSA is a multi-faceted approach to: 1) formalize, capture, and intelligently assist the individual roles in the development life cycle; 2) promote greater user understanding and involvement through informal requirements acquisition, familiar design presentations and frequent prototypes; and 3) create paradigm for development and long term evolution and maintenance of very large software applications, with emphasis on maintaining specification integrity.

The formalizing and capturing of knowledge of all aspects of the life cycle is a major factor in the dramatic improvement in productivity and quality promised by the KBSA. This formalization consists of identifying, relating and modeling the artifacts and processes of a life cycle activity in the knowledge base. It permits the computer to perform the reasoning that provides assistance to human activities as well as automated analysis of requirements and specifications. Formal representation and coordination of all processes and artifacts of development enables more than modeling, analysis and enforcement of high level management procedures and configuration management. It results in three dramatic changes. First, it will enable close management scrutiny, providing the knowledge needed to respond to management queries with dynamic explanations and automate planning and assignment of resources. Second, it is the basis for automation and intelligent assistance such as prompting the user with an agenda of specification details that require elaboration or activities that need to be completed. Third, reuse is inherently provided because of the capturing and recording, in a formal form, of development knowledge such as domain characteristics, artifact attributes, design rationale, and implementation decisions. This reuse does not take place at the customary level of optimized source code, but at the more abstract and effective level of knowledge consisting of requirements, designs, specifications and processes.

Intelligent assistance, supported by representations and views that are familiar, easy to understand and much more abstract than programming or specification languages, will enable users in all roles to work more productively and with a much greater understanding of the evolving system design. The KBSA will be an integrated, distributed environment in which participants collaborate in system development, communicating with familiar tools and notations. All representations whether text, charts, graphics, or other media, will be related via mathematical transformation to the evolving formal, executable specification contained in the knowledge base. Interactions will have global effect, greatly reducing potential design conflicts as changes are rapidly reflected through the knowledge base to all other roles or perspectives. The capability to simulate and execute the specification at any time will enable the continuous validation of the evolving specification by the user.

The KBSA-based development paradigm will be highly iterative with early and frequent opportunity for review and validation. Systems will be valid and "correct" because of the process used to create them. Informal requirements captured through interactions with users and application experts are "formalized" as entities/objects, attributes, relationships, constraints, and actions are added to the evolving description of the system represented in the knowledge base. Elaboration of the formal specification is supported by evolutionary transformations which implement a set of modifications which change the meaning of the specification in well defined ways. The validity of the requirements and resulting specification are continuously checked by the user through textual paraphrases, graphical representations, simulation and execution of the specification. The implementation of the specification, an activity comparable to compilation of programming languages, occurs only after the user is satisfied that it meets his requirements. This derivation of code consists of the application of "correctness preserving" transformations to the already executable specification to achieve the degree of optimization appropriate for the desired cost and performance requirements. Emphasis throughout the KBSA life cycle is on ensuring the consistency and validity of the specification.

The KBSA approach is a departure from the existing software engineering paradigm in that it attempts to formalize all activities as well as products of the software life cycle. By supporting development, evolution, and long-term maintenance of computer software, KBSA captures the history of system evolution. It provides a "corporate memory" of component interaction, design assumptions and rationale, how requirements are satisfied, and explanation of the development process. KBSA accomplishes this through a collection of integrated dedicated facets and an underlying common framework.

KBSA has four main distinguishing features. First, the specification is executable, formal and incrementally defined. The specification is executable like a prototype, allowing the validation of the specification against user intent by actually showing the "running" system. Being formal, the specification is expressed in a language with precise semantics, avoiding the ambiguity of natural language. By supporting incompleteness in the specification detail may be added gradually. The customer does not need to wait until it is complete to review and validate it. Second, the implementation is formal. All decisions made during the implementation are captured. Implementations will be derived using meaning preserving transformations, thus guaranteeing a correct implementation. Third, project management policies will be formally stated and enforced. Project processes will be modeled and enacted defining the relationship between various activities and artifacts (e.g. requirements, specifications, code, test cases, bug reports, etc.) and enforced by KBSA throughout the software development process. Finally, maintenance will be a continuing evolution or merely a continuation of development. It will be done by changing requirements and specifications, rather than by patches to the code since maintenance activities are normally a result of new or better defined user requirements.

To build a KBSA there is a need for specific supporting technologies. These supporting technologies fall into four main categories a wide spectrum language, general inferential systems, domain specific inferential systems, and system integration. A wide spectrum specification language is a single language which provides the user with the ability to capture the formal semantics of the system under development regardless of the level of detail or the stage in the development cycle. A wide spectrum specification language provides uniform expressibility, regardless of what is being described and it must do so in a way which is consistent at all levels, both syntactically and semantically. A general inferential system supports reasoning applicable to all problem domains. Domain specific inferential systems extend general inferential systems to include aspects unique to software development. They focus on the knowledge representation of software development objects and inference rules and how they can be formally represented and used for further reasoning. This category can be further broken down further into three subcategories: formal semantic modeling of the software development environment, knowledge representation and management of software development objects, and specialized inferential systems which incorporate rules that have been tuned to a specific problem or task. System integration technology deals with the inherent compromises necessary to support all facets with uniform interfaces and behaviors .

2.5 Paradigm scenario

The evolutionary and iterative nature of the KBSA is shown in the life cycle model of Figure 2. The iterative process of requirements acquisition, specification elaboration and validation produces a formal specification which is only implemented in code when the user is satisfied that it meets his requirements.

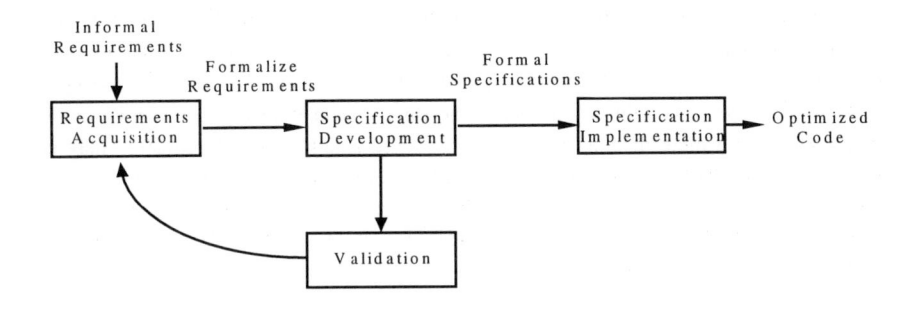

Figure 2

User involvement and productivity are increased by making the life cycle opportunistic and evolutionary. The requirements acquisition process consists of collecting requirements as opportunities arise, using multiple informal and abstract representations including hyper-text, graphics, audio and video [15]. Informal requirements are "formalized" as entities/objects, attributes, relationships, constraints, and actions and are added to the evolving description of the system represented in the knowledge base. The knowledge-based representation of the formal specification enables individual changes to be reflected globally insuring everyone is viewing the same specification and allowing automated analysis of the impact of the change. Graphical presentations, natural language paraphrases, simulation and specification execution provide multiple ways for users to validate the capabilities and developers to demonstrate the consistency of modifications to the specification with their intentions.

Elaboration of the formal specification is supported by evolutionary transformations which implement a set of modifications which change the meaning of the specification in well defined ways [16]. Specification validation will be routine and automatic since the formal specification language will support comprehensive analysis, and may be executed to provide a "rapid" prototype of the system under development. Prototypes will become frequent because they are automatic, supporting continuous validation as the specification evolves. Traceability and consistency with the users needs will be ensured because of the orderly derivation process by which they are created.

Implementation of the specification, an activity comparable to compilation of programming languages, occurs only after the user is satisfied that it meets requirements. Implementation in the KBSA is an automation of formal methods [17]. This derivation of code consists of applying meaning preserving transformations or refinements to the already executable specification to achieve the degree of optimization appropriate for the desired cost and performance. Regions of code to be transformed are selected by the developer. The system then provides the developer with a menu of applicable transformations. The selected transformation is then automatically applied to the identified region of the specification. This process leads to valid implementations since the transformations have previously been proven to be correct. The need for testing is thus greatly reduced or eliminated. Since the final system is merely a rewriting of the specification in a more detailed and functional representation, the specification may be reused. First as the basis for continuing evolution as requirements change, and second as the generic specification for families of systems where only the implementation changes. This will lead to much longer system life, greater portability of applications and data, and produce as a result much greater productivity.

Throughout a development, process mediation, consisting of the coordination of all activities and communications, and a knowledge base relating all artifacts will enable project managers to more effectively plan projects and monitor their progress. Planning and estimation activities will be able to use existing domain libraries of prior developments, the evolving system definition and available resources to produce more accurate cost estimates and schedules. Machine coordination of all activities will enable detailed monitoring of progress and dynamic scheduling and rescheduling of resources as the situation demands. Automated reasoning capabilities will allow identification of "brewing" problems, optimization of resource and schedule allocations and evaluation of processes and procedures. Many of the mechanisms of the KBSA that are used to create programs are also applicable to creating projects since both are merely processes, one computational and one organizational [18]. Automated coordination of all activities and communications will enable management to enforce policy and procedures.

Once developed, systems are not static, continuously evolving to include additional capabilities and move to new environments as users needs and technology change. KBSA provides the opportunity to greatly extend software life because it enables it to be truly "soft" and pliable. In the KBSA paradigm software will evolve after delivery just as it did during development. Requirements will be added (or removed) and the transformation of the new requirements into code will be a process similar to the original development process and will in fact consist of a reuse of much of the original development. Systems will no longer be patched at the code level where the requirements and design have been obscured by optimizations. Instead they will be consistent derivations of code from the updated specification.

2.6 KBSA Program

Since 1983, Rome Laboratory has been developing the Knowledge-Based Software Assistant. The strategy proposed to achieve the goals of the KBSA was to first formalize each stage of the present software life cycle model, with parallel developments of technology and knowledge bases for each particular stage. Supporting technology was to be the subject of concurrent research and development efforts with periodic integration efforts or "builds" to assess progress and identify deficiencies. The general structure guiding this research is shown in Figure 3. Although resource limitations have precluded multiple parallel research thrusts of the magnitude originally proposed, initial products of the program have emerged with the successful completion of efforts which model and automate requirements definition, system specification, performance optimization and project management. Supporting technology has also been developed which will form the core of the KBSA and will be used in merging and managing the activities and processes of the various users. The following paragraphs will provide a brief description of the basic approach of each research effort and resulting products.

The first area to be addressed by the KBSA program was that of project management. In 1984 the definition of a Project Management Assistant formalism and construction of a working prototype began. The goals were to provide knowledge-based assistance in the management of project planning, monitoring, and communications. Planning assistance enabled the structuring of the project into individual tasks followed by scheduling and assigning these tasks. Once planned a project must be monitored. This is accomplished through cost and schedule constraints and the enforcement of specific management policies. PMA also provides user interaction in the form of direct queries/updates and various graphics representations such as Pert Charts and Gantt Charts. PMA is distinguished from conventional management tools because not only does PMA handle user defined tasks, but it also understands the products and implicit relationships among them (e.g. components, tasks, requirements, specification, source code, test cases, test results, and milestones). Contributions of the PMA include the formalization of the above objects, the development of a powerful temporal representation for dependency relationships between software development objects, and a mechanisms for expressing and enforcing project policies. The initial PMA effort was completed in 1986 and subsequent work continued the evolution of PMA, expanding the formalized knowledge of project management to provide enhanced capabilities and to implement PMA as an integral part of a full-scale conventional software engineering environment.

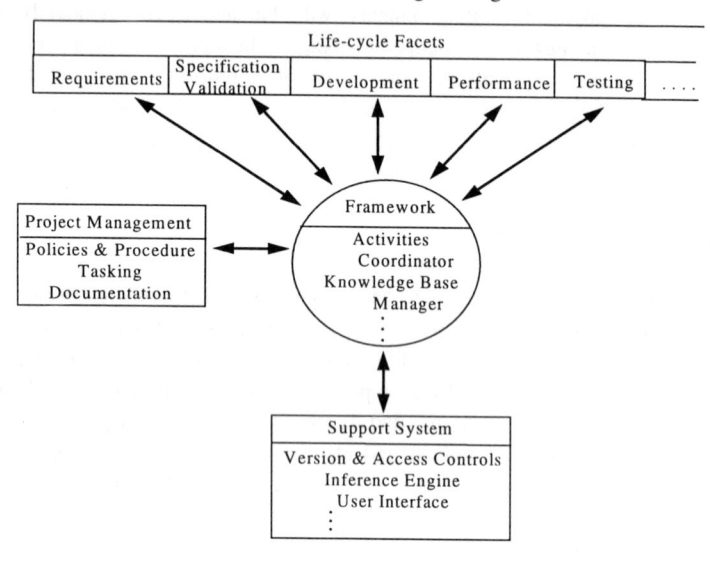

Figure 3

The Knowledge Based Requirements Assistant (KBRA) research began in 1985. Central to the KBRA was the need to deal with the informal nature of the requirements definition process including incompleteness and inconsistency. Requirements can be entered in any order or level of detail using one of many differing views of the application problem. The KBRA is responsible for doing the necessary bookkeeping to allow the user to manipulate the requirements while it maintains consistency among them. Capabilities included in the KBRA are support for multiple viewpoints (e.g. data flow, control flow, state transition, and functional flow diagrams), management and editing tools to organize requirements and the support for constraints and requirements that are not functional in nature through the use of spreadsheet and natural language notations. Other capabilities of the KBRA include analysis to identify inconsistency and incompleteness, and the ability to generate explanations and descriptions of the evolving system. As previously indicated, the primary issue addressed by the KBRA was handling the informality of incomplete user descriptions while building and maintaining a consistent internal representation. This was accomplished through the use of a representation providing truth maintenance support including default reasoning, dependency tracing, and local propagation of constraints. Through these mechanisms the KBRA was able to provide an application specific automatic classification which is used to identify missing requirements by comparing current input against existing requirements contained in the knowledge base.

The KBSA Specification Assistant research also began in 1985. The Specification Assistant facilitates the development of formal executable specifications of software systems. It supports an evolutionary activity in which the system specification is incrementally elaborated as the user chooses among design alternatives. An executable formal specification language combined with symbolic evaluation, specification paraphrasing and static analysis allow early design validation, providing the user an evolving prototype of the system along with English descriptions and consistency checking throughout its design. Specification Assistant capabilities utilize a variety of tools to support the user. One facility peculiar to the Specification Assistant is the support for specification evolution in the form of high-level editing commands, also known as evolution transformations. These commands perform stereotypical, meaningful changes to the specification. They differ from "meaning-preserving" transformations because they are specifically intended to change the meaning of specifications, but in controlled ways. In addition to the top down evolution of specifications supported by the high-level editing commands, the Specification Assistant supports the building of a specification from smaller specifications (i.e.. the reuse of previously defined specifications) with a set of view extraction and merging tools. In 1988 work began to merge the capabilities of the KBRA and Specification Assistant, spanning all the activities needed to derive a complete and valid design from initial user requirements in one system called ARIES.

The purpose of the KBSA Performance Assistant is to guide designers in performance decisions at many levels in the software development cycle. This assistant takes as input a high level program written in a wide-spectrum language and following a combination of automatic, performance-based, and interactively-guided transformations, produces efficient code. It supports the application of a variety of analysis and optimization techniques broken into the two general categories of control optimizations and data optimizations. The control optimizations include finite differencing, iterator inversion, loop fusion, and dead code elimination. Data optimization includes data structure selection, which implements a program's data objects using efficient structures, and copy optimization, which eliminates needless copying of large data objects.

The goal of the KBSA Development Assistant is to support the transformation of formal specifications into low level code. The Development Assistant share many capabilities with the Performance Assistant. It supports the construction of a formal model of the application domain including the specification of the target system's desired behavior and the application of transformations to the specification to produce detailed code. The transformations encode design and optimization knowledge, allowing the user to mechanically make high-level design decisions which the system systematically applies. A facility is also provided which records derivations and provides the basis for future "replay", a fundamental concept of the KBSA.

Research in supporting technology resulted in the definition of requirements for a unifying framework sufficient to support the many varying facets of assistance provided by the KBSA. It would support an object base with a tightly integrated logical inference system, configuration management, activity coordination, and provide a common user interface for the KBSA. It would also provide a common reference for facet development which when followed would allow the sharing of communications and knowledge. One original concept which distinguished the KBSA was that of activity and communication coordination. The goal was to define a formalism with a graphical syntax that could be used to specify and enforce the coordination of the many activities and communications. The result was the development of "Transaction Graphs," a mathematically based formalism for specifying processes. The interrelated problem of change and configuration management was solved at the same time by merging the Transaction Graphs with the previously existing "Artifacts" configuration management system.

Development a system to demonstrate the concepts of the KBSA began 1988. This system provided a demonstration capability for a narrow problem domain and was broad in concept coverage but shallow in functionality . The KBSA Concept Demonstration system combines capabilities from the ARIES, Development, and Project Management assistants and includes example developments from an Air Traffic Control application domain. It allows the demonstration of refinement of requirements and specifications, project management including the automatic creation of tasks as the design progresses, simulation via specification execution, and the automatic generation of Lisp code from specifications. Additional capabilities are provided for examining and manipulating both informal and formal representations of design including hypertext, multiple graphical representations, and English like explanations. The system has successfully demonstrated the usefulness of AI and knowledge-based assistance for the various roles in software development. History mechanisms enabling the retracing of a development tree and agendas where outstanding issues and remaining tasks can be posted greatly assist in organizing development and making it more understandable. The focus of the formal representation of the specification in the knowledge base and the global impact of any change across the development has shown the significant ability to improve productivity. This success in a demonstration system encouraged the next step, building a truly usable KBSA environment, an advanced development model (ADM). The ADM will be the first attempt at integrating the KBSA technologies to form a true working environment. It will provide a robust design and development environment of acceptable performance. The results of this first implementation will be evaluated through development of a moderate sized "real" application. Implementation of the ADM will be accomplished using conventional languages and tools to achieve suitable performance and more widespread user acceptability. Graphical object-oriented design representations to those common in present CASE tools will be used to enhance user acceptability and ease transition to the KBSA.

3. Summary/Conclusion

Although the KBSA is tremendously closer to fruition than true "automatic programming" and much optimism exists as evidenced above, it is an ambitious project and additional research will be needed. Future research must include the continued evolution of existing components and supporting technology, greater emphasis on assistance and promoting design understanding, better abilities to simulate partially specified systems and additional automation in transforming specifications into code. A benefit of the "assistant" approach is that interim capabilities can be used with major impact on productivity and cost. The KBSA is entering a new stage in its research and development life cycle. It may now be used to produce systems and save money while at the same time evolving as it serves as the basis for additional research. Although confidence is now high that significant improvements in productivity are achievable and orders of magnitude improvements in performance have been demonstrated for specific problem domains only actual use and empirical evaluation will validate the early claims of orders-of-magnitude improvements in productivity. With the completion of the ADM in 1996 and its planned use on several significant developments the answer should begin to appear before the end of the decade.

References

[1] B. Boehm and T. Standish, *Software Technology in the 1990's: Using an Evolutionary Paradigm*, Computer, Vol. 16, No. 11, (1983) 30-37
[2] *Software Technology for Adaptable Reliable Systems (STARS) Program Strategy: DoD Report*, ACM-SIGSOFT Engineering Notes, Vol. 8, No. 2 (1983) 56-84.
[3] A. Davis, *Software Requirements: Analysis and Specification*, Prentice Hall, Englewood Cliffs, N.J., (1990)
[4] B.W. Boehm, *Improving Software Productivity*, Computer, Vol. 20, No. 9, (1987) 43-57.
[5] L. Druffel, S. Redwine, Jr. and W. Riddle, *The STARS Program: Overview and Rationale*, Computer, Vol. 16, No. 11, (1983) 21-29.
[6] J. Dean and B. McCune, *Advanced Tools for Software Maintenance*, USAF/RADC Report RADC-TR-82-313, (1982)
[7] A. Barr and E. Feigenbaum, *Automatic Programming*, The Handbook of Artificial Intelligence, Volume II, William Kaufmann, Inc., Los Altos, CA, (1982)
[8] C. Rich and R. Waters, *Readings in Artificial Intelligence and Software Engineering*, Morgan Kaufmann Publishers, Inc., Los Altos, CA, (1986)
[9] J. Voelcker, *Automating Software: Proceed with Caution*, IEEE Spectrum, Vol. 25, No. 7, (1988)
[10] L. Vangelova, *Software Engineering Aid*, Government Executive, Washington, D.C., (1994) 47-48
[11] W. Royce, *Managing the Development of Large Software Systems*, In IEEE WESCON, August 1970. pp. 1-9. Reprinted in Ninth IEEE International Conference on Software

Engineering, Washington D.C.: Computer Society Press of the Institute of Electrical and Electronics Engineers, (1987) 328-38

[12] B.W. Boehm, *A Spiral Model of Software Development and Enhancement*, Proc. IEEE Second Software Process Workshop, ACM Software Engineering Notes, (1986)

[13] M. Rettig, *Cooperative Software*, Communications of the ACM, Vol. 36, No. 4, (1993) 23-28

[14] C. Green et al., *Report on a Knowledge-Based Software Assistant*, USAF/RADC Report RADC-TR-83-195, (1983)

[15] A. Czuchry Jr. and D. Harris, *KBRA: A New Paradigm for Requirements Engineering*, IEEE Expert, Vol. 3, No. 4, (1988) 21-35

[16] W. Johnson, et al, *ARIES: The Requirements / Specification Facet for KBSA*, USAF/RL Report RL-TR-92-248, (1992)

[17] D. Smith, *KBSA Development Assistant*, USAF/RL Report RL-TR-92-26, (1992)

[18] R. Jullig et. al, *KBSA Project Management Assistant*, USAF/RADC Report RADC-TR-87-78, (1987)

KNOWLEDGE-BASED DERIVATION OF PROGRAMS FROM SPECIFICATIONS

THOMAS J. WEIGERT,[*] JAMES M. BOYLE,[†] TERENCE J. HARMER[‡]

and FRANK WEIL[*]

[*]Land Mobile Products Sect.	[†]Math. and Comp. Sci. Div.	[‡]Dept. of Comp. Sci.
Motorola, Inc.	Argonne Natl. Laboratory	The Queens University
Schaumburg, Illinois	Downers Grove, Illinois	Belfast, Northern Ireland

ABSTRACT

We describe an experiment in knowledge-based programming that demonstrates that high-level specifications and program transformation can be used to derive efficient, correct, executable code for a representative communications product software module. Significant advantages of this approach are: The derivation can be carried out automatically; the employed transformations codify programming knowledge so that it can be reused in other applications; and the derived executable code is at least as good as, and in some cases, better than, the hand-written code. These results lead us to conclude that program transformation can be used to derive code that meets the efficiency and reliability requirements for commercial products.

In an appendix we will give an overview of the formalism to write program transformation rules provided by the MOUSETRAP program system which we used in our experiment.

Keywords: Program transformation, knowledge-based software engineering, compilers.

1. Introduction

Programming is a vexing problem. Often, one starts with a fairly simple statement of a problem to be solved or a function to be implemented. One puts a lot of effort into expressing the solution or implementation in a form that can run with adequate speed on the particular computer hardware that is available. Out of this process may come an efficient program for that hardware. Unfortunately, one often loses a number of useful and desirable things on the way to that program.

The first attribute to go is often understandability. To make the code efficient, programming tricks must be used, and each of these tricks obscures the simplicity of the program's original intent.

Correctness also seems to slip rapidly away. Of course, some errors are found in the "debugging" process, but some errors, occasionally disastrous ones, manage to slip into the final product. On the heels of these attributes flees another, modifiability. Because the final program is obscure, often even opaque, it is almost impossible to modify if one wishes to add new features or to reimplement it on new hardware differing in characteristics from the original target.

Finally, many, especially those who are managing program development, notice that large amounts of two other valuable commodities seem to chase after understandability, correctness, and modifiability. These commodities are money and time.

Is there a better way? Could one somehow write a clear original specification of the problem or function and obtain *automatically* an efficient implementation for any of several different hardware platforms?

In the following sections we discuss an experiment in using program transformation to do just that. This experiment represents a small-scale attempt to apply program transformation to the development of code for a commercial software product. Although limited in scope, the experiment had ambitious goals. These goals included demonstrating the use of:

- transcription of clear pseudo-code operations into hardware-specific operations,
- high-level abstract data types,
- separation of concerns (such as correctness from efficiency, function from real-time constraints), and
- high-level specification constructs.

We wished to achieve these goals within the context of producing

- code that is equivalent to, or higher in quality than, that written by hand for the application.

2. The Software Development Process

Our experiment in applying program specification and transformation to software development modifies the later stages of the typical software development process. While the processes used to develop software vary in detail, their idealized form is outlined in the following section.

2.1. Idealized Software Development Process

Starting from informal (and often imprecise) user requirements, system analysts and designers develop a formal requirements specification and a design for the product and its software and firmware. The formal requirements specification models the behavior of the system as input/output relationships over abstract data (possibly maintaining an internal system state), with no attention being paid to physical reality. Design describes the framework of how the functional requirements are achieved. This framework is detailed in terms of an architecture and of processes within the components of the architecture.

Experienced engineers then turn the design representation into pseudo-code. There are two basic goals of this activity: to describe in more detail the actual data on which the system operates, and to describe how the behavior stated in the requirements model can be accomplished based on the data. Additional behavior may have been imposed by the chosen architecture rendering some of the assumptions of the requirements specifications unrealistic. For example, the information between architecture modules may flow over unreliable channels (such as radio links), the chosen processors may not allow the desired amount of parallelism, or a process may have limited ability to listen to multiple simultaneous inputs. Pseudo-code is intended to be a high-level representation of the actual program required for the product. As such, it could be executed for prototyping if suitable tools were available.

The final step in the process is for software engineers to turn the pseudo-code into executable code in the required language, for example, C. In the ideal form of the program development process, it is in this step that "hard reality" enters. The software engineers must take into account the constraints and limitations of the physical components that will execute the program, and they must follow any stylistic conventions that the final program is required to obey.

2.2. Practical Software Development Process

What happens to the later steps of this process in practice? We have observed that the economics of programming dictate that the final step of the process be carried out by inexperienced software engineers. After all, they must be used somewhere, and converting pseudo-code into executable code familiarizes them with the product and with coding practices so that eventually they may be able to work on turning designs into pseudo-code.

In practice, however, the use of inexperienced software engineers leads the experienced engineers to doubt that their pseudo-code will be turned into correct and reliable executable code. Their response is to lower the level of the pseudo-code specification until it is almost at the executable level, leaving the software engineers with little to do except transcription. Of course, the extra work that the experienced engineers do in going from the design to low-level pseudo-code requires additional time on their part, which is relatively expensive.

What happens when we use program specification and transformation in this paradigm? First of all, program transformation automates the transition from pseudo-code to executable code, thereby saving the labor cost of this step. Of course, developing the transformations requires experienced people and costs something. However, once developed, transformations can usually be applied to as many programs as desired. Their cost is thus more nearly proportional to the number of distinct constructs used in the specification rather than to its total length.

Secondly, developing program transformations turns out to be easier if there is more, rather than less, "slack" between the pseudo-code level and the executable code level. Thus, using transformations encourages *raising* the level of the pseudo-code until it is close to the design, thereby reducing the work required of the experienced engineers and its associated cost.

3. What Is Program Transformation?

We do not have space here to give a thorough introduction to program transformation. Recent books edited by Meertens [7] and Pepper [9], as well as an excellent survey article by Partsch and Steinbrüggen [8] contain useful discussions of this approach to deriving programs. In our experiment, we used the MOUSETRAP program transformation system developed by the first author.[1] We sketch briefly how this system applies program transformations in order to facilitate explaining how we were able to carry out our transformational programming experiment. A MOUSETRAP transformation is a *correctness preserving rewrite rule*. Such a rule consists of a *pattern* and a *replacement*, each of which is a program template. A *program template* consists of a sequence made up of terminal symbols from the programming language being transformed and *variables*, which stand for any suitable part of a program without describing its contents in detail. Both the pattern and replacement templates are expressed in terms of the grammar of the programming language being transformed. The grammar provides a convenient way to name the parts of the program that are variable. For example[2], suppose that in some programming language, the program fragment:

```
1  for each i from 1 to n do
2      s := s + a(i).
3  endfor.
```

has the same meaning as the fragment:

---

[1]This experiment was initially conducted using the TAMPER system [2, 4] developed by the second author at Argonne National Laboratory.

[2]Fragments of program code or transformation rules will be typeset in sans-serif font.

```
1  i := 1;
2  while i <= n do
3      s := s + a(i).
4      i := i + 1.
5  endwhile
```

Then a correctness-preserving rewrite rule for MOUSETRAP that transforms *any* fragment having the form of the first one above into the form of the second is:

```
1  ?var: Ident;  ?e1, ?e2: Expr;  ?sl, ?rest: Stmtlist;
2  //.
3  Stmtlist'for each ?var from ?e1 to ?e2 do
4              ?sl
5              endfor.'
6  ==>
7  Stmtlist'?var := ?e1.
8              while ?e1 <= ?e2 do
9                  ?sl
10                 ?var := ?var + 1.
11             endwhile.'
```

Such rules are useful. Indeed, a collection of simple rules such as this one, applied where they match in a piece of pseudo-code, can automate most of the transcription-level programming involved in converting the current low-level pseudo-code into executable code.

However, in order to perform transformations more sophisticated than simple transcription, a program transformation system needs some technique for automatically determining where and in what order to apply transformations. By default, the MOUSETRAP system uses a very simple rule: given a set of transformations that it is to apply to a program, MOUSETRAP applies transformations from the set wherever their patterns match until none can be applied anywhere in the program. (Note that this rule means that, if a transformation applies and its replacement creates new instances to which it or other transformations apply, MOUSETRAP applies those transformations to these instances also.) This application rule turns out to be very powerful because it enables one to write a set of transformations that manipulates a program into a *canonical form*. By using successive sets of transformations to pass a program through a sequence of canonical forms, it is possible to make arbitrarily complex changes in the program [2, 3, 6], for example, to transform it from pure LISP into non-recursive Fortran [2, 4] or C. In such processes, the final program *evolves* from the initial specification in a sequence of tens, or even hundreds, of thousands of transformational rewrites.

It should be clear from this brief discussion that MOUSETRAP transformations are somewhat similar to the rewrite rules used in expert systems. It should not be surprising, then, that the very powerful idea of creating canonical forms can be used to prepare sets of program transformations that constitute an expert system for programming. Such sets of transformations serve as a knowledge base that *codifies programming knowledge*, enabling it to be reused at will in application after application.

4. An Experiment in Program Transformation Applied to a Commercial Product

For our experiment in knowledge-based automated programming we selected the routines from the *Assignable Speakers* module of Motorola's CENTRACOM Series II Plus control center system. We believe that these routines contain a small but representative sample of the functions, operations, and constructs used in communications software.

In outline, the experiment proceeded in the following way:

- We wrote specifications for the routines in *Assignable Speakers* in a high-level pseudo-code language. This language is considerably more abstract than the pseudo-code language used for the routines during the original product development phase.

- We wrote sets of correctness-preserving program transformations that transform the abstract constructs and the abstract data types used in the pseudo-code into equivalent (but less clear) C code.

- We wrote sets of correctness-preserving program transformations that captured domain-specific design and implementation decisions that had been made during the course of the original development of *Assignable Speakers*.

- We applied these transformations to the specifications using the MOUSETRAP program transformation system and produced executable C programs that are as good or better than those developed for use in the product.

In order to produce code of such high quality from specifications as abstract as the ones we used, it was necessary to solve a number of programming problems. In the following sections we discuss the major problems that we tackled and their solutions. We believe that this demonstration of the ability of program transformation to solve such problems makes it credible that real product software can be developed using these techniques.

4.1. Implementing Transcription to Hardware-Specific Operations

As discussed in Sec. 2.1, much of the work done in going from the current level of pseudo-code to executable code involves transcribing a large number of hardware-specific operations from the pseudo-code notation (chosen to some extent for readability) into the hardware-oriented executable notation. One can think of performing the transcription by consulting a "dictionary" of executable equivalents for the pseudo-code notations.

Such a process is simple to automate using program transformations. One just transcribes the equivalences in the dictionary into program transformations. We call the resulting set of transformations "simple rewrites." While they are simple, they are nevertheless numerous; this set of transformations is the largest one that we developed in this experiment.

A typical example of the application of the simple-rewrite transformations occurs in the save_spkr_asgmts routine of *Assignable Speakers*. This routine updates the speaker assignments, which are stored in EEPROM. In the notation of our specification language the update operations are expressed as:

```
1  set_2nd_speaker_destination(speaker_id, channels, current_chnl).
2  if (both_speaker_destinations_are_assignable())
3      then set_1st_speaker_destination(speaker_id, channels, current_chnl).
4  endif.
```

The simple-rewrite transformations are applied to produce the C code:

```
1  w_eeprom(speaker_id, &(_spkr_dest_eeprom_[current_chnl].uns));
2  if (asgn_both_spkr_dest())
3      w_eeprom(speaker_id, &(_spkr_dest_eeprom_[current_chnl].sel));
```

Although this set of transformations is the first we discuss, they are actually applied near the end of the derivation because they remove the last of the abstractions used in the specification. Applying these transformations last reflects a general principle that we use in planning a program derivation: *preserve abstraction for as long as possible*. This principle is important because it helps to maintain flexibility and reusability in the derivation should one later wish to retarget it to a different hardware architecture.

If the only abstractions in our specification were those implemented by the simple-rewrite transformations, our specification would not be very abstract. We turn now to a sequence of examples of increasingly abstract notations that we use in our specifications.

4.2. Implementing High-Level Abstract Data Types

An operation that occurs frequently in the routines of the *Assignable Speakers* module is an iteration over the channel assignments that are stored in the EEPROM. It is useful to hide the obvious hardware dependencies in this statement. For example, in a few years, *Assignable Speakers* may be implemented on hardware that has no EEPROM. Certainly the specification for the process of assigning speakers *need* say nothing about this particular type of hardware; hence, the specification is "better" (more flexible) if it does not.

Iterations over channels can be expressed abstractly by considering the channels to be a *set*, because none of the operations performed in the iterations depend on the channels being ordered or contiguous. Operations that apply to all of the elements of a set can be expressed by a *for each* operator.

Another concept needed to express the operations on channels is that of filter [1]. A filter operates on a predicate and a set and produces the subset of the elements of the original set for which the predicate is true.

An example of the use of these abstractions occurs in the exit_assign_spkr_mode routine of *Assignable Speakers*. This part of the routine turns off each indicator LED for the subset of channels that are assigned to a given speaker identifier:

```
1 for each current_chnl :: Channel in
2    select all chnl :: Channel in elts(channels) such that
3         current_channel_assigned_to_this_speaker_id(speaker_id,
4                                                  spkr_asgmt, chnl).
5    endselect do
6    turn_off_led(current_chnl).
7 endfor.
```

After applying a few sets of transformations, we obtain the C code:

```
1  Int16 current_chnl;
2  ...
3  current_chnl = _low_;
4  while (current_chnl <= hichnl()){
5       if (testbit((Byte*)spkr_asgmt, current_chnl)) {
6            Int16 g1;
7            if (dbmspk(current_chnl, _select_, &g1)) {
8                 ldupd_(g1, _led_off_);
9            }
10       }
11       current_chnl++ ;
12  }
```

This C code *implements* the set operations in a way that is efficient given that the channels are represented as a set of contiguous integers. But recall that we still have the specification expressed in terms of an abstract set, which makes it easy to transform into a new implementation should the hardware representation of channels change.

One may need operations such as *select* and *for each* for several different types of sets and also for other data types such as lists, bags, etc. Clarity is enhanced by using the same operators for all these cases, treating them as polymorphic operators.

If one wishes to write specifications in this way, one must have transformations that compute the types of the operands of such operators in order to enable selection of the implementation appropriate to the type. In addition, such type information is required to construct declarations when the transformations must generate temporary variables not in the specification in order to complete an implementation. Transformations can generate such type information by propagating it from the declarations to the uses of variables, then propagating it through expressions to determine their types, and finally using the information about the types of polymorphic operators to select an appropriate implementation and to generate temporary variables of appropriate type.

4.3. Implementing Separation of Concerns

An important benefit of using abstract specifications and program transformation is that it enables one to *separate concerns* that are necessarily intermixed in ordinary low-level programming. The two concerns that are most obviously intermixed in ordinary C programs are *what the program is supposed to do* and *how to do it efficiently*. However, we discovered that there are several other concerns that are also intermixed in the C code for *Assignable Speakers*.

Some of these concerns are requirements imposed by the "reality" of the implementation. For example, the overall system of which *Assignable Speakers* is a part must meet real-time processing requirements imposed by the need to analyze incoming radio signals. For reasons of efficiency and correctness, it is deemed better to have all routines of the system relinquish control periodically to the scheduling module than to have the scheduling module interrupt running routines to meet the real-time requirements. This approach leads to a requirement that every loop (each of which potentially could execute for a long time) update a "suspend counter" each time through the loop and suspend execution if this counter has exceeded a set maximum.

While this requirement is *vital* to the correct operation of the whole product, it has nothing to do with the specification of *Assignable Speakers*. The requirement for suspending represents a concern that should be separate from this specification. It can be enforced by having a set of transformations insert the suspend counter and its initialization, increment, and testing into *all* potentially long executing loops in the program.

An example of where suspend counting needs to be inserted occurs in the validate_spkr_id routine (as well as several others) of *Assignable Speakers*. The specification for the iteration of this routine is similar to that in exit_assign_spkr_mode. This specification, which is independent of the need for a suspend count, is:

```
1  for each current_chnl :: Channel in
2      select all chnl :: Channel in elts(channels) such that
3          speaker_id_in_second_speaker_dest(channels, chnl, speaker_id).
4      endselect do
5      add_channel_to_speaker_assignment_list(spkr_asgmt, current_chnl);
6      turn_on_led(current_chnl);
7  endfor.
```

The derived C code, obtained after applying the transformation set that inserts operations to implement the suspend count, is:

```
1  Int16 current_chnl;
2  ...
3  Int16 g2 = 0;
4  current_chnl = _low_;
5  while (current_chnl <= hichnl()){
6      if (_spkr_dest_eeprom_[current_chnl].uns == speaker_id) {
7          Int16 g3;
8          modbit((Byte*)spkr_asgmt, current_chnl, _on_);
9          if (dbmspk(current_chnl, _select_, &g3)) {
10             ldupd_(g3, _led_on_);
11             g2 = g2 + 2;
12         }
13         if (g2 >= _max_suspend_) {
14             suspend();
15             g2 = 0;
16         }
17     }
18     current_chnl++ ;
19 }
```

Another "reality" example is related to the manipulation of task pointers. The desire to reuse existing library modules has resulted in a mixture of conventions regarding the responsibility for releasing the task pointer for a task. Some library routines release the pointer; however, code that uses none of these routines (or only routines that do not release the pointer) must release its pointer itself.

The best solution to this problem is to adopt a consistent policy for releasing the task pointer and to accept the necessity to reimplement the library routines. Certainly, the present approach is fraught with danger. It would be easy for the human programmer to lose track of whether a task pointer has been released or not and thereby to release it twice or not at all. There is, however, a worse problem, which is the possibility that the program needs to use the task pointer after calling a routine that releases it. Nevertheless, we did implement transformations that mimic the techniques used in hand coding to cope with this problem.

A policy example is the mandate to use only while loops, not for loops in the C code. It may be convenient to use for statements in the more abstract levels of the code. We have implemented a set of transformations, similar to the example given in Sec. 3, that changes any remaining for loops into while loops, thereby enforcing the transformationally generated code to adhere to this policy requirement. This set of transformations is applied late in the derivation. There are, of course, other policy requirements that we do not discuss here that could be handled, implemented, enforced, or rendered unnecessary by the use of transformations.

4.4. Implementing High-Level Specification Constructs

Another way in which one can increase the abstractness of a specification is to write it using *problem-oriented notations*. Such notations are, of course, too numerous to catalog here. We do, however, discuss one such notation, that for *guarded state tables*, that we implemented for this experiment.

A guarded state table is conceptually similar to a finite state automaton, with the additional requirement that before a state transition takes place based on a state and an event, a guard is checked. Guarded state tables are useful to capture constraints that are not easily represented as part of states or events. Such a table is used in the design for the assignable_spkrs routine of *Assignable Speakers* to control the activation of the various functions of the module. The guarded state table of the design can be transcribed into a linear sequence of states each represented in the form:

```
1 state(state_id, event_id,
2       guard_expression,
3       new_state_id,
4       begin
5              ...actions...
6       end)
```

Using this notation, part of the specification for the assignable_spkrs routine is:

```
1  state(Idle, Depress,
2        test_rc_bit(_rc_busy_) && successfully_entered_assign_speaker_mode(task),
3        Waiting_for_keypad_input,
4        begin
5        end).
6  state(Idle, Depress,
7        true,
8        Idle,
9        begin
10             beep().
11       end).
12 state(Waiting_for_keypad_input, Depress,
13       key_depressed(_shft_key_) && successfully_entered_validate_speaker_id(task),
14       Waiting_for_chnl_sel_or_save,
15       begin
16       end).
17 state(Waiting_for_keypad_input, Depress,
18       key_depressed(_shft_key_),
19       Waiting_for_keypad_input,
20       begin
21             beep().
```

```
22              display_SPR_on_rcp_display(tsk_ptr).
23           end).
24     state(Waiting_for_keypad_input, Depress,
25           true,
26           Idle,
27           begin
28              exit_assign_spkr_mode().
29              relinquish_keypad_and_restore_clock_to_rcp_display().
30           end).
31     ...
```

The derivation of C code from this specification proceeds according to the following steps:

1. Form a state switch by collecting all like states into a switch statement.

2. Once all events for a particular state have been collected, they can be transformed further without regard to the other states.

3. Within each state, form an event switch by collecting all like events into a switch statement. Again, once all the guards for a particular event have been collected, they can be transformed further without regard to the other events.

4. Within each event, form nested conditional expressions to evaluate the guards.

5. Optimize the evaluation of the conditions so that overlapping conditions in the guards are evaluated only once. (Single evaluation of conditions is important if evaluating the condition changes the overall state of the computation.)

6. Define local variables used by the state machine.

7. Optimize the remaining constructs. As a consequence of writing the preceding transformations in simple and general-purpose form, they may leave behind rudimentary switches (having just one alternative) and redundant else clauses. This set of transformations simplifies such constructs.

These transformations derive the following C code from the preceding portion of the specification for assignable_spkrs:

```
1  switch (std_state) {
2      case Idle:
3          switch (task-> tsk_union.trk1.event) {
4              case Depress:
5                  if (! tstrc(_rc_busy_)) {
6                      if (enter_assign_spkr_mode(task))
7                          std_state = Waiting_for_keypad_input;
8                      else beep();
```

```
10                              }
11                         else beep();
13                         break;
14                    default
15                         exit(911);
16                    }
17               break;
18          case Waiting_for_keypad_input:
19               switch (task->tsk_union.trk1.event) {
20                    case Depress:
21                         if (! tstrc(_shft_key_)){
22                              if (get_chrcnt()) {
23                                   g2 = conad();
24                                   g3 = validate_spkr_id(task, g2);
25                              }
27                              if (g3)
28                                   std_state = Waiting_for_chnl_sel_or_save;
29                              else {
30                                   beep();
31                                   g8 = _duplicate_task_pointer_(tsk_ptr);
32                                   clmkpd(g8, 0, _key0_);
33                                   sprdsp();
34                              }
36                         }
37                         else {
38                              std_state = Idle;
39                              exit_assign_spkr_mode();
40                              if (relkpd(_asgn_spkrs_))
41                                   resclk();
43                         }
45                         break;
46                    ...
47               }
48          ...
49     ...
50  }
```

Constructing this code transformationally offers some interesting advantages over writing it by hand. The engineer who implemented this routine ultimately wrote two versions of it in order to find out which was more efficient. In one of these implementations, the outer switch was on states and the inner one on events, as illustrated in the preceding code; in the other, the outer switch was on events and the inner one on states. Of course, constructing the second of these implementations was just as much work as constructing the first.

Using program transformations, constructing the second implementation is almost trivial. Conceptually, it involves nothing more than interchanging the first two sets of transformations. (In practice, these two sets of transformations would require minor modifications, but the remaining sets would be entirely unaltered.) Thus, program transformation can make it easy to explore alternative implementations.

Our emphasis in constructing these transformations was to make them general-purpose. Thus, they contain nothing specific to the *Assignable Speakers* module, and they could be used to implement any module that employs guarded state tables in its specification.

Finally, the use of transformations facilitates analyzing the behavior of the guarded state table specifications themselves (rapid prototyping). It is simple to alter the transformations to insert additional statements that record the states, events, and behavior of the automaton. This information can be analyzed to validate the specification. Then, when the specification is validated, the insertion of additional statements can be removed from the transformations, yielding compact and efficient production code.

5. Conclusion and Future Work

This experiment in knowledge-based programming has demonstrated that high-level specifications and program transformation can be used to derive efficient, correct executable code for a representative commercial product software module. Significant advantages of this approach include:

- The derivation can be carried out automatically.

- The transformations that carry it out codify programming knowledge so that the knowledge can be reused in other applications.

- The derived executable code is at least as good as, and in some cases, better than, the hand-written code.

- The development of the transformations and specifications used in these experiments required only about 5-6 man-weeks. (Moreover, many of the transformations could be reused for other modules, reducing the time to implement those modules.)

These results, together with similar results obtained by two of us [3] that are algorithmically more complex than the examples discussed here, lead us to the conclusion that program transformation could be used to derive code that would meet the efficiency and reliability requirements for commercial products.

The use of program transformation for product software development would offer a number of benefits. The most obvious is the direct cost saving of automating the last step in the program development process.

Other potential cost savings are less obvious. High level pseudo-code specifications of the type used in this experiment are faster to write than the low-level ones currently used. The high-level pseudo-code specifications can be used as a basis for rapid-prototyping, thereby increasing confidence that a problem has been specified correctly. The program transformations preserve the correctness of the specification, thereby reducing or eliminating debugging. The code for the product can be maintained and enhanced at the easily understood pseudo-code level, thereby reducing the cost of enhancements. Finally, in concert these advantages offer the hope of reducing the cost and time to market for a new product.

We are currently undertaking further experiments with the goal of demonstrating that it is worth risking the use of program transformation in the actual software development for a product. These additional experiments are important not only because their successful completion will reinforce the idea that program transformation can produce good executable code, but also because they will lead to the development of a library of transformations that can be reused to reduce the time required to develop new software.

We are in the process of carrying out experiments with commercial software that demonstrate capabilities of program transformation that were not addressed in the experiment reported here. For example, we are exploring automatic selection of implementations for abstract data types based on the usage patterns for the abstract data type accessors and constructors. Another capability that we currently are exploring is construction of highly optimized implementations of time-critical programs using program transformations. Finally, we are working on demonstrating the ability of transformations to construct alternate implementations of modules cheaply, for example, when replacing implementation as subroutines by implementation as tasks. The ability to actually evaluate the relative speeds and costs of such alternative implementations could lead to reduced product cost.

The penultimate step (immediately preceding use of program transformation in the development of a product) in such a series of experiments is to use program transformation to develop code in parallel with conventional methodologies. These experiments will enable us to assess the relative speeds and costs of the two methodologies in real-life situations.

References

[1] R. Bird, L. Meertens, and D. Wile. A common basis for algorithmic specification and development. In *IFIP WG2.1 Working Paper*, number ARK-3. 1985.

[2] J.M. Boyle. Abstract programming and program transformations—an approach to reusing programs. In A.J. Perlis, editor, *Software Reusability*, volume 1, pages 361–413. Addison-Wesley, Reading, 1989.

[3] J.M. Boyle and T.J. Harmer. A practical functional program for the CRAY X-MP. Technical Report MCS-P159-0690, Argonne National Laboratory, Argonne, Illinois, July 1990.

[4] J.M. Boyle and M.N. Muralidharan. Program reusability through program transformation. *IEEE Trans. Software Eng.*, 10(5):574–588, 1984.

[5] J. Goguen, C. Kirchner, H. Kirchner, A. Mégrelis, and J. Meseguer. An introduction to OBJ3. In J.-P. Jouannaud and S. Kaplan, editors, *Proc. Conf. Conditional Term Rewriting*, volume 30 of *Lecture Notes Comp. Sci.*, pages 258–263. Springer Verlag, Berlin, 1988.

[6] R. Kelsey and P. Hudak. Realistic program compilation by program transformation. In *Proc. 16th Ann. ACM Symp. on Programming Languags*, 1989.

[7] L.G.L.T. Meertens, editor. *Program Specification and Transformation*. North-Holland, Amsterdam, 1987.

[8] H. Partsch and R. Steinbrüggen. Program transformation systems. *Computing Surveys*, 15(3):199–236, 1983.

[9] P. Pepper, editor. *Program Transformation and Programming Environments*, volume F8 of *NATO ASI*. Springer Verlag, Berlin, 1984.

[10] D. Wile. Local formalisms: Widening the spectrum of wide-spectrum languages. In Meertens [7], pages 459–483.

A. Appendix: Writing Transformation Rules in the MOUSETRAP Program Transformation System

The MOUSETRAP program transformation system manipulates a source program by the application of transformation rules to the program. A transformation rule states certain semantic relations between two programs (or fragments thereof). The most widely used semantic relations are weakened versions of equivalence. For example, a transformation rule may state that two programs produce identical output given the same input, ignoring erroneous inputs, undefined situations due to non-termination, or the like. Typically, transformation rules do not relate actual programs, but program templates. A program template represents classes of programs. A program can be obtained from a program template by instantiating variables in the template.

Applying a transformation rule means replacing a program (or program fragment) by an "equivalent" program (or program fragment). The resultant program will be just as correct as the original program (the transformation rule promises to preserve correctness) and it is, in some sense, "better" than the original program by some criterion of interest.

The MOUSETRAP program transformation system develops a program by successive application of transformation rules. Typically this may mean that one starts with a specification[3] of the program and obtains the program implementing the specifications as a result of this process. However, program transformation may also result in a transition from a less efficient program to a more efficient program or from a sequential version of a program to a parallel version.

The MOUSETRAP system itself consists of several components:

- The bottom layer is an engine providing order-sorted, conditional term rewriting much like OBJ [5]. Rewrite rules may be formulated by relying on matching modulo equational theories, such as associativity, commutativity, idempotence, and a special list theory. Identities for terms are automatically provided by the rewrite engine. Evaluation strategies are completely flexible.

- Input programs are translated into terms of an appropriate equational theory. A facility is provided which produces parsers, term compilers, unparsers, and definitions of the equational theory of the input language grammar from an extended Backus-Naur notation. Any programming language construct may be given equational properties or identities. The input grammars are restricted to be LALR-1.

- A rule compiler translates transformation rules operating over program templates into rewrite rules operating over terms in its equational theory.

Interacting with the MOUSETRAP program transformation systems mostly means writing transformation rules. The MOUSETRAP system then applies these rules to an input program or program fragment and yields the representation of the resultant program. In this section we will be mainly concerned with the mechanisms provided by the MOUSETRAP program transformation system to write transformation rules easily.

A.1. Simple Transformation Rules

A *transformation rule* is a *correctness preserving rewrite rule* over parse trees. Fig. 1 depicts a simple (and semantically not quite correct) example of a transformation rule which is explained later. Each rule has an optional variable declaration section, a pattern part, an arrow indicating the rewriting relationship, and a replacement part. Comments may be freely interspersed; comments begin with the hash sign and extend to the end of the line. Immediately above the arrow is the pattern. The replacement follows the arrow. Transformations are terminated by a semicolon.

Both pattern and replacement are descriptions of a class of parse trees (we will refer to these as "trees"). In their simplest form, trees are written in terms of the source and target grammars. They consist of the name of a grammatical class or dominating symbol of a production of the respective grammar (in this example,

---

[3]In this context, we will consider a specification to be a program written in a higher-level, maybe even non-executable programming language.

```
1  ?v: Ident;  ?e1, ?e2: Expr;  ?rest: Stmtlist;
2  //.
3  Stmtlist'?v = ?e1;
4           ?v = ?e2;
5           ?rest'              # Two consecutive assignments to the same identifier
6  ==>
7  Stmtlist'?v = ?e2;
8           ?rest'
```

Figure 1: Elimination of double assignment (incorrect)

Stmtlist) and some source text (the text enclosed in ' ... ' quotes). A parse tree is constructed by parsing this source text as deriving from the indicated dominating symbol in the respective grammar. (In this and the following equally contrived examples, we assume that the source text describes parse trees according to a C grammar.) Source and target grammars are not necessarily the same.

A tree may simply be a literal description of a parse tree, or it may be the description of a parse tree with some subtrees left unspecified but restricted to a given grammatical class (in which case there may be many instances matching the pattern). Subtrees are left unspecified through the use of variables standing for subtrees of a particular grammatical class. Variables must be declared before the pattern of the transformation rule. For example,

?e1, ?e2: Expr;

declares that variables ?e1 and ?e2 will represent subtrees deriving from the dominating symbol Expr in the current grammar. Thus the pattern of Fig. 1:

Stmtlist'?v = ?e1;
         ?v = ?e2;
         ?rest'

describes a template for a parse tree representing a Stmtlist (as defined by the grammar of the source language, in this case, C) of the form of two consecutive assignments to the same identifier (denoted by ?v), followed by the list ?rest of other statements (this list may be empty if the grammar for the source text permitted this). A tree may simply be a variable, matching any subtree deriving from the dominating symbol the variable was declared to represent.

If this template matches a subtree of the input program (i.e., of the parse tree to which this transformation rule is applied), the variables in the pattern (here ?v, ?e1, ?e2, and ?rest) are bound to the respective subtrees matched. Then a replacement tree is constructed from the replacement part of the rule and the bindings for the variables previously created. In the example above, the simpler parse tree:

Stmtlist'?v = ?e2;
　　　?rest'

with the variables replaced by the respective subtrees bound to them is constructed. Then the resultant replacement tree is inserted in the input program in place of the matched subtree. In this case, the replacement results in the first assignment to the variable ?v being removed. It should be apparent that pattern and replacement must both derive from the same dominating symbol, for this replacement to be meaningful.

Of course simply matching two trees and performing a replacement using variable substitutions alone will be sufficient only in the most trivial cases. MOUSETRAP provides powerful high level abstractions to allow for convenient representation of transformation rules.[4]

A.2. Qualifiers

A match of a pattern in an input parse tree may be seen as a condition that has to be satisfied in order for a replacement to take place. One can easily imagine additional conditions that have to be met for a transformation rule to be applicable. For example, the transformation rule in Fig. 1 is incorrect since it could be possible that the variable ?v may occur in the expression assigned to ?v in the second assignment statement. If ?e2 contains an instance of ?v we cannot simply eliminate the first assignment if we want to preserve the correctness of the manipulated program.

In Fig. 2 we add the condition to the pattern that the expression ?e2 must not contain an instance of the identifier ?v which was assigned to in the first assignment statement. (Qualifiers may be added to the pattern by the & symbol.) The replacement will only take place in case the additional conditions described by the qualifiers are satisfied.

Various predicates are provided to express additional conditions and are shown in Tbl. 1 below. One can test for equality or containment between ground trees, given the variable bindings currently in place. One can test whether some tree matches another tree or whether an instance of a given tree is contained in another tree. The comparisons or matches are performed modulo any equational theory that may hold for the dominating symbols of the trees in question. Any bindings of variables created in a match performed in a qualifier can be relied upon in subsequent qualifiers and in the replacement. A mechanism is provided to construct additional predicates, should the need arise.[5]

The qualifier in Fig. 2 negates a predicate. Through a negative qualifier (indicated by the ! symbol), no bindings to variables are created.

---

[4]The terminology used subsequently has been gleaned from the TAMPR system.
[5]Predicates are constructed by expressing them in the implementation language (LISP).

Table 1: Tree predicates

t == t'	Tree t equals t'
t >= t'	Tree t' contains t
t <= t'	Tree t contains t'
t ~~ t'	Tree t' matches an instance of tree
t <~ t'	Tree t' contains an instance of tree t
t >~ t'	Tree t contains an instance of tree t'

A.3. Subtransformation

Often it proves useful to be able to perform transformations on a matched subtree in the context of existing bindings for variables. This can be accomplished through the use of subtransformations (i.e., transformations of subtrees of a matched parse tree within a transformation). Subtransformations are another means of constructing a tree: Rather than simply describing a tree by its dominating symbol and the source text to be parsed, a subtransformation takes a tree and constructs a new tree by applying a transformation to it. The original tree as well as the transformation rule applied to it are subject to the bindings of trees to variables that have been made earlier. No bindings made during a subtransformation will persist beyond the scope of the subtransformation with the exception that global variables (see Sec. A.8) can be used to pass information back from the subtransformation.

To continue our example, the transformation rule in Fig. 2 is limited in the sense that it does not manipulate consecutive assignments to the same variable ?v if ?v occurs in the right-hand side of the second assignment ?e2. Through the use of a subtransformation we can recognize those situations and replace the occurrence of ?v in ?e2 by the value it was given in the preceding assignment statement: Before performing the replacement, we rewrite the replacement using a subtransformation. We look through the replacement for an occurrence of a tree bound to ?v, and replace it by the tree bound to ?e1 (which in the input program would have been assigned to the identifier bound to ?v in the first assignment statement). Note that we cannot simply look for a ?v that derives from the grammatical class Ident, but rather we must look for a subtree descending from the grammatical class Expr containing only the Ident bound to ?v in order to be able to replace it by ?e1 which is also an Expr. We parenthesized ?e1 as a safeguard against potential problems with respect to precedence when inserting this expression into ?e2. The braces, "{" and "}", indicate grouping and are mandatory around subtransformations.

The transformation rule in Fig. 3 is not quite correct yet, though. The subtransformation is performed on the whole replacement tree, and thus even the occurrences of the variable bound to ?v throughout the statements following the assignment to it are incorrectly replaced by their earlier value (which was supposed to be overridden

```
1  ?v: Ident;  ?e1, ?e2: Expr;  ?rest: Stmtlist;
2  //. Stmtlist'?v = ?e1;
3             ?v = ?e2;
4             ?rest'
5     & ! ?v <= ?e2
6     ==>
7     Stmtlist'?v = ?e2;
8              ?rest';
```

Figure 2: Elimination of double assignment (correct, but limited)

```
1  ?v: Ident;  ?e1, ?e2: Expr;  ?rest: Stmtlist;
2  //. Stmtlist'?v = ?e1;
3             ?v = ?e2;
4             ?rest'
5     ==>
6     { Stmtlist'?v = ?e2;
7                ?rest' //.
8                Expr'?v'
9                ==>
10               Expr'(?e1)';  };
```

Figure 3: Elimination of double assignment (subtransformation, incorrect)

```
1  ?v: Ident;  ?e1, ?e2: Expr;  ?rest: Stmtlist;
2  //. Stmtlist'?v = ?e1;
3                  ?v = ?e2;
4                  ?rest'
5  ==>
6  {Stmtlist'?v = ?e2;  ' //.
7                      Expr'?v'
8                      ==>
9                      Expr'(?e1)';  } ++ ?rest;
```

Figure 4: Elimination of double assignment (subtransformation, still incorrect)

by the assignment statement)!

We can make use of another method of constructing trees to avoid this problem. The operator ++ takes two trees representing lists and concatenates them into a single tree. As Fig. 4 shows, we can construct the replacement by concatenating the Stmtlist resulting from applying the mentioned subtransformation to the second assignment statement with the Stmtlist bound to ?rest. Now the replacement of ?v by the expression ?e1 occurs only in the assignment to ?v. Unfortunately, this transformation is still incorrect: Even the occurrence of the variable bound to ?v on the left-hand side of the assignment statement is replaced by the expression bound to ?e1![6]

A.4. Scratchpad and additional tree constructors

A scratchpad is a degenerate form of a qualifier in that it will always be trivially satisfied. A scratchpad allows one to bind constructed temporary trees to variables to be used later in the transformation rule.

The difficulty with the transformations of Figs. 3 and 4 was that we want to perform the replacement of occurrences of the variable bound to ?v only within the expression bound to ?e2. In the scratchpad:

```
?new := { ?e2 //. Expr'?v'
                ==>
                Expr'(?e1)'; }
```

we construct a new tree based on the subtree bound to ?e2, in which we replace any occurrence of ?v by the right-hand side of the first assignment statement ?e1 and bind this newly constructed tree to the variable ?new. The newly constructed tree

---

[6]In some programming languages this problem would be ruled out by the grammar in that an Expr would not be allowed on the left-hand side of an assignment statement and, therefore, the subtransformation would not match in those unwanted situations. However, the C grammar does allow this replacement (although the resultant program is illegal due to the static semantics of C).

```
1   ?v: Ident;  ?e1, ?e2, ?new: Expr;  ?rest: Stmtlist;
2   //. Stmtlist'?v = ?e1;
3              ?v = ?e2;
4              ?rest'
5   & ?new := { ?e2 //. Expr'?v'
6                    ==>
7                    Expr'(?e1)'; }
8   ==>
9   Stmtlist'?v = ?new;
10             ?rest';
```

Figure 5: Elimination of double assignment (scratchpad)

will be used as the right-hand side of the assignment statement. In case ?v does not occur in ?e2, ?e2 remains unaltered and the variable ?new simply refers to a copy of ?e2. Fig. 5 shows the modified transformation rule.

By now we have seen two methods of constructing trees from simpler trees: subtransformations and concatenation. Rather than concatenating two trees representing two lists, as in the futile attempt of Fig. 4, we can also add a tree to the front or back of another tree representing a list. The replacement of Fig. 4 could have also been written as:

```
{ Stmt'?v = ?e2; ' //. Expr'?v'
                  ==>
                  Expr'?e1'; } + ?rest;
```

In this case, the modified assignment statement would have been added to the front of the list of statements bound to ?rest. The construct:

new Ident'temp_'

builds a new tree deriving from the dominating symbol Ident which looks like temp_. This construct is only legal for trees that are derived from terminal symbols in the grammar. The appearance of the newly constructed tree results from appending a new number to the printed representation of the given tree. Typically, this construct is used to generate new identifiers. For example, above rule fragment would generate temp_1, temp_2, and so on.

Finally, the names self and parent refer to the subtree that has been matched by the pattern of the transformation rule and to its parent tree, respectively.

A.5. Transformation Sets and Tree Traversal Order

As a further extension, transformations may be grouped into sets which will then be applied together to the same node of a tree. There is no particular order implied

Table 2: Traversal indicators

/.	Single traversal
//.	Repeated traversal
|.	Current node only
\.	Children of current node only

as to which transformation in a set takes precedence should two transformations be applicable to the same subtree. In other words, at each node of a matched tree we attempt to apply all transformations in a rule set in no stated order. Variable declarations hold for all rules in a set. Sets of transformation rules can also be used in subtransformations or scratchpads.

In addition, the mode of tree traversal may be varied. Ordinarily, we want a tree to be rewritten as completely as possible. Therefore, once a match is located, we should try to rewrite the replacement tree again using the same transformation set before it is inserted into the original tree. This procedure guarantees that all applicable rules have actually been applied to the resultant tree. We then describe the resultant tree to be in canonical form. Sometimes we know that no modification can happen to the replacement, or, more severely, sometimes a rewrite sequence might not terminate if performed in the above manner. To allow one to deal with problems of this nature, a mode of traversal is possible in which the replacement tree will not be traversed again (i.e., it is assumed to be already in canonical form). The different modes of rule application and tree traversal are indicated by prefixing the pattern with the traversal indicator symbols shown in Tbl. 2. The standard mode is to continue attempting to apply a transformation rule until a fix-point is reached. Single traversal rules perform a replacement at most once and are not reapplied even if the pattern were to match the replacement tree. Further, there are traversal modes that either do not descend into a tree, but apply the given rule to the current node in the matched tree only, or apply the given rule to the children of the current node only.

A.6. Multiple Grammars

One of the strengths of MOUSETRAP is its ability to express transformation rules in terms of source and target language grammars. Often program transformation also facilitates the transition between different languages (say, a translation from LISP to C). Traditional approaches require that transformation rules are intra-grammatical, i.e., that pattern and replacement belong to the same grammar. This requirement has as a consequence that one must construct a grammar encompassing both source and target languages in order to allow for translation between two different languages. Needless to say, the development of such combined grammars is no trivial undertaking.

```
1  ?v: Ident;  ?e1: Expr;
2  //. Lisp::Stmt'(setq ?v ?e1)'
3      ==>
4      Stmt'?v = ?e1';
```

Figure 6: Moving between grammars

MOUSETRAP allows parse trees constructed from different grammars to coexist. In particular, it is possible to operate on parse trees in which subtrees are parsed according to different grammars. Fig. 6 shows how one might translate a LISP assignment statement (setq) into a C assignment statement. An optional argument can be given to terms describing grammatical classes, as in Lisp::Stmt, indicating that the following text is to be parsed as a Stmt of the LISP grammar. The text in the replacement part is to be parsed according to the default C grammar. After this rule is successfully applied, the parse tree, although constructed according to a LISP grammar, will contain subtrees that belong to a C grammar. The "foreign" C subtrees can be viewed as local formal notations [10], i.e., "ungrammatical" components temporarily introduced into a parse tree. The user is responsible for the eventual removal of these components, which in this case means that the parse tree ultimately belongs to a single grammar.

The grammars according to which the subtrees in a parse tree are constructed have to overlap at some points for a transition to be possible. Only subtrees of the same grammatical class can be substituted for each other. For example, in Fig. 6, we substitute a C Stmt for a LISP Stmt.

A.7. Attributes

Transformational programming often proceeds by analyzing a program fragment for certain properties and in some later step uses the result of this analysis to manipulate that fragment. It would be convenient to have a mechanism that remembers the results of such analysis, or in general, marks nodes in the parse tree. For example, transformations may depend on the type of expressions in the matched tree.

MOUSETRAP allows one to attach attributes (affixes) to nodes in the parse tree. The user is free to use any attribute. The values of attributes can be either parse trees or simple identifiers. Attributes can be tested for equality (in qualifiers), and their value can be retrieved or updated. The notation:

?a.status == ok

is a qualifier which tests whether the status of the subtree matched by ?a is ok. Attributes can be given values in a scratchpad.

The special qualifiers mark and unmark attach and delete a unique attribute to the given node which can be tested via the predicate marked. Unlike ordinary attributes, this attribute is not preserved across transformation rules and serves as a temporary mark only.

A.8. Global Variables

Normally, the lifetime of variables within transformation rules extends only during the match of that rule at a single node in the parse tree. That is, during traversal all variables are initially undefined for each node in the parse tree and become bound through successful matches. These bindings exist only for that particular node.

However, there are occasions where persistence of variables beyond a single node in the parse tree would be advantageous. Imagine a situation where we want to traverse a program fragment and collect all variable declarations in order to insert them at the beginning of the program. Without persistent variables such a task would be quite difficult.

In MOUSETRAP, we will refer to persistent variables as "global." Variables are made global by inserting a declaration at the beginning of a transformation rule. Global variables may be given an initial value. For more detail on the use of global variables, refer to the example below.

A.9. Example: Complex Numbers

In this section, we will work through a small programming problem in order to explain further the style of writing transformation rules. The goal of the transformations developed in this example is to replace arithmetic operations over complex variables by operations over pairs of reals. The purpose of this example is didactic. We will take the grammar of the C programming language to be the grammar of our source text. In other words, we will rewrite program fragments written in C that make use of the datatype "complex number" into fragments where any mention of complex numbers has been replaced by operations over an array of two reals. We will make a few simplifying assumptions on the input program so as to not detract from the essentials of this example.

Transformations often rely on an algebra giving the meaning of the data objects manipulated. For a simplified algebra of complex numbers, consider two sorts: complex and real. Let complex be a binary constructor function for complex numbers taking two reals and composing them into a complex number. Let real and imag be two functions retrieving the real and imaginary part of a complex number, respectively. Eq. (1) explains the construction of complex numbers (let z range over the sort of complex numbers):

$$z = \mathsf{complex}(\mathsf{real}(z), \mathsf{imag}(z)) \qquad (1)$$

Eqs. (2) through (5) define the meaning of arithmetic operators over complex numbers:

$$\mathsf{complex}(r1, i1) + \mathsf{complex}(r2, i2) = \mathsf{complex}(r1 + r2, i1 + i2) \qquad (2)$$

$$\mathsf{complex}(r1, i1)\text{-}\mathsf{complex}(r2, i2) = \mathsf{complex}(r1\text{-}r2, i1\text{-}i2) \qquad (3)$$

$$\text{-}\mathsf{complex}(r1, i1) = \mathsf{complex}(\text{-}r1, \text{-}i1) \qquad (4)$$

$$\mathsf{complex}(r1, i1) * \mathsf{complex}(r2, i2) = \mathsf{complex}(r1 * r2 + i1 * i2, r1 * i2\text{-}r2 * i1) \qquad (5)$$

Finally, the following equations define the accessor functions real and imag:

$$r = \mathsf{real}(\mathsf{complex}(r, i)) \qquad (6)$$

$$i = \mathsf{imag}(\mathsf{complex}(r, i)) \qquad (7)$$

We will use this algebra as a guideline to construct our transformation rules.

We begin by replacing complex variables by the explicit construction of the complex number they represent from its real and imaginary part. Consecutive transformations can then operate on these components separately. We simply write a transformation rule based on Eq. (1), read as a rewrite rule from left to right (i.e., as unfolding the definition of a complex number).[7]

The rule mark-complex initially matches a statement consisting of a declaration section ?decl and a statement list ?sl. Once we matched such a program fragment we remember the statement list as ?newsl and look for declarations of complex variables in ?decl. For every declaration of a complex variable ?z, in a subtransformation, we replace any occurrence of ?z in the statement list ?newsl by its unfolded definition. Finally, the matched statement is replaced by one constructed from the original declaration section ?decl and the possibly modified statement list ?newsl. A similar transformation needs to be provided to deal with variables declared in function headers.

```
 1  # Relies on the simplifying assumption that each
 2  # complex declaration declares one variable only.
 3  ?z: Ident;  ?decl: Decllist;  ?sl: Stmtlist;
 4  global ?newsl: Stmtlist;
 5  /. Stmt'{ ?decl ?sl }'
 6      & ?newsl := ?sl
 7      & for Decl'complex ?z; ' in ?decl do
 8          ?newsl := { ?newsl //. Expr'?z'
 9                              ==>
10                              Expr'complex(real(?z), imag(?z))'; }
11      done
12  ==>
13  Stmt'{ ?decl ?newsl }';
```

---

[7]For the sake of this example, we will assume that each complex variable is declared in a separate declaration. This will allow us to present the relevant rule without assuming detailed familiarity with the input grammar.

The iteration over the program fragment ?decl is a qualifier similar to the mapping functions of programming languages such as LISP or ML: For any instance of the tree Decl'complex ?z; ' within the matched declaration section ?decl, evaluate the qualifiers within the do ... done section. In this particular case, these consist of a single scratchpad and are vacuously satisfied. (The iteration could also have been accomplished by embedding this scratchpad within an additional subtransformation over the declaration section ?decl in which we do not alter the matched declarations.) Note that the variable ?newsl holding the possibly modified statement list had to be declared as global since the impact of the subtransformation on it needs to be preserved outside of the subtransformation. Below we show an alternative way of writing this transformation rule:

```
1  #  Relies on the simplifying assumption that each
2  #  complex declaration declares one variable only.
3  ?z: Ident; ?decl1, ?decl2: Decllist; ?sl, ?newsl: Stmtlist;
4  //. Stmt'{ ?decl1; complex ?z; ?decl2 ?sl }'
5        & ! marked ?z
6        & mark ?z
7        & ?newsl := {?sl //. Expr'?z'
8                     ==>
9                     Expr'complex(real(?z), imag(?z))'; }
10    ==>
11    Stmt'{ ?decl1; complex ?z; ?decl2 ?newsl }';
```

Rather than iterating over the declaration section in the qualifier of the rule, we are looking for the declarations of complex variables in the initial match: a (possibly empty) declaration list, followed by the declaration of the complex variable, followed by another (possibly empty) declaration list. Since this rule is reapplied until a fixpoint is reached (i.e., until no more changes to the tree are made), in effect we are examining the declaration section from left to right and looking for declarations of a complex variable. To avoid reexamining the same declaration over and over, we mark each matched complex declaration and only operate on unmarked complex declarations.

Before we are ready to realize the arithmetic operations, we need to attend to one more detail: There may be cases where operations are intermixed between real and complex numbers. We have two options: We can convert all occurrences of real numbers (and integers, for that matter) into complex numbers by constructing a complex number from the real and setting the imaginary part to 0 and later on eliminate those occurrences that had been converted unnecessarily. Or, we can take into account the possibility of arithmetic operations involving a mixture of real and complex numbers in the transformations realizing those operations.

We will choose the former option.[8] The rule below accomplishes the type conversion; note that we need to check, in a qualifier, whether the expression matched on

---

[8] We have to convert reals to complex numbers only when a real is assigned to a complex

the right-hand side of the assignment is indeed not a complex number. ?n could not stand for an arithmetic expression involving complex numbers since in that case it would not be bound to a variable representing a Term. We can thus simply attempt to match this expression against a Term representing a complex number:[9]

```
1  ?r, ?i: Expr;  ?n: Term;  ?z: Ident;
2  /. ?n
3     & ! Term'complex(?r, ?i)' ~~ ?n
4     ==>
5     Expr'complex(real(?z), imag(?z)) = complex(?n, 0)';
```

We now apply the transformations realizing the arithmetic operations over complex numbers. These transformations are literal translations of Eqs. (2) through (5):

```
1   ?r1, ?r2, ?i1, ?i2: Expr;  ?n: Term;
2   /. Expr'complex(?r1, ?i1) + complex(?r2, ?i2)'
3      ==>
4      Expr'complex(?r1 + ?r2, ?i1 + ?i2)';
5   /. Expr'complex(?r1, ?i1) - complex(?r2, ?i2)'
6      ==>
7      Expr'complex(?r1 - (?r2), ?i1 - (?i2))';
8   /. Expr' - complex(?r2, ?i2)'
9      ==>
10     Expr'complex( - (?r2), - (?i2))';
11  /. Expr'complex(?r1, ?i1) * complex(?r2, ?i2)'
12     ==>
13     Expr'complex((?r1) * (?r2) - (?i1) * (?i2), (?r1) * (?i2) + (?i1) * (?r2))';
```

By now any operation over a complex number has been replaced by an operation over the real and imaginary parts of that complex number. We may be able to simplify some expressions by relying on Eqs. (6) and (7), as exhibited in the transformations below:

```
1  ?r, ?i: Expr;
2  /. Expr'real(complex(?r, ?i))' ==> ?r;
3  /. Expr'imag(complex(?r, ?i))' ==> ?i;
```

Complex variables may also occur as arguments to function calls. Some of those may be handled by special transformations that realize them. Others may simply be implemented as functions that operate on the array of real numbers instead. We then would have to lift the complex number out of the expression in which it occurs, assign it to some temporary variable, and pass that temporary variable into

---

number, but we would then have to provide rules to deal with the special cases of realizing arithmetic operators where one of the arguments is a real number.

[9]We simplified this determination by assuming that there are no functions returning complex numbers.

the expression. We cannot simply pass the real and imaginary parts of the complex number directly, for the same reasons as discussed for the assignment of complex numbers below.

The transformation below examines each statement ?current. In a subtransformation, for each complex number in an expression list within ?current (i.e., as an argument to a function call), a new variable is created (beginning with the letters temp_), a declaration for this variable is created and added to a new declaration section ?tempdecl, an assignment of the complex number to the new variable is created and added to a new statement list ?temp, and finally, the complex number is replaced by the newly created temporary variable. The statement ?current is then replaced by a compound statement consisting of the declaration of all newly introduced variables ?tempdecl, the assignments of the complex numbers to the corresponding newly introduced variables ?temp, and the modified statement itself:

```
1   ?current, ?new: Stmt;  ?r, ?i: Expr;  ?el: Exprlist;  ?z: Ident;
2   global ?temp := Stmtlist''; global ?tempdecl := Decllist'';
3   /. ?current
4       & ?new := { ?current
5                   //. Exprlist'complex(?r, ?i), ?el'
6                   & ?z := newIdent'temp_'
7                   & ?temp := ?temp + Stmt'complex(real(?z), imag(?z)) =
8                                           complex(?r, ?i); '
9                   & ?tempdecl := ?tempdecl + Decl'complex ?z; '
10                  ==>
11                  Exprlist'?z, ?el'; }
12  ==>
13  Stmt'{ ?tempdecl ?temp ?new }';
```

The final step is the implementation of the complex numbers in terms of pairs of reals. This involves the implementation of assignments involving complex numbers, the implementation of the complex numbers themselves, and the implementation of the declarations for complex variables.

When rewriting assignments between complex numbers we need to be careful that the assignments to the imaginary and real parts happen "in parallel," which in effect means that if the real part of an expression to which a value is assigned occurs in the imaginary part of the assigned expression, these values must not be destroyed during a sequential assignment to either components. The first transformation in the rule set below rewrites the complex assignment assuming the mentioned condition is not violated.

If the real and imaginary parts are mutually contained within each other, and no parallel assignment statement is available (as in most sequential programming languages), we need to save the real part in a temporary variable and then perform the two assignments separately.

Finally, we replace any access to the real part of a complex variable by an access to the first element of the array used to implement a complex number, and we replace any access to its imaginary part by an access to the second element of that array. Declarations of variables as complex numbers are replaced by declarations of these variables as arrays of two real numbers (or floats, in the C programming language):[10]

```
1   ?z, ?temp: Ident;  ?r, ?i: Expr;
2   //. Stmt'complex(real(?z), imag(?z)) = complex(?r, ?i);  '
3       & ! Expr'real(?z)' <= ?i
4       ==>
5       Stmt'{ real(?z) = ?r;
6                imag(?z) = ?i;
7                ?z;
8              }';
9   //. Stmt'complex(real(?z), imag(?z)) = complex(?r, ?i);  '
10      & Expr'real(?z)' <= ?i
11      & ?temp := newIdent'tt'
12      ==>
13      Stmt'{ float ?temp;
14               ?temp = ?r;
15               imag(?z) = ?i;
16               real(?z) = ?temp;
17               ?z;
18             }';
19  /. Expr'real(?z)'
20     ==>
21     Expr'?z[0]';
22  /. Expr'imag(?z)'
23     ==>
24     Expr'?z[1]';
25  /. Decl'complex ?z;  '
26     ==>
27     Decl'float ?z[2];  ';
```

We may want to follow these transformations by a set of transformations that further simplifies arithmetic operations by relying on straightforward arithmetic identities:

```
1   ?e: Expr;  ?sl1, ?sl2: Stmtlist;
2   /. Expr'(?e)' ==> ?e;
3   /. Expr'?e * 0' ==> Expr'0';
4   /. Expr'?e + 0' ==> ?e;
```

---
[10]We will, for this example, ignore concerns of allocation on stack or heap.

5 /. Expr'?e - 0'==> ?e;
6 /. Expr'0 - ?e'==> Expr' - ?e';
7 /. Expr'?e * 1'==> ?e;
8 /. Expr'?e / 1'==> ?e;
9 /. Expr'?e / ?e'==> Expr'1';
10 /. Stmtlist'{ ?sl1 } ?sl2'==> Stmtlist'?sl1 ?sl2';

Note that due to the commutativity of the addition and multiplication operators we only need to provide one transformation rule for handling these operators (the second situation, where the identity element is on the left-hand side of the operation, is handled through matching modulo commutativity). The last rule in this set removes the scopes created by compound statements when no variables are declared within the compound statement.

Fig. 8 shows the result of applying the above sequence of transformations to the program fragment shown in Fig. 7. We can easily see how any mention of complex number has been replaced by the corresponding reference to an array of reals (C floats). Notice how the transformations go beyond realizing the datatype via libraries. All operations are open-coded and have been subject to optimizations. The code fragment could now be optimized further. For example, the statements containing only a reference to the variables C_c and temp_278 are dead and could be removed.

```
1  ...
2  { complex A_c;
3    complex B_c;
4    complex C_c;
5    C_c = 24 + C_c;
6    C_c = B_c * C_c-A_c;
7    funcB(C_c + B_c * C_c);
8  }
9  ...
```

Figure 7: A program fragment, initially ...

```
1  ...
2  { float A_c[2];
3    float B_c[2];
4    float C_c[2];
5    C_c[0] = C_c[0] + 24;
6    C_c[1] = C_c[1];
7    C_c;
8    { float tt284;
9      tt284 = C_c[0] * B_c[0] - C_c[1] * B_c[1] - A_c[0];
10     C_c[1] = C_c[0] * B_c[1] + C_c[1] * B_c[0] - A_c[1];
11     C_c[0] = tt284;
12     C_c;
13   }
14   { float temp_278[2];
15     temp_278[0] = C_c[0] * B_c[0] - C_c[1] * B_c[1] + C_c[0];
16     temp_278[1] = C_c[0] * B_c[1] + C_c[1] * B_c[0] + C_c[1];
17     temp_278;
18     funcB(temp_278);
19   }
20 }
21 ...
```

Figure 8: ...and after applying the complex number transformations

AUTOMATIC FUNCTIONAL MODEL GENERATION FOR PARALLEL FAULT AND DESIGN ERROR SIMULATIONS

SHUCHIH ERNEST CHANG, STEPHEN A. SZYGENDA

The University of Texas at Austin,
Electrical and Computer Engineering Department Austin, TX 78712

Received 15 January 1994
Revised 10 May 1994

ABSTRACT

The domain specific automatic programming technique was applied to the design and implementation of an automatic functional model generation system (AFMG) for effectively and efficiently creating simulation models of digital systems. The application domain knowledge, involved in a time-consuming and error-prone process for element modeling, was identified, organized, and incorporated into the AFMG as internal rules to guide several transformation processes for converting various forms of model specification, step by step, into desired functional models. Other design issues, such as incomplete specification handling, different signal representations, multiple-valued model generation, interaction of algorithm and data structure selection, ... etc., were also carefully addressed and resolved during the design process of the AFMG system. Experimental results show that AFMG can significantly reduce the development time of creating functional models for digital logic simulation.

1. Introduction

Digital simulation, which imitates the behavior of a digital circuit with software programs running on a computer, is an essential method for the design verification and the diagnosis of digital systems [1]. Functional models, also called element routines, are created and added into a simulator as the most fundamental and important components in the simulation software. Creating a functional model library is the most difficult and time consuming task in the development of the simulation software, mainly because generating functional models for large digital circuits is not only a complicated and knowledge intensive process but time-consuming and error prone [2,3].

The created models are used by the parallel simulators [4,5], for performing fault simulations [6] and design error simulations [7] in a time based table driven simulation environment [8]. During parallel fault simulation or design error simulation, a good circuit and a fixed number of bad (faulty or error) circuits are simulated at the same time. As shown in Fig. 1, the signal value in a good circuit is packed together in a word with the corresponding signal values of the bad circuits. If the

```
Bit Position:    4   3   2   1   0
              [....|0|0|0|1|0]  word 1
              [....|1|0|1|1|0]  word 2
              [....|1|1|1|0|0]  word 3
Signal Value:    X   E   X   1   0
                 ↑   ↑   ↑   ↑   ↑
                 |   |   |   |   └─── Good Signal Value
                 └───┴───┴───┴─────── Faulty Signal Values
```

Assuming the following signal representations:

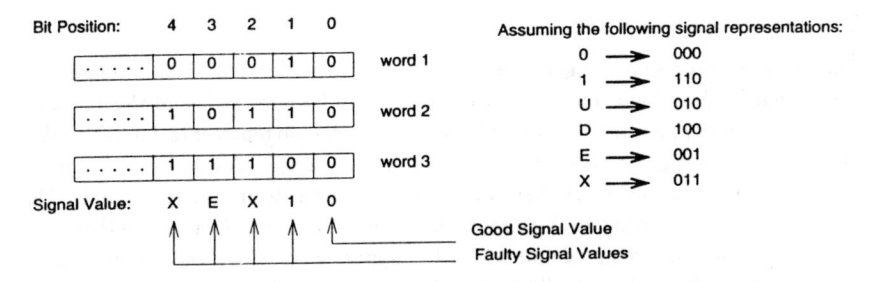

```
0 → 000
1 → 110
U → 010
D → 100
E → 001
X → 011
```

Fig. 1. Example of signal packing for parallel fault simulation.

host computer has an N bit word, $(N-1)$ faults or errors can be simulated in parallel. The number of passes required to test all faults or errors is reduced to the number of total faults divided by $(N-1)$. However, the evaluations required in a parallel model becomes very complicated because of the signal packing.

The automatic functional model generation system (AFMG), which utilized the domain-specific automatic programming approach [9], was designed to generate parallel models more effectively and more efficiently, by providing model designers an easier and more natural way of specifying the desired models. The application domain knowledge of digital simulation and systems modeling, and the domain independent program synthesis knowledge and programming language knowledge, were identified, organized, and incorporated into the system; as internal rules to guide several program transformation processes for converting various forms of input specification, step by step, into the desired simulation code. Other issues, such as: incomplete specification handling, interaction of algorithm and data structure selection, ... etc., were also addressed and resolved during the design process of the AFMG.

2. AFMG Domain Analysis

The AFMG system was proposed to automate the integration of the digital systems modeling process and the software development process [10]. The application domain of the AFMG system is to automate the process of element modeling for parallel simulation. The automatic element modeling process uses two primary knowledge sources: domain-independent programming knowledge, and domain-related knowledge. Domain-independent programming knowledge is knowledge about the syntax of targeted language and data structures. Syntactic knowledge focuses on performance characteristics and implementation mechanisms for data structures and algorithms. The domain-related knowledge is the knowledge about digital system modeling and simulation. How the needed domain knowledge is applied through the element modeling process, for generating element routines, is described in the following paragraphs. The element modeling process can be further divided into three phases: the signal modeling phase, the primitive modeling phase, and the functional modeling phase.

2.1. Signal modeling phase

The software representation of the electrical signal is decided in this phase. In traditional boolean logic, only two logic values, 0 and 1, are used, and this is called two-valued modeling. Under a two-valued signal model, no representation is possible for the voltage range between the lower and upper cut-off voltages. To get a more realistic representation, more logic values must be used to model the signal. The three-valued model, using an X to represent an unknown signal value together with a 0 for the lower cut-off signal and a 1 for the upper cut-off signal, can represent the signals between 0 and 1 by X. The five-valued model can further provide the ability to represent a signal in transition by using: U to represent signals whose voltage is increasing, D to represent signals whose voltage is decreasing, 0 to represent lower cut-off signals, 1 to represent upper cut-off signals, and E to represent all other signal states; which can be errors or unknown. Figure 2 shows an example of representing the electrical signal by two, three, and five value signal models.

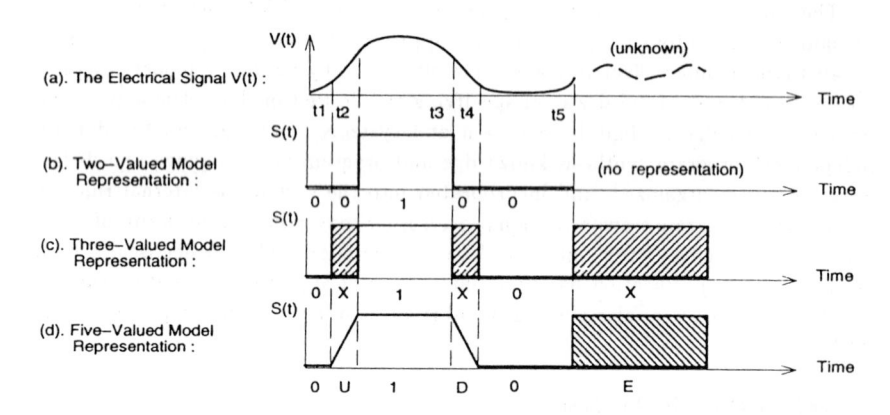

Fig. 2. Modeling the same electrical signal using different signal models.

Selecting a bit representation which correctly models the signals and makes the boolean operations as simple as possible is a very important task during this signal modeling phaser. Figure 3 shows some examples of selecting bit representation in the signal modeling phase. For a two-valued model, one single bit is used to represent the two logic values (0 and 1). For a three-valued model, two bits are used to represent the three logic values (0, 1, and X). For a five-valued model, three bits are used to represent the five logic values (0, 1, U, D, E).

2.2. Primitive modeling phase

Primitive elements are defined as the basic set of elements whose functionality is described by using only the logic states created in the signal modeling phase. For example, two input AND, NAND, OR, NOR, and XOR can be described by only

Two-Valued Model
Two Logic Values: 0, and 1.
One Bit Representation:
    Logic 0 0
    Logic 1 1

Fig. 3(a). One bit representation for two-valued model.

Three-Valued Model
Three Logic Values: 0, 1, and X.
Two Bit Representation:
    Logic 0 00
    Logic 1 10
    unknown X x1
where x represents the don't care value (which can be either 0 or 1)

Fig. 3(b). Two bit representation for three-valued model.

Five-Valued Model
Five Logic Values: 0, 1, U, D, and E.
Three Bit Representation:
    Logic 0 000
    Logic 1 110
    U (up)   010
    D (down) 100
    E (error) xx1
where x represents the don't care value (which can be either 0 or 1)

Fig. 3(c). Three bit representation for five-valued model.

using logic states created in the signal modeling process, and they can be used to build other more complicated hierarchical elements. Truth tables and Karnaugh maps are usually used to describe primitive elements. After the truth table or Karnaugh map description of a primitive element has been specified, one or more transformation processes using this input description will generate the desired element routine code.

Different domain-related knowledge is required during each different stage of the transformation process. The knowledge used to construct the truth table or Karnaugh map, which faithfully represents the element, is essential for processing the element specification provided by the user. The knowledge of superimposing the bit representation (created in the signal modeling phase) onto the truth table to create multiple truth tables, each of which describes logic operations in terms of each individual bit in the bit representation, is critical to the transformation process. Also, the knowledge of mapping a truth table into a Karnaugh map might play another important role during the transformation. Finally, from the performance point of view, the knowledge of why and how to obtain the minimized boolean equations from a truth table or a Karnaugh map is critical to the whole model generating process. In addition to the domain knowledge described above, the knowledge of converting boolean equations into the software code, in a target language, is also essential for this modeling process.

Example 1: OR gate modeling

To model the two-input OR gate shown in Fig. 4(a), using the five value signal model shown in Fig. 3(c), the easiest and the most straight-forward method to specify the input description is to use a truth table, as shown in Fig. 4(b). Under the three bit representation scheme, this truth table is, first, converted into the superimposed truth table shown in Fig. 4(c). Then, the three truth tables, shown in Fig. 4(d), are created by splitting the table of the original three bit representation into three tables of single bit representations, with each new table describing the logic operation of each individual bit of the output signal. The next step is to map each truth table in

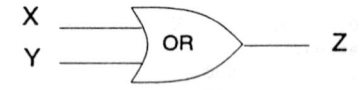

Fig. 4(a). A two-input OR gate.

	Z	Value of X				
		0	1	U	D	E
Value of Y	0	0	1	U	D	E
	1	1	1	1	1	1
	U	U	1	U	E	E
	D	D	1	E	D	E
	E	E	1	E	E	E

Fig. 4(b). Five-valued truth table for the two-input OR gate.

	$Z_1 Z_2 Z_3$	Value of X				
		0	1	U	D	E
Value of Y	0	000	110	010	100	xx1
	1	110	110	110	110	110
	U	010	110	010	xx1	xx1
	D	100	110	xx1	100	xx1
	E	xx1	110	xx1	xx1	xx1

Fig. 4(c). Superimposed five-valued OR gate truth table.

Z_1 Value of Y	Value of X				
		0	1	U	D	E
	0	0	1	0	1	x
	1	1	1	1	1	1
	U	0	1	0	x	x
	D	1	1	x	1	x
	E	x	1	x	x	x

Z_2 Value of Y	Value of X				
		0	1	U	D	E
	0	0	1	1	0	x
	1	1	1	1	1	1
	U	1	1	1	x	x
	D	0	1	x	0	x
	E	x	1	x	x	x

Z_3 Value of Y	Value of X				
		0	1	U	D	E
	0	0	0	0	0	1
	1	0	0	0	0	0
	U	0	0	0	1	1
	D	0	0	1	0	1
	E	1	0	1	1	1

Fig. 4(d). 3 individual truth tables for individual bit operations.

Fig. 4(d) into the Karnaugh map shown in Fig. 4(e). Each of these three Karnaugh maps can be used to generate an optimized logic equation by using switching theory. The three optimized boolean equations are shown in Fig. 4(f). Finally, the optimized equations are translated into their computer language implementation as described in Fig. 4(g).

2.3. *Functional modeling phase*

A functional element is defined as the element constructed by primitive elements or other existing functional elements. Since the components used to construct a functional element are already defined and their corresponding element routines are available, a circuit diagram which details the internal components and their interconnection can be used to specify this desired functional element. With the circuit diagram as input, several transformation processes are used to convert the input into a list of circuit components. At first, the diagram is transformed into its netlist description, which is an internal description of the circuit components and their interconnections. Then a levelizing process, which determines the propagation order

Z_1	Value of X (X1, X2, X3)
	000 001 011 010 110 111 101 100

Value of Y ($Y_1Y_2Y_3$)		
000	0 x x 0	1 x x 1
001	x x x x	1 x x x
011	x x x x	1 x x x
010	0 x x 0	1 x x x
110	1 1 1 1	1 1 1 1
111	x x x x	1 x x x
101	x x x x	1 x x x
100	1 x x x	1 x x 1

Z_2	Value of X (X1, X2, X3)
	000 001 011 010 110 111 101 100

Value of Y ($Y_1Y_2Y_3$)	
000	0 x	x 1 1 x	x 0
001	x x	x x 1 x	x x
011	x x	x x 1 x	x x
010	1 x	x 1 1 x	x x
110	1 1	1 1 1 1	1 1
111	x x	x x 1 x	x x
101	x x	x x 1 x	x x
100	0 x	x x 1 x	x 0

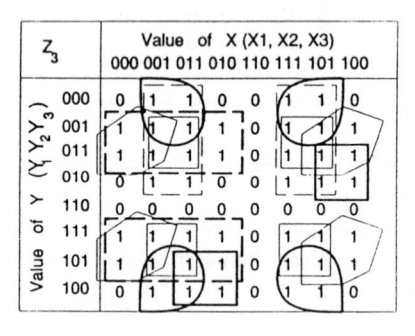

Z_3	Value of X (X1, X2, X3)
	000 001 011 010 110 111 101 100

Value of Y ($Y_1Y_2Y_3$)	
000	0 1 1 0 0 1 1 0
001	1 1 1 1 0 1 1 1
011	1 1 1 1 0 1 1 1
010	0 1 1 0 0 1 1 1
110	0 0 0 0 0 0 0 0
111	1 1 1 1 0 1 1 1
101	1 1 1 1 0 1 1 1
100	0 1 1 1 0 1 1 0

Fig. 4(e). 3 Karnaugh maps corresponding to the individual truth tables in Fig. 4(d).

of all circuit signals inside this circuit net, is applied to transform this net-list into an ordered net-list. The logic level of an element is defined as:

(1) The level of a prime input is assigned to 0.
(2) For any element E, if E has inputs from element E1, E2, ... , Ek, and levels for E1, E2, ... , Ek, are L1, L2, ... , Lk, respectively, then, the logic level of E is defined as:
$$L = 1 + \text{MAX}(L1, L2, \ldots, Lk)$$

$$Z_1 = X_1 + Y_1$$
$$Z_2 = X_2 + Y_2$$
$$Z_3 = \overline{X}_1 Y_3 + \overline{X}_2 Y_3 + X_3 \overline{Y}_1 + X_3 \overline{Y}_2$$
$$+ X_3 Y_3 + X_1 \overline{X}_2 \overline{Y}_1 Y_2 + \overline{X}_1 X_2 Y_1 \overline{Y}_2$$

Fig. 4(f). 3 optimized logic equations for the 5-valued 2-input OR gate.

```
void or2(x1, x2, x3, y1, y2, y3, z1, z2, z3)
int    x1, x2, x3; /* input signal X */
int    y1, y2, y3; /* input signal Y */
int    *z1, *z2, *z3; /* output signal Z */
{
    *z1 = x1 | y1;
    *z2 = x2 | y2;
    *z3 = (~x1)&y3 | (~x2)&y3 | (~y1)&x3 | (~y2)&x3
        | x3&y3 | x1&(~x2)&(~y1)&y2 | (~x1)&x2&y1&(~y2);
}
```

Fig. 4(g). Element routine for the 2 input OR gate.

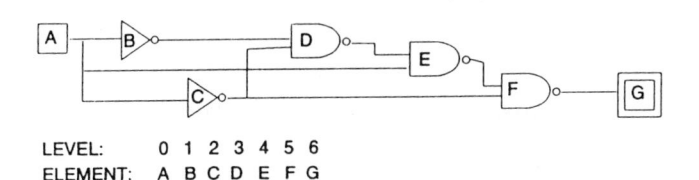

LEVEL: 0 1 2 3 4 5 6
ELEMENT: A B C D E F G

Fig. 5. Example of circuit levelization.

An example of circuit levelization is shown in Fig. 5. The element routine for the functional element can be formed by combining the components element routines, which are already available at this point, according to their orders in the ordered net-list.

There are several different areas of domain knowledge involved in this transformation process. The designers of functional elements need to know the functionality of each component element used to construct functional elements. The knowledge of how to connect the components together is also required to create the circuit diagram. Then, the designers need to know how to convert the circuit diagram into a net-list, while preserving the correctness of the functional description. To levelize the net-list for creating an ordered net-list, the knowledge of the levelization process is required. Finally, the knowledge of generating the desired element routine code according to the levelized ordered list is also essential for this transformation process.

Example 2: XOR gate modeling

The two-input XOR gate shown in Fig. 6(a) is used as a simplified example to illustrate the process of functional modeling. Using the five-valued signal model described in Fig. 3(c), and assuming AND, OR, and NOT gates are primitive models created in the primitive modeling phase, we can use these predefined elements as construction units to build the XOR gate as shown in the circuit diagram described in Fig. 6(b). A transformation process is used to convert the graphic diagram into an intermediate structure which is the net-list described in Fig. 6(c). This net-list has the important information of all predefined component devices and their interconnections. The levelizing process is applied to the net-list to create a levelized net-list shown in Fig. 6(d). According to the order of elements in the levelized net-list, the skeleton of the XOR gate element routine can be created as shown in Fig. 6(e). The target element routine can be derived by replacing each functional block, such as the OR block, by the pregenerated element routine, such as the code generated in Example 1. The final result is shown in Fig. 6(f).

Fig. 6(a). A two-input XOR gate.

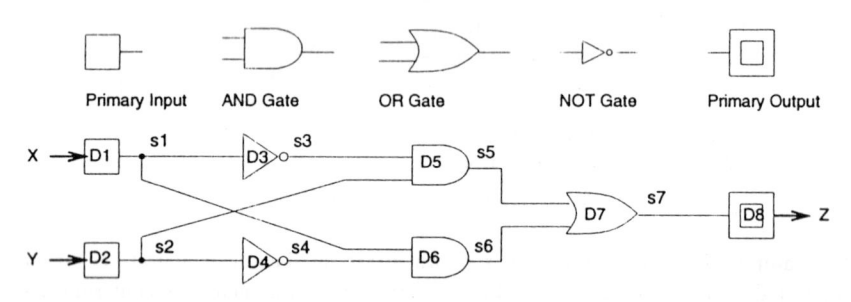

Fig. 6(b). Circuit diagram for the two-input XOR gate constructed by AND, OR, and NOT gates.

```
Device_List((Device, Type)) = ( (D1, PI), (D2, PI), (D3, NOT), (D4, NOT),
                                (D5, AND), (D6, AND), (D7, OR), (D8, PO));
where P1 is primary input, and PO is primary output.
Interconnection((Signal, Pin1, Pin2, ...)) = ( (S1, D1-1, D3-1, D6-1),
               (S2, D2-1, D4-1, D5-1), (S3, D3-2, D5-1), (S4, D4-2, D6-2),
               (S5, D5-3, D7-1), (S6, D6-3, D7-2), (S7, D7-3, D8-1));
```

Fig. 6(c). Net-list generated from the circuit diagram in Fig. 6(b).

```
Levelized_Device_List: (D1, D2, D3, D4, D5, D6, D7, D8)
```

Fig. 6(d). Levelized device list for the circuit diagram in Fig. 6(b).

```
XOR (X1, X2, X3, Y1, Y2, Y3, Z1, Z2, Z3)
int  X1, X2, X3, Y1, Y2, Y3, *Z1, *Z2, *Z3;
{
        Primary Input D1
        Primary Input D2
        NOT  D3
        NOT  D4
        AND  D5
        AND  D6
        OR   D7
        Primary Output  D8
}
```

Fig. 6(e). The skeleton of the generated XOR gate element routine.

```
XOR (X1, X2, X3, Y1, Y2, Y3, Z1, Z2, Z3)
int  X1, X2, X3, Y1, Y2, Y3, *Z1, *Z2, *Z3;
{
      Primary Input D1
      Primary Input D2
      NOT  D3
      NOT  D4
      AND  D5
      AND  D6
      D7z1 = D7x1 | D7y1;
      D7z2 = D7x2 | D7y2;
      D7z3 = (~D7x1) & D7y3  |  (~D7x2) & D7y3
           |  (~D7y1) & D7x3  |  (~D7y2) & D7x3
           |  D7x3 & D7y3
           |  D7x1 & (~D7x2) & (~D7y1) & D7y2
           |  (~D7x1) & D7x2 & D7y1 & (~D7y2);
      Primary Output  D8
}
```

Fig. 6(f). Element routine for the 2-input XOR gate with components replaced by pregenerated code.

2.4. *The domain-independent programming knowledge*

In addition to domain-related knowledge, domain-independent programming knowledge is another primary knowledge sources utilized by the element modeling process.

The language syntax and data structures provided by target language have a direct impact on all design phases during element modeling process. The efficiency of the target language implementation is another general concern of the AFMG project. Optimization of the element routine code becomes indispensable for five value models, because the size of the code is much larger than the size of the code for three value models.

The program-synthesis syntactic knowledge is important for generating the actual target implementation code. For the AFMG design, this knowledge encompasses the C language syntax requirement, and how the language constructs impact performance.

One simple example in C language is the set of constructs to form a loop, such as: the "for" statement, the "while-do" statement, or the "do-until" statement. These three constructs are semantically interchangeable by introducing new temporary variables or by duplicating some code segments, although each of these constructs has specific syntactic requirements, and each is suited towards specific implementation usages. How to select the appropriate loop constructs for various implementation conditions provides an example of this type of knowledge.

The data structure implementation knowledge is knowledge of data structure effectiveness and efficiency in terms of time and space characteristics. The knowledge of the data space characteristics captures important features regarding data types and their definitions and manipulations. The data definition knowledge contains important information regarding the various data sizes, the data alignment and conversions, the static or dynamic memory allocation, the variable scoping rules, and the implementation of abstract data types. The data manipulation knowledge identifies various constraints and methods for manipulating data. This knowledge is used to identify the correct and efficient operators and algorithms to access or pass data.

An example of using the data structure implementation knowledge is in the sequential signal handling, which requires the dynamic (run time) memory allocation and an effective algorithm to keep track of all instances of the sequential signals during every simulation cycle. More details about the sequential signal handling are discussed in Sec. 4.

The data structure knowledge, as well as the program synthesis knowledge, depends on the selected target language. It is used not only to select and refine data structure implementations in transformation processes, but to resolve complex data structure access requirements within the syntactical requirements of the target language.

3. User Interface and Specification Tools

AFMG user interface and specification tools, as shown in Fig. 7, serve to facilitate the communication between users and the AFMG system. Not only should the AFMG front-end provide various types of model specifications, including truth

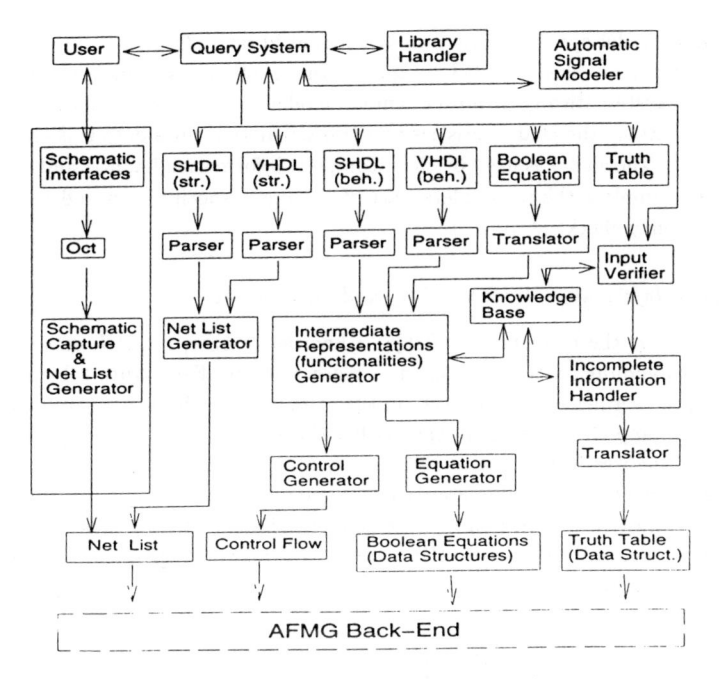

Fig. 7. User interface and specification tools.

tables, boolean equations, behavioral level Hardware Description Languages (HDL), structural level HDL's, and schematic diagrams, as inputs to AFMG but collect and store feedbacks from other AFMG subsystems. The feedbacks include both informative error messages if the input to the back-end was somehow rejected by the transformation process, and performance data and representation objects if the transformation process succeeded. The error messages inform users of reasons of failing, such that users can modify their input specifications interactively and resubmit their requests. The performance data and representation objects are stored in a knowledge base, so that the stored input descriptions and representation objects, corresponding to some well defined and verified simulation model, can be used as a "black box" to construct other more complicated functional models. The user needs to understand only the behavior of the simulation models as specified in their input descriptions, without being concerned about their implementations.

In order to provide a versatile and user-friendly environment, the AFMG front-end as shown in Fig. 7 was designed to contain the following major components:

Interactive Query System (IQS)

By cooperating with the library handler, IQS displays the existing models which can be used to construct new models or stores the generated models and the corresponding model specifications into the model library. During a typical AFMG session, users are asked by IQS to specify the model class name and the model name. By systematically grouping similar models into the same class, users can derive and attach the characteristics and transformation rules onto this class. This class specific knowledge is used by the incomplete information handler and the input verifier to complete the incomplete part of the input specification and to perform the consistency check.

Incomplete Information Handler (IIH) and input verifier

Figure 8 shows the cooperation of IQS, a knowledge base, the IIH, and the input verifier for handling partial or fragmentary descriptions. For a truth table input description, the verifier is used to perform the completeness check and the consistency check. For complete description, the verifier checks whether the description is consistent and invokes IQS to notify users of any input error. If the verifier determines that the description is incomplete, IIH is invoked to fill in the missing information by consulting the knowledge base. If predefined rules exist for the class of the target model, the rules will be used by IIH to derive, as many as possible, of the missing truth table entries.

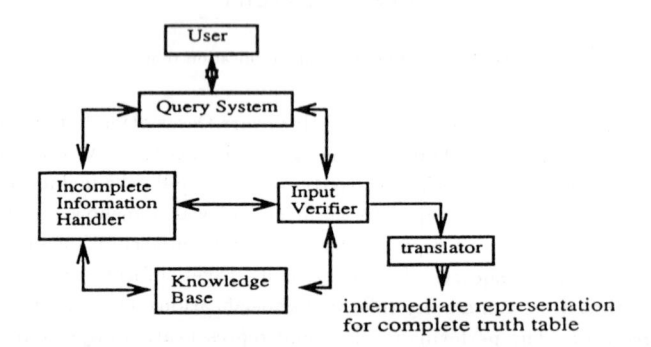

Fig. 8. Handling incomplete input specification.

If there are still some missing entries after all applicable rules in the knowledge base have been used, IQS will ask users to either enter new rules or modify already existing rules for this circuit class. If the rules used by IIH create nondeterministic results users will be notified by IQS of the ambiguities and users have choices to either modify the class rules or just resolve the ambiguities by explicitly entering the missing or ambiguous truth table entries. After the description is completed by IIH and/or the users, the control is again passed to the verifier, which will perform the consistency check on the newly created complete description.

Knowledge base

The knowledge base provides important application domain knowledge to the incomplete information handler for deriving input specifications not provided in the users' incomplete model descriptions. Rules in the knowledge base can also be used by the verifier to validate the correctness of the input descriptions. The rules are organized in a hierarchical structure similar to the class/object structure provided in an object-oriented environment. By carefully grouping similar models into the same class, users can derive and attach the characteristics and transformation rules onto each class. These class specific rules can be inherited by another model class. A class rule inherited from another class can be overridden by a local rule defined with the same rule name. The knowledge base is also used to store rules used by the AFMG back-end to guide the program transformation process, and to perform the data structure and algorithm selection process. The details of the AFMG back-end are described later in this paper.

Schematic interface

The schematic interface provides a higher level, intuitive, and easy-to-use graphics user interface to AFMG users for specifying desired models at a structural level. Figure 9 shows the diagram of the schematic interface subsystem. The schematic interface is built on the OCT tool, which was developed at the University of California, Berkeley, for providing an object-oriented computer aided digital circuit design environment [11,12,13].

Using this schematic graphics user interface, users can select primitive models (such as AND, OR, NAND, NOR, NOT, ... etc.) and other predefined functional models, as construction components, to define a new functional model. In addition to the construction components, users also have to define, by using the schematic graphics user interface, the interconnections among these components. The information of the component elements and their interconnections is used by the schematic interface to create a net-list which will be finally passed to the AFMG back-end for generating the target model.

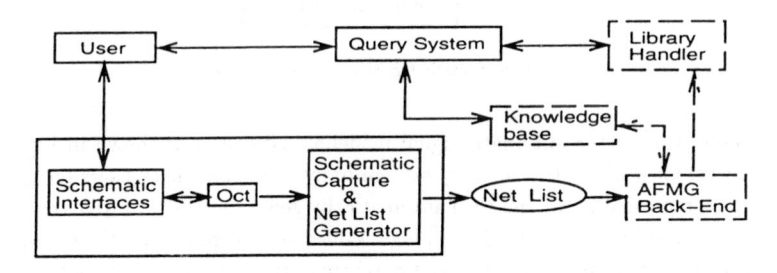

Fig. 9. Schematic interface subsystem.

Automatic signal modeling subsystem

For any k-valued signal model, k distinct logic values can be represented in the computer by sets of bits with at least $\log(K)$ bits in the set. For example, in the three value model, at least 2 bits are needed and we might choose to represent logic 0 (L) by bits 00, logic 1 (H) by 10, and unknown value (X) by 01 or 11. For the five value model using a 3 bit representation, there are 6720 combinations for representing the logic values. This means the generated element routine may have up to 6720 variations. Because a model generated for a specific signal representation can not be directly used by simulations using different signal representations, a rule-based model generation approach, the automatic signal modeling process, was used to achieve a higher level of design automation.

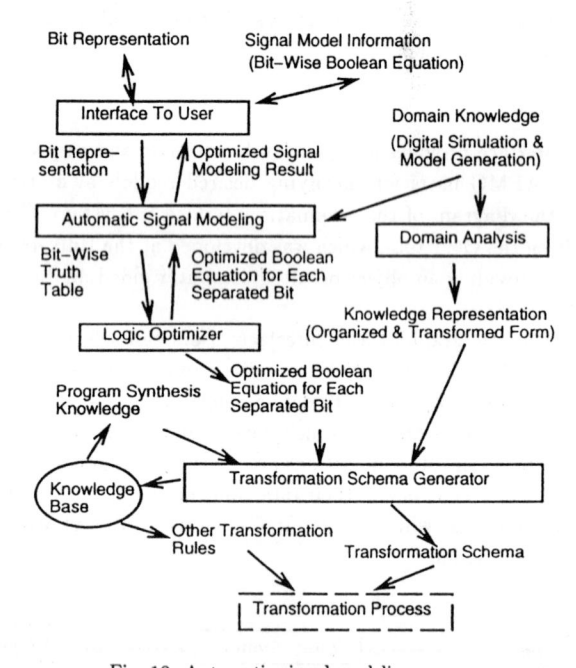

Fig. 10. Automatic signal modeling process.

A description of the automatic signal modeling subsystem is shown in Fig. 10. This subsystem works with the query system to interactively query users for specifying signal representation, and then automatically generates the primitive transformation rules conforming to the desired logic values and bit representations. During the automatic signal modeling process, the schema generator is used to capture and organize important information and store the transformed knowledge as a set of rules. This set of rules is called "transformation schema" and stored in the knowledge base. The transformation schema can be dynamically and efficiently used by AFMG to guide the program transformation process, such that, the model generation process become very flexible in terms of generating models complying with various signal models. More details of the transformation schema are described later.

4. Behavioral Domain Model Generation — Automatic Primitive Modeling

The process of the behavioral domain model generation is shown in Fig. 11. AFMG enhances the concept of primitive modeling by using truth tables, boolean equations, and behavioral HDL's to specify the functional behavior of the desired primitive models. AFMG accepts two types of HDL inputs, VHDL [14] and SHDL (the Simulation Automation System Hardware Description Language), where SHDL is designed and implemented for the purpose of model specification [15,16]. Based on the signal modeling information and the input description type, AFMG chooses appropriate algorithms, data structures, and transformation schema, to perform a series of transformations, which are guided by the internal rules, for converting intermediate representations into simulation models.

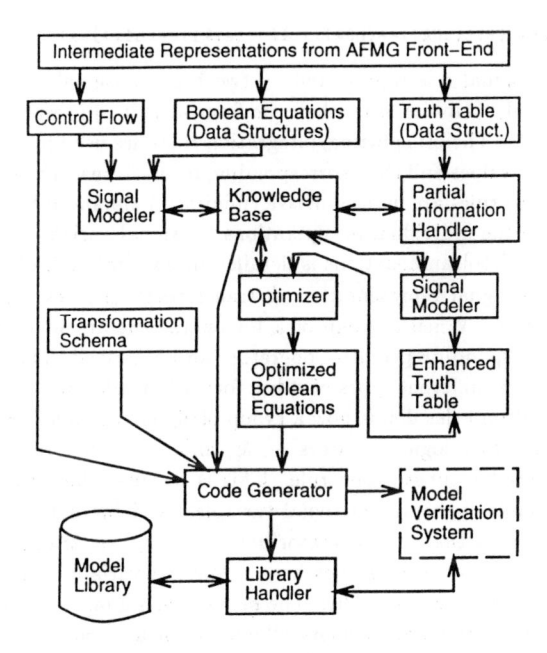

Fig. 11. Behavioral Domain Model Generation (BDMG).

Transformation rules

For each type of input description, domain analysis was applied on the model generation process to identify, organize, and incorporate the transformation rules into the internal knowledge base. These rules include model construction rules, input/output interface rules, "control flow" conversion rules, bit splitting/integration rules, equation derivation/optimization rules, sequential signal handling rules, boolean expression formation/combination rules, and variable substitution rules. The model construction rules define the overall architecture of the generated models. The input (output) interface rules define how to initialize the internal input variables (output parameters) according to the input parameters (output variables). The "control flow" conversion rules provide information for handling control constructs in HDL's, such as: IF statements, FOR loops, REPEAT loops, ... etc. For a 3 value model, if two bits are used to represent each logic state, the following control construct:

```
          REPEAT
             ...
          UNTIL a == b;
can be transformed to the following C code segment:
          do {
             ....
          }
          while (!(((a0&1)==(b0&1)) && ((a1&1)==(b1&1))));
```

In this example, signal a is represented by two bits whose values are (a0&1) and (a1&1), respectively. All bits other than the first bit (bit 0) in words a0 and a1 are used to store faults. The bit splitting/integration rules are used to guide the conversion between logic values and their corresponding bit patterns. The equation derivation (optimization) rules are used to derive (minimize) enhanced boolean equations [16], which define the individual bit operations for signal variables. An example of deriving enhanced boolean equations is detailed in Example 1 (Subsec. 2.2).

Under a k-bit signal representation, k transformation rules are generated for each logic operation. When a group of k bit operations are well defined, the behavior of the corresponding primitive operator can be derived. The transformation schema contains two different types of rules; formation rules and variable substitution rules. Formation rules define how a group of signal variables are concatenated together using bit-wise logic operators $\sim$, &, and |. Each formation rule has a corresponding variable substitution rule. This is because the formation rule uses the deferred variables to represent signal variables, and the deferred variables will not be instantiated by the code generator until the very last transformation phase. The variable substitution rule defines how a deferred variable is substituted by a run-time signal variable dynamically. The sequential signal handling rules handle the required dynamic run-time memory allocation for all feed-back signals inside a sequential circuit.

The behavioral HDL parsers generate two types of intermediate representations, control flows and boolean equations, which are handled by AFMG using control flow conversion rules, boolean expression formation/combination rules, and variable substitution rules. The boolean expression combination rules are used to convert IF statements into enhanced boolean equations. Figure 15 describes the rules for combining boolean expressions, and an example of using the rules is described in the last paragraph of this section.

Sequential signal handling

Dynamic run-time memory allocation is required for all feed-back signals inside a sequential circuit. During the first time an instance of any element routine is called by the simulator, this element routine must allocate run-time memory for all of its internal state variables. This process needs to be invoked at run time instead of compile time, because there might be more than one instance of the same element routine being used in a simulated system.

Between various calls to an element routine, the dynamically allocated memory is used to reference, update, and store the internal state of every instance of this element routine. Right before a call to an instance of a model is completed, the model has to store the internal state, defined by internal state variables, back to the run-time memory for the current model instance. In addition, the model also needs to pass the run-time memory pointer back to the simulator, and the simulator needs to store this state information in its internal table. For subsequent calls to the same instance of an element routine, the simulator must provide the run-time memory pointer as an input parameter to the element routine, so that, the element routine can use this pointer to restore its previous state. This state variable handling process is also required for generating sequential models from all other kinds of input descriptions.

Truth table descriptions

Example 1 shows a simplified scenario of applying a series of transformations to generate the desired models from truth tables. For the actual design and implementation, more issues were taken into considerations, such as: automatic signal modeling, optimization alternatives, and sequential circuit handling. The result from automatic signal modeling is used to guide the selection of appropriate data structures, and to derive rules which are used in the code generation process for signal initialization and input/output interfacing. The input interface rules define how to initialize the internal input variables according to the input parameters. The output interface rules define how to construct the output parameters according to the internal output variables. For generating the optimized boolean equations, AFMG can use various techniques including the McCluskey's minimization algorithm [17]. The design considerations of optimization alternatives and automatic signal modeling are also required in the transformation processes for other forms of model descriptions.

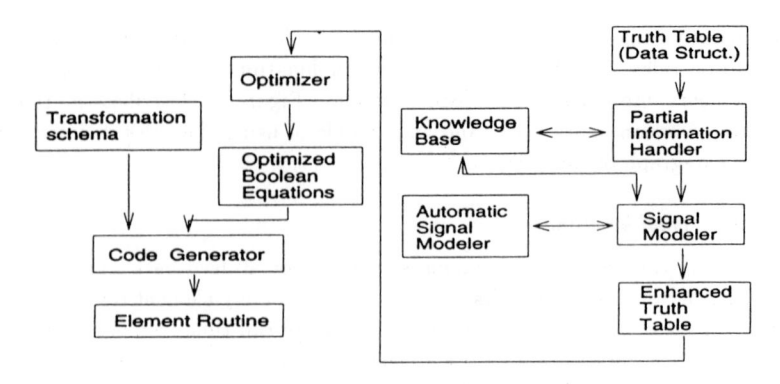

Fig. 12. Model generation from a truth table.

Boolean equation descriptions

Formation rules and variable substitution rules are used in handling boolean equation descriptions. A boolean equation assigns the value of a boolean expression to a variable which can be either an output signal, an internal variable, or a state variable. A boolean expression contains one or more operands and one or more operators. An operand can be an input signal, an internal variable, or a state variable. An operator can be a logic operation such as a logic AND, a logic OR, or a logic NOT. For example, the behavior of an XOR gate can be described by the boolean equation:

$$y\ =\ \sim a\ \&\ b\ |\ a\ \&\ \sim b$$

The above compound boolean expression, which contains more than two operands and more than one operator, is converted to a series of primitive boolean expressions. Each of these primitive expressions is a 2-operand AND expression, a 2-operand OR expression, or a NOT expression. The XOR equation will be transformed into a series of equations, such as:

$$c\ =\ \sim a$$
$$d\ =\ \sim b$$
$$e\ =\ c\ \&\ b$$
$$f\ =\ a\ \&\ d$$
$$y\ =\ e\ |\ f$$

Since these equations only involve primitive expressions, the transformation schema, which defines the logic behavior of the 3 primitive operators, can be used to generate enhanced equations for the target element routine.

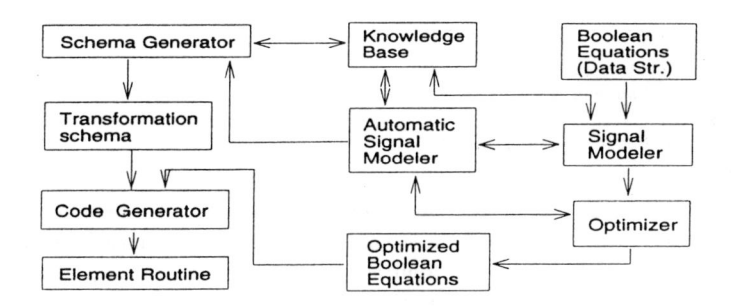

Fig. 13. Model generation from Boolean equations.

Behavioral HDL descriptions

There are two types of intermediate representations, control flows and boolean equations, generated by the AFMG front-end. The boolean equations contained in an HDL description are handled using the same transformation approach as described in previous paragraphs. The control flows in an HDL description may contain assignment statements, IF statements, FOR loops, and REPEAT loops. The assignment statement is handled exactly in the same way as the boolean equations. Some control flows, such as the IF statements, can also be converted into boolean equations [18,19], which can be further transformed into enhanced equations. Then, all the enhanced equations are merged with other control flows, which are handled by using the control flow conversion rules, to form the core code segment. Finally, the core code segment is used by the code generator to synthesize the element routine.

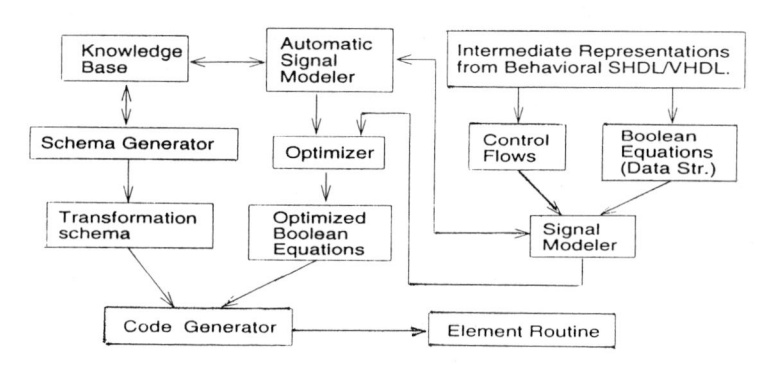

Fig. 14. Model generation from behavioral HDL's.

For example, an IF statement looks like:

```
IF  condition_expression  THEN
       x = boolean_expression1
ELSE
       y = boolean_expression2
END IF
```

If 3 logic states and 2 bit representations are assumed, the "condition_expression" is transformed to an enhanced condition expression, which is merged into the enhanced equations derived from equations "x = boolean_expression1" and "y = boolean_expression2" to create the following enhanced equations:

```
x0 = (enh_cond_expr)&(enhanced_expression10) | ~ (enh_cond_expr)&x0
x1 = (enh_cond_expr)&(enhanced_expression11) | ~ (enh_cond_expr)&x1
y0 = ~ (enh_cond_expr)&(enhanced_expression20) | (enh_cond_expr)&y0
y1 = ~ (enh_cond_expr)&(enhanced_expression21) | (enh_cond_expr)&y1
```

where the enhanced equations "x0 = enhanced_expression10" and "x1 = enhanced_expression11" are derived from the boolean equation "x = boolean_expression1", and the enhanced equations "y0 = enhanced_expression20" and "y1 = enhanced_expression21" are derived from the boolean equation "y = boolean_expression2".

The algorithm of the transformation process shown in Fig. 15 is used to generate the enhanced boolean expressions from the condition expressions with a 3 bit signal representation. A condition expression has the form:

$$a == b \quad \text{or} \quad a\ != b$$

where a and b can be any signal constant (such as: L, H, U, D, E, or X), or a signal variable. Function N is a negation function, which returns the string "$\sim$" if the input parameter is '0'. The operators '+' and '^', used in Algorithm Gen_Enh_Expression, represent the string concatenation operator and the bit-wise XOR operator, respectively.

If the logic signal H is represented by bits 110, the IF statement

```
IF  a==H  THEN
       x = p
ELSE
       y = q
END IF
```

is transformed into the following C code:

```
x0 = (a0&a1& ~ a2)&p0 | ~(a0&a1& ~ a2)&x0;
x1 = (a0&a1& ~ a2)&p1 | ~(a0&a1& ~ a2)&x1;
x2 = (a0&a1& ~ a2)&p2 | ~(a0&a1& ~ a2)&x2;
y0 = ~(a0&a1& ~ a2)&q0 | (a0&a1& ~ a2)&y0;
y1 = ~(a0&a1& ~ a2)&q1 | (a0&a1& ~ a2)&y1;
y2 = ~(a0&a1& ~ a2)&q2 | (a0&a1& ~ a2)&y2;
```

```
FUNCTION N(bit)
    IF bit_rep(bit) == '0'
        RETURN the string "~"
    ELSE
        RETURN an empty string
    END IF
END FUNCTION
ALGORITHM Gen_Enh_Expression(expression)
    IF expression looks like "a == c"  /* c is a signal constant */
        enh_expression = N(c0) + "a0&" + N(c1) + "a1&" + N(c2)+ "a2"
    ELSE IF expression looks like "a != c"  /* c is a signal constant */
        enh_expression = N(~c0) + "a0&" + N(~c1) + "a1&" + N(~c2) + "a2"
    ELSE IF expression looks like "a == b"  /* a, b are variables */
        enh_expression = "~(a0 ^ b0)&~(a1 ^ b1)&~(a2 ^ b2)"
    ELSE IF expression looks like "a == b"  /* a, b are variables */
        enh_expression = "(a0 ^ b0)&(a1 ^ b1)&(a2 ^ b2)"
    END IF
    RETURN enh_expression
END ALGORITHM
```

Fig. 15. Generating the enhanced Boolean expression.

5. Structural Domain Model Generation — Automatic Functional Modeling

A structural domain model is a hierarchical simulation model which contains embedded lower level models. The functional modeling process, shown in Example 2, is a simplified example of structural domain model generation (SDMG). By selecting the construction components and specifying their interconnections, a new model can be built by SDMG. A lower level component element routine, used as a construction unit, can be a model generated from truth tables, boolean equations, behavioral HDL's, structural HDL's, or high level schematic diagrams. Once a model is generated by the SDMG, it can be used as a new construction component to build the more complex functional models. This hierarchical partitioning approach makes it possible to define and generate simulation models of large digital systems [16,20].

SDMG accepts three kinds of input descriptions: SHDL, VHDL, and schematic diagrams. These inputs are converted by the automatic functional modeling subsystem, as shown in Fig. 16, into net-lists, which contain the information of components and their interconnections. Figure 6(c) shows a simplified example of the net-list. The loop detection process is essential for identifying sequential circuits. If a loop is found, a "state variable" is automatically defined for the detected feedback loop, and some sequential circuit specific transformations need to be done during the

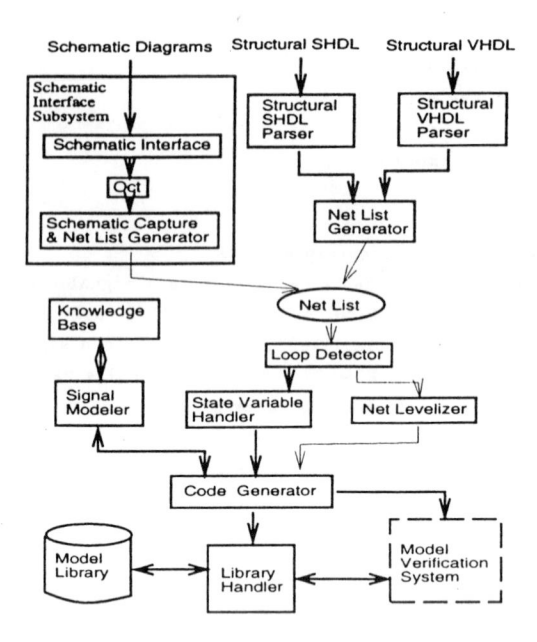

Fig. 16. Structural Domain Model Generation (SDMG).

levelization process and the code generation process. The levelized net-list, the component element routines, and appropriate transformation rules described above are used by SDMG to synthesize the targeted simulation model.

Loop detection

The loop detection process builds an internal directed graph from the net-list. All components, including primary inputs and primary outputs, are represented by nodes. All connection wires between components are represented by directed edges. If an output of component A is fed, as an input, into component B, then there would be an directed edge starting from node A and ending at node B. The loop detection algorithm, as described in Fig. 17, performs the Depth First Search technique [21], for all primary inputs.

Net list levelization

The levelization process is used to determine the propagation order of all circuit signals inside the simulated circuit net. The original net-list is transformed, by this levelization process, into an ordered net-list, such that, the generated C code can follow the ordered list, to derive a correct and efficient signal evaluation sequence.

```
ALGORITHM Loop_Detection
loop_found = FALSE;
FOR every primary input i DO {
        FOR every node n DO
                visited[n] = FALSE;
        dfs (i);
        if (loop_found)
                break;
}
END ALGORITHM

PROCEDURE dfs(n: node)
IF (loop_found)
        return;
visited[n] = TRUE;
FOR all nodes j adjacent to n DO {
        if (visited[j]) {
                loop_found = TRUE;
                break;
        }
        dfs (j);
}
END PROCEDURE;
```

Fig. 17. Algorithm for detecting loops.

For combinational models, if all signals are evaluated according to the ordered net-list, it is guaranteed that every component element is evaluated after all of its input signals have already been evaluated. This approach makes the generated model more efficient, because the model only has to evaluate its components once to derive its output signals. The algorithm shown in Fig. 18 can be used to effectively levelize combinational circuits.

For levelizing sequential circuits, a different algorithm, which takes the delays of the internal elements into account, is used to make sure the internal signals are handled properly. This new algorithm, shown in Fig. 19, evaluates any element output as soon as one of its input signals is changed, while the combinational algorithm, shown in Fig. 18, defers the evaluation until all input signals are ready. Based on the new levelization method, the generated sequential element routine can detect the transient output changes and the intermediate signal changes caused by the different component delays. However, this also means that a steady state, for a sequential element, may not be reached by one iteration of the signal evaluation, while only one iteration of the signal evaluation is required for a combinational element levelized by the combinational algorithm. Therefore, the signal evaluation process in a generated element routine needs to be embedded in a loop, such as a "do-while" loop. After each iteration of the loop, the new values of the signal

```
ALGORITHM levelize_combinational_circuit
    FOR all element e DO {
            e.level = -1;
            e.visited = FALSE;
    }
    FOR all primary output po DO
            levelize(po);
    Sort all element e by e.level in ascending order;
END ALGORITHM
PROCEDURE levelize(e)
    IF (e.visited)
            IF (e.level == -1)
                    quit; /* not a combinational circuit */
            RETURN (e.level);
    e.visited = TRUE;
    IF (e has no input element) {
            e.level = 0;
            RETURN (0);
    }
    temp = 0;
    FOR all input element i of e DO
            temp = MAX(temp, levelize(i));
    e.level = temp;
    RETURN (e.level);
END PROCEDURE
```

Fig. 18. Algorithm for levelizing combinational circuits.

variables are compared with the values obtained from the previous iteration, to check whether a steady state has been reached. More details of handling sequential circuit are described in the next section.

Code generation

As described in Fig. 16, the code generator uses the information provided by the signal modeler, the loop detector, the net levelizer, and the sequential signal handler. The signal modeler expands the signals, the input parameters, and the output parameters, according to the signal representation and the word length determined in the signal modeling phase. The other three components give the code generator the domain knowledge and the refined representation of the desired model.

There are two different approaches for generating element routine code at the structural level. One approach is to expand the hierarchical structure using lower level elements. The other approach does not "flatten" the high level model into the lower level primitive elements. Instead, the second approach dynamically links the low level element, and treats the component elements as predefined functions. The "circuit flattening" approach may run faster, because there is no overhead of function calls. However, the code size of a flattened simulation model, of a complex system, can exceed acceptable limits. In addition to the advantage of generating

```
ALGORITHM levelizing_sequential_circuit
    FOR all element e DO {
        e.total_delay = INFINITE;
        e.visited = FALSE;
    }
    FOR all primary input pi DO
        calculate_delay(pi);
    Sort all element e by e.total_delay in ascending order;
END ALGORITHM
PROCEDURE calculate_delay(e)
    IF (e.visited)
        IF (e.total_delay == INFINITE) {
            copy e to fe;
            add fe to feedback element list;
        }
        ELSE
            RETURN (e.total_delay);
    ELSE {
        e.visited = TRUE;
        IF (e has no input element) {
            e.total_delay = 0;
            RETURN (0);
        }
        temp = INFINITE;
        FOR all output element oe of e DO
            temp = MIN(temp, calculate_delay(oe));
        e.total_delay = temp + e.element_delay;
        RETURN (e.total_delay);
    }
END PROCEDURE
```

Fig. 19. Algorithm for levelizing sequential circuits.

models with a much smaller code size, the "dynamic linking" approach can generate high level models even when the lower level component models are not yet available. After the high level model was generated, the component models, at the lower level, can be defined, generated, and even modified, without changing or regenerating the high level model. The dynamic linking approach is selected and used by the SDMG code generator.

The generated model code has the structure shown in Fig. 20. Each generated model is a C function with several lines of comments. The function has three parameters; the input, the output, and the state pointer. The declaration part of the generated function declares variables, signals, and pointers. Right after the declarations, there is code for handling state variables. The main evaluation block appears after the declarations part and the initialization part. Within this block, the component elements are evaluated according to the order of the levelized netlist. For a sequential circuit, the main evaluation block is in a loop and will be executed repeatedly, until the feedback signals converge or the preset iteration limit

```
Comments for the generated model
Model_Name(inputs,  outputs,  state_pointer)
{
        Declarations of variables
                Inputs;
                Outputs;
                Signals;
                Feedback signals;
                Temporary variables;
        Declarations of pointers for included sequential elements;
        If (state_pointer == NULL) /* first call */
                Allocate memory for internal states;
        Initialize variables and signals
                using input parameters and state_pointer;
        REPEAT
                Evaluate internal elements in order;
                IF (feedback signals are stable)
                        BREAK loop;
        UNTIL (loop counter reaches preset value);
        Set output parameters;
        Save states of feedback signals and sequential elements;
}
```

Fig. 20. The code structure of the generated element routine.

is reached. Finally, the output parameters are set, and the values of the state variables are saved in the run-time dynamic memory.

6. Experiment Results

AFMG was tested on a UNIX based SUN4/110 workstation with the memory size of 32M bytes. In the experiment, the unknown value X was selected as the default value for unspecified table entries and the initial signal values. The Behavioral Domain Model Generation subsystem (BDMG) was tested for generating primitive models from truth tables, boolean equations, and behavioral HDL's. Table 1 shows some statistics including abstraction level, code size, model generation time, and model execution time.

The model generation time and the model execution time were measured by the "time" UNIX command and/or the "times" C library function. In both cases, the system time (used by the UNIX kernel processes) and the user time (used by the AFMG processes) are added together to obtain the total time used. The generation time shows that the AFMG system can generate primitive element routines in seconds or minutes, for reasonably sized circuits. Generated models were tested by the tester program or the simulators (PARSIM [4] and P-SIM [5]). The generated primitive element routines have the model execution times in the neighborhood of micro-seconds, which are comparable to the model execution times of manually written element routines.

Table 1. Results of the behavioral domain model generation.

Circuit	Descr. Type	Level	Logic Values	Code Size	Gen. Time	Exec. Time
2 input NAND gate	Truth Table	Gate	3-Valued	25 lines	1.3 sec	3.1 usec
3 input decoder	Truth Table	Func.	6-Valued	82 lines	603 sec	16.3 usec
2 input NOR gate	Truth Table	Gate	6-Valued	29 lines	1.6 sec	3.1 usec
2 input XOR gate	Boolean Eq.	Func.	3-Valued	36 lines	0.15 sec	4.9 usec
4 input NAND gate	Boolean Eq.	Func.	6-Valued	69 lines	0.6 sec	16.7 usec
4 input decoder	Boolean Eq.	Func.	6-Valued	950 lines	3.0 sec	467 usec
RAM	HDL	Beh.	3-Valued	49 lines	0.12 sec	2.3 usec
J-K flip-flop	HDL	Func.	6-Valued	206 lines	0.2 sec	16.7 usec
ALU	HDL	Beh.	6-Valued	85 lines	0.2 sec	3.2 usec

Table 2. Results of the structural domain model generation.

Circuit	Included Elements	Code Size	Gen. Time	Exe. Time
Multiplexer	7 gates	150 lines	0.28 sec	0.12 msec
J-K flip-flop	8 gates	300 lines	0.32 sec	0.27 msec
Counter 7493	1 gate, 4 JKFF	149 lines	0.23 sec	0.38 msec
D flip-flop	6 gates	227 lines	0.27 sec	0.18 msec
Counter 74163	29 gates, 4 DFF	711 lines	0.55 sec	3.13 msec
Decoder 74154	21 gates	483 lines	0.53 sec	0.61 msec
ALU 74181	63 gates	1406 lines	0.92 sec	2.17 msec

Using the Structural Domain Model Generation subsystem (SDMG), many 6 valued structural models were generated. Table 2 shows the statistics for some experiments performed on a SUN4/110.

The generation time shows that the SDMG system can generate element routines within seconds. This result demonstrates the substantial advantage of the automatic model generation approach, especially by comparing to the many man-hours of effort which are usually required by the more error prone manual approach.

7. Conclusion

Generating simulation models is a knowledge intensive, time consuming, and error-prone task in implementing a digital logic simulator. To ease this problem, AFMG has been designed to generate simulation models in an efficient and reliable way, by utilizing the domain specific automatic programming technique. The application domain knowledge of logic simulation and model generation was identified, organized and incorporated as internal rules into the AFMG to guide the program transformation processes. Several program transformation processes were designed and implemented within AFMG to transform, step by step, the different forms of model specifications into element routines which are the target language implementations of the functional models. Using AFMG, the process of developing functional models becomes a much easier task than writing the high level device specifications. The desired models can be 3-valued, 5-valued, or 6-valued with 2-bit or 3-bit signal representations which can be flexibly specified by AFMG users. Using the concept of transformation schema, AFMG provides the flexibility of generating models conforming to the user specified signal representations. AFMG also provides the capability of incomplete specification handling and input verification. The AFMG back-end is composed of two subsystems: BDMG and SDMG, BDMG is designed to automate the primitive modeling phase which creates primitive element routines to be used in simulation or in constructing more complicated functional models. SDMG automates the functional modeling phase and provides a powerful hierarchical partitioning approach to synthesize the complex simulation models. As a result of providing an effective and efficient way to create simulation models, AFMG can significantly reduce the development time and improve the quality of the digital system design process.

References

[1] S. A. Szygenda and E. W. Thompson, "Modeling and Digital Simulation for Design Verification and Diagnosis," *IEEE Trans. Computers*, December 1976, pp. 1242-1253.
[2] S. A. Szygenda, "Simulation of Digital Systems: Where We Are And Where We May Be Headed," in *Computer Aided Design*, pp. 41-54, 1979.
[3] M. Bloom, "Behavioral Models Take the Pain out of System Simulation," in *Computer Design*, February 15, 1987, pp. 38-46.
[4] S. Kang and C. H. Han, "PARSIM Manual," Electrical and Computer Engineering Department, The University of Texas at Austin, June 1990.
[5] C. Ang, "A Technique for Parallel Design Error Simulation," The University of Texas at Austin, Master Thesis, August, 1993.
[6] Y. Levendel and P. R. Menon, "Fault Simulation," in Fault Tolerant Computing, D. K. Pradhan, ed., Prentice-Hall, pp. 184-264, 1986.
[7] S. Kang and S. A. Szygenda, "Modeling and Simulation of Design Errors," in ICCD, 1992.
[8] S. A. Szygenda and E. W. Thompson, "Digital Logic Simulation in a Time-Based Table-Driven Environment, Part 1: Design Verification," in *Computer*, March 1975.
[9] D. Barstow, "Domain-Specific Automatic Programming," in *IEEE Trans. on Software Engineering*, pp. 1321-1336, November 1985.
[10] C. H. Han, S. Kang, and S. A. Szygenda, "AFMG: Automatic Functional Model Generation System," in *4th Annual IEEE International ASIC Conference for Digital Logic Simulation*, September 1991.
[11] R. Spickelmier, "Policy Guides for Oct Tools Distribution 3.0," Electronics Research Lab., University of California, Berkeley, Mar. 1989.
[12] R. Spickelmier, P. Moore, and A. Newton, "A Programmer's Guide to Oct," Electronics Research Lab., University of California, Berkeley, Feb. 1990.
[13] D. Harrison, "Symbolic Editing with VEM", University of California, Berkeley, Mar. 1989.

[14] R. Waxman, "The VHDL (IEEE Standard 1076) Language Features Revisited," IEEE Compcon 1988, Digest of Papers, San Francisco, pp. 41–47, Spring 1988.
[15] C. H. Han, and S. A. Szygenda, "Automatic Generation of Functional Element Routines from Behavioral Descriptions," in *Simulation and Modeling Conference*, 1992.
[16] S. E. Chang, "The Design of a Multiple-Valued Automatic Model Generation System for Digital System Simulations," The University of Texas at Austin, Ph.D. Dissertation, May 1994.
[17] E. J. McClusky, "Logic Design Principles," Prentice-Hall, 1986, pp. 194–236.
[18] Y. Levendel and P. R. Menon, "Fault Simulation," in *Fault Tolerant Computing*, D. K. Pradhan, ed., Prentice-Hall, pp. 184–264, 1986.
[19] K. Wu, "Synthesis of Accurate and Efficient Functional Modeling Techniques for Performing Design Verification of VLSI Digital Circuits," The University of Texas at Austin, Ph.D. Dissertation, 1979.
[20] C. H. Han, S. A. Szygenda, and Benyu Fan, "Structural Domain Model Generation for Digital Logic Simulation," in *Summer Computer Simulation Conference*, 1992.
[21] A. Aho, J. Hopcroft, and J. Ullman, *Data Structures and Algorithms*, Chapter 6, "Directed Graphs," Addison-Wesley Publishing Corp., 1983.

Visual Reverse Engineering using SPNs for Automated Diagnosis and Functional Simulation of Digital Circuits

J.Gattiker and S.Mertoguno

Binghamton University, Dept. EE/AAAI lab, Binghamton, NY 13902

Abstract

This paper presents a new methodology, which leads to the development of a prototype system for the automated diagnosis and functional simulation of digital circuits. The methodology presented here is based on an automated visual reverse engineering process, which assists the extraction of the functional behavior of a digital circuit and for the detection and diagnosis of functional faults and defects on it. The visual information extracted from a PCB (printed circuit board) or a digital circuit is represented by SPN (stochastic Petri-net) forms in order to maintain both its structural and functional characteristics.

1. INTRODUCTION

The automatic visual inspection for detection of defects and diagnosis of faults in printed circuit boards (PCBs), before and after the mounting process, has significantly improved the quality of these products. The visual inspection process starts with imaging the PCB. Segmentation and feature extraction provides primitive features then used for matching to expected features of the PCBs. The electronics industry is active in applying these techniques to PCBs, integrated circuits, photomasks, and hybrid circuits. Performance is an issue when inspecting PCBs for defects. Some defects such as faulty connections, are easily detected visually, whereas other faults cannot be diagnosed from strictly image information. To pinpoint a faulty component and replace it, electronic technicians are required. This process can be automated to increase performance and reduce cost. This automation is not the simple visual inspection, but the automation of the entire visual reverse engineering process.

Visual reverse engineering is the process of analyzing a manufacturers item (PCB, chips, etc.) to determine its component parts, the connection between them, and by combining this information the functionality of the system [9,10,13,20]. The visual inspection processes can be divided into two stages: visual detection of defects, and generation of the functional behavior of the circuit. Visual reverse engineering of circuits has not been fully explored in the literature, although some publications for PCBs exist [9-11].

PCBs are inspected extensively before the insertion of components and the soldering process to isolate defects such as shorts, opens, over-etching, under-etching, mouse bites, pad size violations, and spurious metals. A huge variety of PCB inspection algorithms have been proposed [1-11], which fall into two approaches: reference comparison where the scanned image is compared to that of an ideal board to identify defects, and design-rule verification where the inspection is simplified to dimensional verification [1]. Existing techniques for fault diagnosis include standard image processing techniques such as morphological operators [4], shape extraction [5], primitive feature comparison [6], up to extraction of a high-level representation of circuit features [7].

Most of the activities to date have been in the inspection of bare PCBs. Practically nothing has been done toward system integration of various PCB manufacturing processes. The visual inspection process alone, without the development and acquisition of domain knowledge, has severe inherent limitations. Visual comparison of images can only take the process so far. Without retaining information regarding previously inspected images and the correlation and interrelation of defects, and without background knowledge on functionality of primitive units and the ability to assess system behavior, the process of reverse engineering: modeling, inference, and hypothesis, is impossible. The benefits of adding contextual information, functional modeling capability, and knowledge acquisition and manipulation capability in the inspection process will be to allow the manufacturing inspection system to pinpoint the source of the problem, that is, the electromechanical part or sub-process that produced the defect [9,10].

2. THE DETECTION DIAGNOSIS METHODOLOGY

We consider that a circuit can be represented by a directed graph with attributes, where nodes represent modules of the circuit and arcs represent connections and relationships among the modules. The generation of the graph, however, requires the visual extraction of the circuit's structural description. Thus, an automated visual reverse engineering process is needed to do it.

2.1 Visual Reverse Engineering

Definition: Vision reverse engineering is a process which extracts information from a given system via visual processes, and describes it as structural and functional features by using an associated knowledge-base.

The visual inspection of a circuit, using a vision camera, will assist significantly on the development of the circuit's functional simulation. In particular, it will visually inspect the circuit for possible defects, such as open circuit, bridges, etc. if any, and inform the user with appropriate messages [11]. In case that no defects will be detected, the visual reverse engineering begins. More specifically, the vision camera scans the entire circuit area by following a recursive pattern (Zeta, or Peano) which retains the region neighborhood property [12], see figure 1. At the ith scanning step of the recursive scanning process, the camera focus on the ith subregion Wi and processes it as an independent image. Thus, a set of image processing and recognition tasks are applied on Wi for the extraction of the functional behavior of the ith potion of the circuit. In particular, a smoothing process is applied to remove noise from the digital image-region Wi due to illumination. An edge detection and segmentation process is performed to define the boundaries and colored sub-areas of Wi. Features extraction and recognition processes are used to identify the circuit's functional components and their connectivity included in Wi. At this point, an attributed graph Gi is developed, by using each component's functional status (i.e. true table) and its connections with other components, or I/O with other subregions. The functional behavior of each component is provided from a Database, which holds the functional behavior of several circuits' components. Thus, the extracted circuit's components represent the nodes of the graph and the connection lines represent the arcs. The functional behavior of this particular portion of the circuit is described by the use of a predicate calculus approach. At this point, the same process is repeated on the next region Wi+1 scanned by the recursive algorithmic pattern. The graph, Gi+1, extracted from the sub-region Wi+1 is synthesized with Gi and a new graph Gi,i+1 is produced [13]. The synthesis of graphs is continued by following the recursive algorithmic scanning pattern until the entire circuit area to be covered. Thus, this reverse process will automatically produce the circuit's simulation required by the next step.

The vision reverse engineering system can further be divided into a number of processes related to image processing, pattern recognition, image understanding, and knowledge representation. Generally the input of this system is digital image which was acquire using a digital video camera and image grabbing device. The digital image is stored in the memory for further processing. Filtering is often performed on the image in order to remove any undesirable noise. Image segmentation, edge detection, line normalization processes are applied in a pipelined manner to define (isolate) several important part of the system. At this point a feature extraction process is used to isolate and represent the system parts in a coded form. This sequence of processes is followed by pattern recognition, in order to determine the type of these features, such as straight lines, curved lines, angles, etc, and their characteristics. Thus, attributed graphs are formed from these features. Initially the sizes of these graphs are small, and they represent a region with the same color or light intensity. In the next step, these graphs are correlated, producing new graphs which may represent larger areas. These graphs are compared with the contents of the knowledge base. Each recognized graph corresponds to a specific system's component. Additional correlation processes are performed again over all of the un-recognized graphs, in an effort to generate larger (higher) recognizable graphs. This processes are followed by

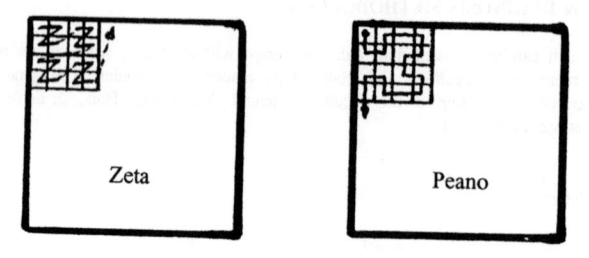

Figure 1

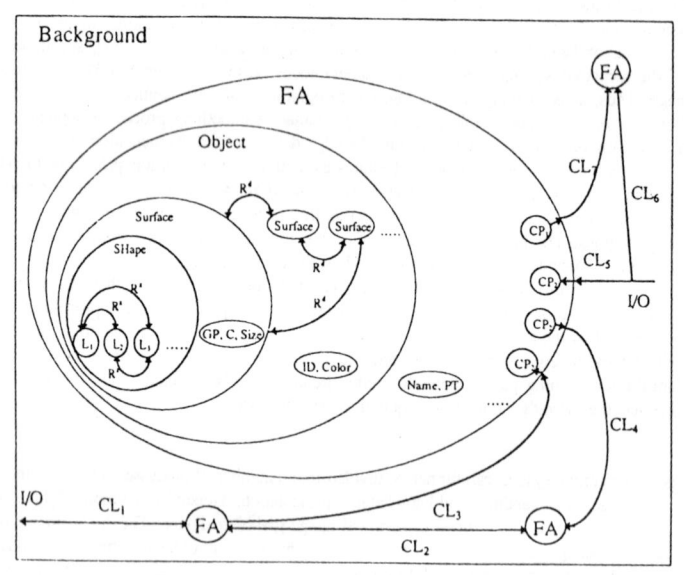

Figure 2 : The highest Graph Level of a PCB

object recognition again. The sequence of larger graph generation (correlation) and object recognition are performed repeatedly until almost none of the graph left unrecognized. This hierarchical abstraction and representation of the circuit into graph form is shown in figure 2.

When a graph matching (object recognition) process is successful, the functional behavior of the corresponding component of the system is extracted, and represented by a stochastic petri-net (SPN) model, as shown in the following section. Figure 3. shows the global architecture of a visual reverse engineering system. The input of the system comes from two ends, the user end and the image end. In the image end, images that has been captured by the camera and undergone image processing, are transformed into graph form. These graphs are fed to either graph database (GDB) or the graph matching unit, or both, depending on the task that the whole system carries on. The processed images are also send to be stored in the image database (IMGDB). From the user end, the query produced by the user will be passed to the query analyzer and transformed into a graph form.

The graph matching processor is the heart of this system. One of its function is to recognize graph (by comparing it with the graph database), access the appropriate functional database, and send them both to the stochastic petri-net generator (SPNG). SPNG in turn will combine the functional and structural (graph) model into stochastic petri-net (SPN) version of reverse engineering model (RE-M).

The natural language processing of query into the graph form and petri-net form is shown in figure 4. Any user's query is analyzed by a query analyzer unit with its own inference engine and knowledge database to generate the graph version of it. This graph version of the query is then sent to the graph matching processor to be recognized. Should it be necessary the recognized query graph can be transform into the SPN version, by the adding its functional description to it. In particular, the use of SPN modeling is necessary to maintain the structural and functional characteristics of the circuit.

3. CONVERGING FUNCTIONAL & STRUCTURAL KNOWLEDGE WITH SPNs

A single scheme that can model both structural or descriptive knowledge as well as functional or direct-model knowledge will be necessary in a system that is to model state as well as properties. Traditional knowledge base methodologies, which can be generally represented by predicate calculus and inference, is suitable for modeling systems where the inference will be a successive application of rules to a database, adding to the database as inference is performed [14]. However, when modeling a system with a changing state, a knowledge tool capable of modeling state, synchronization, parallelism, concurrency, and time is necessary. Petri nets are an excellent tool for system modeling, and it is also possible to implement the function of traditional knowledge bases using a Petri net knowledge representation methodology.

Petri net techniques for modeling and analysis of systems have been well developed [15]. The use of Petri nets for representation of traditional knowledge forms has also been investigated [16,17]. The general approach of a converged structural and functional knowledge representation methodology supports reasoning using knowledge that changes according to the state of some internal model. This model changes either through the passage of time (e.g. the evolution of the model), through the addition of new facts that cause the model to change, or through internal implications made through structural (traditional) inference. Some investigation of a knowledge representation system using Petri nets consistent with these properties can be found in [18].

A system for converged knowledge representation will be necessary where reasoning about structural properties is required for the functional model. For example, a description of a digital circuit is both dependent on models of components for a model of the overall system. However, the structural properties are, in an actual circuit, inseparable from this description. The connectivity, physical distance, functional groupings (e.g. integrated circuits in a larger design), are critical for defining the physical implementation. This convergence is required for defining both the functionality of the system as well as the physical properties. Once this convergence is in place, the circuit is truly defined. The implication of physical properties, such as faults or distance, to the function can be investigated, and vice-versa.

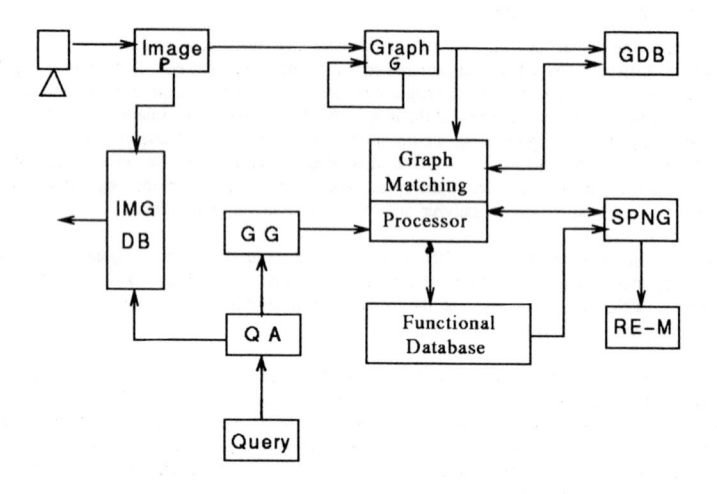

Figure 3

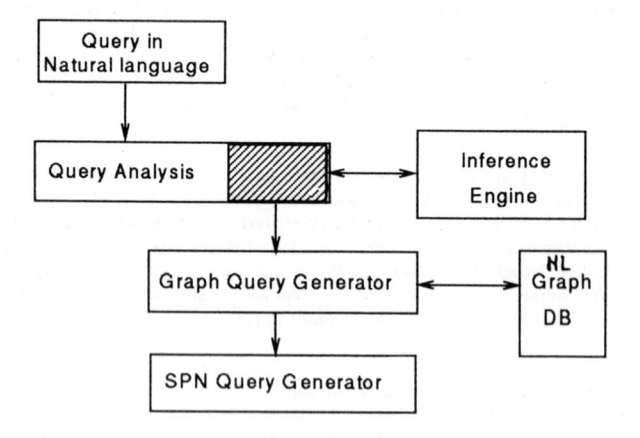

Figure 4

3.1. Definitions and Primitive for SPN Knowledge Representation

Knowledge representation using Petri nets is based on the following correspondences:
1) Knowledge atoms ==> places
2) Rules ==> transitions
3) Instantiation ==> Tokens

There are several problems with this simple representation scheme. To avoid these problems, which will be discussed briefly below, we have developed the knowledge representation primitive shown in Fig.5. Since this net is a primitive, knowledge representation can be described with the shorthand as shown in Fig.6.

This primitive has several features desired for representation of structural knowledge: representation of negative instantiation, detection of contradiction, and avoiding redundant inference. One of the significant features is the capability of representing both positive and negative instantiation of a concept. By this feature, and the corresponding automatic detection of contradiction, both detection of database inconsistency and the predicate calculus technique of including the negative of a clause and detecting subsequent contradiction for proving a clause are enabled without needing development of specific mechanisms for this purpose. The problem of redundant inference, which is coupled with the problem of persistence of knowledge, is addressed by the interconnection mechanism. The problem in the simplistic Petri net representation of one place per concept is that once an instantiation is used for inference, it is consumed by the transition and is lost. On the other hand, if that instantiation is allowed to multiply without bound, without external intervention every inference will be made multiple times. The primitive shown in Figure 5 distributes one copy of the instantiation to a place corresponding to each transition that can use that knowledge for inference. That way, it enables every possible inference, but for only one firing of a particular rule.

3.2. Mapping Graphs onto SPNs for Knowledge Processing

Knowledge can be represented in graph form where nodes in the graph correspond to objects in the world, and the edges of the graph are the relations between these objects. It is important to transfer this knowledge form into the SPN, where the SPN will model also the functional behavior of the system.

The graph knowledge form is defined by:
N_i = the set of nodes in the graph
R_{ij} = the set of relations between the nodes.
The graph form is then characterized by clauses of the form:

$$N_i\text{-}R_{ij}\text{-}N_j\text{-}R_{jk}\text{-}N_k\text{-}R_{ki}\text{-}N_i$$

The mapping onto the SPN is as follows:
$N_i = \{P_{xi}\}$ Graph nodes are mapped into a set of transitions, each representing either features of that node/object, and/or possible functional states.
$R_{ij} = \{t_{ij}\}$ Relations are represented by a set of transitions connecting the places of the node.

In order the structural and functional characteristics of a circuit to be maintained, a generic SPN module is used, as shown in figure 7. More specifically, if a structural query is issued the SPN module activates the structural part of SPN, if however, a functional query is issued then the functional part of SPN is activated. In case that both structural and functional part are requested, then an extra token is needed to activate them simultaneously.

4. AN ILLUSTRATIVE EXAMPLE

In this section, we provide a detailed illustrative example in order to show the visibility of the methodology. The selected example is a digital circuit from which a small portion is extracted by a Zeta scanning, as shown in figure 8. Table 1 shows the primitive graph elements used for the generation of the attributed graph, figure 9. Table 2

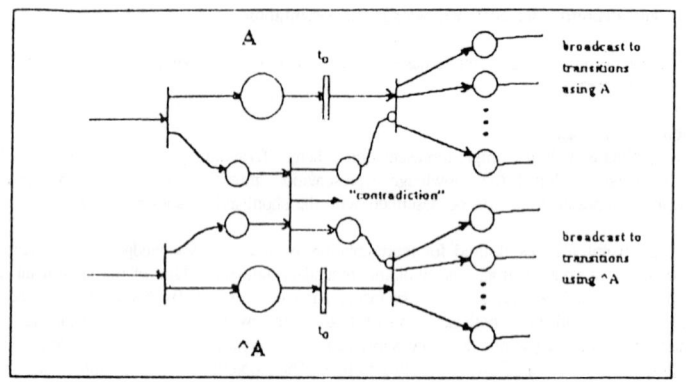

Figure 5.: Details of Petri net representation primitive

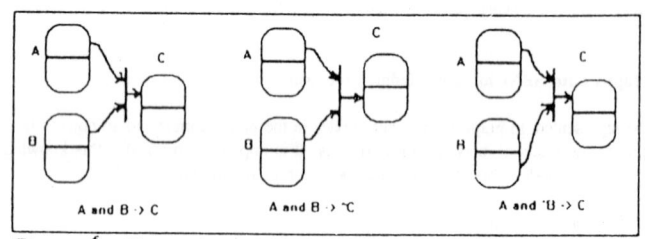

Figure 6: Examples of representing production rules with Petri Net primitives

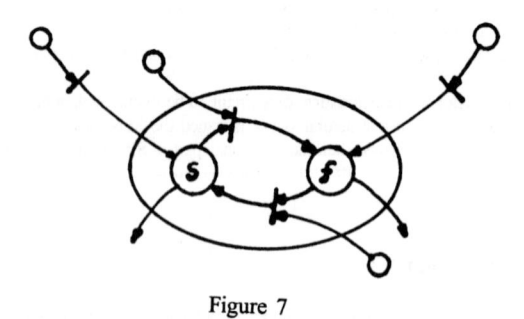

Figure 7

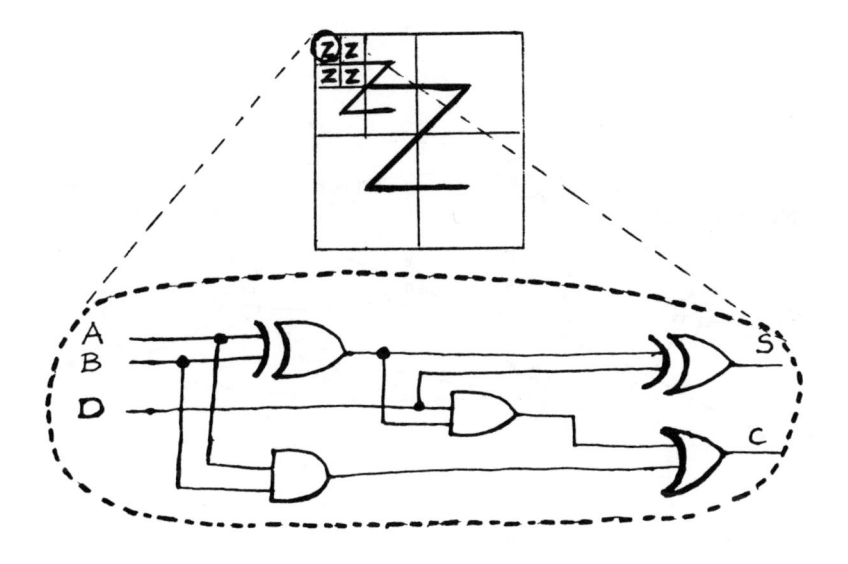

Figure 8: A digital circuit

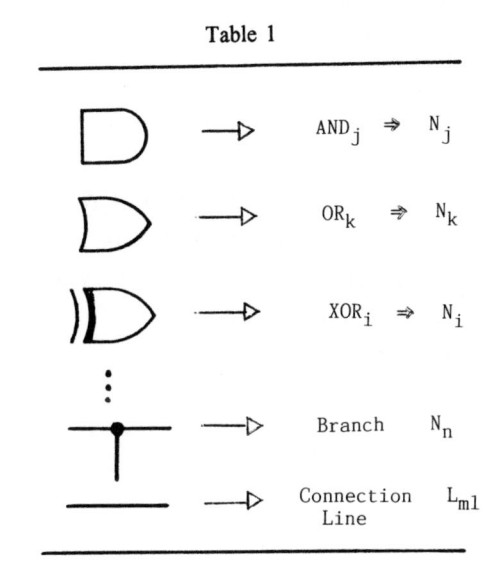

Table 1

AND symbol	→	AND_j ⇒	N_j
OR symbol	→	OR_k ⇒	N_k
XOR symbol	→	XOR_i ⇒	N_i
Branch symbol	→	Branch	N_n
Line	→	Connection Line	L_{ml}

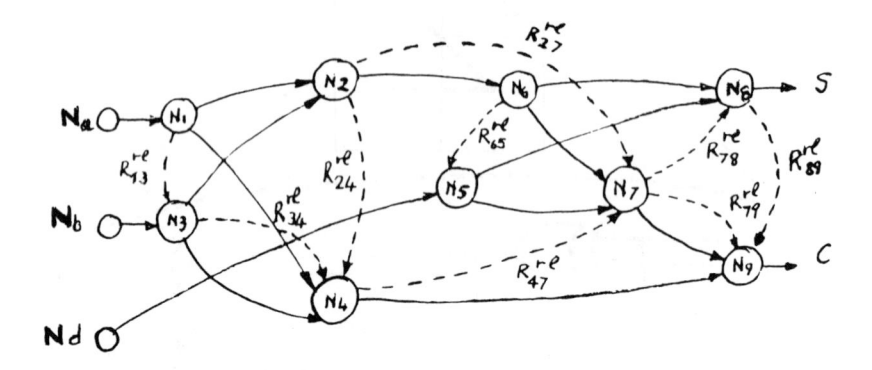

N1 represents a *branch b_1* node
N2 represents an *XOR_1* node
N3 represents a *branch b_2* node
N4 represents an *AND_1* node
N5 represents a *branch b_3* node
N6 represents a *branch b_4* node
N7 represents an *AND_2* node
N8 represents an *XOR_2* node
N9 represents an *OR* node

$$G = \{N_a R_{a1}^c N_1 R_{12}^c N_6 R_{68}^c N_8\} \# \{N_1 R_{14}^c N_4 R_{49}^c N_9\} \# \{N_b R_{b3} N_3 R_{32}^c N_2\} \# \{N_3 R_{34}^c N_4\} \#$$
$$\{N_d R_{d5}^c N_5 R_{58}^c N_8\} \# \{N_6 R_{67}^c N_7 R_{79}^c N_9\} \# \{N_5 R_{57}^c N_7\} \#$$
$$\{N_1 R_{13}^{rl} N_3\} \# \{N_3 R_{34}^{rl} N_4\} \# \{N_2 R_{27}^{rl} N_7\} \# \{N_6 R_{65}^{rl} N_5\} \# \{N_2 R_{27}^{rl} N_7\} \# \{N_4 R_{47}^{rl} N_7\} \#$$
$$\{N_7 R_{78}^{rl} N_8\} \# \{N_7 R_{79}^{rl} N_9\} \# \{N_8 R_{89}^{rl} N_9\}$$

 # represents the systhesis operator

Figure 9

Table 2

$$R_{12}^{c} = connection[(b_1(O_1), XOR_1(I_1)), L_{12}]$$
$$R_{26}^{c} = connection[(XOR_1(O), b_2(I)), L_{26}]$$
$$R_{68}^{c} = connection[(b_2(O_1), XOR_2(I_1)), L_{68}]$$
$$R_{14}^{c} = connection[(b_1(O_2), AND_1(I_1)), L_{14}]$$
$$\vdots$$
$$R_{57}^{c} = connection[(b_4(O_2), AND_2(I_2)), L_{57}]$$

$$R_{13}^{rl} = relative\ location[N_1\ above\ N_3]$$
$$\vdots$$
$$R_{89}^{rl} = relative\ location[N_8\ above\ N_9]$$

$\{L_{12} R_H^P L_{32}\}$ P = Parallel
$\{L_{14} R_{H,V}^P L_{34}\}$ H = Horizontal
    $\vdots$ V = Vertical
$\{L_{49} R_H^P L_{79}\}$

$$L_{ij} = \{GP, length, orientation, curvature\}$$

If A=1 token in A
If A=0 token in $\overline{A}$

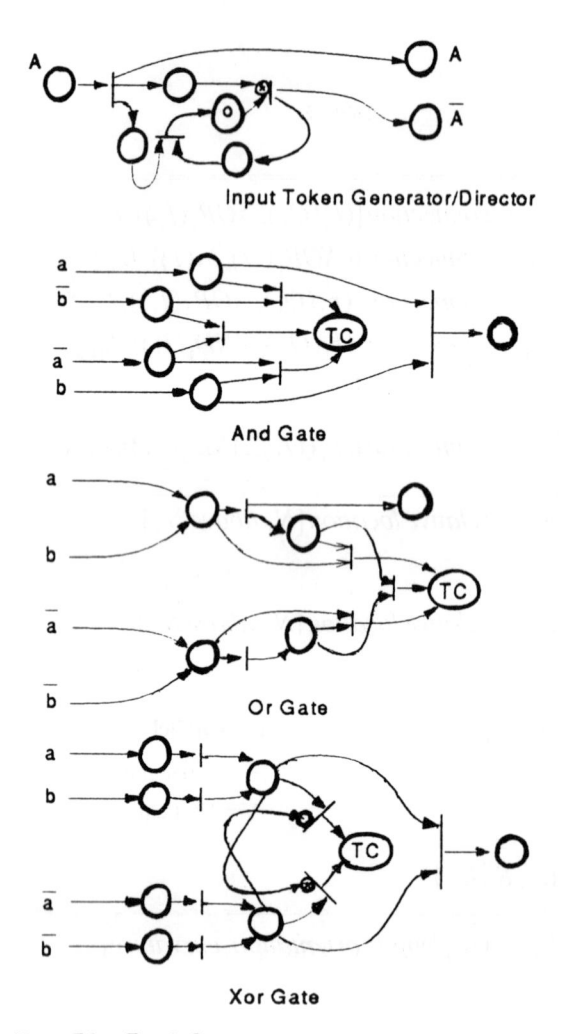

Input Token Generator/Director

And Gate

Or Gate

Xor Gate

Note: TC = Trash Can

Petri-Net representation of the nodes

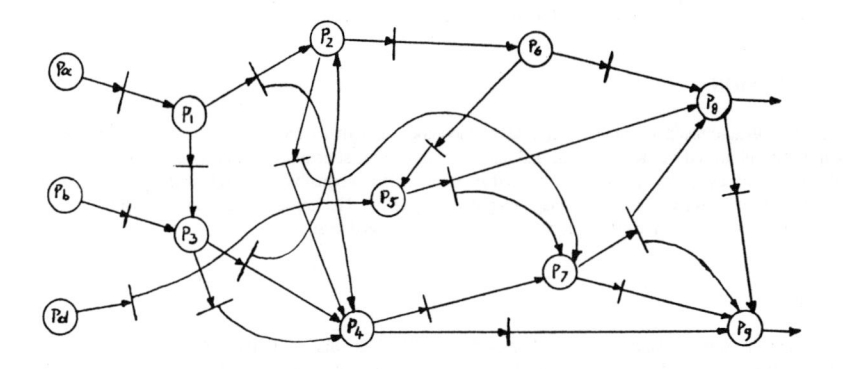

Figure 10: Structural SPN model with color tokens

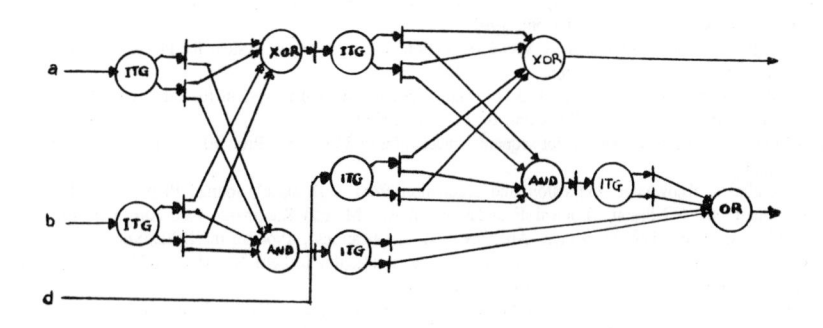

Figure 11: Functional SPN model

includes the relationships among the circuit components. At this point the graph is converted into an SPN module, which includes both the structural and functional SPN models of the circuit.

ACKNOWLEDGEMENT
The authors wish to express many thanks to Professor N.Bourbakis for his guidence and fruitful suggestions in the areas of visual reverse engineering and automated functional simulation.

5. CONCLUSION

In this work an automated functional diagnosis methodology of digital circuits was presented. The methodology was based on the automated visual extraction and representation of the structural and functional characteristics of a given digital circuit by using the appropriate knowledge bases. This a promisible approach with application to real PCBs under the consideration of some extra image preprocessing effort. The impact of this approach will be very important on the electronic circuits repairing process by reducing time and cost.

References

1. R.T.Chin, A survey: Automated visual inspection 1981-87 Journal CVGIP,41,1988
2. Proc. SPIE Conf. on Integrated Circuit Methodology Inspection and Process Control, Vol.775,1987
3. Proc. SPIE Conf. on Machine Vision Applications for IC Inspection, Vol. 2423, 1995
4. O.Z.Ye and P.E.Danielson,"Inspection of printed circuit boards by conductivity preserving shrinking",IEEE Trans PAMI, 10,737-742,1988
5. J.R.Mandeville,"Novel method for analysis of printed circuit images,IBM Journal R&D Dec.1985,73-86
6. G.A.W.West,"A system for automatic visual inspection of bare printed circuits boards,IEEE Trans. SMC,14,1984
7. A.M.Darwish and A.K.Jain,"A rule-based approach for visual pattern inspection,IEEE T-PAMI,10,1988
8. C.S.Fahn et.al."A topology based component extractor for understanding electronic circuit diagrams, Journal CVGIP, 44,1988,119-138.
9. N.Bourbakis and CV Ramamoorthy, Specifications for the development of an expert system for automated visual reverse engineering, IEEE Test Conf. 1991
10. A.Mog/zadeh and N.Bourbakis,"A visual inspection diagnosis expert system for PCBs: reverse engineering, IEEE Conf. on TAI 1993, Boston, MA 396-403.
11. A. Mog/zadeh and N.Bourbakis,"A 3-D visual inspection diagnosis system for damaged VLSI boards: VLSI reverse engineering", Proc. SPIE Conf. on EI, Feb. 1993,CA
12. N.Bourbakis, SCAN a language for sequential accessing of 2-D arrays",Proc.IEEE Workshop on LFA, Aug.1986, Singupore
13. N.Bourbakis,"Knowledge extraction and acquisition from 2-D space", Journal PRAI,vol.9,1,1995
14. M.Genesereth, N.Nilsson, "Logical Foundations of AI", Morgan Kaufmann, Palo Alto, CA, 1985.
15. T. Murata, "Petri Nets: Properties, Analysis, and Applications", Proc. of the IEEE, vol.77, no.4, 1989.
16. G.Peterka, T.Murata, "Proof Procedure and Answer Extraction in Petri Net Model of Logic Programs", IEEE Trans. on Software Engineering, vol.15, no.2, 1989.
17. C.Lin, A.Chaudhury, A.Whinston, D. Marinescu,"Logical Inference of Horn Clauses in Petri Net Models", IEEE Trans. on Knowledge and Data Engineering,vol.5, no.3,June 1993.
18. J.Gattiker and N.Bourbakis, "Representing Structural and Functional Knowledge Using SPN", IEEE Int'l Conf. on Software Engr. and Knowledge Engr., Maryland, June 1995.
19. A.Gal, O.Etzion, " Maintaining Data-Driven Rules in Databases", IEEE Computer, January 1995.
20. J.Gattiker, S.Mertoguno, A.Mogzadeh and N.Bourbakis,"Visual reverse engineering using SPNs for automatic testing and diagnosis of digital circuits", Proc. IEEE Conf. on ATC-95, Atlanta,GA, Aug.1995

THE IMPACT OF AI IN VLSI DESIGN AUTOMATION

MOHAMMAD MORTAZAVI & NIKOLAOS BOURBAKIS
AAAI Laboratory
Department of Electrical Engineering
State University of New York
Binghamton, NY 13902-6000

ABSTRACT

This chapter discusses several method based on of knowledge expert systems with application to automation of Very Large Scale Integrated Circuit Systems (VLSI) design environment. The impressive progress on VLSI technology and the shifting from design of VLSI chips to Ultra Large Scale Integrated ULSI circuit systems made the use of design automation to a necessity. The developments of computer-aided design (CAD) programs (tool) helped significantly to the automation of VLSI design. Each tool were able to solve different stages of task design more effectively. However, by integrating and putting these tools together into one package, the effectiveness and functionality of CAD programs decreased drastically. To overcome the problems at different design stage, the researchers introduced artificial intelligent (AI) techniques in VLSI design automation. AI techniques such as knowledge-based and expert systems, at first try to define the problem and then choose the best solution from the domain of different possible solutions. Different AI techniques and their performances used for VLSI design automation are discussed in this paper.

Keywords: Artificial Intelligent, VLSI Design Automation, Knowledge-Based Expert Systems, Synthesis Systems, Design Environments, and Evaluation Tools

1. Introduction

Since the invention of transistor in 1950's, the development of Integrated Circuit (IC) technology has focused on integration of more complex transistors into one chip. During the last three decades, fabrication of IC technology has gone through four generations: Small Scale Integration (SSI); Medium Scale Integration (MSI); Large Scale Integration (LSI); and Very Large Scale Integration (VLSI) and yet furthering the IC technology to new era of Ultra Large Scale Integration (ULSI). However, the transition from tens and hundreds (SSI) of transistors to thousands and higher (MSI) can not be accomplished manually. This is mainly due to the time and money constraints. As the IC complexity increases the needs for CAD tools would become volatile. In particular, with the existence of current technology containing of millions of transistors (VLSI), necessitated the designers to be heavily dependent on design automation process. CAD software provide solution different stages of the VLSI design such as logic synthesis, layout synthesis, chip behavior simulation, circuit simulation, optimization, and so forth. The purpose of using CAD system is to help designer perform more complex task designs as well as to arrange large design data appropriately. These tools also help to check and verify the validity of a design. Thus, an efficient CAD system must have the following components [Ryc88, Pre88, Bou88, and Bus88]:

. *Design Automation (DA)*. The resulted output from these tools produce automatically different solutions. The structure of these tools consists of the following levels (more detailed explanation of DA will be discussed later):
(i) <u>Synthesis</u> The synthesis systems produce new design data from a higher level specification. In general, synthesis systems are divided into three categories:

    (a) <u>High-Level synthesis</u> converts high level , the systems specification and behavioral aspects of a circuit which is described as programs into structural (topological) designs. The topological designs are in form of interconnected set of RTL (Register Transferred Level) components such as multiplexer, ALU, and registers.

    (b) <u>Logic-Level synthesis</u> converts a structural design into optimized combinational logic and maps that logic onto IC library of cells.

    (c) <u>Layout-Level synthesis</u> converts the IC cells into the physical layout.
(ii) <u>Analysis</u> level checks the validity of the design.

(iii) *Optimization* level improves the final layout and the overall performance of the chip.

.Interaction. It provides to the user the ability to communicate with the computer more efficiently. This mandates the development of an interface which can allow the designer to interact with the computer during design cycle.

. Database. The computer used for the CAD system must rapidly access to large amounts of design data. This requires to create and maintain a reliable database.

The first generation of CAD tools in the field of design technology for electronic systems were mainly modeled by numerical analysis formulation and generating algorithmic solutions for those problems. These solutions were coded in the form of procedural languages such as BASIC, FORTRAN, or PASCAL. The conventional procedural programming approach were successfully applied to many physical design tasks. However, as the circuit design became more complex, the time and cost of achieving the acceptable results became very crucial.

Advancement in the VLSI technology lead into new challenges for development of CAD tools. One of the basic challenges in design automation was to solve the problems of unmanageable computational tasks, which had grown too large for previously used algorithmic solutions. The complexity of a task depends on many details and degrees of freedom. Hence, performing an exhaustive search for an algorithmic solution lead to combinatorial explosion. The second challenge lied on the problems with "ill-defined" tasks, in which their foundations were not based on any kind of mathematical theory and difficult to be formulated into an algorithmic solution. Some of these sub-problems are too complex so it is almost impossible to generate an algorithmic solution [Ryc88].

In result, CAD tool developers started to look for new techniques capable to solve more complicated design tasks. Some of the techniques lied within artificial intelligence (AI) area which made it possible to address some of the shortcomings of conventional (procedural) programming. Section one briefly explains the VLSI design automation process. Then section two looks at the different techniques from AI environment such as expert systems, knowledge-based systems, heuristic algorithms, and learning algorithms used in VLSI design process. Section three will discuss how AI tools can be applied to VLSI DA. Finally some concluding remarks about AI tools for VLSI DA will be described.

1.1 VLSI Design Automation

Due to its nature the VLSI layout design process, is naturally hard to be automated. There are, however, various levels of abstraction of designing a VLSI chip.

By giving an abstract specification of an object(chip) to be designed, a design automation system generates the physical design automatically and verifies that the design agrees with the specification. Design automation tools are defined as those computer-based tools that assist through automation of procedures and would not be performed manually. These tools enable designs which are too large or complex to deal with, reduce time of design process, improve quality (performance and reliability), and reduce costs. Design automation tools can be generalized as a class, known as CAD (Compute-aided design) tools.

CAD systems provide tools to synthesize, analyze and verify portions of the design but require active participation of the engineers. An engineer guides the tools at various stages of design until the completion. The engineer must provide the degree of creativity (the fitness function), and the CAD system provides different solutions for tedious problems [Bus88].

1.2 Top-down Design Methodology For DA Process

A typical VLSI design process can be shown by a flow chart in the form of top-down design fashion [She93]. It consists of the following phases (see Figure 1):

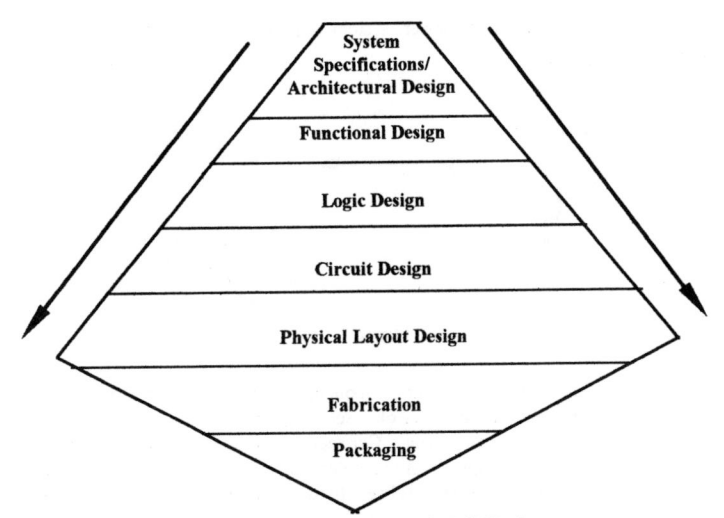

Figure 1. Various Levels of VLSI Design

1.2.1. System Specification- At this stage, the engineer must know the system specification of the design for creating different levels of representations of the system. The designer must consider factors such as performance (bandwidth, bus widths, external signal wave forms), functionality, size, speed, and power of the VLSI system.

1.2.2. Functional Behavior- The designer considering the behavioral conditions of the system such as timing diagram and relationships of units among sub-units. This can help designer to reduce the circuit complexity of the prior phases as well as improving the overall design process (see Figure 2).

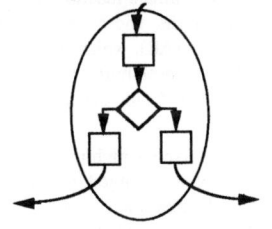

Figure 2

1.2.3. Logic- In this step, the functional behavior can be translated into a logic structure in the form of a textual, schematic or graphic description. Logic structures usually can be represented in the form of Boolean expressions. These expressions can be minimized to achieve an optimized logic design which corresponds to the functional design.

$$Z1 = \overline{A} * (B \oplus C)$$

1.2.4. Circuit- At this stage, the logic structure is represented in the form of circuit diagrams. Boolean expressions are transformed into a circuit representation (see Figure 3). At this stage the parameters which must take into account are speed, power, system requirements, and electrical behavior of each component of the circuit.

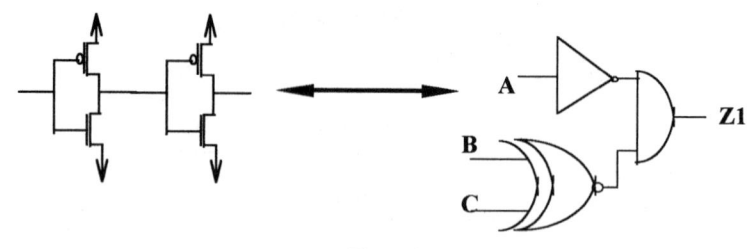

Figure 3

1.2.5. Physical Layout In this step, each circuit component is converted into a geometric representation called as a layout. This representation consists of a set of geometric patterns which functions as the intended logic for the corresponding component. Connections between components are also shown as geometric patterns. Physical layout design is very complex process and the most time consuming step in the VLSI design process. Therefore this stage has been divided into several sub-steps (see Figure 4):

(i) Partitioning - This task partitions the layout into smaller parts. The main goal is to partition the circuit layout into parts such that the size of the components is within prescribed ranges, and the number of connections between the components is minimized. The output partitioning is a set of blocks along with the interconnections required between blocks. These interconnections are referred to as a netlist. Different partitionings correspond to different circuit implementation. Therefore, a good partitioning can improve significantly the circuit performance and reduce layout costs.

(ii) Floorplanning - This step is required with selections of good layout alternatives for each block. Floorplanning is the placement of flexible blocks, which have fixed areas but unknown dimensions. The floorplanning problem is to determine the approximate location of each block (module) in a rectangular area. The blocks are rectangular and their lengths, widths, and location of each block are determined. Floorplanning usually deals with the aspect ratio (AR), which is the ratio of width to its length. There is usually an upper and lower bound on the AR. Floorplanning techniques are typically used in hierarchical manners. An important step in floorplanning is to decide the relative location of each module.

(iii) Placement - In placement, each module is fixed, which means it has fixed shape and terminals. The goal of placement is to find an optimum area arranged for the blocks that allows completion of interconnections among the blocks. Usually the placement is done in two phases. At the first phase, an initial placement is constructed. At the second phase, the initial placement is evaluated by alternative cost functions and iteratively improved until the minimum area is achieved and corresponds to design specifications. The only difference between floorplanning and general block placement is the freedom of cell's interface characteristics.

(iv) Routing - At this stage, routing is divided into two levels of routings: global routing and detailed routing. The purpose of global routing is to decompose a large routing problem into small, manageable (detailed routing). This decomposition is accomplished by finding a rough path for each net in order to reduce the chip size, shorter wire length, and evenly distributed the congestion over routing area. On the other hand, the detailed routing is carried out by a two-layer Manhattan model [Had77] with reserved layer, where horizontal wires are routed in one layer and vertical wires are routed in the other layer. To connect a horizontal and vertical segment, a contact (via) must be placed at the interconnection point.

(v) Compaction - In this stage a layout is optimized by minimizing the number of vias and compacting the area. It is the task of confining the layout in all directions in which the total area is reduced. If the area is compressed, the wire lengths is shortened. Therefore the signal delay and critical path between each component of the circuit may be shortened. It is also important that by making smaller area occupied by a IC or transistor, more ICs can be placed on a wafer which results in reduction of the cost of IC manufacturing (See Figure 4).

Figure 4

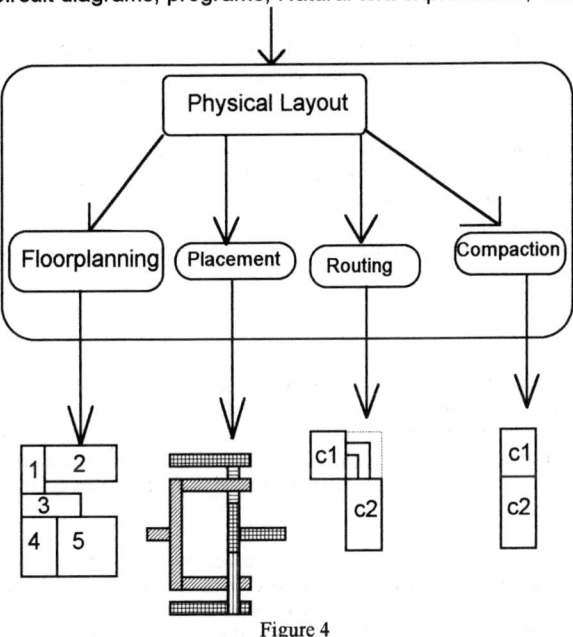

1.2.6. Design Verification- The layout is verified and tested to satisfy design and layout rules and ensuring that the layout meets the system specification and fabrication requirements and requiring two steps:
. *Design rule checking-* Verifies all the geometric patterns meeting the design rules imposed by the fabrication process
. *Circuit extraction-* taking place after checking the layout for design rule violations and removing them.

1.2.7. Fabrication Line- Making and testing a prototype before the mass production and consisting of several steps (see Figure 5):
. preparation of wafer
. deposition
. diffusion of various materials on the wafer

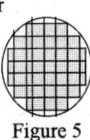

Figure 5

1.2.8. Packaging- Testing the packaged chip after the wafer being fabricated and diced to ensuring the system specifications met and functioned properly .

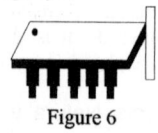

Figure 6

It is also important that during the design automation process, different kinds of errors are being generated. The design must be reconfigured and modified to overcome these problems. Some of these errors are:

. *Electrical Parasitic* may occur when an electrical schematic converted it into a layout.
. *Propagation Delay* for bus signals may not be acceptable, due to choosing excessive length of the buses. These excessive bus lengths will generate a capacitive and resistive loads in which makes the propagation delay varied from specified design
. *Radial defect* distribution on VLSI wafers can occur on chips near the outer edge, which are more defective than chips in the interior of the wafer.

1.3. Classification Of AI Tools

As it was mentioned in the previous section, it is essential to integrate all of the CAD tools levels into one design environment. However, prior design environments showed that the performance of these tools, when functioned under one integrated CAD tool, will become very inefficient to generate a maskable layout. Therefore, the use of AI techniques helps to improve the CAD tools performance effectively. In order to integrate a set of CAD tools into one design automation system, there are three major problems which must be considered [Ade90]:

(i) The *control problem* involves in determining which portion of the design should work on next, and which CAD tool must be applied to that portion. To overcome this problem, a design scheme must be introduced. The control strategy appoints to the environment which CAD tool is appropriate for further elaboration of the current design. This selection is done by the scheduler, which interprets descriptions of the design tasks and control strategy. The scheduler requires an inferencing mechanism. This is because of choosing which CAD tool to be used next. There are two kinds of inferencing mechanism available:

(a) **Logical theorem proving** which is provided by *resolution theorem provers (RTP)*. This mechanism has been used since early 60's. The major program which its mechanism is based on RTP, is called PROLOG. This RTP provides logical reasoning facilities using a data base of facts. Both the facts and inferences are described in a language that resembles the predicate calculus. Predicates can be applied to objects to express true or false relationships among the objects., however, in a very large database, the VLSI design becomes very complex. The inference mechanism will create unnecessary rules which may not be used at all.

(a) **Rules Matching and Firing** which are provided by a new concept called *Knowledge-Based Expert Systems (KBES)*. KBES is an AI programming system in which rules, known as productions, match against data in a database. There will be more explanation on these two elements individually and integrated tool together in a design environment.

(ii) The Design Detail Problem is involved in controlling, storing, and manipulating a very large amount of information that describes a VLSI design space. There are many levels of design representation, as well as many competing designs. Thus, this design detail must be stored efficiently and must be represented in a proper relationships among the CAD tool data files.

For each level of competing design in the design space, its origin and the history of major decisions that led to the design must be represented. The design space emanates from an initial design specification written by the designer in the design environment hardware description language (HDL).
Each design is classified in a form of the following decisions[Ade90]:

(a) Specification An original or an altered specification for the entire design.
(b) Temporary Decision--A decision point in the design process.
(c) Permanent Decision A temporary decision that is now permanent.
(d) Choice A design point representing one of the options for a temporary or permanent decision and stored as a child of the decision.
(e) Elaboration- An elaboration of the parent design with more design data synthesis.
(f) Sub-goal -A design point whose children represent several sub-problem that must be solved to complete the design.

(iii) The Programming Abstraction Problem concerns with the difficulty of describing the tasks performed by human designers into a CAD system. Conventional programming languages and environments make it very hard to describe design tasks. This leads of using a better descriptive mechanism, such as LISP.

Study shows the types of knowledge in the human design process can be categorized as the following [Ack88]:

(a) Primitive knowledge: This is the profound knowledge which is realized by engineering and circuit theory. For example, principles of device physics that define the operation of the basic electrical components have to be considered. The models of transistor operations and transmission line theory must be known. Even though, this could principally be used to design a complete circuit, it is seldom utilized directly. This is due to the time constraint required to complete a design process. Therefore, it is usually abstracted at higher levels.

(b) Heuristic Knowledge: This knowledge which is developed to lead the design process at higher levels of abstraction. At the higher level, this knowledge yields either to proceed in larger design steps or to cut the design search space. For example, a CMOS gate can only store 0 or 1 logic level. This is a heuristic knowledge that obtained by applying transistor theory to the circuit configuration. To appropriately apply this knowledge, is up to the engineer's understanding of the problem. Engineers use heuristics to see effects of trade-offs between design parameters.

(c) Procedural Knowledge- Both primitive and heuristic knowledge are used to generate procedures. They indicate how to accomplish a particular design task. Logic minimization algorithm is an example of procedural knowledge.

(d) Design Experience: This knowledge is based on the results of past that provided better results. An experienced designer can used many tricks from previous problem solving experiences to obtain a better result. Previous solutions from different problems will give insight into tackling a new problem with an efficient manner. It is also important, design experience can make a distinction the difference between an acceptable and a better solution.

(e) Control Knowledge: The knowledge mentioned in above are classified as domain knowledge. They describe properties of the design domain and straight forward algorithmic techniques for improving the state of the design. Control knowledge determines how and when to apply the various pieces of domain knowledge. This kind of knowledge is sometimes referred to as problem solving knowledge. Control knowledge can hierarchically decompose the design problem into manageable design spaces and search for an acceptable solution.

It should be noted that other types of knowledge, such as experience, control knowledge, and special-case heuristic can be easily automated. The design knowledge is used in an exploratory fashion, that is according to incomplete plans that guide the design through a large, constrained solution space. Mapping this type of knowledge into algorithmic control structures leads to an over simplified model of the design process that must be extended with many pieces of special case code before the program can execute any useful design. The knowledge gets buried in the control structures of the program, which become increasingly more complicated as new knowledge is added in the form of modified code. In the use of such tools results in limitations of these techniques to a narrow class of problems. They become unable to adapt to new knowledge or situations that do not match their simplified model of the design process.

The limitations of the algorithmic approach can be seen in building a silicon compiler. A silicon compiler is a VLSI synthesis tool for converting functional specification to layout using techniques borrowed from compiler technology. Even though a number of such tools have built efficiency, there are each quite limited in their domain of application and rely heavily on knowledge in the form of pre-defined layout cells or generators.

2. Roles Of AI In VLSI Design Automation

2.1 Expert Systems in Design Automation Process

The field of Artificial Intelligence (AI) concerned with understanding of human intelligent behavior and building an artificial (computing) systems that simulate intelligent behavior in performing some set of tasks. In 1960's, AI research focused mainly on general principles and techniques such as logical reasoning, search heuristic that could be applied to many problems. Although, these techniques were used with some success in a number of abstract areas like theorem proving, they were found to be inefficient for complex real-world tasks. In 70's the focus shifted to developing techniques for capturing, storing and using domain-specific knowledge to guide and minimize the search process. This led to the development of a new class of AI tools known as expert systems.

An expert system can be defined as a computer program that [Wei84]
(i) provides solution to special problems which usually requires human expertise; and
(ii) reaches its solution using a model of computation based on human expert reasoning.

Several programming languages such as Lisp, Prolog, OPS5, and Smalltalk have been used for development of expert system tools in last decade. These languages are used to facilitate in manipulation of symbolic information.

For instance, Lisp features in data representation, dynamic storage allocation (particularly used in link-list structures), flexible variable binding, the processing of programs by other programs, and the computation of recursive functions (such as trees and graphs).

Prolog is used for logic programming that describes objects and their relationship with each other using a set of logic clauses (i.e., IF---->Then statements). Prolog is best known as backward chaining. This process continues searching for the facts (in which case the goal is satisfied) or the search space is exhausted (the goal is not satisfied).

OPS provides a working memory, and an interpreter that selects and fires rules according to a simple priority function. Working memory can be modified when the rules are fired. The solutions are reached as more rules are being fired, This means that the process is moving from known facts towards a conclusion. This process is known as forward chaining.

2.2 Rule-Based Systems

Knowledge-based expert systems are implemented as production or rule-based systems. Rule-based system consists of three components (see Figure 7):

. **Working Memory** used for intermediate results. This is when attempting to solve each individual problem, the system stores in the working memory any new facts which are established, goals to be satisfied, and any intermediary results. This is a global database that contains the objects which describe the current state of the problem. All problem specific data is help in working memory.

. **Knowledge Base** consist of two parts, known as the rule part and the fact part, in which rules and facts are stored, respectively. The knowledge base sometimes is called as rule memory which contains a set of production rules that specify how the objects in the working memory might be modified in order to advance towards a solution. Rules typically consist of two parts: -antecedent (or IF clause), which specifies conditions under which the rule should fire and a **consequent** (or THEN clause), which specifies what action to perform on the working memory.

. *Inference Engine is* used for the part of the expert system. It deals with the selection and execution of rules. This engine provides control by determining which rules are able to fire, selecting one according to some strategy and then executing the consequent of that rule.

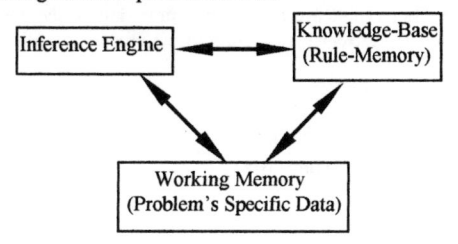

```
        ┌──────────────────┐      ┌──────────────────┐
        │ Inference Engine │◄────►│  Knowledge-Base  │
        │                  │      │  (Rule-Memory)   │
        └──────────────────┘      └──────────────────┘
                   ▲                        ▲
                   │                        │
                   ▼                        ▼
                  ┌──────────────────────────┐
                  │     Working Memory       │
                  │ (Problem's Specific Data)│
                  └──────────────────────────┘
```

Figure 7. Rule-Based System or Production System[Ack88]

At first, the working memory generates description for the problem to be solved. Then the inference engine produces rules one at a time to transform these descriptions. This process continues until either there is no more rules to be fired, or a rule detects a halt condition and informs the inference engine to be stopped.

2.3 Heuristic Algorithms in VLSI DA

CAD tools are classified into two types. The first type is to assist human designers to manipulate a layout. For example, a layout editor allows the designer to manipulate a layout and may perform some functions such as routing and design rule checking. The second type of CAD tools is designed to perform some task on the layout automatically. The major accomplishment has been focused on partitioning the entire problem into smaller problems. These are still computationally hard and proven to be non-deterministic polynomial time complete (NP-Complete) or NP-hard. Therefore, the best available algorithms which assures optimal solution requires more number crunching and CPU times. As the circuit complexity increases the CPU times grows exponentially.

This resulted the researchers focused on development on design and analysis of heuristic algorithms for partitioning, placement, routing and compaction. Many of these algorithms are based on computational geometry and graph theory.

Heuristic algorithms are also used in allocation (data path synthesis) of high-level synthesis systems. Allocation are involved in assignment of operation, variable, and communication path to a piece of hardware. Heuristic algorithms select one element of operation or variable at a time to allocate and assign to hardware. The selection of allocation is based on different criteria. Heuristics algorithms are fast and produce desirable results [Wal91].

2.4 Learning Algorithms in VLSI DA

Learning has not been used properly to CAD tool. In order to use different IC library, it is essential to redesign higher level components and take advantage of the new library. For example, if the previous library had only 2-bit adder while the new library has 4-bit adder, the designer must revise the carrry-look-ahead function. This revision can be done through learning algorithms. Learning can also be used in optimization at layout, logic, and system levels [Cam91].

Some Genetic and Neural Network schemes are considered as a learning algorithms. Genetic is a searching scheme which emulates the natural process of evolution as a mean of progressing toward the optimum[You95].

Neural Networks are mostly used for an on-chip learning purposes. An ANN is a network of many very simple processors units, each possibly having a small amount of local memory. Each processing unit is characterized by an activity level, an output value, a set of input connections, a bias value, and a set of output connections. Each of these aspects of the unit is represented mathematically by real numbers.

Hardware support for an on-chip learning greatly increases the silicon area of the neural system and affects the formation of the architecture. Consequently, based on the their particular applications, many researchers found it more efficient to leave the learning procedure to the host and spare the entire chip area for massively integrated and extremely fast and compact processors [Alh95].

3. AI Tools In Design Automation

In this section, some AI tools for CAD VLSI systems which have been proposed and built are described. These tools are classified into three types:

3.1 Synthesis Tool Systems

Synthesis systems generate new design data from a system specification. There are many interesting synthesis tool systems in which few interesting will be mentioned.

R1, one of the earliest design synthesis tool systems which was applied successfully in VLSI CAD systems, was developed as a prototype expert system at Carnegie Mellon University (CMU). This system takes a set of components such as disks, CPUs, floating point, etc. as input and generates diagrams. These diagrams describe what additional components (such as controllers, power supplies, etc.) are needed and how these components can be placed in order. R1 was coded in OPS5. Design orders are controlled by number of state variables called goals. Rules are accordingly arranged into sets of different goals and sub-goals.

DAA (Design Automation Assistant) is another synthesis design program in which used in an automating a data path [Kow88]. Development of DAA was based on a rule-based expert system approach. The input of DAA accepts functional description of system specifications and applies dataflow and hardware allocation for providing a technology independent hardware description that realizes the data path. This synthesis consists of three stages. First, the functional descriptions such as memory arrays, constants, and I/O register, shown in the form of ISPS (Instruction Set Processor Specification) language, are transformed a dataflow format. This is known as the Value Trace (VT). At this stage, DAA starts with initial hardware assignment. Then, at the second stage, DAA produces a hardware solution to the design which requires optimization. Finally, at the third stage, DAA attempts to improve and optimize the design by applying design heuristics that are used by design experts. DAA was successfully applied to design a simple microprocessor MCS6502 and more complex system the IBM 370. Figure 8 shows the general tasking orders of DAA.

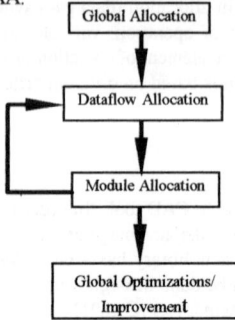

Figure 8. DAA tasks [Kow88]

Another major synthesis tool is called WEAVER [Joo85] (An Expert Channel Router). This was the first CAD tool using blackboard model as the architecture of the expert system. The blackboard architecture provides a problem solving schemes using cooperations of different experts where placed around a blackboard. For instance, if a problem is given to the blackboard, various experts attempt to solve the problem in an opportunistic fashion. After a solution can be provided by a certain expert to portion of the problem, that expert, activates and insert the appropriate results to the blackboard for other experts to work on the rest of problem.

The main objective of using Weaver is to develop an automated route wiring (termed nets) on a chip. This routing automation performs either channel routing or switch-box routing by employing multiple optimization metrics such as wiring length, vias, congestion, pre-assigned nets and single layer availability of pins. Channel routing is referred as wires entering from two opposite sides of a rectangular region. On the other hand, switch-box routing is referred as wires entering from all four sides of a rectangular region. The problem that WEAVER tackles is how best to route together a set of terminals placed on the boundary of a rectangular region using a specified number of routing layers. Finding the optimal solution is an NP-complete problem. Therefore heuristic searches are used to limit the searching space. These heuristic algorithms do not perfectly complete routing. These are due to the uses of :
  (i) simple metrics,
  (ii) oversimplified model of the routing process, and
  (iii) are not designed to deal with special cases.

WEAVER's (blackboard model) architectural programming consists of three major partitions (see Figure 9).
  (i) The problem partition consists of the design data objects such as pins, nets, wires, etc.
  (ii) The decision partition lists suggestions made by the various experts (for example, which net should be routed next).
  (iii) The scratch-pad partition provides restricted areas for each knowledge source to perform internal book keeping.

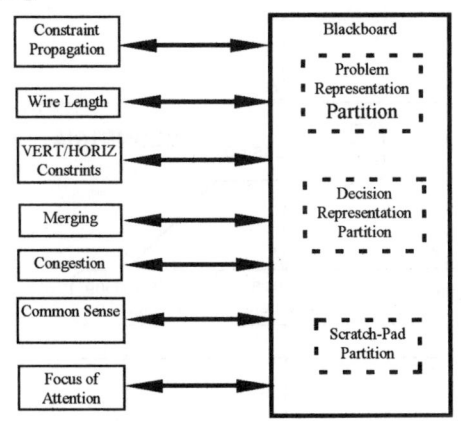

Figure 9. Weaver Architecture [Joo85]

GEOMETRIA [Bou88 and Mor94] is another synthesis tool in which is developed for physical layout level such as partitioning, floorplanning, placement, and compaction. More specifically, the GEOMETRIA structure supports automatically three different VLSI tasks. The first task is a floor planning process of macro-blocks placed effectively in a specified area [Mor94]. The second task is a design/synthesis process which produces the layout of a desirable circuit. The third task is the block reshaping process which minimizes the placement area by using a compaction schemes [Bou94]. The floorplanning part of the methodology is based on the hierarchical cooperation of two context-free languages (SCAN and GEOMETRIA). To achieve an acceptable planning, the SCAN language defines the partitioning of the floor area and the global acquisition strategy (scan patterns) for the placement of the macro blocks. On the other hand, GEOMETRIA language deals with the local synthesis of the blocks under the constraints superimposed by global scan patterns. The transistor layout synthesis part of the methodology is also based on the GEOMETRIA language. More specifically it accepts various user's inputs such as stick diagram, circuit schematics, Boolean expression, netlists, or natural language text expressions, and produces automatically the desirable as well as compacted VLSI layout. The major characteristic of GEOMETRIA is that it is an automated process for VLSI layout placement and synthesis of blocks at various levels of integration (See Figure 10). The implementations of GEOMETRIA methodology are done in C++ using object-oriented-paradigm.

In particular, for the layout synthesis case the user has the option to provide to GEOMETRIA model one of the following possible inputs:

(1) Stick diagram (SD);
(2) Electronic circuit diagram (ECD);
(3) Boolean Logic diagram (BLD);
(4) Boolean Expression (BE);
(5) Netlists (NLS);
(6) Natural language text expressions (NLTE) ;as shown in Figure 11.

When a user inputs one of the first three cases (SD,ECD, and BLD) through a vision camera, the GEOMETRIA user interface activates the reverse engineering package. The reverse engineering package has the ability to recognize ,and analyze the diagram received by a vision camera as 2-D digital image [Bou95].

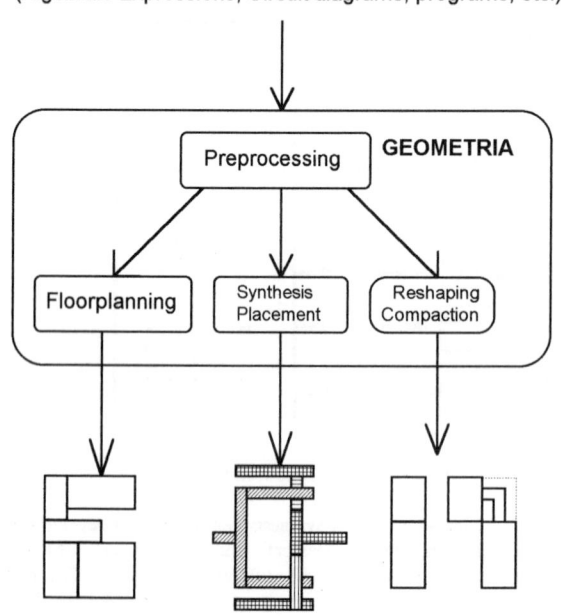

Figure 10. GEOMETRIA Tasks

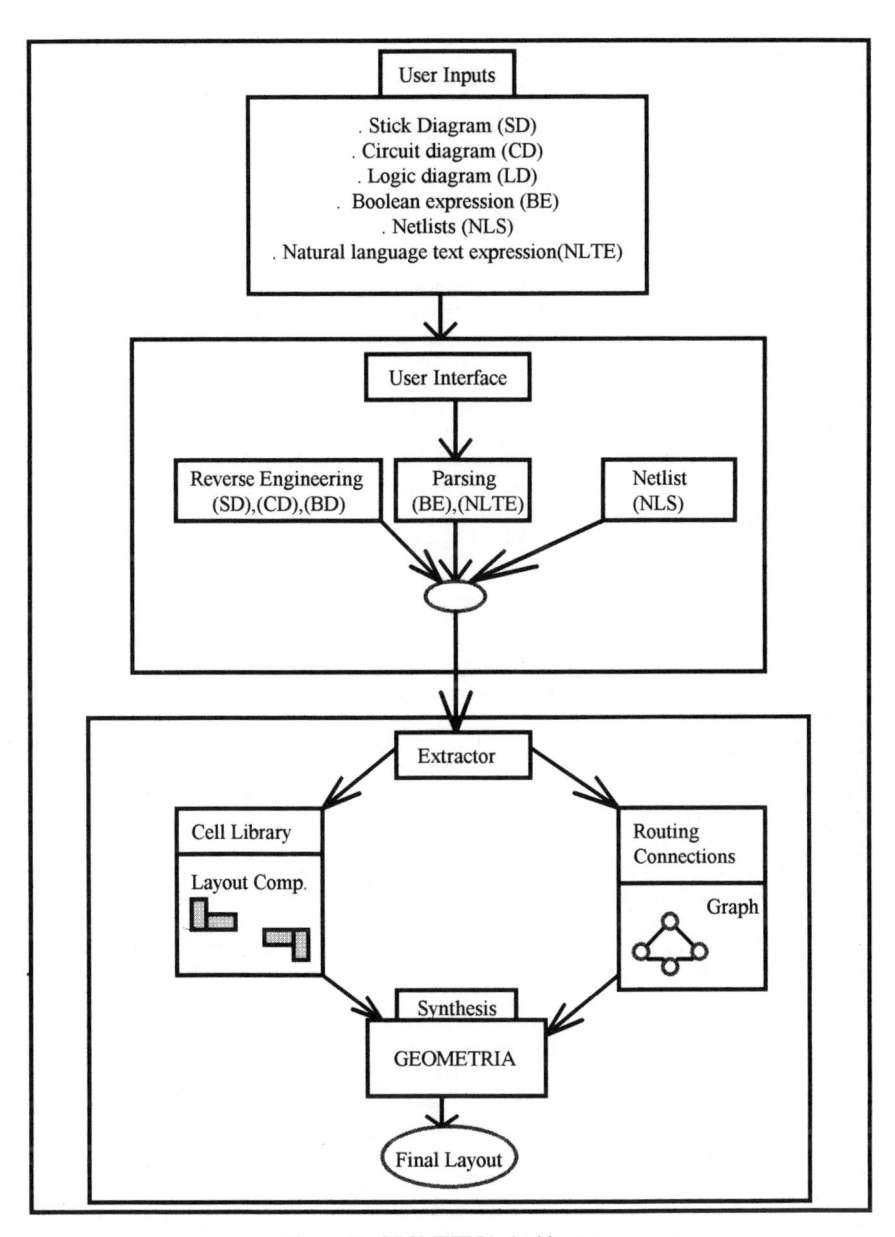

Figure 11. GEOMETRIA Architecture

Table 1 summarizes some other important synthesis tool systems with their some highlights features.

Name/Affiliation	Function
ADPS- Case Western Reserve University, USA	- scheduling and data path allocation
ALPS/LYRA/ARYL- Tsing Hua University, Taiwan ROC	- scheduling and data path allocation
BDSYN- UC Berkeley, USA	- FSM synthesis from DECSIM language for multilevel combination-logic realization
BECOME- AT&T Bell Labs, USA	- FSM synthesis from C-like language for PLA, PLD and standard cell realization
BOLD	- logic optimization
BRIDGE- AT&T Bell Labs	- High-level synthesis FDL2-language descriptions
CADDY - Karlsruhe University, Germany	- behavioral synthesis using VHDL as the input/output language, based on data-flow analysis; automated component selection (allocation), scheduling, and assignment. Different architecture styles are supported, such as multiplexers Vs busses and two-phase Vs single phase clocks.
CALLAS - Siemens, Germany	- high level, algorithmic and logic synthesis (contains CADDY)
CAMAD-Linkoping Univ, Sweden	- scheduling, data path allocation and iteration from a Pascal subset
CARLOS - Karlsruhe University,	- multilevel logic optimization for CMOS realizations
CATHEDRAL- Univ. of Leuve, Phillips and Siemens, Belgium	- synthesis of DSP-circuits from algorithm descriptions
CATREE- Univ.of Waterloo, Canada	- scheduling and data path allocation
CHARM-AT&T Bell Labs	- data-path synthesis
CMU-DA (2)- Carnagie-Mellon University (CMU), USA	- behavioral synthesis from ISPS
CONES- AT&T Bell Labs	- FSM synthesis, produces 2-level logic realizations (truth-table)
DAGAR-Univ of Texas, Austin	- scheduling and data-path allocation
DELHI- IIT	- design iteration, scheduling and data path allocation
DESIGN AUTOMATION ASSISTANT(DAA)- CMU	- expert system for data path synthesis
ELF-Carleton University, Canada	- scheduling and data path allocation
EUCLID- Eindhoven University of Technology, Netherlands	- logic synthesis
EXLOG-NEC Corp., Japan	- expert system, synthesizes gate level circuits from FDL descriptions
FACE/PISYN-General Electric (GE), USA	- FACE: high-level synthesis tools and a tool framework, PISYN: synthesis of pipelined architecture DSP systems (mostly)
FLAMEL- Stanford University,	- data path and control-logic synthesis from Pascal description
HAL-Carleton University, Canada	- data path synthesis
HARP- NTT, Japan	- scheduling and data path-allocation from FORTRAN
HYPER- UC Berkeley	- synthesis for real-time applications (scheduling, allocation, module binding, controller design)
IMBSL/RLEXT-Univ. of Illinois	- data-path allocation, RTL-level design
LSS (Logic Synthesis System) - IBM, USA	- logic synthesis and optimization from many RTL-languages
MAHA - USC, USA	- data path synthesis
MIMOLA- University of Dortmund, Germany	- scheduling, data-path allocation and controller design
OLYMPUS/HERCULES	- behavioral synthesis from C-language (HERCULES), logic and

- Stanford University	physical synthesis
SEHWA - USC	- pipeline-realizations from behavioral descriptions
SIEMENS' SYNTHESIS SYSTEM - Siemens	- partitioning, data path allocation and scheduling
SIS (formerly MIS (II/MV)) - UCBerkeley	- synthesis and verification system for sequential logic
SOCRATES- GE, and Univ. of Colorado, USA	- expert system - logic optimization and mapping for different technologies
SPAID- University of Waterloo	- DSP-synthesis for silicon compiler realizations
SYNFUL- Bell-Northern Research, Canada	- RTL and FSM synthesis for a production environment
SYSTEM ARCHITECT'S WORKBENCH- CMU	- behavioral synthesis
UCB'S SYNTHESIS SYSTEM - UC Berkeley	- transformations, scheduling and data path allocation
V COMPILER- IBM	- scheduling and data path allocation from V-language
VSS- UC Irvine, USA	- transformations, scheduling and data path allocation from VHDL to MILO
YORKTOWN SILICON COMPILER - IBM T.J.Watson Research Center	- data path synthesis, logic synthesis etc.

Table 1. Summary of Synthesis Tools

3.2 Evaluation/Analysis Tools

These tools are used to analyze the functionality, performance, testability, and verify the correctness of the behavior of the designed system. Unlike synthesis tools, evaluation tools play a passive role in the design process [Dan88]. A brief review of some evaluation-type CAD tools that associated with AI methodologies is provided here.
One of the earliest and most successful applications of AI techniques to an evaluation task was the EL system. EL examines the direct current (dc) analysis of an IC using a set of rules written in a language called ARS. These rules are based on fundamental circuit laws such as Kirchoff's laws and Ohm's law or even more developed complex device models as transistors.
For instance, information given about a circuit are stored as a set of assertions (known facts). When a fact is asserted to the data base, some rules from circuit analysis will be fired. When an action fires, it will either make a new assertion or detect a contradiction. The new generated assertions are based on linear constraint propagation. Contradictions are based on a non-linear devices represented as piece wise linear operators, and the system must predicate the operating region of a device. Wrong predicts result in a contradiction and the system investigates the constraints that led to a contradictions. EL uses a scheme called as dependency directed backtracking to return only to those statements which have caused the contradictions. After the incorrect statement is detected, a new operating point will be chosen and forward analysis continues.
DIALOG is another evaluation tool based on AI techniques [DeM83 & DeM85]. This knowledge-based design tool provides assistance in criticizing MOS VLSI circuits developed by the VLSI Systems Design group at IMEC in Belgium. DIALOG contains information on worst and best design practices for a limited design styles such as clocking schemes and lambda rules for a given fabrication technology. This information consists of logic configurations, noise margin, charge sharing, non-static CMOS gates, capacitive loads, etc. unlike conventional simulators, using knowledge-based schemes in DIALOG can aid to identify more design errors. This is because of the conventional simulators lack of global knowledge on the nature of the certain design to be analyzed. Table 2. shows the highlights some knowledge-based analysis tools.

Name	Function
Critter	Functional Timing Constraint Propagation
Debugger	Verify Correct nMOS Transistor Connectivity
VeryFun	Functional Correctness of IC Logic
Prove	Verify Logic against Functional Specification
TLTS	Analysis and Redesign of Transmission Line Circuits
DFT	Verify LSSD Design for Testability
PLA-ESS	Select PLA Test Strategy
Amber	Area Estimation Assistant for Custom Layout
Leap	A learning Apprentice System
Hitest	A test generation system
FP	Functional Partitioning for Test Generation

Table 2. Summary of Knowledge-based Evaluation Systems [Ack88]

3.3 Design Environment/Management System Tools

These system tools will manage the overall design process. They will decide which action is appropriate to be taken next in order to advance the state of the design. One of the major problems using the set of CAD tools under an environment for complex VLSI system design is that of CAD tool integration. This is mainly due to three reasons :

(i) *Continuos growth in complexity of CAD tools used during the design process.*
(ii) *Lack of standard interface/interconnection between each CAD tool makes the integration inefficient. This results in spending more time in converting the output of one CAD to the input of another CAD tool.*
(iii) *The sheer size of design details is so enormous such that the designer has no choice but to depend on the computer to keep up and verify the design database.*

In addition, the requirement to handle many tools, CAD systems should be able to adapt to different design environments. Not only is essential for a CAD system being able to handle different IC technologies, such as Metal Oxide Semiconductor (MOS), Complementary MOS (CMOS), bipolar, gallium arsenide, etc., but is also important the system can adapt quickly to new changing design rules and new IC technologies. It is also important to support higher level systems design specification, such as circuit schematics, programming expressions, hardware description languages, etc.

Besides the CAD issues mention above, another major problems with the power of CAD systems is user interface. For instance, the interface must be able to span all of the system facilities such that a set of single commands is all needed for learning. It is necessary for the interface to be flexible with group and individual needs as well as to be interpreted into programming languages such as Lisp and Prolog.

The last important characteristics of a CAD system in a design environment must be independent of any kind of platform. The rapid growth workstations technology, it is no longer needed to tie a system to one machine. It is rather more important to provide portability of the CAD system so it can run any machine [Rub91].

One of the earliest integrated CAD tools based on knowledge-based techniques acting as management process was Palladio. Palladio methodology proposed that design must be followed by a parallel refinement of structural and behavioral perspectives or view. In particular, perspectives are described as a particular level of abstraction in forms of either basic components or a set of rules defining their interconnections. More specifically, structural perspectives (SP) are used to specify the logical interconnection. For example, SP could be used to represent a set of clocked storage registers, or it could define a circuit at the transistor level. On the other hand, behavioral perspective are used to describe changes of circuit state over time. They can express rules relating signal behavior.

In Palladio, both behavioral and structural perspectives are orderly linked by design procedure and rule-based logical simulator. Design procedures produces structure from behavior such as PLA generation. A rule-based logical simulator provides structure to be simulated and verified against its modeled behavior. Different programming techniques were used to implement different parts of the CAD system. Rule based programming was used to incrementally constructed expert design aids. Data oriented programming was used to propagate design constraints through the environment. The logical-reasoning programming was for behavioral specification and simulation, and finally, object-oriented programming was used for structural specification. Figure 12 shows the architectural systems design of Palladio.

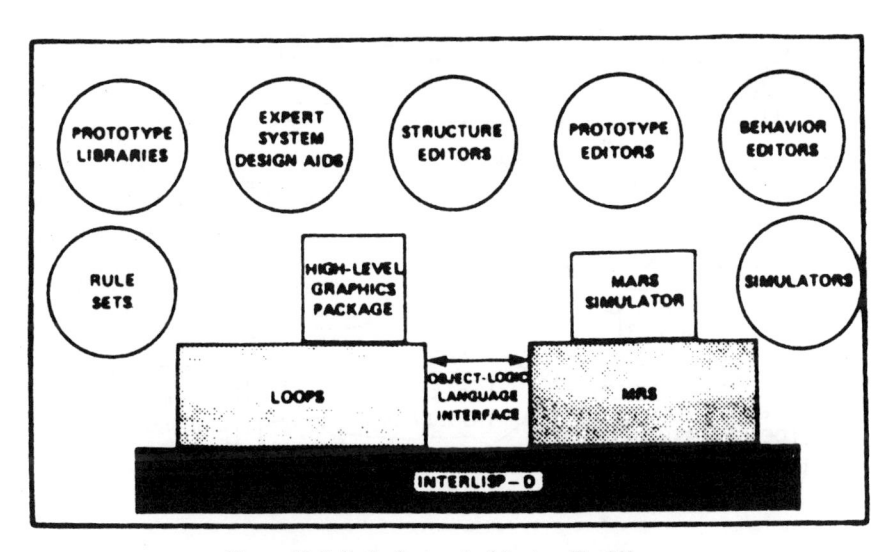

Figure 12. Palladio System Architecture [Bro83]

There are some problems with Palladio that is worthwhile to mentioning. Palladio is not a good production environment in terms of execution speed, its development did result in a basic understanding of the nature of the tool integration problem. One of the fundamental problems of Palladio design approach was that its difficulty of mapping its hierarchical representation of a chip to some other intermediate forms. It can not handle or adapt large number of other CAD tools. The major bottleneck in the Palladio environment caused by limited representation of using the CAD tools in the design environment.

Another attempt of developing VLSI design environment was Ulysses. It addressed major issues of integrating any arbitrary CAD tools. Ulysses was capable of allowing integration of a series of CAD tools under a design environment and automatically codify and execute the design methodology. It could handle complete interaction between various CAD systems. Ulysses would allow the designer to interrupt anytime for the purpose of halting or re-designing during the design process. Moreover, if a CAD tool failed functioning or the designer changed some parameter in the design specification, Ulysses was capable of automatically backtrack and reinstate the consistency of a design floorplan.

Furthermore, Ulysses employed the concept of blackboard model in order to provide a mechanism for communication and to handle efficiently interaction between various CAD systems. The control of the blackboard was supported by the Ulysses scheduler. The blackboard methodology was used and developed in Ulysses more effective than the WEAVER's knowledge-based synthesis. This was due to the limitation of generating WEAVER's rule-base system.

Design activities in Ulysses were described by a language called scripts. Script implements high level representation of a design task. It accommodates knowledge of the CAD tool execution sequence. It provides reasoning for each step in the sequence. Finally, script takes the controlling of how the output of a tool could be used as the input to another tool. The Ulysses' architectural design environment is shown in Figure 13.

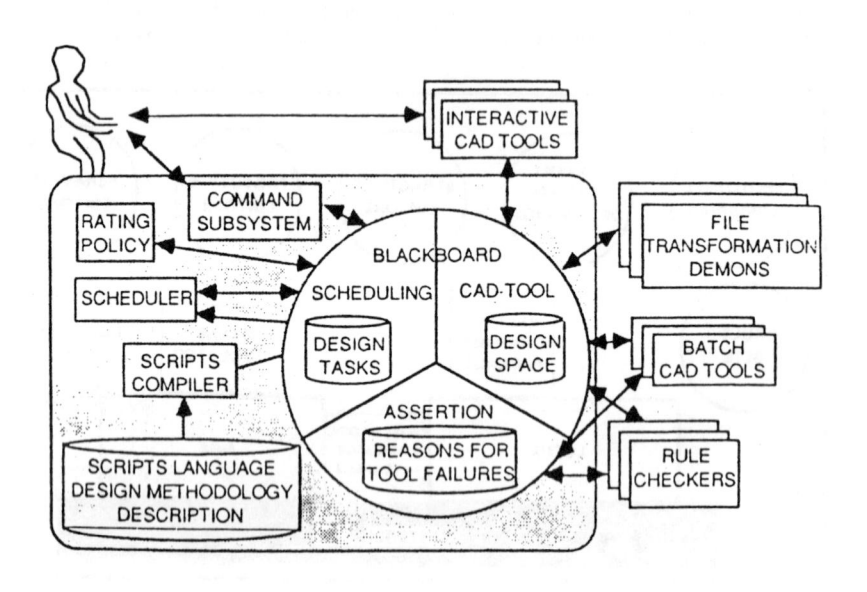

Figure 13. The Ulysses Design Environment Architecture [Bus88]

However, there were some limitations with Ulysses. The inference mechanism in Ulysses was very slow. This is because of the assumptions were made for rule firing implementation were simple.
Another important knowledge-based design environment systems for VLSI design automation that is called SHEDIO and is under the development. The SHEDIO environment is a multilevel frame based which evaluates and optimizes the topological level of system's component. This is composed of a digital system which examines a pre-defined set of algorithms that helps the system execute efficiently. Furthermore, the SHEDIO environment tries to optimize the layouts by means of minimizing the length of the connections among the system components and at the same time attempts to minimize the occupied chip area and the critical path of the chip.
General configuration of the SHEDIO environment is shown in Figure Z. The inputs of SHEDIO are dependent on parameters such as architectural system design, system's components, bus organization, algorithms, micro-code or Boolean expressions, dimensions of the chip area, IC technology, functional speed, and power dissipation. In order to accomplish the primary objectives of SHEDIO environments which are interactive and recursive, we must to be concerned with two important tasks:
One important feature of the SHEDIO environment is the interactive expert system used here. The SHEDIO environment accepts natural language descriptions for a particular system, converts them into internal SPN graph representations and at the other end produces a variety of VLSI layout form of the system with various performances simulated by ARTEMIS [Bou95]. The major disadvantage of SHEDIO is the slow processing of rules due to heavy work loads.

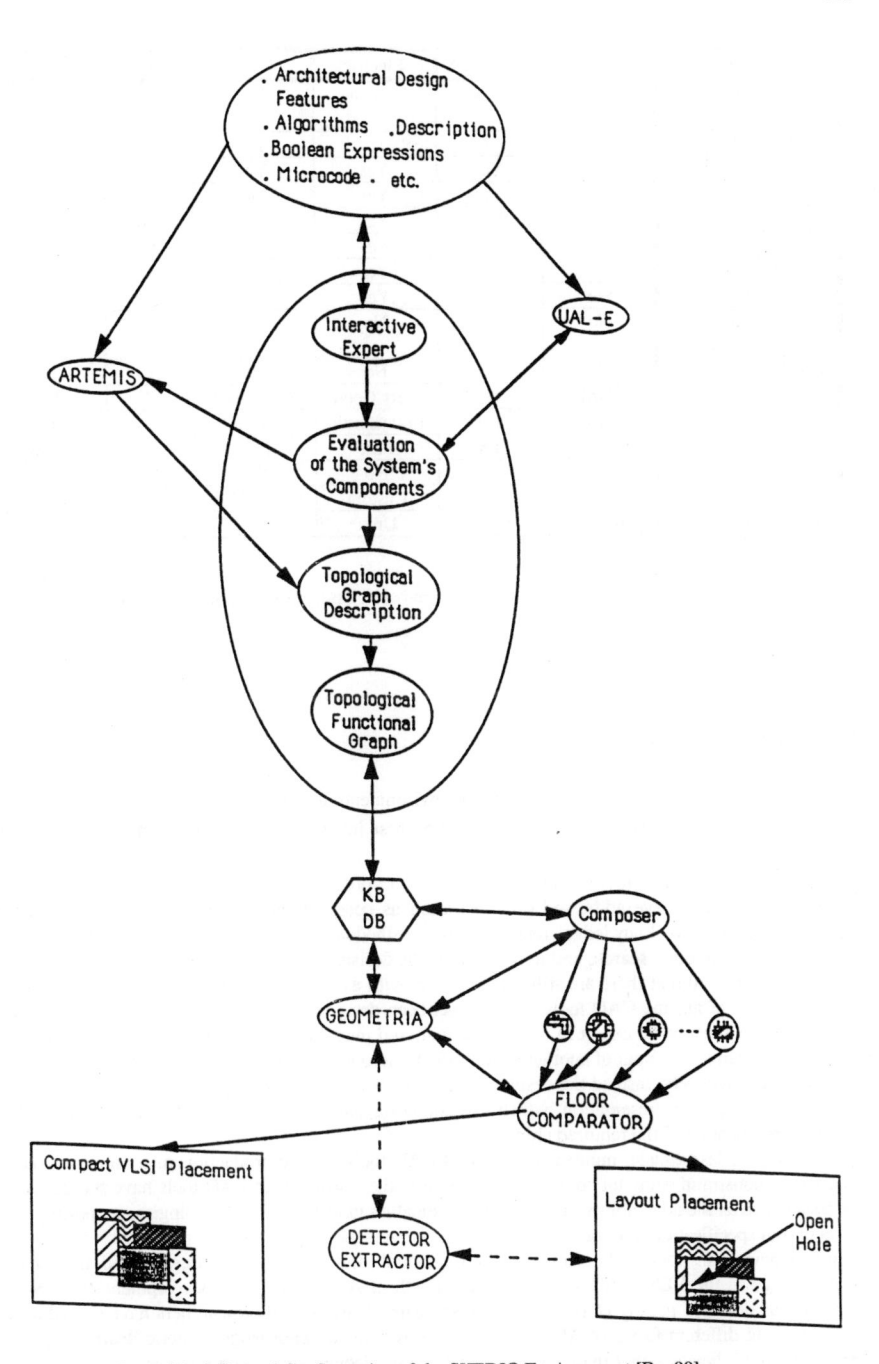

Figure 14. A General Configuration of the SHEDIO Environment [Bou88]

Feature	Palladio	ULYSSES	Electric	DWB	ADAM	SHEDIO
Existing Tool Integration	No	Yes	Re-write	Yes	No	In Progress
Networking	No	?	?	Yes	?	?
Schematic Capture	Yes	Yes	Yes	Yes	Yes	Yes
Layout Synthesis	No	Yes	Yes	No	No	Yes
Layout Editing	No	Yes	Yes	Yes	No	Yes
PLA Generation	No		Yes	No	No	Yes
Layout Compaction	No	Yes	Yes	No	No	Yes
Simulation	Yes	Yes	Yes	Yes	Yes	In Progress
Test Vector Generation	No	?	No	Yes	No	In Progress
Reliability Analysis	No	?	No	No	No	No
Ease of Use	Hard	Yes	Very Good	Very Good	?	Yes
User Acceptance	Not Good	Yes	Enthusiastic	Enthusiastic	?	Yes
Speed	Very Slow	Slow	Fast	Fast	?	slow
File Translators	No	Yes	No	Yes	No	?
HDL	Custom	Yes	Custom	Hiwire UCL	Custom	Yes
Operating System	Xerox 1100	Unix	Unix	Unix	?	Unix

Table 3. Summary of Knowledge-based Design Environment tools [Bus88]

4. Concluding Remarks

Concluding, highlights of different design environments for automation of VLSI design have been introduced. The most important problem with all of these integrated CAD tools are not being efficient in speed execution.

AI techniques have provided some solution to the problem arising from VLSI CAD community. Heuristic knowledge have addressed problems such as computationally inflexible and ill-defined task designs through new knowledge representation methodology, CAD tool architectures, and new approaches in planning, search, and non-deterministic decision-making.

A study showed that there are still some barriers with using of the CAD tools in an integrated form [Bus88]: First of all, the CAD tools have not yet been designed to work together. This means that the VLSI designer must be an expert in using each tool, and must know the connectivity between the output of a CAD system to the input of another CAD system. Secondly,. there is no mechanisms to automate and run the tools effectively, instead the designer must run everything. Finally, the tools do not conform to a simple design space representation, which can contain a subset of all IC layouts that might satisfy the systems specification of the required chip.

During the last decade, number of AI-Based CAD tools applied production rules; however the time and money constraint were not considered. Therefore, the researches in CAD tools have started focusing on issues such as the adaptability of the design automation to different IC technologies, supporting higher level design specification, the user interface, the user interactions, platform independence, etc. rather than building new CAD tool architectures. Chart one indicates the serious influence and impact of AI methodology in the VLSI CAD systems. The growth of AI techniques in development of VLSI CAD systems has not been greatly considered in recent years. Instead, CAD researchers have started to focus on employing different forms of AI techniques such as heuristic knowledge, generic, learning, planning and searching schemes rather than developing a new knowledge-based expert systems.

Number of Articles Published in IEEE Tran. on CAD

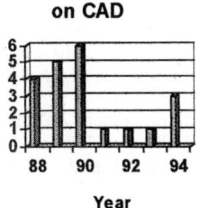

[bar chart showing values for years 88, 90, 92, 94 on Y-axis 0–6]

Chart 1. This data was taken from Proceeding of IEEE Transaction on CAD since 1988.

5. References

[Ack88] Ackland,B.D., *Knowledge-Based Physical Design Automation*, Chapter in Physical Design Automation of VLSI Systems, by Preas,B & Lorenzetti,M., Bejamin Cummings Pub., 1988.
[Aco86] Acosta, R., M. Huhns, and S. Liuh, *Analytical reasoning for digital system synthesis*, in Digest Int. Conf. on CAD, pp. 173-176, 1986.
[Ade90] Adeli,H., *Knowledge Engineering* , Vol I & II, McGraw Hill Company, 1990.
[Afs86] Afsarmanesh, H., D. Knapp, D. McLeod, and A. Parker, *Information management for VLSI/CAD*, in Proc. Int. Conf. on Computer Design, pp. 476-481, 1986.
[Alh95] Alhabibi, B.A. and Bayoumi,M., *A Scalable Analog Architecture for Neural Networks with On-chip Learning and Refreshing* , Fifth Great Lakes IEEE Symposium on VLSI, pp 33-38,1995.
[Ash91] Ashar,P., *Optimum and Heuristic Algorithms for Approach to Finite State Machine Decomposition*, IEEE Trans. on CAD of IC and Sys, pp296-310, Mar. 1991.
[Bal89] Balakrishnan, M. et al. *Integrated Scheduling and Binding: A Synthesis Approach for Design Space Exploration* , Proc. of the 26th DAC, pp. 68-74, June 1989
[Bar86] Bartlett, K., *Synthesis and Optimization of Multilevel Logic Under Timing Constraints* , IEEE Trans. on Computer-Aided Design, Vol 5, No 10, October 1986
[Bat93] Batalama, S., *Heuristic Single-Row Router Minimizing Inter-street Crossings*, IEEE Trans. on CAD of IC and Sys, pp.946-955, Jul. 1994.
[Ben84] Bending, M. *Hitest: A knowledge-Based Test Generation System* IEEE Design & Test1, 2 pp. 83-92, 1984.
[Ber90] Berkelaar, Michel R.C.M. and Theeuwen, J.F.M., *Real Area-Power-Delay Trade-off in the EUCLID Logic Synthesis System* , proc. of the Custom Integrated Circuits Conference 1990, Boston MA, USA, pp 14.3.1 ff
[Ber89] Berstis, V: *The V Compiler: Automatic Hardware Design* , IEEE Design & Test, pp. 8-17, Apr. 1989.
[Bor94] Borah,M., *An Edge-Based Heuristic for Steiner Routing*, IEEE Trans. on CAD of IC and Sys, pp.1563-1568, Dec. 1994.
[Bou94] Bourbakis N.G. and M.Mortazavi, *GEOMETRIA: A formal language for VLSI layout representation and manipulation* ,TR- 1993, Int. Journal on Computer languages submitted
[Bou95] Bourbakis, N.G., *Understanding Natural language text expression by using SPN graphs*, Int. J. Artificial Intelligent with tools, Vol. 4, 1995.
[Bou91] Bourbakis N.G. & Ramamoorthy CV *Specification for the Development of an Expert Tool for the Automatic Optical Understanding of Electronic Circuits: VLSI Reverse Engineering*, IEEE VLSI Test Symp. 1991.
[Bou88] Bourbakis, N., Saviddes,I., *Specifications of A Knowledge-Based Environment for VLSI System Architectural Design*, in IEEE Workshop on languages for Automation, 1988.
[Bra86] Brayton, R. *Multiple-level Logic Optimization System* , Proc. of IEEE ICCAD, Santa Clara, Nov. 1986
[Bra88] Brayton, R.K., et al *The Yorktown Silicon Compiler* ,Silicon Compilation, pp. 204-311, Add. Wesley, 1988.
[Bro83] Brown,H, Tong,C., and Foyster, G. *Palladio: An Exploratory Environment for Circuit Design* IEEE Computer 16 (Dec. 1983).
[Bus88] Bushnell, M.L., *Design Automation: Automated Pull-Custom VLSI Layout Using the ULYSSES Design Environments*, AP. , 1988.
[Cam91] Camposano,R. and Wolf, W. *High-Level VLSI Synthesis*, Kluwer Academic Pub., 1991.
[Cam89] Camposano, R. *Synthesing Circuits From Behavioral Descriptions* IEEE Trans. on CAD, V8, 2, Feb. 89.

[Chu89] Chu, C-M. et al.: *HYPER: An Interactive Synthesis Environment for Real Time Applications*, Proc. of ICCD '89, pp. 432-435, October 1989
[Dan88] Daniell, J., Dewey, A.M., and Director,S.W., *Artificial Intelligence Techniques: Expanding VLSI Design Automation Technology*, Chapter 9 of Expert Systems for Engineering Design by Rychener, M., AP, 88.
[Dan91] Daniell,J., *Object-Oriented Tool Integration Methodology for CAD tool Control*, IEEE Trans. on CAD of IC and Sys, pp.698-713, Jun. 1991.
[Dar84] Darringer, J. et al. *LSS: A System for Production Logic Synthesis*, IBM Journal of Research and Development, vol. 28, No. 5, pp. 272-280, Sept 1984.
[DeM88] De Micheli, G.: *HERCULES - A System for High-Level Synthesis*, Proceedings of the 25th ACM/IEEE Design Automation Conference, pp. 483-488, IEEE 1988.
[DeM90] De Man, H.: *Architecture-Driven Synthesis Techniques for VLSI Implementation of DSP Algorithms*, Proceedings of the IEEE, Vol. 78, NO. 2, pp. 319,February 1990
[Dem83] DeMan, H., Darcis, L., Bolsens, I., Reynaert, P., and Dumlugol, D. *A Debugging and Guided Simulation system for MOS VLSI Design*, IEEE Conf on CAD, pp. 137-138, 1983.
[Dev89] Devadas, S. *Algorithms for Hardware Allocation in Data Path Synthesis*, IEEE Trans. on CAD, pp. 768-781, July 89.
[Dew86] Dewey, A. And Gadient, A. *VHDL Motivation.* IEEE Design and Test of Computers 3,2, 1986.
[Den90] Deng, A.C, *Generic Linear RC Delay Modeling for Digital CMOS Circuits*, IEEE Trans. on CAD of IC and Sys., pp.367-376, Apr. 1990.
[Elt89] El-Turky, F., *BLADES Knowledge-Based Analog Circuit Design Environment*, IEEE Trans. on CAD of IC and Sys, pp.680-692, June 1989.
[Fun93] Funabiki,N., *Neural Network Approach to Topological Via Minimization Problems*, IEEE Trans. on CAD of IC and Sys, pp.770-779, Jun. 1993.
[Geb88] Gebotys, C.H.: *VLSI Design Synthesis with Testability*, Proc. of the 25th DAC, pp. 16-21, June 1988
[Geu87] de Geus, A.J., *The Socrates Logic Synthesis and Optimization System*, Design Systems for VLSI Circuits, pp. 473-498, Martinus Nijhoff Publishers, 1987.
[Gir85] Girczyc, E.F. et al.: *Applicability of a Subset of Ada as an Algorithmic Hardware Description Language for Graph-Based Hardware Compilation*, IEEE Trans. on CAD, pp. 134-142, April 1985.
[Gos92] Gosh,A., *Heuristic Minimization of Boolean Relations Using Testing Techniques*, IEEE Trans. on CAD of IC and Sys, pp1166-1172, Sep.1992.
[Gut92] Gutberlet P., Mueller J., Kraemer H., Rosenstiel W.: *Automatic Module Allocation in High-level Synthesis*, Proc. of 1st EURO-DAC, 1992
[Gup93] Gupta,A, *Knowledge-Based System-Level Synthesis Tool for Small Computer Systems Using Hierarchical Select-and-Interconnect Method*, IEEE Trans. on CAD of IC and Sys, pp.473-48, Aug. 1993.
[Had75] Hadlock, F.O., *Finding a Maximum Cut of a Planar Graph in Polynomial Time*, SIAM Journal of Computing, 4, no.3, pp. 221-225, September 1975.
[Har89] Harjani,R., *Hierarchically Structured Framework For Analog Circuit Synthesis*, IEEE Trans. on CAD of IC and Sys, pp.1247-1266, Dec. 1989.
[Har89] Haroun, B. *Architectural Synthesis for DSP Silicon Compilers*,IEEE Tran on CAD,pp431-447,V8,4,Apr 89.
[Hay94] Haykin, S. *Neural Networks: A Comprehensive Foundation* Macmillan Publishing Company, NY 1994.
[Hon89] Hong Y.-S., *Ordering of Columns in one-dimensional Logic Array to Minimize Necessary Number of Tracks*, IEEE Trans. on CAD of IC and Sys, pp.547-562, May 1989.
[HU90] HU, Y. H., *Gate Matrix Layout Algorithm GM_Plan, Based on Artificial Intelligence Planning Techniques*, IEEE Trans. on CAD of IC and Sys, pp.836-845, Aug. 1990.
[Joo85] Joobbani, R. *WEAVER An application of Knowledge-Based Expert Systems to Detailed Routing of VLSI Circuits* Ph.D. Th., CMU, EE & CE Department, June 1985,
[Jun94] June-K R., *Exact and Heuristic Algorithm for the Minimization of Incompletely Specified State Machines*, IEEE Trans. on CAD of IC and Sys, pp.167-177, Feb.1994.
[Kna91] Knapp,D., *ADAM Design Planning Engine for Managing Digital Design Process*, IEEE Trans. on CAD of IC and Sys, pp.829-846, Jul. 1991.
[Kna89] Knapp D.W.: *Manual Rescheduling and Incremental Repair of Register Level Data Paths*, Proc. of ICCAD '89, pp.58-61, November 1989.
[Kos90] Koster, M. et al.: *ASIC Design Using the High-Level Synthesis System CALLAS: A Case Study*, Proc. IEEE (ICCD '90), pp. 141-146, Cambridge, Massachusetts,Sept. 17-19, 1990
[Kow88] Kowalski, T.J. *The VLSI Desig Automation Assistant: An Architecture Compiler*, Silicon Compilation, pp. 122-152, Addison-Wesley, 1988
[Kow84] Kowalski, T., *The VLSI Design Automation Assistant: A Knowledge-Based Expert System* Ph.D. Th., CMU, EE & CE Dept., 1984

413

[Kun93] Kunz,W., *Accelerated Dynamic Learning for Test Pattern Generation in Combinatorial Circuits*, IEEE Trans. on CAD of IC and Sys, pp.684-694, May 1993.
[Lee89] J-H: et al.: *A New Integer Linear Programming Formulation of the Scheduling Problem in Data Path Synthesis*, Proc. of ICCAD89, pp. 20-23, November 1989.
[Lis88] Lis, J. et al.: *Synthesis from VHDL*, Proc. ICCD'88, pp. 378-381, October 1988.
[Mao90] Mao,W., *DYTEST Self-Learning Algorithm Using Dynamic Testability Measure to Accelerate Test Generation*, IEEE Trans. on CAD of IC and Sys, pp.893-898, Aug. 1990.
[Mar90] Marwedel, P. *Matching System And Component Behavior in MIMOLA Synthesis Tools*, Proc. of EDAC90 pp. 146-156, March 1990.
[Mat88] Mathony, H-J.: *CARLOS: An Automated Multilevel Logic Design System for CMOS Semi-Custom Integrated Circuits*, IEEE Transactions on Computer-Aided Design, Vol 7, No 3, pp. 346-355, March 1988
[Min75] Minsky, M. *A framework for representing knowledge.* ,The Psychology of Computer Vision, NY, 1975.
[Mor94] M. Mortazavi and N.G. Bourbakis, *A generic floorplanning methodology*, IEEE conf. on Autotestcon-94, CA, Sept 1994
[Pap90] Papachristou, C.A. et al.: *A Linear Program Driven Scheduling and Allocation Method Followed by an Interconnect Optimization Algorithm*, Proc. of the 27th DAC, pp. 77-83, June1990.
[Par86] Parker, A.C. *MAHA: A Program for Data Path Synthesis*, Proc. 23rd ACM/IEEE Design Automation Conference, pp. 252-258, IEEE 1986.
[Par86] Park, N. *SEWHA: A Program for Synthesis of Pipelines*, Proc. 23rd ACM/IEEE Design Automation Conference, pp. 454-460, IEEE 1986.
[Pau89] Paulin, P.: *Force-Directed Scheduling for the Behavioral Synthesis of ASIC's*, IEEE Transaction on Computer-Aided Design, pp. 661, Vol. 8, No. 6, June 1989.
[Pen88] Peng, Z.: *CAMAD: A Unified Data Path/ Control Synthesis Environment*, Proc. of the IFIP Working Conference on Design Methodologies for VLSI and Computer Architecture, pp. 53-67, Sept.1988.
[Pre88] Preas, B. and Lorenzetti, M., *Physical Design Automation of VLSI Systems*. Benjamin Cumming Pub. Company, 1988.
[Raj89] Raj. V.K.: *DAGAR: An Automatic Pipelined Micro-architecture Synthesis System*, Proc. of ICCD '89, pp. 428-431, October 1989.
[Rob83] Robinson, G. D. *Hitest - Intelligent Test Generation.* Proc. IEEE Int. Test Conference, pp311-323, 1983.
[Rob89] Robach ,C., *Knowledge-Based Functional Specification of Test and Maintenance Programs*, IEEE Trans. on CAD of IC and Sys, pp.1145-1156, Nov. 1989.
[Ros91] Rosenstiel, W., Kraemer, H.: *Scheduling and Assignment in High-Level Synthesis*, in 'High-Level VLSI-Synthesis' R. Camposano, W. Wolf Ed. Kluwer, 1991
[Rub91] Rubin, S.M., *A General-Purpose Framework for CAD Algorithms*, IEEE Comm.Mag., pp56-62, May 1991.
[Rub87] Rubin, S.M., *Computer Aids for VLSI Designs*, Addison Wesley, Reading, MA 1987.
[Ryc88] Rychener, M.D., *Expert Systems for Engineering Design.* , Academic Press, 1988.
[Sch90] Scheichenzuber, J. et al.: *Global Hardware Synthesis from Behavioral Dataflow Descriptions*, Proc. of the 27th DAC, pp. 456-461, June 1990.
[Sen92] EM Sentovich,KJ Singh,L Lavagno,C Moon,R Murgai,A Saldanha,H Savoj,PR Stephan,RK Brayton,A Sangiovanni-Vincentelli *SIS:A System for Sequential Circuit Synthesis*, TR-UCB/ERL M92/41, May 92
[She93] Sherwani,N., *Algorithms for VLSI Physical Design Automation*, Kluwer Academic Publishers, 1993.
[Sim89] Simoudis, E., *Knowledge-Based System for Evaluation and Redesign of Digital Circuit Networks with Signal Integrity Problems*, IEEE Trans. on CAD of IC and Sys, pp.302-315, Mar 1989.
[Sim90] Simoudis,E., *Knowledge-Based Transmission-Line Troubleshooting System TLTS for Redesign of Circuits on PCB*, IEEE Trans. on CAD of IC and Sys, pp.1047-1062, Oct. 1990.
[Smi89] Smith, W.D. et al.: *FACE Core Environment: The Model and it's Application in CAE/CAD Tool Development*, Proc. of the 26th DAC, pp. 466-471, June 1989.
[Str86] Stroud, C.E.: *CONES: A System for Automated Synthesis of VLSI and programmable logic from behavioral models*, Proc. of IEEE ICCAD, Santa Clara,Nov. 1986.
[Swi90] Swinkles, G.M., *Schematic Generation by Knowledge-Based System*, , IEEE Trans. on CAD of IC and Sys,pp.1289-1306, Dec. 1990.
[Tan89] Tanaka, T. et al.: *HARP: Fortran to Silicon*, IEEE Trans. on CAD, pp. 649-660, June 1989.
[Tho88] Thomas, D. *The System Architect's Workbench*, Proc. of the 25th ACM/IEEE DAC, pp. 337-343,1988.
[Tho87] Thomas, D.: *Linking the Behavioral and Structural Domains of Representation for Digital System Design*, IEEE Transactions on Computer-Aided Design, pp. 103-110, Vol. 6, No. 1, January 1987
[Tri87] Trickey, H. *Flamel: A High-Level Hardware Compiler*, IEEE Tran. on CAD, Vol 6, No 2, March 1987.
[Tse88] Tseng: *Bridge: A Versatile Behavioral Synthesis System*, Proc. of 25th ACM/IEEE DAC, pp. 415-420, 1988
[Wal91] Walker, R.A. and Camposano,R., *A Survey of High-Level Synthesis Systems*, Kluwer Academic Pub. 91.

[War90] G. Ward, *Logic Synthesis at BNR: A SYNFUL Story*, Proceedings Canadian Conference on Very Large Scale Integration, October 1990.
[Wat87] M. Watanabe, et al.,: *EXLOG: An Expert System for Logic Synthesis in Full-Custom VLSI Design*, Proc. of 2nd Int. Conf. Application of Artificial Intelligence, August 1987.
[Wei84] Weiss, S, and Kulikowski, C, *A Practical Guide to Designing Expert Systems*, Rowman & Allenheld, Totowa, NY, 1984.
[Wei88] Wei, R-S.: *BECOME: Behavior Level Circuit Synthesis Based on Structure Mapping*, Proc. of 25th ACM/IEEE Design Automation Conference, pp. 409-414, IEEE, 1988
[Woo90] Woo, N-S.: *A Global, Dynamic Register Allocation and Binding for a Data Path Synthesis System*, Proc. of the 27th DAC, pp. 505-510, June 1990.
[You95] Youssef,H., Sait,S.,Nasser,K., and Benton,M.S, *Performance Driven Standard-Cell Placement Using the Genetic Algorithm*, Fifth Great Lakes IEEE Symposium on VLSI, pp 124-127,1995.

THE AUTOMATED ACQUISITION OF SUBCATEGORIZATIONS OF VERBS, NOUNS AND ADJECTIVES FROM SAMPLE SENTENCES

FERNANDO GOMEZ
Department of Computer Science
University of Central Florida
Orlando, FL 32816
USA

ABSTRACT

An expert system consisting of learning algorithms that acquire syntactic knowledge for a parser from sample sentences entered by users who have no knowledge of the parser or English syntax is described. It is shown how the subcategorizations of verbs, nouns and adjectives can be inferred from sample sentences entered by end-users. Then, if the parser fails to parse the sentence, say *Peter knows how to read books*, because it has limited knowledge or no knowledge at all about "know," an interface, which incorporates the acquisition algorithms, can be activated and "know" be defined by entering some sample sentences, one of which can be the one which the parser failed to parse[1].

1 Introduction

One of the stumbling blocks of natural language processing (NLP) applications is that of syntax acquisition. If the parser used by the NLP system fails to parse a user's sentence, then the entire application system collapses. Formal knowledge about syntax is a form of expertise that most people do not have. Moreover, the encoding of this knowledge into the parser is not an easy matter either. The ideal situation would be one in which end-users would communicate the syntactic knowledge to the knowledge acquisition system by typing some sample sentences. This is the approach that is explored in this paper. Algorithms for the acquisition of syntactic knowledge from sample sentences entered by end-users, who have no formal knowledge of syntax or the parser, are described. We will refer to these learning algorithms as the interface to the parser, or simply the interface. This interface is one of the main components of SNOWY [6, 5].

The philosophy underlying our parser is that the core of the syntax of a language does not rely on "general" rules, but rather on lexical items, especially verbs and

---

[1]This research is being in part funded by NASA-KSC Contract NAG-10-0058.

nouns, and to a lesser degree, adverbs and adjectives. This view was presented and argued for by Gross [7] and his associates, who concluded that "for French, ... the phenomena that would be called lexical by Chomsky are the rule, while the ones he termed transformational are quite rare." As a consequence, our parser is a lexically driven algorithm that incorporates mechanisms to handle "general" syntactic phenomena such as long distance dependency, embedded clauses, passive sentences, syntactic ambiguity, etc. However, the enormous syntactic diversity of most words in English is handled by storing the syntactic rules in the lexical entries. The parsing of a sentence is driven by these syntactic usages, which are retrieved by the parser as they are encountered in the sentence. Thus, a word like "news" may be followed (among other things) by a clause introduced by the conjunction "that." Or a word like "way" may be followed by an infinitive clause. This behavior is idiosyncratic to "news" and "way." Likewise, the verb "know" may be followed by "how," but "want" followed by "how" will result in an ungrammatical sentence.

Formalizing the syntax of words like "know," "be," etc. in a transition diagram or in a grammar is not easy for people with no knowledge of linguistics. In fact, our experience has shown that even college graduates have a lot of trouble in capturing the syntax of words in any formal mechanism. In this paper, we show how the syntax of words can be inferred from sample sentences entered by end-users. Then, if the parser fails to parse the sentence, say *Peter believed that Mary read the books*, because it has limited syntactic knowledge or no knowledge at all about "believe," the interface can be activated and "believe" defined by entering some sample sentences. The appendix contains a sample session indicating how this interaction proceeds. The reader may want to read that sample session prior to delving into the paper. This paper is organized as follows: Section 2 gives a very brief overview of our parser. Sections 3 to 6 describe the algorithms and the acquisition of usages for verbs, adjectives and nouns. Section 7 explains the problems with ambiguity; section 8 describes the related research and section 9 gives our conclusions.

2 Overview of the Parser

This section describes only those features of the parser necessary to understand the interface acquisition algorithms. The parser [4] is a top down algorithm that uses a "grammar" indicating the structure of the sentence *up to the main verb*. There are two main stacks in the parser. One stack, called *non-terminals*, contains the non-terminal symbols, and a stack, called *actions*, contains the actions to be executed when a non-terminal is reduced from the *non-terminals* stack. The algorithm consists of two steps. In the first step, the *non-terminals* stack is initialized with one rule of the "grammar" and the *actions* stack is initialized with the actions for those non-terminals. In the second step, the symbols in the stacks are manipulated. By way of example, consider the following productions of the initial "grammar:"

```
S --> SUBJ VERB
    | INF VERB
    | INF SUBJ VERB
```

The first rule of the "grammar" says that an English sentence consists of a subject followed by the inflected form of a verb. (The non-terminal VERB refers always to the inflected form of the verb). Whatever follows the verb depends on each concrete verb and, consequently, is not part of the "grammar". When the non-terminal VERB is reduced, its syntactical "usage" will be retrieved from the parser's lexicon entry and pushed on the stack. Similarly, the nouns in the subject may have special usages, in which case those usages will be pushed on the stack during execution of the NP (noun phrase) procedure. The second production of the "grammar" says that an English sentence is formed by an infinitive followed by a verb (*To read is good.*), while the third production says that an English sentence is formed by an infinitive followed by a subject and a verb. In this case, the infinitive introduces a purpose clause as in "To read Mary went to the library."

The syntactic usage of a word is represented in the lexicon by a list of the form (a1 a2 ... ai), meaning that the word may be followed by just one of the ai elements. If an asterisk (*) belongs to the list, it means that none or one ai may follow the word. If any ai is a list, it means that the word may be followed by the first element in ai. This first element in turn may be followed by any number of elements in ai. The usage of "eat" is represented by (obj *), which means that "eat" may or may not take an object. Consider the usage of "want:"

```
want obj
want inf
want subj inf
```

This usage is represented as (inf (obj (inf *))), which corresponds to the structure (a1 a2), where a1 matches inf and a2 the list (obj (inf *)). The meaning of this representation is that "want" may be followed by an infinitive or object, and the object in turn may be followed by an infinitive or nothing. A transition diagram representing the usage of "want" and "like" is depicted in Figure 1. The actions corresponding to this usage are:

```
(same-subj (obj (diff-subj *)))
```

There is a one-to-one mapping between the symbols in the usage and the symbols in the actions. The symbols in the actions are names of functions that will be executed when their matching symbols in the usage are reduced. Note that the representation of the usage of "want" assumes that the NP following the verb is a direct object. However, this assumption will be corrected by the action *diff-subj* if the NP is followed by an infinitive. This action will make the NP the subject of the embedded clause introduced by the infinitive. An alternative representation of the usage of "want"

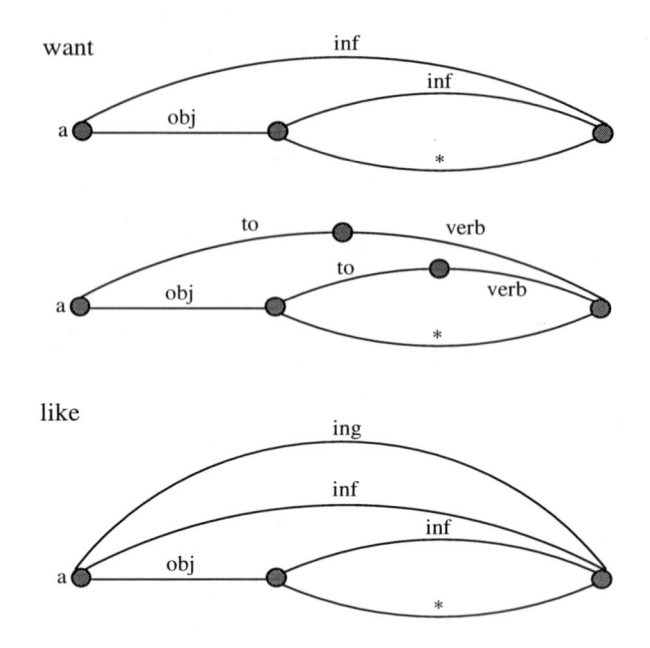

Figure 1: Transition network representing the usages of "want" and "like." For "want," two possible representations are depicted.

is shown in Figure 1. In that case, the action *same-subj* would be associated with the first instance of "to" and the action *diff-subj* with the second occurrence of "to." These actions could take care of the verb following "to." However, we have opted for taking care of infinitives in the lexical analyzer function called *advance*. This function takes care of the entire verb sequence.

The key idea in the algorithm is the function *coalesce*, which selects a non-terminal from the *non-terminal* stack on the basis of the senses of the word being scanned. The selection of the correct arc or non-terminal is determined by matching the non-terminal or non-terminals on the top of the *non-terminals* stack to the senses in *wordsenses*. For instance, *coalesce* selects an arc labeled *inf*, if "inf" belongs to *wordsenses*; an arc labeled *verb*, if "verb" belongs to *wordsenses*; an arc labeled *that*, if the current word is "that," etc. We use the feature *descr* to mark all those words which can form part of a noun group. Thus, adjectives, nouns, articles, pronouns and the -ing form of verbs are marked with the feature *descr*. Hence, the function *coalesce* selects a non-terminal labeled *subj*, or *pred* (predicate) if *descr* belongs to the senses in *wordsenses*. *Coalesce* selects the non-terminal *obj* or *io* (indirect object) if *descr* belongs to *wordsenses* and the marker *nominative* does not belong to it. If *coalesce* selects a non-terminal in this way, we say that a *match* between the non-terminal and the senses in *wordsenses* takes place.

If the top of the *non-terminals* stack contains a single non-terminal, and a *match* exists between that terminal and the senses in *wordsenses*, *coalesce* returns true; otherwise it returns false. If *coalesce* returns true, the top of the *actions* stack is executed, which is the action corresponding to the non-terminal on top of the *non-terminals* stack. Every action, when it is executed, will pop both stacks and, if appropriate, will push a usage and a corresponding action on the *non-terminals* and *actions* stacks, respectively.

Let us now explain the behavior of *coalesce* when the top of the stack is a list. Refer to Figure 1. We call the node marked *a* the *initial* node and the arcs coming from the *initial* node *initial* arcs. The *initial* arcs for the usage of "want" are *inf* and *obj*. Suppose that the top of the stack contains (a1 a2), where a1 is a single non-terminal and a2 is a list. The function *coalesce* will choose between a1 and the first element in a2. In other words, what *coalesce* does is to coalesce the *initial* arcs for the usage of a word. If the usage of "want" is being considered, *coalesce* will try to select the arc labeled *obj* or the arc labeled *inf*. If *coalesce* selects one arc, it will discard the remaining *initial* arcs. It may happen that, in some cases, *coalesce* will fail to coalesce the *initial* arcs. In that case, the saving of some arcs is needed. See the report [4] for a detailed description of this.

For example, consider the sentence *Peter wanted Mary to read a book*. When the non-terminal *verb* is reduced, the usage and actions for "want" will be pushed on the *non-terminals* and *actions* stacks, respectively. Then *coalesce* will be activated with (inf (obj (inf *))) on top of the *non-terminals* stack and "Mary" as the current word. *Coalesce* will return true, pop the stack and push "obj (inf *)" on it. *Coalesce* will perform a similar change in the *actions* stack. Namely, it will pop it and push "obj (diff-subj *)" on it, so that the correspondence between non-terminals and actions is kept.

Relative clauses and prepositional phrases are handled in the following way. When the function that parses noun groups is ready to exit, it pushes (prep np pp rel ing-rel *) onto the *non-terminals* stack, and (preposition rel-np partc-rel relatives ing-rel-act *) onto the *actions* stack. The non-terminal *prep* stands for preposition, *rel* stands for "which," "who" and "whom." The non-terminal *pp* stands for past-participle. Then, it is up to *coalesce* to select the correct arc. *Coalesce* will select the non-terminal *prep* if the marker *prep* belongs to the senses in *wordsenses*; *ing-rel* if the marker *ing* belongs to *wordsenses*; "that" if the the current word is "that;" *rel* if *rel* belongs to the senses in *wordsenses*; *np* if the marker *descr* belongs to *wordsenses*; or *pp* if the marker *pp* belongs to *wordsenses*. The selection of any of these non-terminals is followed by the selection of its corresponding action.

The *non-terminals* stack holds two special symbols: $ and +. The dollar sign indicates that the stack is empty. The + sign is used to indicate the end of the non-terminals for a sentence or an embedded clause. When the parser finds the + sign, it knows that a sentence or clause has been parsed, and therefore, that the structure which it is building for that sentence or clause is complete. The + sign is put on the stack by the initialization step and whenever an embedded clause is found by the action corresponding to a non-terminal.

The following detailed example will give the reader a feeling of how the parser works. Consider the sentence *The book Peter donated vanished.*

```
(1)  (subj verb + $)->(the book peter donated vanished)
(2)  ((prep np pp rel ing-rel *) verb + $)->(peter ...)
(3)  (subj verb + verb + $)->(peter donated vanished)
(4)  ((prep np pp rel ing-rel *) verb + verb + $)->(donated ...)
(5)  (verb + verb + $)->(donated vanished)
(6)  (obj (io for to *) + verb + $)->(vanished)
(7)  (io for to *) + verb + $)->(vanished)
(8)  (+ verb + $)->(vanished)
(9)  (verb + $)->(vanished)
(10) (+ $)->nil
(11) ($)->nil
```

When we refer to the stack, we will always mean the *non-terminals* stack. The initialization step has initialized the non-terminals stack to "subj verb + $" and the *actions* stack (not shown in the example) with the actions corresponding to those non-terminals. Also, a stack called *structs* contains the name of the structure (a gensym) being built for the parse of the sentence. So *structs* is equal to (g01). Line 1: Coalesce matches *subj* to the senses of the current word, "the," and returns true. The action corresponding to the non-terminal *subj* calls the noungroup function which returns "the book." Also the noungroup function inserts in the stack (prep np pp ....). The action *subj* fills the slot subject with "the book" in the structure in top of structs, g01.

Line 2: *Coalesce* matches the non-terminal *np* to "peter," returning true. The action corresponding to that non-terminal, namely *rel-np*, is executed. This action creates the slot rel in g01 and fills it with "the book," for use later; pushes "subj verb +" on the non-terminals stack; and creates a new structure, say g02, and pushes it on *structs*. The stack *structs* contains (g02 g01). Line 3: Coalesce matches the non-terminal *subj* to "peter," returning true. The action for *subj*, after calling the noungroup function, builds the slot subject in the structure in top of structs, namely g02, and fills it with "peter." Line 4: Coalesce does not match any of the arcs to "donated" and returns false. The main algorithm checks to see if * is one of those arcs and, consequently, pops the stack. Line 5: Coalesce returns true. The action for *verb* opens the verb slot in top of *structs* and fills it with "donated," and pushes the usage for "donated" on the stack. Line 6: Coalesce returns false since it cannot match "obj" to "vanished." The main algorithm checks for the non-terminal obj on the top of the stack. Since that is the case, it then searches the structures being built for a slot called rel. In this case, it finds it in the structure g01. Then, it makes the content of slot rel the object of the structure being built (g02); changes the name of the slot from rel to rela, so that it cannot be used again; and fills the rela slot with the name of the structure in top of structs, namely g02. At this point the contents of the structures are:

```
g01 = (subject (the book) rela (g02))
g02 = (subject (peter) verb (donated) object (the book))
```

Line 7: Coalesce returns false. The main algorithm pops the stack since the asterisk belongs to the non-terminals in the top of stack. Line 8: When the main algorithm finds + on top of *structs*, it pops the non-terminals stack and the *structs* stack. Then, the gensym g02 is popped from structs, which is left only with g01. Line 9: Coalesce matches *verb* to "vanished" and the action for *verb* creates and fills the slot verb with "vanished." Note that the structure is g01, since g02 has been popped. The structure g01 will contain at that point:

```
g01 = (subject (the book) rela (g02) verb (vanished))
```

Line 10: the structure g01 is popped from the stack. Line 11: the dollar sign and the end of the sentence are reached, consequently accepting the sentence.

3 The Acquisition of the Syntactic Usages

One easy way to define a syntactical usage for a new word is to search the lexicon for another word which behaves the same as the word one is defining and to copy its definition into the new one. This method, of course, is limited by the word usages currently defined in the lexicon. Moreover, there are many words that do not fall into any category. An even more serious problem with this method is that most end-users are unable to determine if a word has the same syntactic distribution as another one.

It is clear that, if one wants to help end-users, it is imperative that the usage be built from sample sentences posed by the users themselves. In the case of verbs, one could think of attacking this problem by attempting to parse the user's sentences with all distinct verb usages in the lexicon, and picking that usage which succeeds for all sample sentences. Again this method will be limited by the usages presently defined in the lexicon, but a more serious limitation with this method is that it simply does not work. This may be illustrated with a very simple example. Let us assume that a user tries to define the verb "read" and he/she types the sample sentences *Mary read* and *Mary read the book*. The first sentence will be successfully parsed with the usages (obj *) and (*). The usage (obj *) means that the verb may take or not take an object. The second sentence, *Mary read the book* can be parsed with the following usages (among others): (obj), (obj *) and (obj (inf *)). From these usages and the two sample

0. Repeat steps 1 to 4 until the user stops entering sample sentences.

1. Input user's sample sentence.

2. Select segment usages by examining the senses of the word immediately following the one the user is defining.

3. Put these segments in the list Segment-Usages.

4. Repeat steps 4.1 to 4.3 until the list Segment-Usages is empty.

   4.1 Parse the user's sentence with the first usage in the list Segment-Usages.

   4.2 If the parser succeeds, save the usage in the list Successful-Segment-Usages.

   4.3 Remove the first usage from Segment-Usages.

5. Merge the segment usages in Successful-Segment-Usages into a single usage.

Figure 2: Algorithm for selecting the relevant usages of a word.

sentences above, only one usage, namely (obj *), is the correct one. The interface could rule out some of these usages by generating sentences from the successful usages and asking the user if the sentences are grammatical. For instance, the interface could generate the following sentence from the third usage: *Mary reads Peter to go to the movies* and ask the user whether the sentence is correct. This method could work for some cases. However, it may produce a tedious interaction with the user. It also has the problem that a user may reject a sentence as ungrammatical for semantic reasons rather than for syntactic reasons. Most people will consider the sentence *Rocks eat ideas* as ungrammatical although the sentence is syntactically correct.

The solution to this problem is to parse the user's sample sentences with pieces or segments of usages. What a segment usage is will be discussed later. For now, let us observe that the usage (inf (obj (inf *))) is composed of the following segments "obj," "inf," "obj inf." This is to say a verb may be followed by object, or infinitive, or object plus infinitive. Of course, a verb may be followed by many other things, which will be discussed in the next section. The basic idea is to parse the user's sentence with these segment usages, save all segment usages which successfully parse the sentence and finally merge those usages into the final usage for the word. The algorithm is described in Figure 2 above.

Suppose that a user is defining "read" and he/she types the sentences *Mary read* and *Mary read a book*, and let us further assume that the segment usages in the list

Segment-Usages are: (obj), (inf), (*), (obj inf). In this case, the parser will succeed when it parses the first sentence with the usage (*), and the second sentence with the usage (obj). The list Successful-Segment-Usages will contain ((obj) (*)) in step 5 of the algorithm. This list will be passed to the function *merge*, which will form a single usage for "read" *based on the two sentences* provided by the user. Let us consider each one of these steps separately.

4 Learning the Subcategorization of Verbs

First, we discuss the acquisition of the syntactic usage for verbs. The list below contains the segment usages for verbs that are implemented in the current version of the interface, and also example sentences illustrating each usage. All these segment usages are stored in the list Segment-Usages. For each one of these segments, there is a corresponding action, which is not listed here.

1. (*)   He read.
2. (adj)   He looks sad or He is sad.
3. (obj)   He read the book
4. (obj adj)   He kept her warm.
5. (obj bare-inf)   I saw him fall.
6. (obj inf)   I want Mary to read a book.
7. (obj ing)   I saw her reading a book.
                               (obj + present part.)
                               I stopped him talking all night.
                               (obj + gerund)
8. (obj io)   I gave Mary the book.
9. (obj verb)   I know Mary likes Peter.
10. (obj adverb inf)   She showed him how to read.
11. (obj particle)   He gave smoking up.
12. (obj adverb subj verb)   He showed Mary how Peter proved the theorem.
13. (inf)   Peter wants to read.
14. (prep)   She believes in you.
15. (particle)   She gave up.
16. (ing)   Peter likes flying planes.
17. (poss ing)   I enjoyed his coming.
18. (adverb inf)   He knows how to read.
19. (adverb subj verb)   He knows why Mary likes books.
20. (that subj verb)   He knows that Mary likes books.

The segment (*) is for intransitive verbs. The segment usage (obj inf), as has been explained, will succeed in sample sentences like *I want Mary to read*. Note that that usage will fail in the sentence *I want Mary* because "Mary" is not followed by an infinitive. The usage (obj bare-inf) is handled like the (obj inf). The NP following the verb is taken as a direct object, but if a bare infinitive is found, the obj is made the subject of the bare infinitive. The segment usage (obj io) handles bitransitive verbs by the usual procedure of assuming that the NP following the verb is the object, but if that NP is immediately followed by another one, the action for the non-terminal io will change the object slot to io, and will make the second NP the object. That segment usage will not handle indirect objects introduced by the prepositions "to" or "for." The segment (obj verb) handles verbs which take a clause as direct object e.g., *Mary knows Peter likes her*. This is a done in a way similar to the usage (obj inf). The parser will take the NP following the verb to be the object, but if a verb is found then the the object will become the subject of the verb. Note that it is the usage (obj (verb *)), which makes possible this parsing. The usage (obj (verb *)) is the result of merging the segments (obj) and (obj verb). The usages (particle) and (obj particle) deals with particles. The interface infers that a preposition is a particle if the user's sentence ends in a preposition. For instance if the user types *Bears rolled over*, or *Peter gave cigarettes up*, the interface will infer that "over" and "up" are particles. Adverbs that may be used as particles are marked in the dictionary. Then, the interface recognizes these adverbs in sentences like *Peter found out that Mary like him*, or *Peter found out the truth*. (See appendix for an example.)

The segment usages (prep) and (ing) are for verbs which are followed by a prepositional phrase and a gerund, respectively. In the case of (ing), there is a problem with ambiguity. If a user is defining the verb "like" and she/he types the sentence *Mary likes flying planes*, the parser will return an ambiguous parse for this sentence, and also for the sentence *Mary sells flying planes*. Ambiguous cases are discussed in section 6. The two segments which start with "adverb" handle sentences like *Mary knows how to read books* and *Mary understood how Peter entered the house*. Finally, the last segment covers that-clauses like *I think that Peter saw Mary*. There is no need to try a user's sample sentence defining a verb with all segment usages, since the word immediately following the one she/he is defining constrains the relevant usages to a small subset. When a user types a sample sentence the senses of the word following the one being defined are examined. If the following word is unknown, he/she is asked to define that word prior to proceeding with the one presently being defined.

5 Merging the Segment Usages into a Final Usage

For the sample sentences *Peter wants a book*, *Peter wants Mary to buy a book* and *Peter wants to buy a book*, the parser will succeed with the following segment

1. Make the list Successful-Segment-Usages a set.
2. Move all singleton subsets of Successful-Segment-Usages to the list Singleton-Sets.
3. Repeat steps 3.1 up to 3.4 until the set Successful-Segment-Usages is empty.
   - 3.1 Call the first subset of Successful-Segment-Usages FIRST. Move FIRST and all subsets of Successful-Segment-Usages whose first element is the same as the first element in FIRST to the list L.
   - 3.2 If there is more than one set in L, eliminate the first element of all subsets of L, except from one. Form the union of all subsets of L and put the result back in L.
   - 3.3 If the first element in the first subset of L is a member of any set in Singleton-Sets, remove that set from Singleton-Sets and insert-the-asterisk in L. (The function insert-the-asterisk yields (a (b c d *)) if L is equal to (a b c d). In Lisp this is defined as (list (car l) (append (cdr l) '(*))).)
   - 3.4 Put L in Final-Usage. Set L to nil.
4. Move all sets in Singleton-Sets - if any left - to Final-Usage.
5. Return Final-Usage as the final usage for word.

Figure 3: Algorithm for merging the relevant usages of a word into a deterministic representation.

usages: (obj), (obj inf) and (inf), respectively. When the function *merge* is called it will merge those segment usages into the final usage for "want," namely (inf (obj (inf *))). The role of *merge* is to transform a representation which could be highly non-deterministic into a deterministic one, making it possible for *coalesce* (see section 2) to select deterministically the correct arc. The algorithm for *merge* is described in Figure 3. (Successful-Segment-Usages is a list containing all segment usages which successfully parsed the user's sample sentences).

Let us explain each one of these steps and illustrate them with examples. Let us assume that Successful-Segment-Usages contains ((a) (a f) (a k) (a m) (h m) (d)). Step 1 is necessary because two or more sample sentences can be parsed using the same usage. For instance, a user may type *Peter wants the red book* and *Peter wants books* as examples of two usages of "wants." In that case, the list Successful-Segment-Usages will contain (obj) twice. We could tell the user that his/her second sample sentence

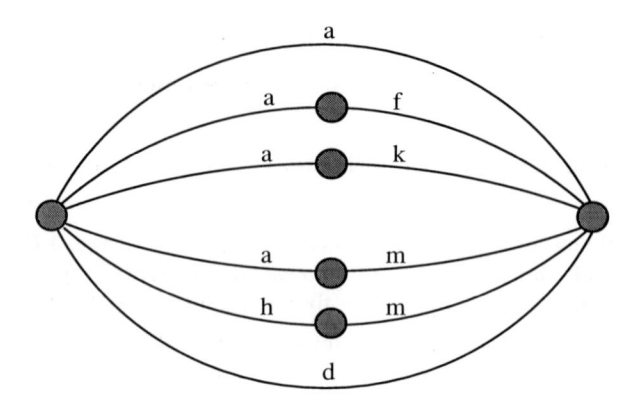

Figure 4: Transition diagram representation of input ((a) (a f) (a k) (a m) (h m) (d)) for the function *merge*.

does not show a new usage of "wants," and teach her/him a little bit of syntax along the way. We may do that in later versions of the interface. Step 2 is intended only to facilitate the next two steps. After executing step 2, the list Successful-Segment-Usages is equal to ((a f) (a k) (a m) (h m)) and the list Singleton-Sets is equal to ((a) (d)). After executing 3.1 the list L is equal to ((a f) (a k) (a m)) and the list Successful-Segment-Usages is equal to ((h m)). After executing step 3.2, L is equal to (a f k m). After executing step 3.3, L is equal to (a (f k m *)) and Singleton-Sets is equal to ((d)). After executing step 3.4, Final-Usage is equal to ((a (f k m *))) and L is equal to nil. In the second iteration (h m) is put into Final-Usage. The asterisk is not inserted because (h) does not belong to the Singleton-Sets. In step 4, (d) is moved to the Final-Usage whose final content becomes ((a (f k m *)) (h m) d). Figure 4 depicts the example above in a transition diagram, and Figure 5 depicts the output of *merge*.

6 Learning the Subcategorization of Nouns and Adjectives

It is very easy for an end-user to understand that in order to define a verb she/he needs to type some sample sentences using the verb. However, a completely different situation occurs in the case of nouns and adjectives. We mention that some nouns and adjectives have special "constructions" and provide some examples. This does not work in many cases, however, because many end-users do not distinguish between "that" as a conjunction and as a relative. It is common for a user to define,

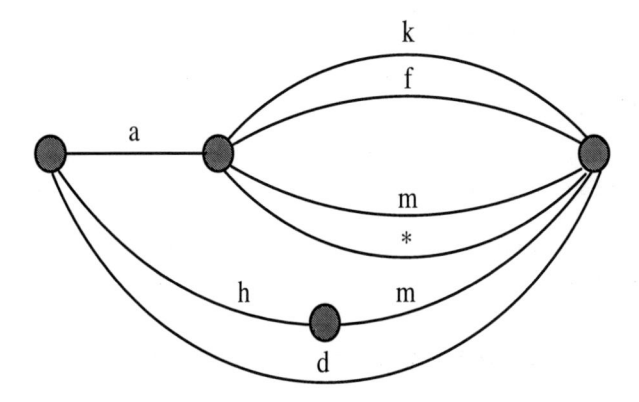

Figure 5: Transition diagram representation of merge's output for input depicted in Figure 4.

say, "announcement" or "ability" as a noun without providing any sample sentences indicating its usage. Then, one week later he/she is surprised that the parser fails to parse the sentences *The announcement that Peter was coming pleased her* and *Her ability to write is unique*. The problem with reminding users that certain nouns may be followed by a *that-clause* is that the next time they are defining a noun, say, "news," they may type the following sample sentence *The news that Peter brought was sad*. This does not cause any problem for the interface, because it is very easy to prevent this type of mistake by parsing the sentence with "that" as a relative in case the other usages for nouns fail.

This problem is even worse in the case of adjectives, because some adjectives in combination with *be* distribute over a large domain of different usages. The data we have gathered with the interface has indicated to us that the best way to deal with this problem is by telling the users not to be overly concerned about the definition of adjectives or nouns until the parser fails to parse a sentence. Then, they should activate the interface with the *same* sentence that failed (The automatic activation of the interface after the parser fails has not yet been implemented.). Because the user's sentence may contain several user-defined words, they must redefine the one that they suspect caused the problem. An aspect that is presently under design is to let the parser determine which word caused the problem and correct it by automatically activating the interface with the failed sentence. The following three usages for nouns are implemented:

```
(inf)              Peter has the ability to read fast.
(that-clause)      The fact that Mary came pleased him.
```

```
(it be np subj verb)  It's a pity Kafka died so young.
```

In the third usage, the head noun of the np (*pity*) is the *noun* being defined by the user. The following usages for adjectives have been implemented:

```
(be adj inf)            It was impossible to read at the store.
(be adj prep np inf)    It was nice of you to help him.
                        It was impossible for Peter to leave.
(be adj that-clause)    I am sad that Mary left.
(be adj prep)           I am sorry about your tree.
```

7 Ambiguity and Other Problems

In his/her sample sentence a user may include any word which is defined in the lexicon, or he/she may define a word and then use that word in defining other words. For instance, a user may define "duck" by saying that it is a noun and a verb. Then, in defining the verb "see," he/she may type the sample sentence *I saw her duck*. The interface will try this sentence with all segment usages which begin with obj. The parser will succeed with the three segment usages (obj), (obj bare-inf) and (obj io). A user may type the sample sentence *I took her flowers* for which the parser will succeed with the segments (obj) (obj io). In an earlier version of the interface, if a user's sentence was ambiguous he/she was asked to choose another sample sentence. In our present implementation, the user's sample sentence is modified and he/she is asked to judge if the altered sentence is grammatical. The modification of the sentences follow very simple rules. A lexical entry which may be a pronoun or an adjective is replaced with a proper noun. For instance, "her" will be replaced in *I saw her duck* with "Mary" resulting in *I saw Mary duck*. Then, the user will be asked if this sentence is grammatical. In the sentence *I like flying planes*, the parser will succeed with the segment usages (obj) and (ing). In this case, the rule used is to modify the user's sentence by deleting every word following the -ing *form*. In this example, the altered sentence will become *I like flying*.

The unrestricted use of adverbs by end-users caused some problems in an earlier stage of the interface. The problem was solved by marking in the dictionary those adverbs and prepositions which may act as particles, and also by encoding knowledge in the parser about a large class of adverbs including words such as "yesterday," "later," "there," "home," ...

Time NPs may also pose some problems because they may appear almost anywhere in the sentence. For instance, a user may type a sentence like "Popeye ate spinach every day," which will result in building an indirect object for "ate." Of course, this problem can easily be solved by asking the users not to construct sample sentences having nouns referring to time, such as "day," "hour," etc. Because the present implementation of the parser is able to handle time NPs, we are asking the user to tell us if the noun they are defining is a noun referring to time. If they do that, the parser is able to determine that "every day" in the sentence above is a time NP.

8 Related Work

Carter's work [3] has a similar aim as this, but is based on elicitation techniques. Work on transportable natural language interfaces [1, 8] bears only an indirect relation to our work, because those authors designed their systems as front ends for data base systems. The system described in [9] is a intended for the acquisition of semantic knowledge, not syntactic subcategorizations, and is designed on principles very different of those presented in this paper. As indicated in the introduction, the view which underlies our approach to syntactic knowledge can be traced back to a paper by Gross [7]. He and his associates realized, after building a transformational-generative grammar of French containing 600 rules, that the task could not be achieved without abandoning the main theoretical tenets of transformational-grammar. This realization came about after applying these 600 rules to over 12,000 lexical items. Gross found out that, for every syntactic transformational rule, one can find some lexical items which obey the rule and others which don't. Our syntactic knowledge acquisition algorithms are a consequence of this point of view: "General" syntactic rules are not the target of the acquisition algorithms, but rather verb patterns, adjective patterns, etc. This is clearly in contrast to the ideas presented in [2].

From a learning point of view, our approach is a knowledge-intensive one. Our learning algorithms learn because they already know a lot. This knowledge is in the parser, which covers a considerable portion of English syntax, in the form of a few "general" syntactic rules and a considerable number of syntactic word usages.

Finally, the way in which our algorithm acquires the syntax of new words is somewhat similar to the way in which a non-native speaker of a language acquires the syntax of new words. She/he, we hypothesize, starts by mastering a set of basic and "general" syntactic rules (e.g. passive, relative clauses ...) and, then, she/he begins to face the enormous irregularity of English syntax by learning the syntax of individual words. Then, one day he/she starts building sentences with the verbs "want" and "expect" such as:

Mary expected to read a book.
Mary wanted to read a book.

Then, when in future days she/he says,

Mary was expected to read a book.
*Mary was wanted to read a book.

he/she finds, with unpleasant surprise, that "Mary was wanted to read a book" is not grammatical, and that she/he needs to learn a usage for "want" and another for "expect." Now, when she/he had thought that the two verbs could be put in "the same sack"!

9 Conclusion

We have presented an algorithm that acquires syntactic subcategorizations for verbs, nouns and adjectives from sample sentences entered by end-users. The algorithm synthesizes the rules in a way which allows the parser to parse the sentences deterministically. The interface is implemented in Allegro Common Lisp and is running on Symbolics and SPARC workstations. The time the interface takes to acquire a subcategorization for a given word depends, naturally, on how complex the syntax of that word is. However, the average time that the interface takes in processing a user's sentence describing one syntactic usage of a word is about 4 seconds. In most cases, the user has barely finished typing the sentence, when the interface comes back asking him/her for a new sentence describing a different syntactic usage. All these tests have been performed using interpreted code, not compiled code. The interface is now being used on a daily basis by SNOWY's users, and has performed beyond our most optimistic expectations. Moreover, it is helping us to gather highly interesting information about naive users' intuitions about English syntax. In that regard, we are using the interface as the main component of a intelligent tutoring system to help 7th and 8th graders to learn to analyze parts of speech. In this application, the children type English sentences, many of them taken verbatim from the exercises in their textbooks, and observe the interface identifying the subject, direct object, predicate nominative, predicate adjective, embedded clause, adverbial clauses, etc. Of course, the interface finds this out from the sentences typed by the children, not from a pre-stored set of sentences. In order to test this aspect of the interface, we are in the process of transporting the parser and the interface to PCs running Allegro Common Lisp.

References

[1] B. Ballard and N. Tinkham. A grammatical framework for transportable natural language processing. *Computational Linguistics*, (1984), **10**, No. 2, pp. 81-96.

[2] R. C. Berwick. *The Acquisition of Syntactic Knowledge*. MIT Press, Cambridge, Massachusetts (1985).

[3] D. Carter. Lexical acquisition in the core language engine. CRC-012, Cambridge Research Center, SRI International (1989).

[4] F. Gomez. WUP: a parser based on word usage. UCF-CS-90-1, Department of Computer Science, Orlando, FL 32751, (1990).

[5] F. Gomez and C. Segami. Knowledge acquisition from natural language for expert systems based on classification problem-solving methods. *Knowledge Acquisition*, (1990), **2**, pp. 107-128.

[6] F. Gomez and C. Segami. Classification-based reasoning. *IEEE Transactions on Systems, Man and Cybernetics*, (1991), **21**, No. 3, pp. 644-659.

[7] M. Gross. On the failure of generative grammar. *Language*, (1979), **55**, No. 4, pp. 859-883.

[8] B. Grosz, D. Appelt, P. Martin and F. Pereira. TEAM: an experiment in the design of transportable natural-language interfaces. *Artificial Intelligence*, (1987), **32**, No. 2, pp. 173-243.

[9] U. Zernik and M. Dyer. The self-extending phrasal lexicon. *Computational Linguistics*, (1987), **13**, No. 3-4, pp. 308-327.

Appendix

Sample Session with the Interface

The output of the system is in capital letters. The input by the user is in lower case letters. Explanatory notes to the session are preceded by the expression "**REMARKS."

```
Script started on Sat May  1 04:25:08 1993
barros 1% cl
Allegro CL 4.1 [SPARC; R1] (6/9/92 16:51)
Copyright (C) 1985-1992, Franz Inc., All Rights Reserved.
(start)
(TYPE SENTENCE/DONE)
they believed mary ate the apples.
(I DO NOT UNDERSTAND * BELIEVED *)
(SUBJ VERB + $)->(THEY BELIEVED MARY ATE THE APPLES)
(VERB + $)->(BELIEVED MARY ATE THE APPLES)
```

     **REMARKS: PARSER FAILS BECAUSE IT DOES NOT KNOW
     ''BELIEVED.''

```
ERROR: UNKNOWN VERB
PARSING-ATTEMPTS-COMPLETED
(TYPE SENTENCE/DONE)
done
```

    **REMARKS: INTERFACE IS LOADED AND ACTIVATED IN NEXT
    LINE BY USER IN ORDER TO DEFINE ''BELIEVED.''

```
USER(3): (in interface.l)
; Loading /usr2/faculty/gomez/parser/interface.l.
USER(4): (interf)
PLEASE TELL ME IF THE WORD YOU WANT TO ADD IS A VERB. YES/NO.
yes
PLEASE ENTER IN A LIST THE PRESENT TENSE, THE SIMPLE PAST
AND THE THE PAST PARTICIPLE FORMS OF THE VERB. E.G. (READ READ READ)
(believe believed believed)
PLEASE ENTER A SENTENCE USING THE SIMPLE PAST.
peter believed Mary ate the apples.
DO YOU WANT TO TYPE ANOTHER SENTENCE USING THE SAME WORD.
BUT SHOWING A DIFFERENT USE OF THE WORD? YES/NO.
yes
PLEASE ENTER A SENTENCE USING THE SIMPLE PAST.
peter believed that mary went to spain.
```

```
DO YOU WANT TO TYPE ANOTHER SENTENCE USING THE SAME WORD? YES/NO.
yes
PLEASE ENTER SENTENCE USING THE SIMPLE PAST.
peter believed in mary.
DO YOU WANT TO TYPE ANOTHER SENTENCE USING THE SAME WORD?
yes
PLEASE ENTER A SENTENCE USING THE SIMPLE PAST.
peter believed mary.
DO YOU WANT TO TYPE ANOTHER SENTENCE USING THE SAME WORD?
no
BESIDES ** (IN) **
ARE THERE ANY OTHER PREPOSITIONS WHICH MAY immediately FOLLOW
   **believe**?
IF MOST PREPS CAN FOLLOW IT, WRITE "MOST." IF ONLY THOSE
WRITE "ONLY." IF OTHERS WRITE THOSE PREPS IN A LIST.
HELP/MOST/ONLY/LIST CONTAINING OTHER PREPS ?
only
(DO YOU WANT TO STORE THE NEW WORD IN THE FILE USER-LEX? YES/NO)yes
(DO YOU WANT TO ADD ANOTHER WORD?YES/NO)no

            **REMARKS: PARSER IS ACTIVATED BY USER AGAIN.

USER(5): (start)
(TYPE SENTENCE/DONE)
they believed that mary ate the apples in her car.
(SUBJ VERB + $)->(THEY BELIEVED THAT MARY ATE THE APPLES IN HER CAR)
(VERB + $)->(BELIEVED THAT MARY ATE THE APPLES IN HER CAR)

**REMARKS: NOTE THE USAGE BUILT FOR ''BELIEVE'' NOW ON TOP OF STACK.

((THAT IN (OBJ (VERB *))) + $)->(THAT MARY ATE THE APPLES IN HER CAR)
(SUBJ VERB + + $)->(MARY ATE THE APPLES IN HER CAR)
((PREP NP PP REL ING-REL *) VERB + + $)->(ATE THE APPLES IN HER CAR)
(VERB + + $)->(ATE THE APPLES IN HER CAR)
((OBJ *) + + $)->(THE APPLES IN HER CAR)
((PREP NP PP REL ING-REL *) + + $)->(IN HER CAR)
((PREP NP PP REL ING-REL *) + + $)->NIL
(+ + $)->NIL
(+ $)->NIL
($)->NIL(PARSED! WANT TO SEE STRUCTURES? Y/N)y

**REMARKS: OUTPUT OF PARSER.
```

```
STRUCTS
USED-STRUCTS
G7589
(SUBJ ((PRON THEY)) VERB ((MAIN-VERB BELIEVE BELIEVED) (TENSE SP)) OBJ
     ((G7635)))
G7635
(SUBJ ((PN MARY)) VERB ((MAIN-VERB EAT ATE) (TENSE SP)) OBJ
     ((DFART THE) (NOUN APPLES)) PREP (IN ((ADJ HER) (NOUN CAR))))
(TYPE ANY LETTER TO CONTINUE)
(TYPE SENTENCE/DONE)
done

**REMARKS: INTERFACE IS ACTIVATED AGAIN

USER(6): (interf)
PLEASE TELL ME IF THE WORD YOU WANT TO ADD IS A VERB. YES/NO.
yes
PLEASE ENTER IN A LIST THE PRESENT TENSE AND THE SIMPLE PAST
AND THE THE PAST PARTICIPLE FORMS OF THE VERB. E.G. (READ READ READ).
(find found found)
PLEASE ENTER A SENTENCE USING THE SIMPLE PAST.
peter found the truth.
yes
DO YOU WANT TO TYPE ANOTHER SENTENCE USING THE SAME WORD,
BUT SHOWING A DIFFERENT USE OF THE WORD? YES/NO.
yes
PLEASE ENTER A SENTENCE USING THE SIMPLE PAST.
peter found that mary went to spain.
DO YOU WANT TO TYPE ANOTHER SENTENCE USING THE SAME WORD?
yes
PLEASE ENTER A SENTENCE USING THE SIMPLE PAST.
mary found peter happy.
DO YOU WANT TO TYPE ANOTHER SENTENCE USING THE SAME WORD?
yes
PLEASE ENTER A SENTENCE USING THE SIMPLE PAST.
peter found out that some apples are green.
DO YOU WANT TO TYPE ANOTHER SENTENCE USING THE SAME WORD?
no
DO YOU WANT TO STORE THE NEW WORD IN THE FILE USER-LEX? YES/NO.
DO YOU WANT TO ADD ANOTHER WORD?YES/NO.
no
```

USER(7): (on find)

**REMARKS: USAGE BUILT FOR FIND:

(((((CTGY VERB) (ROOT FIND) (TENSE PS)
    (USAGE (((OUT THAT) (OBJ (PRED *)) THAT)))
    (ACTIONS (((PARTICLE THAT-CLAUSE) (OBJ (PRED *)) THAT-CLAUSE))))))

REMARKS: INTERFACE IS ACTIVATED AGAIN
USER(8): (interf)
PLEASE TELL ME IF THE WORD YOU WANT TO ADD IS A VERB. YES/NO.
yes
PLEASE ENTER IN A LIST THE PRESENT TENSE, THE SIMPLE PAST
AND THE PAST PARTICIPLE FORMS OF THE VERB. E.G. (READ READ READ).
(learn learned learned)
PLEASE ENTER A SENTENCE USING THE SIMPLE PAST.
peter learned the truth.
(DO YOU WANT TO TYPE ANOTHER SENTENCE USING THE SAME WORD.)
yes
PLEASE ENTER A SENTENCE USING THE SIMPLE PAST.
peter learned how mary ate the apple.

DO YOU WANT TO TYPE ANOTHER SENTENCE USING THE SAME WORD?
yes
PLEASE ENTER A SENTENCE USING THE SIMPLE PAST.
peter learned to eat apples.
DO YOU WANT TO TYPE ANOTHER SENTENCE USING THE SAME WORD?
yes
PLEASE ENTER A SENTENCE USING THE SIMPLE PAST.
peter learned how to eat apples.
DO YOU WANT TO TYPE ANOTHER SENTENCE USING THE SAME WORD?
yes
PLEASE ENTER A SENTENCE USING THE SIMPLE PAST.
peter learned that apples are good for you.
DO YOU WANT TO TYPE ANOTHER SENTENCE USING THE SAME WORD?
no
DO YOU WANT TO STORE THE NEW WORD IN THE FILE USER-LEX? YES/NO.
yes
DO YOU WANT TO ADD ANOTHER WORD?YES/NO.
no
USER(9): (on learn)

```
**REMARKS: USAGE BUILT FOR LEARN:

(((((CTGY VERB) (ROOT LEARN) (TENSE PS)
    (USAGE (((HOW (INF (SUBJ VERB +))) OBJ INF THAT)))
    (ACTIONS
     (((ADVERB-ACT (INF-AFTER-ADV (SUBJ VERB))) OBJ SAME-SUBJ
       THAT-CLAUSE))))))
USER(10): ^D
Really exit lisp [n]?
; Exiting Lisp
barros 2% exit
script done on Sat May  1 03:27:26 1993
```

SECTION : 4
PLANNING

GENERAL METHOD FOR PLANNING AND RENDEZVOUS PROBLEMS

KAREN I. TROVATO
Philips Laboratories - Philips Electronics North America Corporation
345 Scarborough Road
Briarcliff Manor, New York   10510 USA

ABSTRACT

This chapter describes a general method for representing and solving planning problems. The framework has well defined subcomponents to simplify the problem using transforms, define costs over permissible motions, define illegal regions in the transformed space, and efficiently find optimal motions. The method fully exploits any a-priori information and also provides a method to augment this information efficiently at runtime.

The method is powerful in that it can be used in arbitrarily high dimensional spaces, and has been used to solve non-holonomic problems efficiently and with ease. Examples will be given for moving robotic equipment optimally while avoiding obstacles, for automatically maneuvering a vehicle around obstacles, and for determining alternative rendezvous locations for machinery based on the separate constraints (time, space, fuel, etc.) of each.

1.0 Introduction

Diverse types of planning problems share a number of attributes. They often have a goal or set of goal states, a current state, some measure of success, and limits on how the system can progress from the starting state to any other state. The objective is to provide an optimal path to the nearest goal while avoiding the limits, which can be hard obstacles, movement based on rules of the game or even kinematic constraints. This chapter describes a framework that can be used to represent such problems, and then gives examples where it has been applied.

2.0 Framework

There are several elements that are required in the framework: a problem state-space, a neighborhood, a cost metric on the neighborhood, a transformation of the obstacles/forbidden regions, and a goal set. Each of these elements are described in further detail below along with a simple example of each. The general idea is that by employing these elements it is possible to transform the problem into an easier space in which to solve the problem. The solution is then transformed back to the original 'task' space where it is carried out.

The state of the system must be characterized by one or more parameters which can uniquely describe that state. For example, the two joint angles of a two link robot arm uniquely describe the position of the robot. A test for the chosen parameters is to ask 'Do these parameters give a unique state of the system?'.

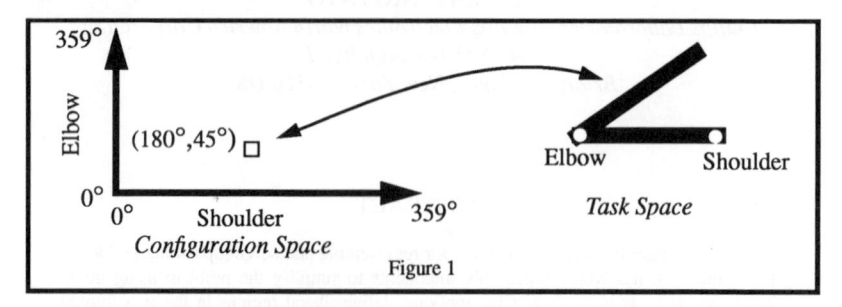

Figure 1

The span of legal ranges for the parameters defines the problem state-space. The reformulation of the problem in this space is called *configuration space*[1] in robotics and is sometimes abbreviated CS. This is an easier space in which to solve the problem than the task space because the status of the entire system can be characterized by a single state within the configuration space. Continuing with the robot example, a robot that has shoulder and elbow joints that are fully revolute, would have a configuration space as in Figure 1. A single state describes a specific pose of the robot Each state is specified by the values assigned to the parameters. Figure 1 also shows the mapping between a state and the corresponding pose in task space.

The *neighborhood* in the configuration space is a set of possible motions for moving 'legally' from one state to another assuming no obstacles. It describes the fundamental motions of the system, and is depicted by an arrow showing the direction of a particular motion. Therefore if the example robot has one motor that is switched between the shoulder and elbow joint so that only one can move at a time, then the neighborhood of possible motions is as in Figure 2.

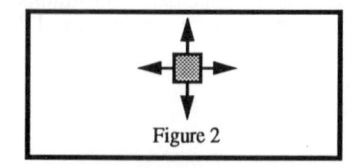

Figure 2

This neighborhood represents one unit of motion in the horizontal direction corresponding to a unit change in the shoulder angle, and one unit of motion in the vertical direction corresponding to a unit change in the elbow angle. This is a simplified example of the neighborhood. If the joints could be moved simultaneously then four diagonal arrows would be seen as well.

$$M(\theta_1,\theta_2,\delta\theta_1,\delta\theta_2) = \sqrt{(L_1\delta\theta_1)^2 + (L_2\delta\theta_2)^2 + 2L_1\delta\theta_1 L_2\delta\theta_2\cos(\theta_1-\theta_2)}$$

Figure 3

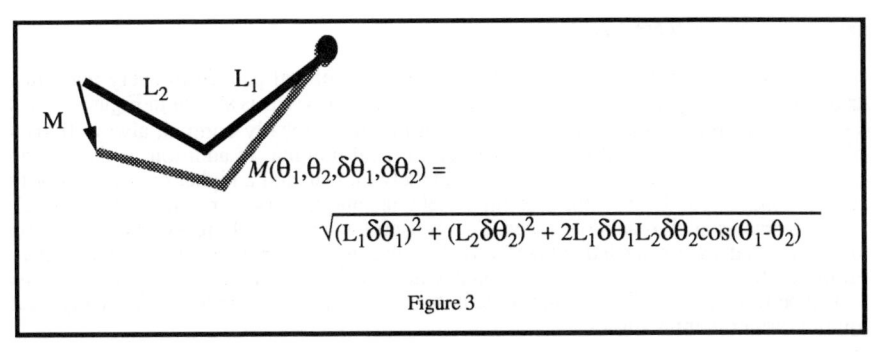

A cost can be associated with each motion, resulting in a cost-weighted neighborhood relative to a given state. The *cost metric*[*] is usually expressed as a function of the 'current' state and the neighbor state, or equivalently, the current state and the difference to achieve the neighbor state. The metric need not be memoryless, that is, it may have cost factors that are a function of prior motions. Admissible heuristics also may be employed as part of the metric. The metric can be used to implement an optimality criterion, for example, a metric M that will minimize the distance travelled by the 2-link robot's end effector (giving the straightest possible motion in the task space) can be expressed as in Figure 3.

The *transformation* of obstacles from task space to configuration space can be very difficult, particularly for high-dimensional spaces with complex shaped obstacles. Others[2,3] have examined such cases, but no fast, general transform methods currently exist. Fortunately in problems such as point-to-point routing, rendezvous and strategic planning, 2-DOF (degrees of freedom) robot arm path planning and vehicle maneuvering, the transformation can be performed in real time.

Goal states are a specification of acceptable target locations. These may be supplied by a higher level task planner[4], or by manual specification (pointing to a grid on a map for example).

The *starting state* must be known in order to issue commands to the controllable entities. It is not required for the planning process unless a heuristic (using the starting state) is employed.

Once the elements of the framework are in place, the cost-weighted neighborhood is simultaneously propagated from all goal states through the configuration space in an A*[5], least-cost-first manner until the space is filled (or at least until the start state has been reached). In most machine based problems it is preferable to propagate waves from the goal rather than the start, although wave propagation can clearly be performed in the reverse direction (with appropriate reverse calls to the metric function). Details for an efficient cost-wave propagation (CWP) technique are given in a previous paper[6]. The cost wave propagation marks each state with the cost to the goal and a pointer toward the adjacent

---

[*]. The word 'metric' is used here more loosely than the strict mathematical definition. The cost must be >=0, but the triangle inequality and commutative properties need not be enforced.

state leading to the nearest goal.

The result of the cost wave propagation is a globally optimal path from every reachable state to the nearest goal. The total solution is a field of arrows as is shown in Figure 4. This is different from potential field techniques[7], in that it is globally optimal, always finds a solution if there is one, and never gets stuck in local minima. In addition, if there is no solution to get to the goal, then this is immediately known, and a higher control authority can be summoned. Unlike the potential field technique however, there are optimal-direction arrows at all reachable states. This is useful when controlling a device or issuing commands that are not carried out perfectly. For machines in the field, unexpected slippery terrain or failing gear mechanisms could cause such problems. In this event, if the controlled device falls off the optimal path, then a new path can be found from any state without further computation.

An example of a cost-wave propagation solution for the robot arm example is shown in figure 4. The black region in the configuration space represents all of the configurations where the pose of the robot arm results in a collision with the round obstacle. For each state in the configuration space there is at least one arrow that shows the next pose (setpoint) that the robot should achieve. By following these setpoints the robot will move optimally and purposefully to the goal. Note that the imperfections in the end-effector travel shown in the task space of Figure 4 are due in part to the motion required for obstacle avoidance, but also to discretization error.

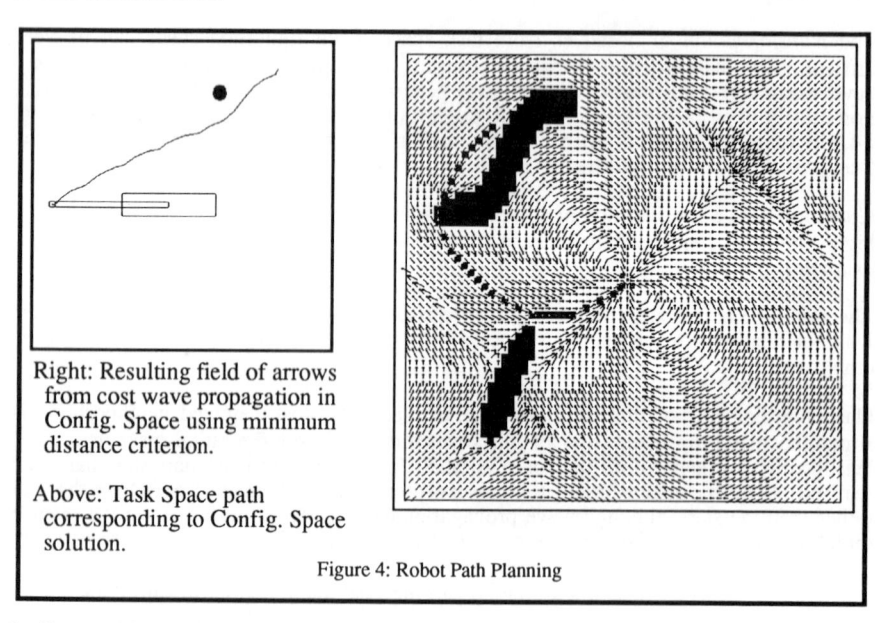

Right: Resulting field of arrows from cost wave propagation in Config. Space using minimum distance criterion.

Above: Task Space path corresponding to Config. Space solution.

Figure 4: Robot Path Planning

In live problems, the environment is not static but ever changing. Obstacles and goal opportunities arrive and depart over time. To keep pace with the changing environment, the planner must adapt the plan as quickly as possible. In the current implementation, a 10,000

(100 x 100) state configuration space with an 8-connected neighborhood can be computed from scratch in 595 ms (~1/2 second) on a Sun-4 (SPARCstation IPX). A similarly connected 90,000 (300 x 300) state configuration space can be computed 6177ms. Higher connected neighborhoods require more time, as will large numbers of states. Even though these times seem fast enough, more complex problems are easy to create.

Since global environments and the resulting plans often change piecewise it is more efficient to update only the changed areas and the subsequently affected areas. In these situations, a method called Differential A* [8,9] can be used. This has two advantages. Not only does the computation take less time, but the unaffected regions can continue to be used for control. Thus if an incident takes place at the opposite end of the area of interest from a controlled entity, then the controlled entity can continue its work while the incident region is recomputed. Another application is when information about the environment is being assimilated while the machine is operating. In this case the machine can, in a simple sense, learn and adapt to the environment quickly. For example, an intelligent vacuum cleaner might detect the layout of furniture while it is cleaning so as to create a better vacuuming pattern the next time.

Another improvement on the basic framework is to create a more concise representation of the freespace by grouping adjacent free configurations into hyper-parallelepipeds of various sizes[10]. Alternatively, the 'swept bubble' concept [11] can divide recursively smaller spaces according to obstacle occupancy. Swept bubbles also provide a uniform geometric representation of the machine and the obstacle environment which provides quick intersection tests required for high-dimensional planning.

The framework that has been described allows the computation of quite diverse problems simply by modifying the relevant elements in the framework. This has proven quite powerful, especially when computing higher dimensional spaces.

A variety of machine examples follow that describe what modifications were necessary to each element that result in the desired plans. Finally, the framework is applied to the rendezvous problem which can be used to predict worst-case enemy strategies and efficient coordinated responses.

3.0 Algorithm Analysis / Accuracy

Discretizing configuration space causes some error to be introduced. The amount of error for Euclidean distances in 8-connected and 16-connected neighborhoods has been studied[12,13]. The 8-connected neighborhoods are the horizontal, vertical and diagonal transitions in configuration space. A 16-connected neighborhood additionally includes the 8 knights-moves. By using 8 connected neighbors in a Euclidean metric there is a 4% error from the optimal solution. By using 16 connected neighborhoods there is a 1.4% error. Therefore a trade-off between increasing discretization and increasing neighborhood size can be made.

The worst case timing for this algorithm is $O(N \log N)$ for N states in the configuration space. Wave propagation is a one pass method with no computation for obstacle states. Therefore, obstacle strewn environments give faster results than those with no obstacles.

For robot examples, where the number of states is determined by the resolution R, and by the number of degrees of freedom D, $N=R^D$. For i neighbors, Time $\propto$ i R^D log R. Even though this is exponential, many practical problems have still been solved in acceptable time by accommodating the most essential factors of each specific problem. Some key factors that improve computation time are described next.

The nature of the solution method makes it more efficient for spaces with more obstacles, which are also typically problems that are the most difficult to perform by inspection. The resolution R is a tunable parameter that can be adjusted for regions of greater sensitivity, and has even been adjusted dynamically[10,11]. For robotic applications, D may be adjusted by observing that most machines have the greatest volume of motion in only 3 or 4 degrees of freedom, whereby the remaining degrees of freedom may be treated in a fixed configuration until the final manipulation is required. Finally, the number of neighbors i may be adjusted for a slight gain in computing time, although this may not be desirable because of the loss in accuracy. The trade-off between the time, space and accuracy must be made separately for each implementation.

4.0 THE FIRE EXIT PROBLEM

The objective is to provide exit directions for inhabitants of an unfamiliar building (conference center, hotel, etc.). Observe that currently in a fire, we are all trained to blithely follow the red exit signs even if they lead directly into the blaze of a fire. Oddly enough, the fire/emergency systems today can specifically pinpoint the problem areas, but cannot convey that information to the inhabitants.

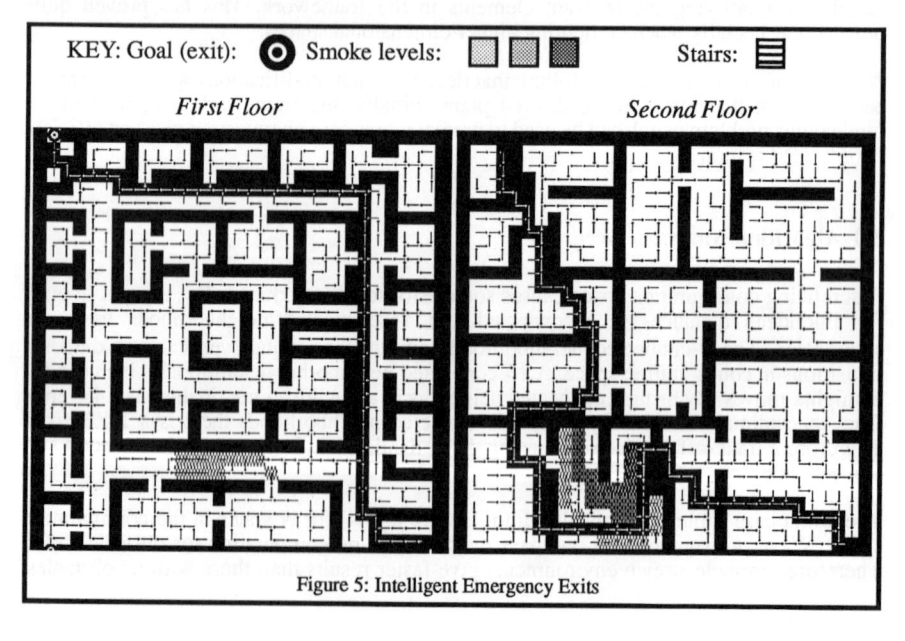

Figure 5: Intelligent Emergency Exits

Directing people out can be performed using the framework described. The directions to inhabitants are envisioned as lighted arrows at about knee height distributed about the building at regular intervals and at logical junctions. The directions should lead the inhabitants out of the building by the most direct route, but by avoiding the problem areas where possible.

The building in the example shown in Figure 5 has two floors, two exits, and one set of stairs. The first floor is on the left, and the second floor is on the right. The exits are in the upper left and lower left corners of the first floor. The stairs are in the lower right corner and lead to the second floor.

$$M(x_1,y_1,x_2,y_2) = \sqrt{(x_2-x_1)^2 + (y_2-y_1)^2}$$
Formula 1

The configuration space is therefore three dimensional, and has a 'solid' floor between the two floors at all states not representing stairs. There is no difficulty in performing obstacle transformations. Walls, and the solid floor are the only forbidden regions. The exits are the goals. The cost metric given in Formula 1 is the minimum distance approximated by a straight line between 6 adjacent neighbors (up,down,right, left, forward, back). Additional penalties can be imposed by adding extra cost at regions where smoke (or other) alarms have been activated. In this example, there are three types of alarm. The darker color represents a more severe alarm. From all locations in the building, a path is found to the nearest exit. If there is no exit from a particular location (perhaps because of being blocked in by an impassable alarm) causing people to be trapped, then this would also be apparent to the emergency personnel monitoring the system.

While this problem initially seems specific to fire exits, it is also applicable to other planar point to point routing problems. Some examples are routing over varied terrain, selecting optimal roads for personal or emergency use, and routing automated guided vehicles on a factory floor.

5.0 The Robot Path Planning Problem

The objective is to plan a path for the robot to follow from the current position to the goal position without hitting anything. It should move optimally in terms of the minimum energy, where the energy is a function of the mass of each link. The goal is for the arm to be stretched out to the left ($180°,180°$), where the angle is measured absolute from a $0°$ reference pointing right.

Figure 6 contains the path computed from a starting state of about ($45°,100°$). In this case, the robot can move both the shoulder and elbow simultaneously using a 16 connected neighborhood.

The configuration space for this problem is two dimensional as described previously. The span of angles for the shoulder and elbow joints gives the axes and limits of the configuration space. The cost metric imposed on each of these motions is governed by the equation in Figure 6. The transform gives the two black areas in the figure which represent forbidden poses. They are computed in real-time based on a function[14]. Based on only this information, a path can be found from any reachable starting state to the goal by reading out joint states from the configuration space and sending them as setpoint commands to the robot.

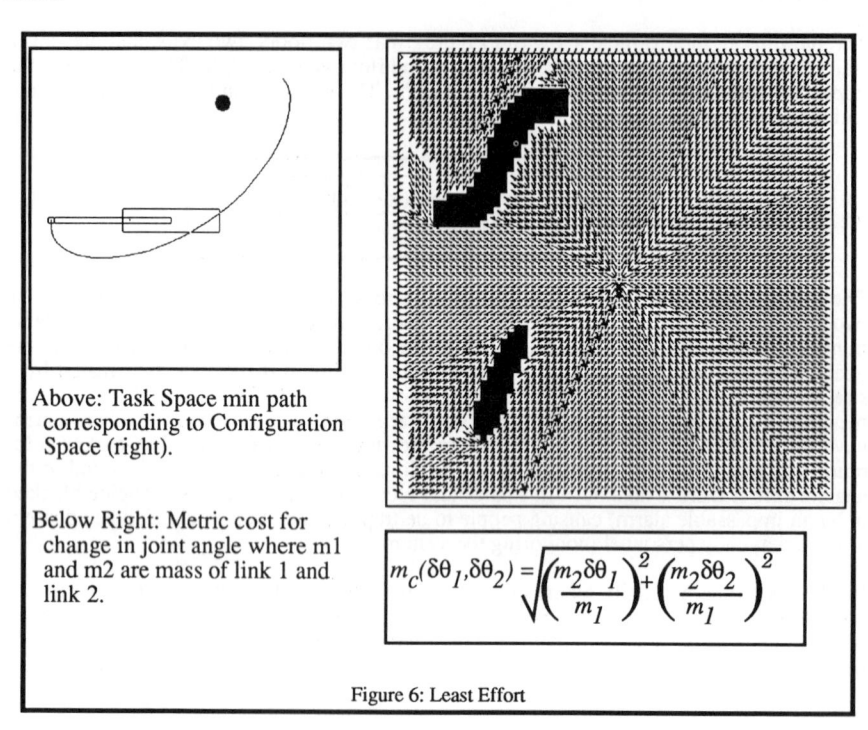

Above: Task Space min path corresponding to Configuration Space (right).

Below Right: Metric cost for change in joint angle where m1 and m2 are mass of link 1 and link 2.

$$m_c(\delta\theta_1, \delta\theta_2) = \sqrt{\left(\frac{m_2\delta\theta_1}{m_1}\right)^2 + \left(\frac{m_2\delta\theta_2}{m_1}\right)^2}$$

Figure 6: Least Effort

6.0 The Vehicle Maneuvering Problem

The objective is to plan a path for a vehicle, such as a car, to follow from the current position to the goal position without hitting anything. The configuration space is best characterized in 3 dimensions by the x,y location of the vehicle and its orientation. It should move optimally in terms of the minimum distance that a specific point on the car travels. The goal in this example is to be parallel parked between two other cars. The vehicle's body dimensions and steering capability is known in advance. The control parameters are clearly the direction of the steering wheels and the forward/backward motion of the drive wheels. Since the planning parameters and the control parameters are different, this problem is non-holonomic.

Left: Trace of possible motion in 2-D
Right: Trace of possible motion in 3-D Configuration Space

Figure 7: Neighborhood for a Vehicle

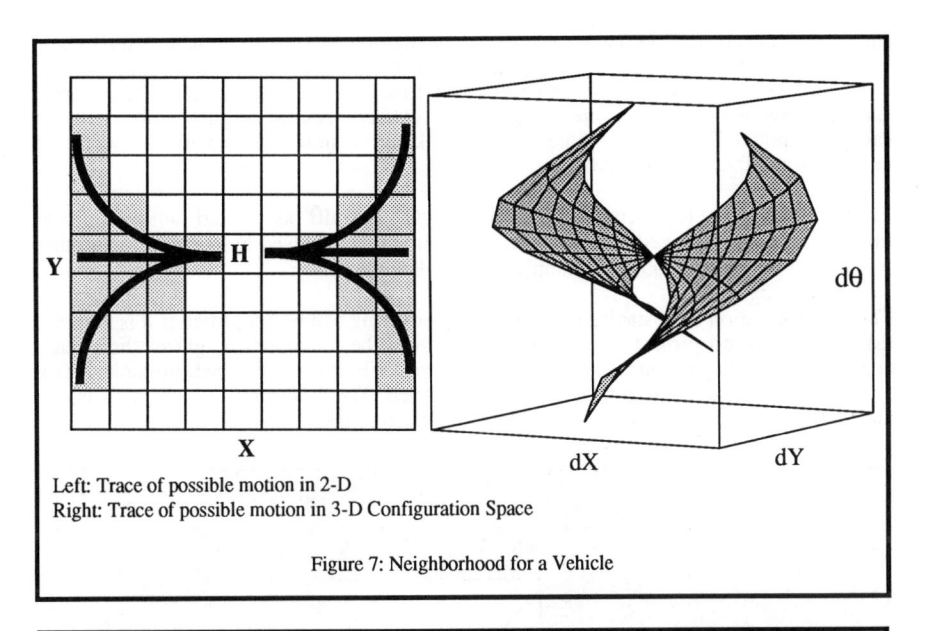

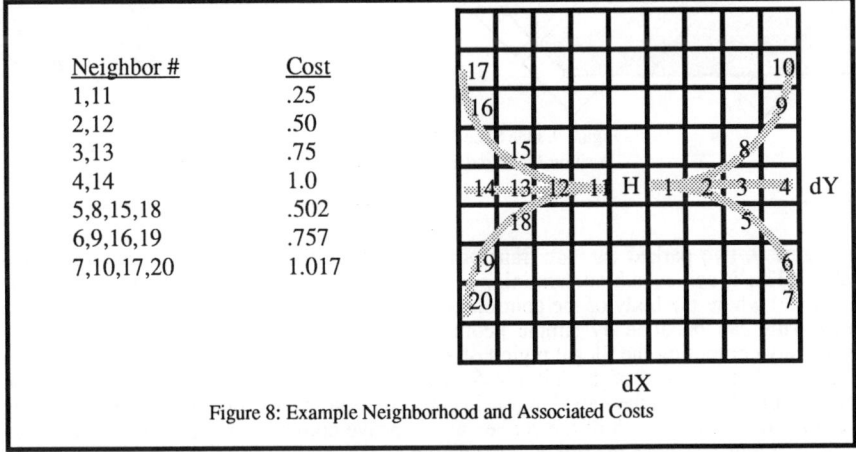

Neighbor #	Cost
1,11	.25
2,12	.50
3,13	.75
4,14	1.0
5,8,15,18	.502
6,9,16,19	.757
7,10,17,20	1.017

Figure 8: Example Neighborhood and Associated Costs

The neighborhood is interesting and different from the previous examples. The vehicle is kinematically constrained. It must follow a particular pattern for each forward/backward motion given a fixed steering wheel orientation. By using the center of the rear wheels as the reference point on the car, the hard-right, hard-left and straight motions (for forward and reverse) result in a bowtie shaped trace if drawn on the ground as the car drives. It is depicted in Figure 7 on the left. In configuration space however, the bowtie has 3 parameters, the change in x and y, and the change in orientation of the vehicle. This 'twisted bowtie' is shown in Figure 7 on the right. The bowtie describes only the partial motion of the vehicle which results in a fixed circular arc based on the turning radius. Therefore if distance travelled is the measure along an arc between specific neighbors, the metric is easily computed. For clarity, Figure 8 uses the planar version of the bowtie to specify the neighbors and their respective costs.

Each neighbor in the neighborhood has a dx, dy, dθ associated with it. In an implementation, improved performance and accuracy is achieved by computing and then storing a separate neighborhood template for each angle θ.

The transformation of obstacles for vehicle maneuvering can be simplified if it is assumed that the downward projection of any obstacle from the maximum height of the vehicle determines the illegal zone. In the vehicle maneuvering examples, each obstacle is first enclosed by a rectangular area and then transformed. The controlled vehicle itself is also approximated by a rectangle to simplify the intersection test.

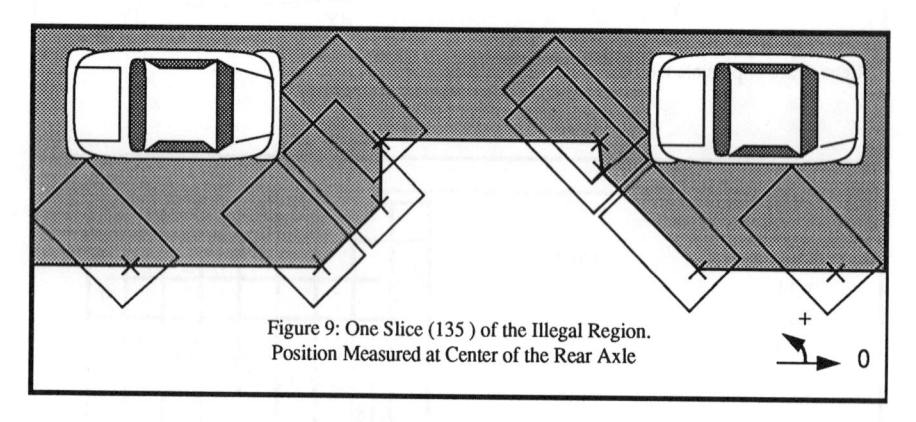

Figure 9: One Slice (135) of the Illegal Region.
Position Measured at Center of the Rear Axle

In Figure 9, two parked cars are transformed based on the dimensions of the controlled vehicle. This is achieved by determining the outer region (as measured by the center of the rear axle) where the body of the controlled vehicle would intersect the parked cars or the curb. This can be done by simple geometry, and is most efficiently implemented by separately computing the illegal region of x,y for each possible angle. The transformation of one angle is a 'slice' of the full transformation along the x,y plane. For each angle θ of the controlled vehicle, the illegal region in configuration space is determined by computing the transformation at each of the convex and concave corners and filling in the forbidden region. Figure 10 shows the full transformation of the configuration space

The *goal position* is simple to transform. If the goal is to be parallel parked between the two parked cars, in a specific orientation (e.g. parked pointing left) then the transform of the goal x,y coordinate of the car and orientation can be given directly. Any number of goals can be accommodated using the CWP technique. If there are several parking spots to choose from, then each could contain at least one goal. In the current example, only one goal is used.

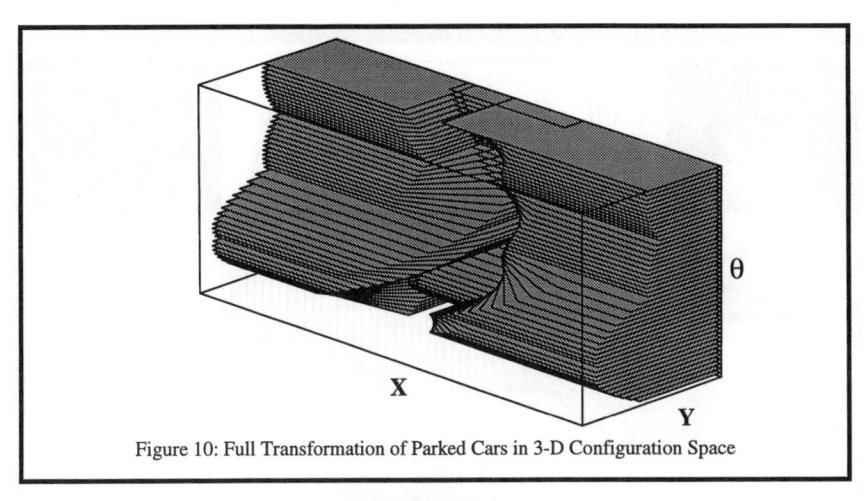

Figure 10: Full Transformation of Parked Cars in 3-D Configuration Space

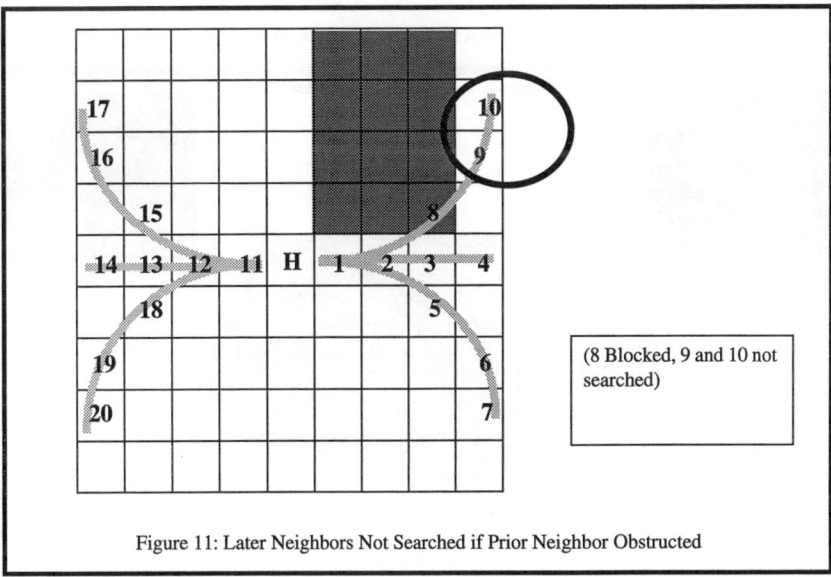

Figure 11: Later Neighbors Not Searched if Prior Neighbor Obstructed

(8 Blocked, 9 and 10 not searched)

Once the framework is in place, the cost wave propagation method can be used with one modification. The original CWP method propagated waves by repeatedly evaluating the neighborhood of the least cost state in the wavefront. No particular ordering of the neighbors was imposed.

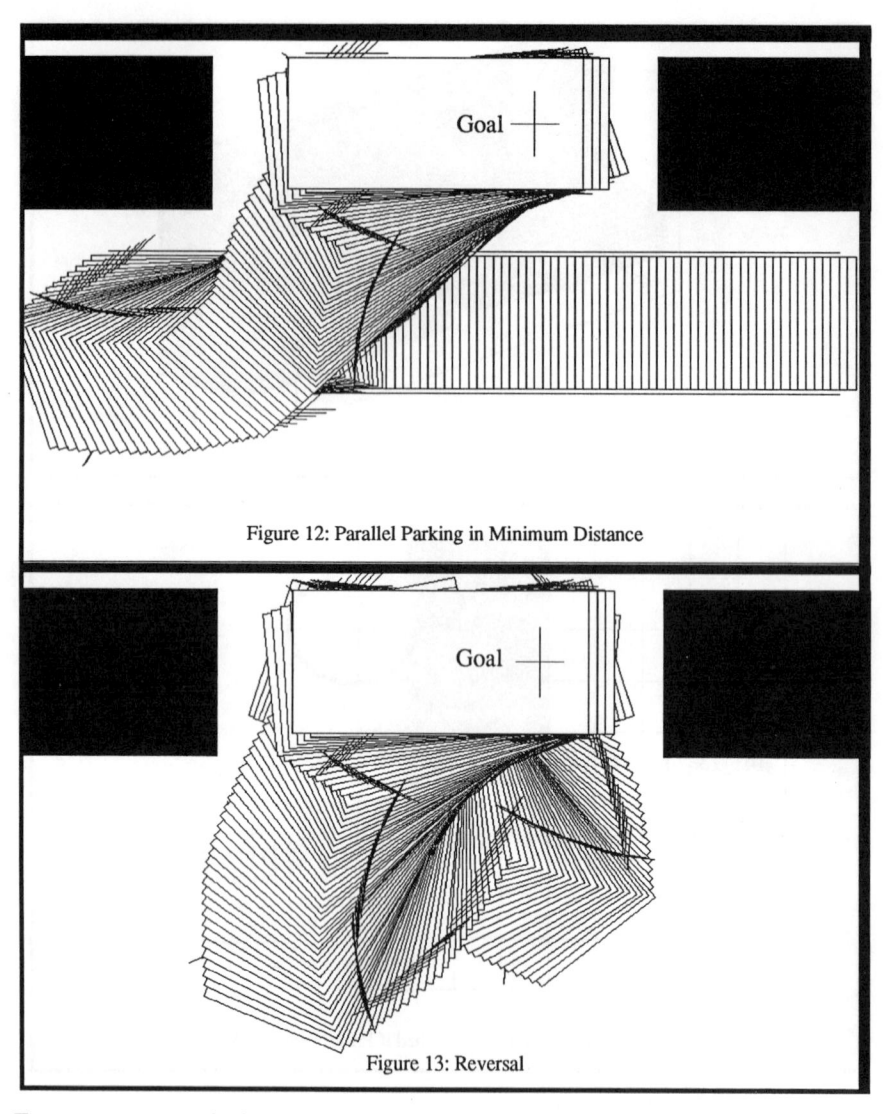

Figure 12: Parallel Parking in Minimum Distance

Figure 13: Reversal

To propagate waves in the vehicle maneuvering example however, the neighborhood must be evaluated so that the nearer neighbors along a particular direction are evaluated first. If a near neighbor is blocked by an obstacle state, then the remainder of the neighbors along that direction are not explored (see Figure 11). This feature is necessary to avoid collisions with convex corners because otherwise it is possible to step over obstacles when evaluating far neighbors.

The result from the cost wave propagation is a direction arrow in each reachable configuration state indicating the steering wheel orientation and transmission direction (i.e. forward or reverse) for the vehicle. By following the arrows from any starting state, the minimum distance path will be travelled to the goal. Figure 12 shows the path resulting from a minimum distance criterion starting from a position parallel to the rear parked car. Figure 13 shows the maneuver necessary if the controlled vehicle is started in the parking spot but facing backwards. Of course, parallel parking is not the only maneuver that can be computed.

7.0 A Forward-Only Constraint

If the above method were used to make a right turn, it would likely lead to a maneuver with several transmission changes, because it is the actual minimum distance solution. This is obviously undesirable in ordinary traffic. This could be solved by greatly penalizing changes of transmission, although in this case it is simpler to eliminate the neighbors corresponding to the reverse direction entirely. This has the additional benefit of speeding the wave propagation considerably. All other aspects of the setup and control are the same. Figure 14 shows a simple right hand turn constrained to allow only forward motions/

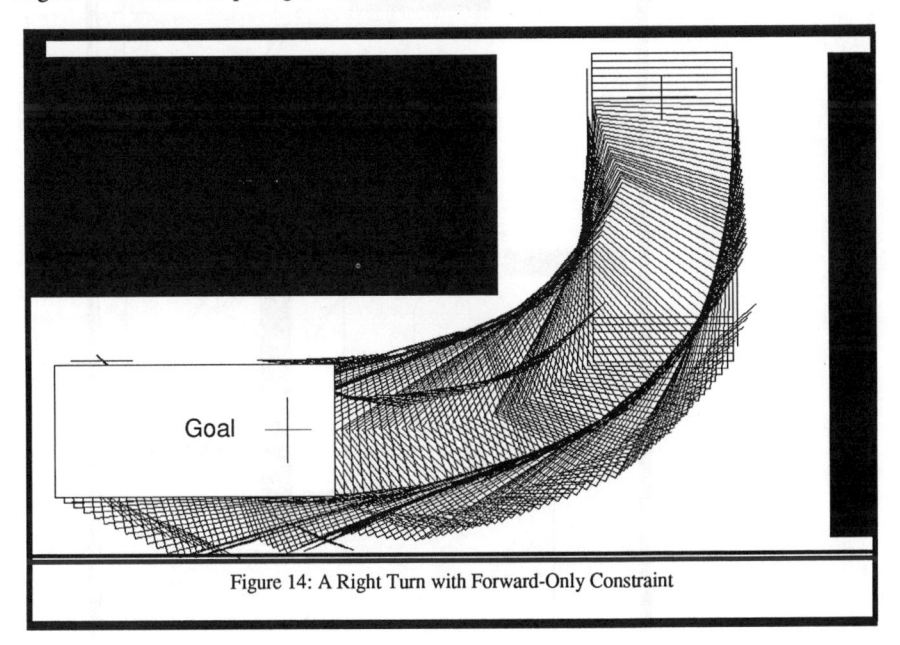

Figure 14: A Right Turn with Forward-Only Constraint

8.0 Use of the System for Complex Maneuvers and Larger Areas

Complex maneuvers are required when there are many perhaps odd shaped obstacles or a restricted space. The problem is set up and computed in the same way as for simpler arrangements, however the sensing problem is more difficult. In the case where the vehicle is performing anything much more than parallel parking, inexpensive sensors will not suffice. This is particularly true if complex obstacles must be sensed, or a large area is used for maneuvering. Even if the environment is known, the position and orientation of the vehicle must be reported in real time. This is perhaps the most difficult problem, since open-loop control is only accurate for a few motions. Only if the vehicle is performing in a restricted location, such as warehouse, airport, ship, or commercial establishment, can fixed sensors be placed efficiently to track or locate the vehicle. Global positioning with differential correction may be of use in some outdoor applications, although this is an area for future research. Other than the sensing problem, there is no difficulty in computing an optimal path around numerous obstacles.

Figure 15: Transform of 8 obstacles and 4 containing walls for a car at +135 degrees

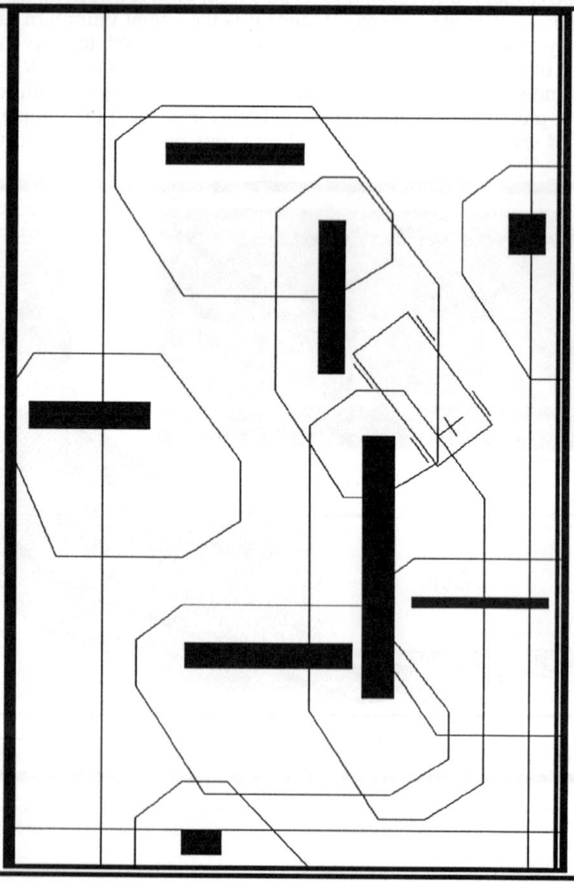

The setup and computation follow directly from the previous example. Given the vehicle's dimensions and steering capability, the neighborhood can be computed automatically. The cost used is the distance the cross-hair marker travels. If the obstacles and vehicle are approximated by rectangles, then the transform resulting from a single angle of the vehicle is as shown in Figure 15. The region around an obstacle represents the closest that the cross-hair of the vehicle can get to the obstacle at the current orientation of the vehicle. The illegal region in configuration space is determined by computing the transformation at each of the convex and concave corners and filling in the forbidden region. More complex shapes are certainly possible. For each possible angle of the car, a transform is computed. The transforms for all angles of the car can be performed in real-time. The optimal path is generated by performing cost wave propagation from the goal (or goals). The waves avoid the transformed obstacle regions, and eventually fill the reachable 3-D configuration space. As mentioned before, the more obstacles that are present, the faster the solution is computed. By following the steering and transmission commands from location to location the path in Figure 16 is found.

Figure 16:
Minimizing Distance
Travelled While Avoiding
Obstacles

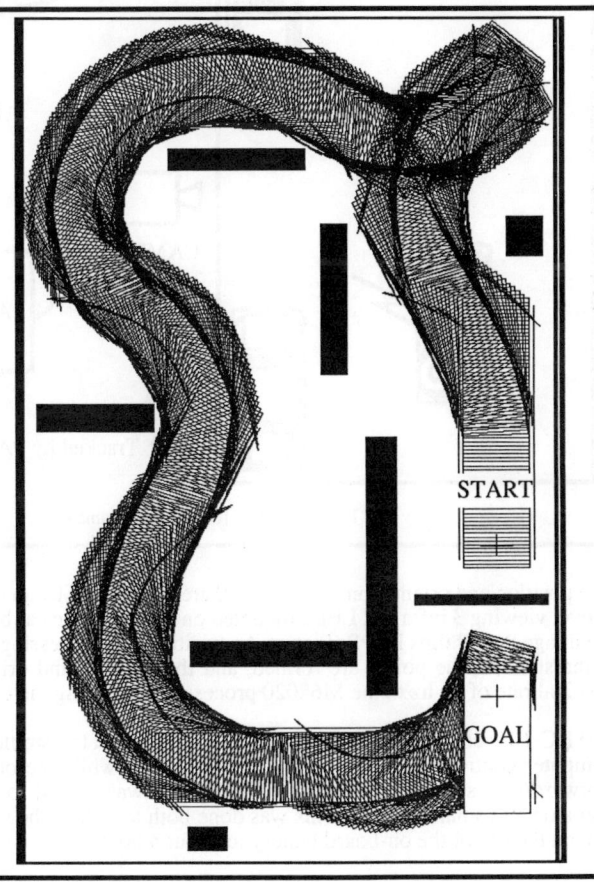

9.0 A Radio Controlled Example

The simulated path of the vehicle shown in Figures 12, 13 and 14 have been executed in a testbed environment shown in Figure 17. In addition to these, many more complex maneuvers constrained by the four walls were accomplished. A Motorola 68020 based multi-processor system called SPINE (Structured Processor Interactive Networked Environment), developed at Philips Laboratories for real-time experiments, is used to coordinate and control the activities of the car. A stock 1/10 scale RC-10 radio controlled car is used as the controlled vehicle. The body dimensions and steering ability are used as input for planning the motions. The testbed also contains two car bodies as obstacles.

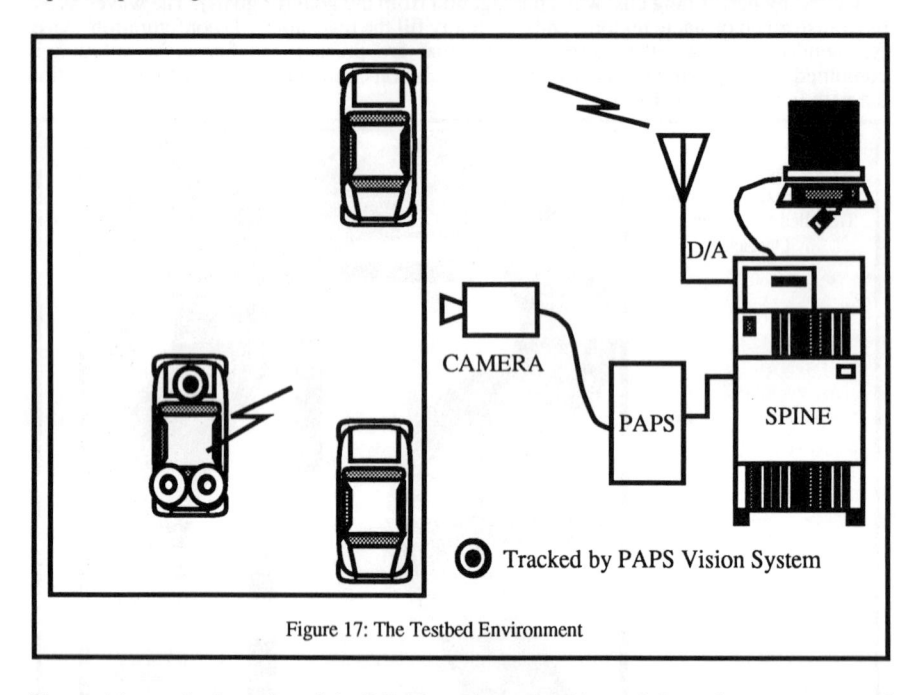

Figure 17: The Testbed Environment

The position and orientation of the RC-10 are determined by an infra-red camera, mounted above, viewing 3 infra-red LEDs mounted on the top of the car body. The camera transfers the image to a Philips PAPS (Picture Acquisition and Processing System) where the image is thresholded, the points are refined, and the position and orientation are reported at a nominal rate of 20 hz to the M68020 processor controlling the car.

The RC-10 is controlled from a stock FM radio-control transmitter, fitted with D/A taps for computer control. One tap controls steering angle while the other controls transmission (forward/reverse). The drive-train of the RC-10 was altered so that it has a geared down rear axle and smaller motor. This was done both to reduce the speed of the vehicle and to extend the life of the on-board battery to about 4 hours.

The setup of the parked cars is input via a graphical user interface (GUI) on a Sun 3/160. In an expanded implementation, the location of the parked cars could be determined by sensors. The goal and selection of optimization criterion are also given in the GUI. Once the configuration space is computed, it is downloaded to the 4Megabytes of common memory in SPINE.

From the common memory area, the controller can read the direction arrows indicating the proper control of the vehicle from any currently sensed position. In practice, a few enhancements to the previous control schemes (i.e. for the robot) are needed. These are needed because the direction arrows in the configuration space tend to point in wildly different directions for even minor variations in state, thus a smooth gradient field is not available for control.

The first enhancement is, since neighbors are arcs of limited length, any sequence of like-kind neighbors are concatenated to determine the next setpoint for control. For example, if the path contains two directives that both result in forward motion with wheels at 45 degrees, then the setpoint should be at the end of the farthest 45 degree neighbor. This allows some smoothing of the discretization errors that necessarily occur when planning a continuous problem in a discrete space.

The second is that an arc is struck from the current active position through the next x,y,θ setpoint and the vehicle is PID controlled through that point. This helps correct for an imperfect mechanical system or minor sliding on the surface.

The third enhancement is that a control stopping criterion must be provided. The vehicle is considered to have reached the setpoint when the vehicle's position crosses the line that runs through the setpoint perpendicular to the setpoint orientation. This line allows for some mechanical or frictional error, while still ensuring a fairly robust hand off between setpoints.

Fourth, once the vehicle has crossed the line defining the stopping criterion, an evaluation is made. If the current location is within reasonable range of the setpoint, then control resumes with the next neighbor in the path, otherwise the path is determined anew from the current location. As an alternative to this, an acceptability test can be made while control is underway. If the current position is wildly out of track with the setpoint, such as if we had reached in and flipped the car around, the controller can abort the motion and continue from a newly determined starting position.

To use the testbed once the system is loaded, the RC-10 is placed anywhere within the physical boundaries, and a 'start button' is pushed. The car then drives to the goal.

Anyone interested in a free videotape (specify PAL or NTSC) showing the vehicle maneuvering testbed may contact the author at the address given at the beginning of the chapter or electronically at kit@philabs.philips.com .

10.0 High Speed Vehicle Maneuvering

Initially, the planning for a scenario any larger than that of the complex region in the previous section would seem impractical, not only because of the real-time sensing issues, but also because the planning would be too slow and require too much memory. Since maneuver planning determines the proper reaction in the sensing-reaction control loop, it is important to be able to perform this quickly to get adequate 'reaction time'. Fortunately, quick planning in a relatively small space can be performed for this problem.

In the situation where a vehicle is moving along a highway, the configuration space should be considered as the space moving <u>relative</u> to some other object or marker. Previously, all configuration spaces were considered to be relative to a fixed origin.

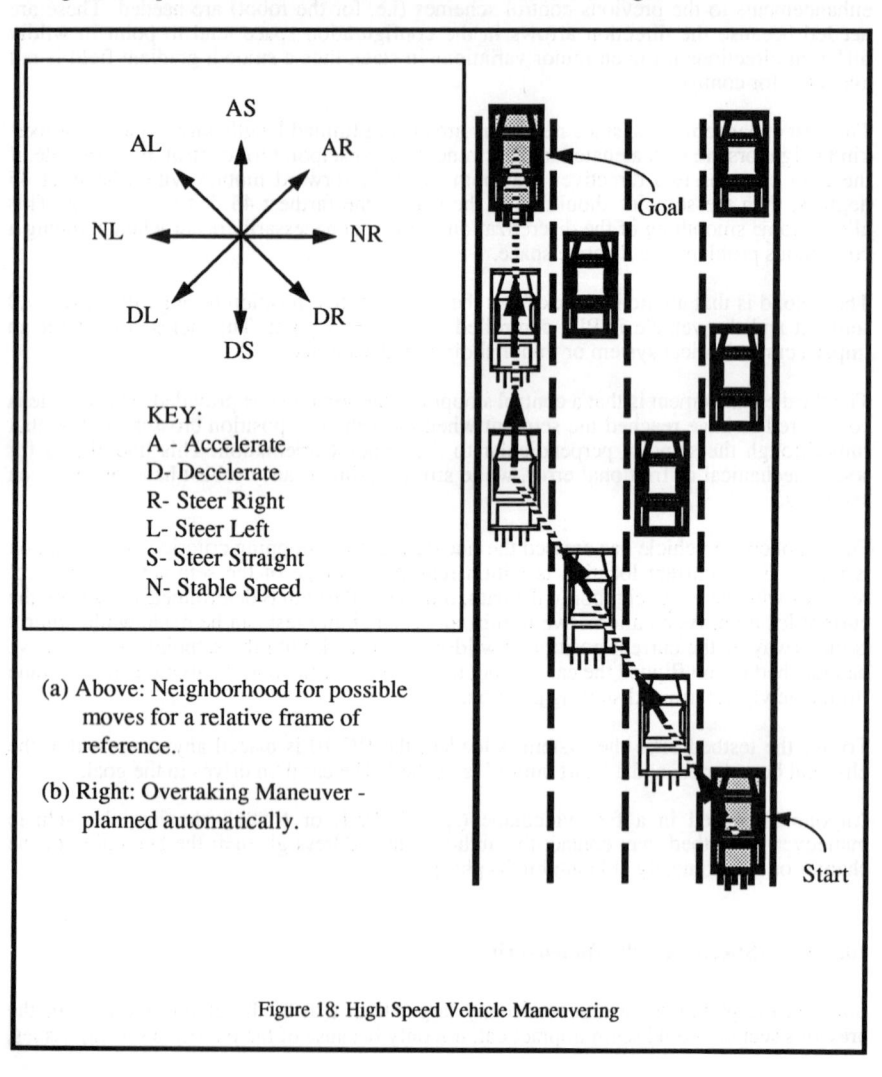

KEY:
A - Accelerate
D - Decelerate
R - Steer Right
L - Steer Left
S - Steer Straight
N - Stable Speed

(a) Above: Neighborhood for possible moves for a relative frame of reference.

(b) Right: Overtaking Maneuver - planned automatically.

Figure 18: High Speed Vehicle Maneuvering

Looking at the 'neighborhood of permissible motions' for a high speed vehicle, it is discovered that the vehicle can in fact move sideways, relative to other vehicles, simply by steering. Also, forward and backward motion can be achieved by accelerating or decelerating. This is a delightfully simple neighborhood for planning purposes, as shown in Figure 18(a). Figure 18(b) gives the results of straightforward path planning in a relative space for vehicles moving at higher speeds. The goal is to pass the other cars. At each state there are directives such as "accelerate and move one lane to the left". Speed limits can be incorporated naturally by changing the neighborhood as a function of the current speed. Since the computation is very quick, any change in the relative positions of the other cars can be accommodated in a short time.

11.0 Planning the Coordination of Multiple Actors - Synergistic Planning

In the previous examples, only single machines are under the control of the planning system without regard to the requirements of other machines or to the collective synergy that can be obtained. The same general planning framework that has been outlined previously may be used to find locations for rendezvous or simultaneous requirements of multiple machines or actors. Each actor may have different cost measures for efficiency, different neighborhood constraints, and different starting and ending requirements. These may all be factored into a single problem so that predictive, optimal motions can be computed.

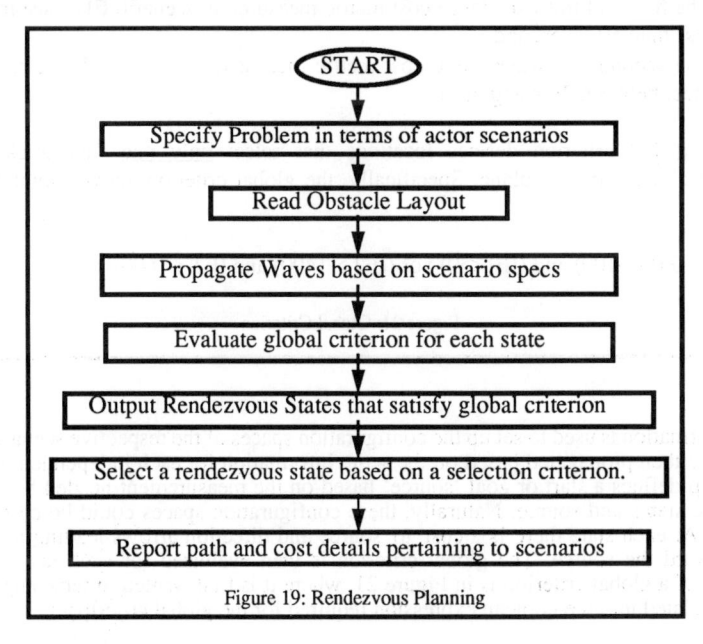

Figure 19: Rendezvous Planning

There are several steps required to perform this task which are listed in Figure 19. The first step is to specify the problem in terms of scenarios for each of the actors. A scenario can be as simple as the information in Figure 20. Next, the obstacle layout is sensed.

Scenario a1   /* actor a: scenario 1 */
Start   /* tells whether to propagate away or towards the 'source' */
        3,2   /*equivalent locations that the actor could start from */
        1,3   /* coordinates are listed <row>,<column> */
Metric
        Min_distance /* for Neighborhood, each move is cost=1 */
                /* horiz and vert. moves only */

Figure 20: A Scenario

At least one of two fuel trucks and an airplane must rendezvous but:
- The plane (fuel cost metric measured in scenario A) must still have at least 100 gallons of fuel in its tank,
- The first fuel truck (distance cost metric measured in scenario F1) must travel less than 50 miles, and
- The second fuel truck (time cost metric measured in scenario F2) can only meet between 2pm and 4pm.

The global criterion defines locations that select <u>only one</u> fuel truck for rendezvous with the plane. Specifically, the global criterion above might look like:

G=(A>100) and ((F1<50) xor ((F2 > 1400) and (F2 < 1600)))

Figure 21: Global Criterion

This information is used to set up the configuration spaces of the respective scenarios. Cost waves are then propagated based on the setup information for each independent scenario. The setup defines a start or goal 'source' based on the measurement needed between the reachable states and source. Naturally, these configuration spaces could be computed in parallel. At each state there is a cost_to_source and direction arrows leading to the next state toward the source. The global criterion is then evaluated for each task state. An example of a global criterion is in Figure 21, where it is first written in terse english, and then translated into the boolean expression required for the global criterion.

The task states are the regions where any possible meeting might take place between the actors. They may naturally map or transform into more than one configuration state in each of the configuration spaces. By evaluating the global criterion at each of these transformed task states, a set of satisfying (rendezvous) configuration states results. Since many states can be workable for the rendezvous, a 'best choice' can be selected from the set, based on a selection criterion. For example, of the rendezvous points possible, choose the one that will cost the least in terms of total dollars spent by all of the actors.

The specific movement/action details for each actor can then be read out from the respective scenario-based configuration spaces. This gives coordinated motion for all the actors to achieve a shared mission.

11.1 Example Coordination Problem

The following is a problem involving the coordination of two actors, A and B. Candidate rendezvous points are to satisfy the following global criteria:

1) the distance A travels from the start to the rendezvous state must be between (2,8],
AND
2) the distance A travels in total must be precisely 7,
AND
3) the time B travels from the start to the rendezvous state must be less than 4,
AND
4) The fuel spent by B from the rendezvous state to the goal must be at least 4,
AND
5) the sum of the (distance travelled by A + the fuel spent by B) must be less than 9,
AND
6) the sum of the (distance travelled by A + the fuel spent by B) must be greater than 6.

Scenarios representing the needed measurements for each actor can be enumerated as follows:

A: distance travelled away from the potential starting states
A: distance travelled from the rendezvous to the potential goal states
B: time travelled from the potential starting states
B: Fuel spent from the rendezvous to the potential goal states

For each of these four requirements, scenarios can be written for the specific start and goals. They are as follows:

Scenario A1 /* actor A: scenario 1 */	Scenario B1/* actor B: scenario 1 */

Start
  3,2
  1,3
Metric
  Min_distance
/* H&V moves. Each move
  is cost=1 */
	Start
	  4,8
	  4,9
	Metric
	  Min_time
	/* H&V moves. Each move
	  is cost=1 */

Scenario A2/* actor A: scenario 2 */	Scenario B2/* actor B: scenario 2 */

Goal
  4,8/
  1,7
Metric
  Min_distance
/* H&V moves. Each move
  is cost=1 */
	Goal
	  3,2
	  1,3
	Metric
	  Min_fuel
	/* H&V moves. Each move
	  is cost=1 */

The boolean expression corresponding to these constraints is:

(A1 > 2) and (A1 <= 8) and (A1+A2 = 7) and (B1 < 4) and (B2 > 4) and (A2+B2 < 9) and (A2 + B2 > 6)

Each of the above scenarios correspond to a cost wave propagation in a configuration space. For this example, the obstacle layouts are assumed to be the same for each actor-scenario. The four resulting spaces might be as in Figure 22.

Once these spaces are completed, the global criterion is evaluated for each state these actor-scenarios have in common (all in this case). The candidate rendezvous states that meet all the constraints will have a 'true' boolean result. For this example, the states are highlighted in Figure 23.

461

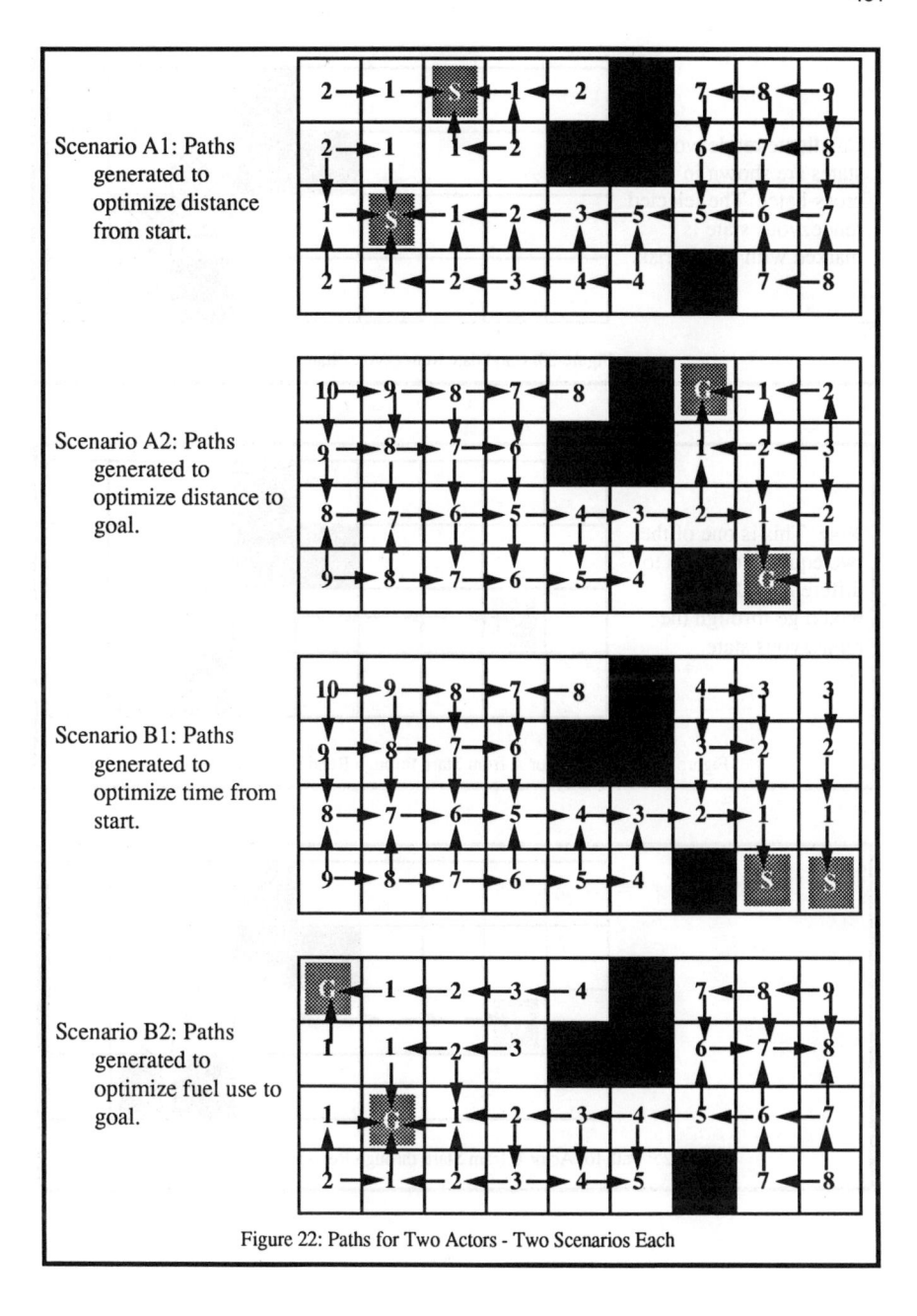

Scenario A1: Paths generated to optimize distance from start.

Scenario A2: Paths generated to optimize distance to goal.

Scenario B1: Paths generated to optimize time from start.

Scenario B2: Paths generated to optimize fuel use to goal.

Figure 22: Paths for Two Actors - Two Scenarios Each

Candidate rendezvous states are shown in cross-hatch. The selected rendezvous state is marked with an asterisk.

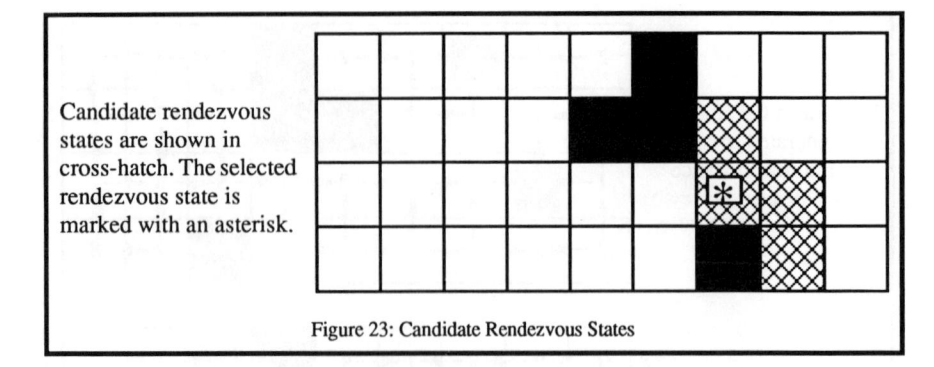

Figure 23: Candidate Rendezvous States

Note: This is one of the two equivalent paths to different goals. Both would go through the rendezvous state.

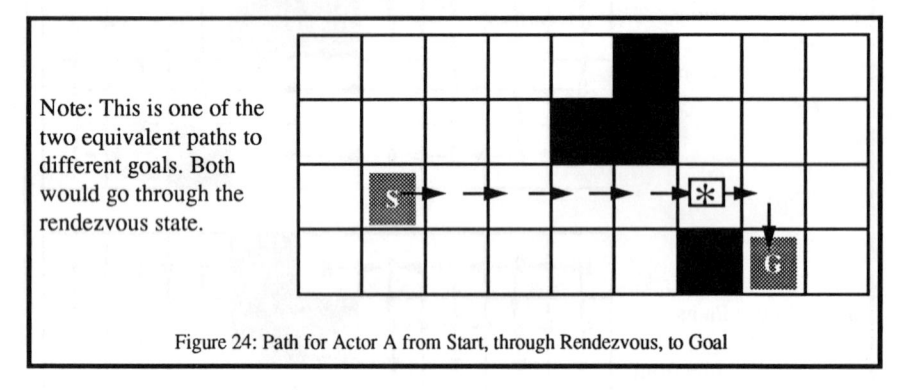

Figure 24: Path for Actor A from Start, through Rendezvous, to Goal

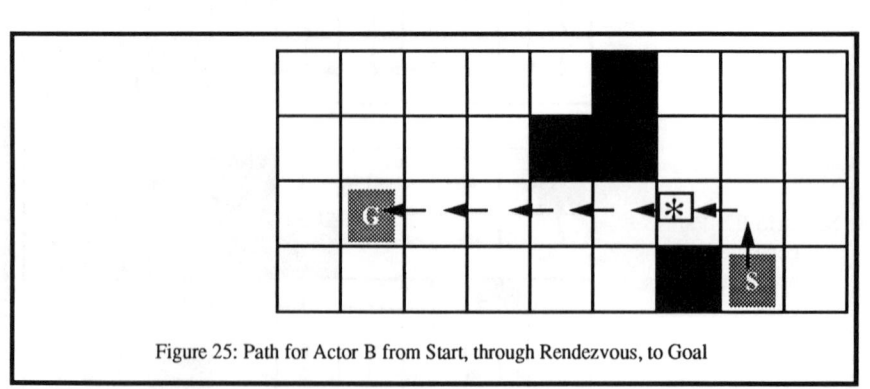

Figure 25: Path for Actor B from Start, through Rendezvous, to Goal

463

If it is assumed that any candidate rendezvous state is acceptable, then a state may be selected at random for implementation. It is often the case however that one of the candidates is preferable to the others. In this case a selection criterion can be used to find the best candidate, for example one with the least implementation cost.

Once a state is selected, then the moves for each of the actors can be read from the respective scenario-based configuration spaces. For example if the state marked with the asterisk (*) is selected, then Actor A would proceed as in Figure 24 and Actor B would proceed as in Figure 25.

11.2 Coordination of Fleet Trucks

The above synergistic planning method is general in that it is equally applicable to many domains. To show the breadth of the types of problems that can be handled, I will enumerate a few using a sample scenario.

Suppose we own a national fleet of delivery trucks which have irregular delivery sites. Some trucks haul double-trailers and some haul single trailers. Some of the truckers are independent and have licenses to haul only specific cargo in specific states. Some trucks have limits on the weight they can haul. We can then compute answers for the following types of questions:

- What locations are feasible for exchanges between specific trucks?
- How shall they be coordinated so that all of the deliveries arrive on time?
- How can certain truck overhead be minimized by selecting the most time-direct routes for each?
- If a central depot for a regular trucking schedule is to be established, where should it be so as to be central to all trucks given their limitations?
- If a road is closed due to weather, traffic, construction, etc., how can coordination take place to manage the exchange between two trucks elsewhere?

11.3 Brief Analysis

While most methods would require a computation for a product of the state spaces, this method uses the sum of the state spaces by simply overlapping them with a global and selection criterion.

11.4 Other Applications

The military is another area where coordination of forces is critical. In a defensive or offensive strategy, one may place opposing actors into the equation, assuming that they will use the most opportune strategy available. This computational tool identifies the weak areas of each strategic side, and the timing of a strategic arrangement with the most favorable outcome (e.g. fewest casualties). This is also described in a previous paper[14].

12.0 INVITATION

Most of the information given in this chapter is covered by issued and pending patents. Anyone interested in collaborating or directly licensing these techniques is encouraged to contact the author at the given address.

13.0 CONCLUSION

A framework has been presented which can encapsulate a machine or agent's fundamental behavior and automatically compute intelligent maneuvers. By transforming the problem into a configuration space, path planning for a point can be used to yield optimal collision free maneuvers. It is general and efficient for many types of problems. A nice property is that the nature of the method makes it more efficient for spaces with more obstacles, which are also typically problems that are the most difficult to perform by inspection.

Many examples are given in this chapter including a solution for a safer fire exit system which provides the best route out of a building for all inhabitants based on the sensed alarms. This can be implemented by installing intelligent arrow-lights at regular intervals throughout the building which point toward the safest exit. This example is important because it can also be used for other planar problems such as optimal point-to-point highway selection and rough terrain routing.

Another example is a robot that moves optimally from one place to another without hitting obstacles. The specific machine movement can be incorporated into a neighborhood which ensures that the resulting optimal path is feasible for a specific machine. Because the path is computed automatically, only higher level directives or goals are needed for the machine to complete complex tasks.

The maneuver for a vehicle can be computed in much the same way as for the robot. The main difference is that the machine movement is more complex for a non-holonomic problem. The result is a car that can derive complex maneuvers (such as parallel parking) based on the obstacle layout. A testbed implementation has shown that the computed path is indeed controllable.

High speed vehicle maneuvering can be performed by using a relative frame of reference and by using the neighborhood that maps acceleration and deceleration to forward and backward motion, and higher speed left and right steering to left and right motion. This makes the problem simple to solve in two dimensions, which in turn gives relatively fast reaction time.

Finally, multiple machines or actors can be coordinated to satisfy their individual and collective constraints. Given this general framework, more efficient trucking distribution can be planned, as well as problems in other domains such as strategic defense. It is possible to evaluate the scope of threats and determine the appropriate temporal response.

By continuing to challenge this framework with diverse problems, ever-evolving facets help shape its domain. New challenges are therefore always welcome.

REFERENCES

1. T. Lozano-Pérez and M.A. Wesley, "An Algorithm for Planning Collision-Free Paths Among Polyhedral Obstacles", Communications of the ACM, **No. 10**, 1979. pp. 560-570.

2. L. Kavraki, "Computation of Configuration-Space Obstacles Using the Fast Fourier", 1993 Int. Conf. on Robotics and Automation, **Vol. 3**. pp. 255-261.

3. D. Lyons and A. Hendriks, "Planning for Reactive Robot Behavior", 1992 IEEE Int. Conf. on Robotics and Automation, Nice, France.

4. J. Pearl, "Heuristics: Intelligent Search Strategies for Computer Problem Solving", Addison Wesley, Reading, MA, 1984.

5. L. Dorst and K. Trovato, "Optimal Path Planning by Cost Wave Propagation in Metric Configuration Space", SPIE Conference on Advances in Intelligent Robotics Systems, Nov. 1988.

6. O. Khatib, "Real Time Obstacle Avoidance for Manipulators and Mobile Robots", International Journal of Robotics Research, **Vol 5** #1. pp. 90-98, Spring 1986.

7. K. Trovato, "Differential A*: An Adaptive Search Method Illustrated with Robot Path Planning for Moving Obstacles and Goals, and an Uncertain Environment", International Journal of Pattern Recognition and Artificial Intelligence **Vol. 4** #2. pp 245-268. World Scientific Publishing Company.

8. U.S. Patent Serial # 4,949,277. "Differential Budding: Method and Apparatus for Path Planning with Moving Obstacles and Goals", K.Trovato and L.Dorst. Issued August 14, 1990.

9. R.A. Brooks, T. Lozano-Perez, "A Subdivision Algorithm in Configuration Space for Findpath with Rotation," IEEE Transactions on Systems, Man and Cybernetics, **Vol. 15** #2, 1985, pp. 224-233.

10. R. Featherstone, "Swept Bubbles: A Method of Representing Swept Volume and Space Occupancy", Philips Public Document. MS-90-069. 1988.

11. G. Borgefors, "Distance Transformations in Arbitrary Dimensions", Computer Vision Graphics and Image Processing **Vol.27**, pp. 321-345.

12. L. Dorst and A.W.M. Smeulders, "Length Estimators for Digitized Contours", Computer Vision Graphics and Image Processing **Vol. 40**, pp. 311-333.

13. W. Newman, "High-Speed Robot Control in Complex Environments", Ph.D. Thesis, MIT, October, 1987.

14. K. Trovato, "General Method for Strategic Assessment and Planning Problems", Sixth Joint Service Data Fusion Symposium 1993, **Vol. 1** Part 1, pp. 305-335, June 1993.

LEARNING TO IMPROVE PATH PLANNING PERFORMANCE

PANG C. CHEN
Sandia National Laboratories
Albuquerque, NM 87185, USA
E-mail: pchen@cs.sandia.gov

ABSTRACT

In robotics, path planning refers to finding a short, collision-free path from an initial robot configuration to a desired configuration. It has to be fast to support real-time task-level robot programming. Unfortunately, current planning techniques are still too slow to be effective, as they often require several minutes, if not hours of computation. To remedy this situation, we present and analyze a learning algorithm that uses past experience to increase future performance. The algorithm relies on an existing path planner to provide solutions to difficult tasks. From these solutions, an evolving sparse network of useful robot configurations is learned to support faster planning. More generally, the algorithm provides a speedup-learning framework in which a slow but capable planner may be improved both cost-wise and capability-wise by a faster but less capable planner coupled with experience. The basic algorithm is suitable for stationary environments, and can be extended to accommodate changing environments with on-demand experience repair and object-attached experience abstraction. To analyze the algorithm, we characterize the situations in which the adaptive planner is useful, provide quantitative bounds to predict its behavior, and confirm our theoretical results with experiments in path planning of manipulators. Our algorithm and analysis are sufficiently general that they may also be applied to other planning domains in which experience is useful.

1. Introduction

One of the most important problems in robotics is path planning, which in known environments refers to finding a short, collision-free path from an initial robot configuration to a desired configuration. Path planning algorithms have to be fast (ideally within seconds) to support real-time task-level robot programming. Accordingly, path planning has received much attention [21,17] and there are now a number of implemented path planners based on a variety of approaches. However, the practicality of these planners in general has been hampered by their time-consuming search, as they often require several minutes, if not hours of computation.

To improve the performance of these planners and hence increase their practical value, we present a learning algorithm that uses past experience to increase future performance. Our work is motivated by the observation that robots often perform multiple tasks in virtually the same environment. In such environments, the total planning time can be amortized and significantly reduced by reusing the computation results for one task to plan for another. One example is Sandia National Laboratories'

Remote Radiation Survey and Analysis (RRSAS) project, which is to automatically inspect nuclear waste transport casks for radiation using a robot.[16] Without hitting any obstacle, the robot is to maneuver through the workspace and sample a set of randomly chosen points on the cask. For this series of tasks, learning is possible because the workspace is unchanged during inspection, therefore past experience is useful. Learning is also feasible for this problem because there are only a small number of different kinds of movements that the robot needs to learn.

Thus, we make the underlying assumption that similar tasks are to be performed repeatedly before the robot environment is changed. To perform each task in this stationary environment, our algorithm primarily uses a fast but necessarily incomplete planner that responds quickly and can solve simple tasks. For difficult tasks, however, the algorithm relies on a more complete but slower planner to provide solutions. The algorithm learns from these solutions, building an evolving sparse network of useful robot configurations that guides and supports fast planning. More generally, the algorithm provides a speedup-learning framework in which a slow but capable planner may be improved both cost-wise and capability-wise by a faster but less capable planner coupled with experience.

We demonstrate the effectiveness of the algorithm via both mathematical analysis and experimentation. Our focus is on the fundamental algorithmic behavior of learning as opposed to other equally important issues of knowledge representation, solution abstraction, and implementation. Thus, we present the algorithm in its abstract form so that we may provide an in-depth theoretical analysis for understanding its behavior. Our analysis begins with some general quantitative relationships governing the learning process, followed by a specific case analysis illustrating these results. To achieve predictive power while preserving some generality, we next study the algorithm under models with additional simplifying assumptions. Using these models, we derive global quantitative bounds on planning cost and capability in terms of training time. We show that the reliance of the improved planner on the original slow planner is at most inversely proportional to the training time. We also characterize the situations in which learning is useful and prescribe the amount of training required. Finally, in our empirical evaluation of the algorithm, we validate our theory and use it to gain insight into several experimental results. Not only can we explain the observed data using our theory, but we can also use it to predict unobservable quantities such as the maximum achievable speedup.

After studying our algorithm for the fundamental stationary case, we next extend it to handle incrementally changing environments. In this more general environment, we assume that for each robot task, the obstacles are stationary, but may slowly change their configuration or shape over the course of the robot performing many tasks. One example application is manufacturing of evolving products in which the design changes made to a product will cause incremental changes to the robot environment. Another example is waste-site remediation in which wastes are typically removed one by one, resulting in a slowly changing environment. Our algorithmic extension consists of two experience manipulation schemes: For minor environmental change, we use an object-attached experience abstraction scheme to increase the flexibility of the learned experience; for major environmental change, we use an on-demand experience repair scheme to retain those experiences that remain valid and useful. With these modifications in place, we show that the learning algorithm is indeed able to adapt to its working environment, provided that the frequency of change is sufficiently low.

2. Related Work

As mentioned in the introduction, a large amount of research has been done on robot path planning, most of which deals with solving one-time problems in stationary environments.[2,3,6,15,20,22,27] Most implemented path planners have been developed for mobile robots and manipulators with a few degrees of freedom (dof). There are some that are designed for many dof manipulators based on random[2] (Brownian motion), sequential[15] (backtracking with virtual obstacles), or parallel[3] (genetic optimization) search. For mobile robots, there is also some work on solving one-time problems in time-varying environments that contain moving obstacles with known trajectories.[12,19,29] All of these planners, however, typically require minutes of computation for mobile robots, and tens of minutes for 6 dof manipulators. Further, little work has been done for changing environments[1] in which movable obstacles remain relatively stationary during sequences of tasks, as opposed to time-varying environments with constantly moving obstacles.

For solving several problems in stationary environments, there are a few other path planners that incorporate learning: some[14,26] take a higher-level, reasoning approach, while others[18,25] take a lower-level, memory-based approach similar to ours. Learning can be done incrementally, or in phases which some consider as preprocessing.[18] To decrease the effective cost of solving each problem, all of these works maintain a network (roadmap) of useful robot configurations (landmarks) and employ some sort of a local planner for moving through the network. Algorithmically, there are some differences between ours and that of other memory-based approach.[18,25] First, we assume and use the same distribution of tasks (problems) for both training and subsequent problem-solving. In other works,[18,25] a uniform problem distribution is used for training. Second, we assume the existence of a fairly reliable, albeit slow, global planner to act as a teacher, whereas they do not. Thus, while their algorithms may be more general, they may also require more training time to compensate for the lack of solutions when local planning fails.

The work presented here is the culmination of three years of research.[5,8,9] The algorithm for stationary environments is initially presented[5] with some general but preliminary analysis on the learning process. Later, the algorithm is extended to cope with incrementally changing environments.[8] Most recently, a deeper analysis for the fundamental stationary case is developed.[9] Overall, the most significant difference between all of the aforementioned work and ours is that we aim to provide a theoretical foundation for algorithm analysis to: 1) better understand and predict our experimental results; and 2) suggest similar analysis techniques that others may apply to better understand their algorithms.

3. Algorithmic Framework

Given an arbitrary but fixed environment, let task (u, w) be defined as finding a collision-free path to move the robot from configuration point u to w. We assume that there are initially two path planners available: **fast** and **slow**. Both return true (1) if successful, false (0) otherwise. The **fast** planner is required to be fast, symmetric, and only locally effective, i.e., it should have a good chance of success if u and w are close to each other. Any greedy hill-climbing method using a potential field[2] or sliding[13] approach should be sufficient to implement **fast**. The **slow** planner, on the other hand, is required to be much more globally effective than **fast**, and hence may be very slow. It is the performance of this planner that we wish to improve with our speedup learning algorithm. Notice that this 'planner' can even be the human operator himself.

In our learning algorithm, we retain the global effectiveness of **slow** by calling it whenever necessary, while reducing the overall time cost by calling **fast** whenever possible. We plan paths for arbitrarily shaped robots or arbitrarily jointed manipulators by planning for a point robot in the corresponding configuration space (C-space). To utilize **fast** fruitfully, we remember significant intermediate robot configurations learned from the solution paths of **slow**. These subgoals (landmarks) represent fully specified robot configurations and are stored in memory V, with connecting edges E (indicating successes of **fast**) maintained so that complete solution paths may be regenerated through applications of **fast**. The subgoals V can be thought of as 'trail-markers' in that each marker can be traced to one another through the trails E. We call the connected network of trail-markers the experience graph $G = (V, E)$. (In contrast, the graphs constructed by other algorithms[18,25] are not necessarily connected since the availability of planner **slow** is not assumed.) Ideally, G is to be used by **fast** to achieve most tasks without the help of **slow**. If **fast** is incapable of achieving a task through G, **slow** is called. If **slow** is also incapable of finding a solution, then we simply skip to the next task. Otherwise, we learn from the solution of **slow** by abstracting (or compressing) it into a chain consisting of a short sequence of intermediate robot configurations that **fast** can use later to achieve the same or similar tasks.

Incidentally, instead of representing the trail-markers as fully specified robot configurations, we can try to be more sophisticated and remember a more general subgoal representing subspaces of configurations. With a more general subgoal representation,

```
Algorithm  Adapt(Fast, Slow)
    u ← current position; v ← u; G ← ({v}, ∅);
    do forever
        w ← goal;
        if (not Fast(u, w; G, h)) then
            if (not Slow(u, w; G, h)) then continue;
            ρ ← Abstract(Slow[u, w; G, h]);
            G ← Learn(G, ρ);
        endif
        execute(Fast[u, w; G, h]); u ← w;
    enddo
end.
```

Fig. 1. A learning algorithm for improving path planning.

the experience graph can be more compact and powerful in solving new tasks. However, taking this research direction requires us to delve into deeper knowledge representation issues instead of focusing on the basic learning process behavior. Hence, we restrict our subgoals to specific configurations to gain simplicity, which allows us to develop a more rigorous understanding of the learning processes within the framework, and hopefully will provide further insight into the more sophisticated ones.

Formally, the learning algorithm Adapt is shown in Figure 1. In the algorithm, u is the current robot configuration, and w is the next goal configuration. To access G, we maintain two pointers: $\hat{u}$ and $\hat{w}$, each of which points to a vertex of G that is known to be reachable with one call of fast from u and w, respectively. We may view these pointers as tethers. The algorithm is based on two planners: Fast and Slow, which are in turn based on fast and slow, respectively. Both Fast and Slow have task (u, w) as arguments, and graph G and a heuristic vertex ordering function h as parameters. Planners Fast and Slow attempt to achieve (u, w) using G as guideline. Since in stationary environments, G forms a connected component and the tether from u to $\hat{u}$ stays valid, the planners only need to check the reachability of w from a known reachable configuration v of G. For planner Fast, we use Fast(·) to denote the predicate that Fast is successful, and Fast[·] to denote the path found when Fast succeeds, and similarly for Slow.

The algorithm for Fast(·) is simple: Search the vertices of G in order according to heuristic h, and find a vertex v satisfying fast(v, w). If v exists, then set $\hat{w} ← v$, and return success; else return failure. To generate Fast[·] once Fast(·) succeeds, first trace a shortest sequence of vertices Γ in G from $\hat{w}$ back to $\hat{u}$, so that $\Gamma_1 = \hat{u}$, and $\Gamma_k = \hat{w}$ for some $k \geq 1$. Then, set $\Gamma_0 = u$, $\Gamma_{k+1} = w$, and output fast$[\Gamma_j, \Gamma_{j+1}]$ for j going from 0 to k.

The algorithm for Slow(·) is even simpler: Call slow(v, w) with v being the best vertex in G according to h, and set $\hat{w} \leftarrow w$. To generate Slow[·] once Slow(·) succeeds, simply output slow[v,w].

The two nested if-statements present Adapt as essentially Fast backed up by Slow. Learning occurs when Fast fails but Slow succeeds. In this case, we apply Abstract to condense the solution path slow[v,w] into a short chain ρ of trail-markers with each edge traversable by fast. We do not specify how Abstract is to be implemented, only that it return a short chain efficiently. In practice, this is a reasonable assumption, since a typical task consists of only 3 smooth motions: departure, traversal, and approach. Moreover, the abstraction can be implemented efficiently by locating the markers with binary search on a discretized solution path. After digesting work experience slow[v,w] into ρ, we next incorporate it into our repertoire G to achieve incremental learning. Here, there are many ways to implement Learn, ranging from connecting all feasible edges between ρ and G to simply connect ρ to G at v. The tradeoff is between time cost and solution quality. The more edges we attempt to introduce, the more time we will take, but the more choices of solution paths we will have. We do not specify how Learn is to be implemented, just that it needs to augment G with enough edges of ρ to ensure a solution path for reaching w if it were to be requested again.

We illustrate the learning algorithm with a simple example designed to capture the key aspects of the cask inspection problem. Consider a point robot in a two-dimensional workspace with an open disk in the center as obstacle. Let the goal points be uniformly distributed on the boundary of the disk, with the robot positioned initially at one of these points. Let fast implement a go-straight procedure, with fast(u,w) returning success iff w is visible from u, and fast[u,w] returning the line segment $\overline{uw}$. Let slow implement a greedy 2-step go-straight procedure, with slow(u,w) returning success iff there is a point v visible from both u and w, and slow[u,w] returning the shortest path from u to w through such v. To complete the specification, let the heuristic used in Fast and Slow be $h = h_1$, with h_1 ordering the vertices of G according to the distance to w, starting with the closest point first. We call the above specification $\mathcal{E}_1$.

Figure 2 illustrates Adapt with a series of snapshots. Frame (1) shows the initial setting with the robot at position u, and G initialized to the single vertex v_1. The first goal indicated by w is shown in Frame (2). Since Fast is unable to plan using only fast and G, Adapt then calls Slow. Using h, Slow chooses to extend from v_1 to w, since v_1 is the only vertex in G. The path produced by slow(v_1,w) consists of the line segments $\overline{v_1v_2}$ and $\overline{v_2v_3}$ in Frame (3). This path is then abstracted into the chain connecting v_1 to v_2 and v_2 to v_3. The result of augmenting G is that G now becomes the 3-vertex chain. Using this augmented G, Fast is now able to produce a path from u to w, which consists of the segments $\overline{uv_1}$, $\overline{v_1v_2}$, $\overline{v_2v_3}$, and $\overline{v_3w}$, with $\overline{uv_1}$ and $\overline{v_3w}$ being null segments. With the first task accomplished, the next task is shown in Frame (4). This time, Fast is able to solve the task without the help of Slow.

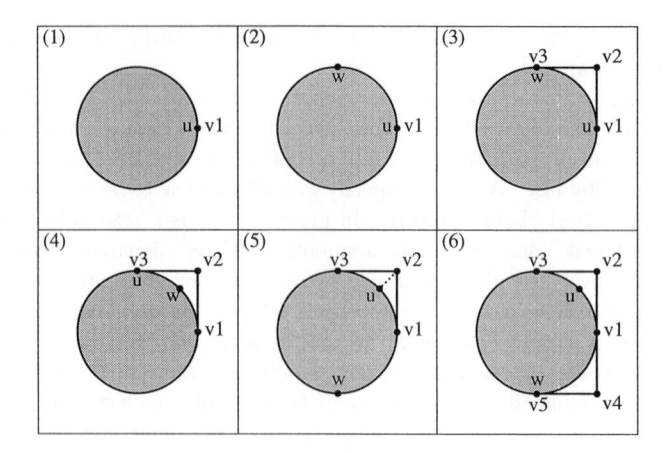

Fig. 2. Snapshots of Adapt.

Using h, it first picks v_2 to test for fast(v_2, w). Accordingly, fast succeeds; hence, Fast sets $\hat{w} = v_2$ and produces the path with segments $\overline{uv_3}$, $\overline{v_3v_2}$, and $\overline{v_2w}$. The result is shown in Frame (5), with the next task indicated. This time, Fast is not able to accomplish the task, so Slow is called. Using h, Slow chooses to extend from v_1, since it is the closest point to w. The resulting path produced by slow(v_1, w) is shown in Frame (6). The chain produced connects v_1 to v_4, and v_4 to v_5. After G is augmented with this chain, Fast is able to solve the third task by producing the segments $\overline{uv_2}$, $\overline{v_2v_1}$, $\overline{v_1v_4}$, $\overline{v_4v_5}$, and $\overline{v_5w}$.

4. General Analysis

In this section, we provide some general analysis to better understand the performance of Adapt. A specific case analysis follows in the next section to illustrate the general results derived here. The techniques developed here should also be useful in analyzing other types of probabilistic learning.

In studying the speedup-learning framework provided by Adapt, two performance measures are of interest: efficiency and capability. To quantify, we assume that the problems are drawn randomly and independently from a distribution (as in PAC-learning[24]) on some configuration space (C-space) S. We do not require slow to be complete; we do require that it have a success probability σ in solving a random task. We assume that only slow, fast, and Learn have costs, each being a constant. (The cost of Abstract can be absorbed into the cost of Learn.) To normalize, let 1, r, and c be the respective costs of slow, fast, and Learn. (Both r and c are typically $\ll 1$.) We use subscript n on a program variable to denote its value at the end of the nth loop. Thus, G_n denotes the experience graph G after Adapt has been trained

Table 1. Random variables of interest.

A_n	The probability that Adapt will need to call slow in solving problem $n+1$, i.e., $1 - \mathbf{E}(\mathsf{Fast}(u, w; G_n, h))$, the probability that a random goal w will not be Fast-reachable via G_n.
I_n	0-1 variable indicating the invocation of Slow when w_{n+1} is not Fast-reachable.
L_n	The probability that a random goal w not Fast-reachable via G_n will now be Fast-reachable via G_{n+1}, assuming that $I_n = 1$.
K_n	The number of times that Slow has been called at the end of the nth loop.
E_n	The cost of Fast in solving problem $(n+1)$.
F_n	The cumulative cost of Adapt after n steps of training.

with n problems. We are interested in both the speedup that Adapt has over the plain iterations of slow, and the capability of Adapt as it increases with training. We are also interested in the performance of Fast, which is Adapt without the backup of Slow after some training. Hence, we analyze the relationships between the random variables in Table 1. These variables are important in characterizing the performance of Adapt. In particular, how fast A_n, the failure probability of Fast after n steps of training, goes to zero determines how well learning takes place.

To analyze these random variables, we use standard techniques in conditional probability theory.[30] Let superscript (n) on an operator denote the conditional operator given A_n. Table 2 summarizes the basic relationships between the random variables (Theorems 1,2,3,4) as well as the major results on estimating the failure probability A_n (Theorems 7,9,10). With respect to Adapt, Theorems 1 and 2 measure time cost; Theorem 3 measures space cost; and Theorem 4 measures capability. Since all these measures depend critically on A_n, we need to analyze A_n carefully. Our analysis is based on the learning rate L_n, which is a key quantity governing the learning process. The expected learning rate $\mathbf{E}(L_n \mid A_n) = \mathbf{E}^{(n)} L_n$ is also important in determining the success of Adapt. Thus, Theorems 7 and 9 explore the consequences when the expected learning rate has a lower bound, while Theorem 10 explores the same when the learning rate has an upper bound.

We now begin our analysis in detail. Readers uninterested in these details may skip to the next section.

Theorem 1 *The average number of calls that* Adapt *will make to* Slow *after n steps of training is*

$$\mathbf{E} K_n = \sum_{0 \le j < n} \mathbf{E} A_j.$$

Proof This is an immediate result of $K_n = \sum_{0 \le j < n} I_j$, and $\mathbf{E} I_j = \mathbf{E}\mathbf{E}^{(j)} I_j = \mathbf{E} A_j$. ∎

Theorem 2 *The average planning cost of* Adapt *per problem after n steps of training is*

$$\mathbf{E} \Delta F_n \stackrel{\text{def}}{=} \mathbf{E}(F_{n+1} - F_n) = (1 + \sigma c)\mathbf{E} A_n + \mathbf{E} E_n. \tag{1}$$

Table 2. Summary of general analysis.

Thm	Condition	Implication
1		$\mathbf{E}K_n = \sum_{0 \leq j < n} \mathbf{E}A_j$
2		$\mathbf{E}F_n = (1 + \sigma c)\mathbf{E}K_n + \sum_{0 \leq j < n} \mathbf{E}E_j$
3		$\mathbf{E}\|V_n\| = \Theta(\mathbf{E}K_n)$
4	$\mathbf{E}A_n \leq \epsilon\delta$	$\Pr(A_n \geq \epsilon) \leq \delta$
7	$\mathbf{E}^{(n)}L_n \geq \alpha$ $\bar{\alpha} \stackrel{\text{def}}{=} 1 - \alpha < 1$	$\mathbf{E}A_n \leq A_0 \bar{\alpha}^n$ $\mathbf{E}K_n \leq A_0(1 - \bar{\alpha}^n)/\alpha \leq A_0/\alpha$
9	$\mathbf{E}^{(n)}L_n \geq \alpha A_n^r$ $r > 0, \ \alpha > 0$	$\mathbf{E}A_n \leq (\alpha r(n+1))^{-1/r} \exp\left(\dfrac{0.52 - \ln n}{2rn}\right)$
10	$L_n \leq \beta A_n$ $0 < \beta < 1$	$\left(\dfrac{1-\beta}{\beta}\right)\ln(A_0/\mathbf{E}A_n) \leq \mathbf{E}K_n \leq \ln_{1-\beta}(\mathbf{E}A_n/A_0)$

Consequently, the average cumulative cost of Adapt *after n steps of training is*

$$\mathbf{E}F_n = (1 + \sigma c)\mathbf{E}K_n + \sum_{0 \leq j < n} \mathbf{E}E_j. \tag{2}$$

Proof The cost for ΔF_n is obvious since in addition to E_n, a cost of $(1 + \sigma c)A_n$ is required to call slow with probability A_n and Learn with probability σA_n. The second equation follows immediately from Theorem 1. ∎

Theorem 3 *The average amount of memory required by* Adapt *as measured by* $\|V_n\|$ *after n steps of training is* $\Theta(\mathbf{E}K_n)$.

Proof Obvious since by our initial assumption on the abstraction process, each call to Slow generates at least one and at most a constant number of vertices. ∎

Theorem 4 *Suppose that* Adapt *is to be trained with random instances until the probability of further* Fast-*failures becomes less than ϵ with confidence at least $1 - \delta$, i.e., until* $\Pr(A_n \geq \epsilon) \leq \delta$. *Then the number of training steps required is at most* n_t, *with n_t being the smallest n satisfying* $\mathbf{E}A_n \leq \epsilon\delta$.

Proof If $\mathbf{E}A_n \leq \epsilon\delta$, then $\Pr(A_n \geq \epsilon) \leq \delta$ because $\mathbf{E}A_n \geq \Pr(A_n \geq \epsilon)\epsilon$. ∎

To interpret this result, suppose that we would be satisfied if Adapt can solve 90% = $1 - \epsilon$ of the problems using only Fast. Then with n_t steps of training, where n_t is the smallest n satisfying $\mathbf{E}A_n \leq 0.001$, we can guarantee with $1 - \delta = 99\%$ confidence that the trained Adapt will be adequate.

The four theorems above show that the expected performance of Adapt depends critically on the behavior of $\mathbf{E}A_n$, whether it is terms of time, space, or accuracy. We now examine the behavior of A_n more closely.

Lemma 5 $A_{n+1} = A_n - L_n I_n$.

Proof If $I_n = 0$, then no change will be made to the experience graph, implying $G_{n+1} = G_n$, and $A_{n+1} = A_n$. If $I_n = 1$, then the graph will be augmented to solve more problems whose probability is measured by L_n. ∎

Lemma 6 For $j \geq 1$, the jth moment of A_n under Adapt satisfies
$$\begin{aligned}\mathbf{E}A_{n+1}^j &= \mathbf{E}A_n^j - \mathbf{E}A_n(A_n^j - \mathbf{E}^{(n)}(A_n - L_n)^j)\\ &\leq \mathbf{E}A_n^j - \mathbf{E}A_n(\mathbf{E}^{(n)}L_n)^j.\end{aligned}$$
In particular, for $j = 1$, $\mathbf{E}A_{n+1} = \mathbf{E}A_n(1 - \mathbf{E}^{(n)}L_n)$.

Proof To derive the equation, it suffices to prove that
$$\mathbf{E}^{(n)}A_{n+1}^j = A_n^j - A_n(A_n^j - \mathbf{E}^{(n)}(A_n - L_n)^j),$$
which follows from the fact $\mathbf{E}^{(n)}I_n = A_n$ so that during the $(n+1)^{\text{th}}$ loop, A_{n+1} will remain unchanged from A_n with probability $1 - A_n$, and decrease by L_n from A_n with probability A_n.

To derive the inequality, observe that
$$X^j + Y^j \leq (X + Y)^j$$
for $X = A_n - L_n \geq 0$, $Y = L_n \geq 0$, and $j \geq 1$. Hence, we have
$$\begin{aligned}\mathbf{E}A_{n+1}^j &\leq \mathbf{E}A_n^j - \mathbf{E}A_n(A_n^j - \mathbf{E}^{(n)}(A_n^j - L_n^j))\\ &= \mathbf{E}A_n^j - \mathbf{E}A_n(\mathbf{E}^{(n)}L_n^j),\end{aligned}$$
yielding the result. ∎

Theorem 7 Suppose that Adapt has at least an expected learning rate of $\mathbf{E}^{(n)}L_n \geq \alpha = 1 - \bar{\alpha}$, for some positive α. Then $\mathbf{E}A_n \leq A_0 \bar{\alpha}^n$ and
$$\mathbf{E}K_n \leq A_0(1 - \bar{\alpha}^n)/\alpha \leq A_0/\alpha.$$

Proof Obvious from unfolding the first moment recurrence of Lemma 6. ∎

Lemma 8 Any positive decreasing sequence $\{a_n\}_{n\geq 0}$ with $a_0 \leq 1$ that satisfies the inequality
$$a_n \leq a_{n-1}(1 - \alpha a_{n-1}^r) \tag{3}$$
for some positive α and r, has an upper bound of
$$a_n \leq (\alpha r(n+1))^{-1/r} \exp\left(\frac{0.52 - \ln n}{2rn}\right), \tag{4}$$
for $n > 0$.

Proof Let $a_n = (\alpha r(n+1))^{-1/r} e^{b_n}$. Then Inequality (3) becomes
$$e^{b_n} \leq (1 + 1/n)^{1/r} e^{b_{n-1}} \left(1 - e^{rb_{n-1}}/rn\right).$$
Taking the logarithm of both sides and Taylor-expands the RHS yields
$$b_n \leq b_{n-1} + \frac{1}{r}\left(\frac{1}{n} - \frac{1}{2n^2} + \frac{1}{3n^3}\right) - \frac{e^{rb_{n-1}}}{rn},$$
which further simplifies to
$$b_n \leq \left(1 - \frac{1}{n}\right) b_{n-1} - \frac{1}{rn^2}\left(\frac{1}{2} - \frac{1}{3n}\right).$$
Multiplying both sides by n and letting $c_n = nb_n$ yields
$$c_n \leq c_{n-1} - \frac{1}{rn}\left(\frac{1}{2} - \frac{1}{3n}\right).$$
Finally, the recurrence unfolds and simplifies to
$$c_n \leq (\pi^2/9 - \ln n - \gamma)/(2r) \leq (0.52 - \ln n)/(2r),$$
which yields the lemma. ∎

Theorem 9 *Suppose that* Adapt *has at least an expected learning rate of* $\mathbf{E}^{(n)} L_n \geq \alpha A_n^r$, *for some positive* α *and* r. *Then the average* Fast-failure *probability of* Adapt *after* $n > 0$ *steps of training has an upper bound of*
$$\mathbf{E} A_n \leq (\alpha r(n+1))^{-1/r} \exp\left(\frac{(0.52 - \ln n)}{2rn}\right).$$

Proof From Lemma 6, we have for $n > 0$ that
$$\begin{aligned}
\mathbf{E} A_n &= \mathbf{E} A_{n-1}(1 - \mathbf{E}^{(n-1)} L_{n-1}) \\
&\leq \mathbf{E} A_{n-1}(1 - \alpha A_{n-1}^r) \\
&\leq (\mathbf{E} A_{n-1})(1 - \alpha \mathbf{E}^r A_{n-1}),
\end{aligned}$$
where the last inequality comes from the fact that $\mathbf{E} X^r \geq \mathbf{E}^r X$ for $X \geq 0$. The desired upper bound can be obtained by identifying $\mathbf{E} A_n$ with a_n of Lemma 8. ∎

Theorem 10 *Suppose that* Adapt *has a maximum learning rate of* $L_n \leq \beta A_n$ *for some positive* $\beta < 1$, *and that the initial probability of failure is* $A_0 = \mathbf{E} A_0 > 0$. *Then* $\mathbf{E} K_n$ *has, in terms of* $\mathbf{E} A_n$, *the bound of*
$$\beta^{-1}(1-\beta) \ln(A_0/\mathbf{E} A_n) \leq \mathbf{E} K_n \leq \ln_{1-\beta}(\mathbf{E} A_n/A_0),$$
which implies that $\mathbf{E} A_n$ *has, in terms of* $\mathbf{E} K_n$, *the bound of*
$$e^{-(\beta/(1-\beta))\mathbf{E} K_n} \leq \mathbf{E} A_n \leq (1-\beta)^{\mathbf{E} K_n}.$$

Proof We first prove the lower bound for $\mathbf{E}K_n$. From Lemma 5, we have

$$A_{n+1} = A_n(1 - L_n I_n/A_n) \geq A_n(1 - \beta I_n) > 0.$$

Therefore, $A_{n+1}^{-1} \leq A_n^{-1}/(1 - \beta I_n) = A_n^{-1}(1 + \beta I_n/(1 - \beta))$, implying that

$$\begin{aligned}
\mathbf{E}A_{n+1}^{-1} &\leq \mathbf{E}A_n^{-1}(1 + \beta I_n/(1 - \beta)) \\
&= \mathbf{E}A_n^{-1}(1 + \beta A_n/(1 - \beta)) \\
&\leq (\mathbf{E}A_n^{-1})(1 + \beta \mathbf{E}A_n/(1 - \beta)).
\end{aligned}$$

Unfolding the recurrence above yields $\mathbf{E}A_n^{-1} \leq A_0^{-1} \prod_{0 \leq j < n}(1 + \beta \mathbf{E}A_j/(1-\beta))$. Taking the logarithm of both sides and using the fact that $\mathbf{E}A_n^{-1} \geq 1/\mathbf{E}A_n$ gives us

$$\begin{aligned}
\ln(A_0/\mathbf{E}A_n) &\leq \sum_{0 \leq j < n} \ln(1 + \beta \mathbf{E}A_j/(1-\beta)) \\
&\leq \sum_{0 \leq j < n} \frac{\beta}{1-\beta} \mathbf{E}A_j = \frac{\beta}{1-\beta} \mathbf{E}K_n,
\end{aligned}$$

which yields the desired lower bound.

To obtain the upper bound of $\mathbf{E}K_n$, we simply unfold $A_{n+1} \geq A_n(1 - \beta I_n)$ into

$$A_n \geq A_0 \prod_{0 \leq j < n}(1 - \beta I_j).$$

Taking the logarithm of both sides yields

$$\begin{aligned}
\ln A_n/A_0 &\geq \sum_{0 \leq j < n} \ln(1 - \beta I_j) \\
&= \sum_{0 \leq j < n} \sum_{i \geq 1} (\beta I_j)^i/i \\
&= \sum_{0 \leq j < n} \sum_{i \geq 1} \beta^i I_j/i \\
&= \ln(1 - \beta) \sum_{0 \leq j < n} I_j.
\end{aligned}$$

The upper bound follows by taking the expectation of both sides and using the fact that $\mathbf{E} \ln A_n \leq \ln \mathbf{E}A_n$. ∎

5. A Specific Case Analysis

To demonstrate our general theorems, consider again the simple two-dimensional environment $\mathcal{E}_1$ introduced earlier in illustrating the algorithm. In this environment, G is a chain wrapping around the obstacle with the boundary vertices on the circle. Points that are **Fast**-reachable are exactly those points of the arc covered by G and delimited by the two boundary vertices. Let C be the arc covered by G, and let the circumference be of unit length. Then probability A_n is exactly 1 minus the length of C_n.

To evaluate L notice that since we are using heuristic $h = h_1$ to guide our vertex selection for graph extension, we will always choose the boundary vertex closest to goal w as $\hat{w}$ in calling Slow when w lies outside C. Moreover, $\mathrm{arc}(w, \hat{w})$ outside C contains exactly those points that would not be Fast-reachable after learning w. Hence, probability L_n is exactly the length of $\mathrm{arc}(w_{n+1}, \hat{w}_{n+1})$ when $I_n = 1$, where I_n indicates the case of w_{n+1} lying outside C. Because w_{n+1} is uniformly distributed, we then have L_n uniformly distributed on the interval $[0, A_n/2]$. Consequently, Adapt has a maximum learning rate of $L_n \leq A_n/2$, and an expected learning rate of $\mathbf{E}^{(n)} L_n = A_n/4$.

Theorem 11 *The average number of calls that* Adapt *will make to* Slow *under* $\mathcal{E}_1$ *is* $\mathbf{E} K_n = \Theta(\ln n)$.

Proof Using Theorem 9 with $\alpha = 1/4$ and $r = 1$, we obtain $\mathbf{E} A_n = O(1/n)$. Applying Theorem 1 with this fact gives us immediately the upper bound of $\mathbf{E} K_n = O(\ln n)$. Applying the first inequality of Theorem 10 with $\beta = 1/2$ gives us the lower bound of $\mathbf{E} K_n = \Omega(\ln 1/\mathbf{E} A_n) = \Omega(\ln n)$. ∎

Define the *inefficiency* of Adapt to be the ratio of $\mathbf{E} K_n$ over n. From Theorem 3 and the theorem above, we see that although the inefficiency of Adapt under $\mathcal{E}_1$ does approach 0, the memory requirement is actually unbounded. Fortunately, we can avoid this problem by seeking an Fast-planner with less than 100% accuracy. We simply stop the learning process after a training period as prescribed by the following theorem.

Theorem 12 *For* Adapt *under* $\mathcal{E}_1$ *to obtain, with confidence* $1 - \delta$, *a* Fast*-planner that can solve all but ϵ fraction of the tasks, at most* $n_t \leq 5.2/(\epsilon \delta)$ *steps of training are necessary.*

Proof From Theorem 9, we have $\mathbf{E} A_n \leq 4 e^{0.26}/(n+1)$. Applying Theorem 4 with this fact yields the result. ∎

Incidentally, if we have a convex m-sided polygon instead of a disk as the obstacle, Adapt would be able perform much better. Once Adapt learns to reach a particular point of a side, every point on that side becomes reachable. Therefore, for Adapt to learn in this environment $\mathcal{E}_2$, at most m calls to slow and $O(m)$ number of trail-markers are required (cf. Theorem 7).

Let us now reconsider the unbounded memory problem that Adapt faces under $\mathcal{E}_1$. The difficulty is that its set of reachable goals can only approach, but never be the desired set. We can rectify this situation by modifying Adapt so that it learns more in the beginning. The modification allows the desired set to be learned completely so that slow learning does not occur at the end. Figure 3 shows the modified algorithm, Modapt, that satisfies our need.

```
Algorithm Modapt(Fast, Slow)
    u ← current position; v ← u; G ← ({v}, ∅);
    do while training
        w ← goal;
        if (not Fast(u, w; G, h')) then
            if (not Slow(u, w; G, h')) then continue;
            ρ ← Abstract(Slow[u, w; G, h']);
            G ← Learn(G, ρ);
        endif
        execute(Fast[u, w; G, h']); u ← w;
    enddo
    do forever
        w ← goal;
        if (not Fast(u, w; G, h)) then continue;
        execute(Fast[u, w; G, h]); u ← w;
    enddo
end.
```

Fig. 3. A modified algorithm to speed up learning.

In Modapt, we separate the initial training period that uses heuristic h' from the later working period that uses h. Ideally, h' should be more stringent than h, i.e., $\mathsf{Fast}(u, w; G, h') \leq \mathsf{Fast}(u, w; G, h)$, so that $\mathsf{Slow}(u, w; G, h')$ can be called more often. For $\mathcal{E}_1$, we use h_2 that not only orders the vertices as in h_1, but also rules out vertices v whose best potential path from u to w through v has length exceeding a certain value, say half the circumference. Using $h' = h_2$, we can force $\mathsf{Fast}(u, w; G, h')$ to fail on cases where w lies on C, but the length of $\mathrm{arc}(u, w)$ covered by C exceeds half the circumference. Subsequently, Modapt would be able to call $\mathsf{Slow}(u, w; G, h')$, so that G can be extended to cover the complement of C in one step. Once the entire circle is covered by G, Modapt can stop learning and use $\mathsf{Fast}(u, w; G, h)$ to reach every goal on the circle.

To analyze Modapt under $\mathcal{E}_1$, again let A_n be the probability of Fast-failure using G_n and h. Let $J_n = \lceil \hat{w}_{n+1} = w_n \rfloor$, with $\lceil \cdot \rfloor$ denoting the indicator function: 1 if the predicate is satisfied; 0 otherwise. Let $X_n \leq 1/2$ be the length of the shorter arc connecting w_{n+1} and $\hat{w}_{n+1}$ when Fast fails using G_n and h', and 0 otherwise. Define $\bar{a} = 1 - a$, and $b = a - 1/2$.

Lemma 13 *For $0 < x \leq 1/2$, the density functions $f_j(x, a)$ for X_n at x given $A_n = a$ and $J_n = j$ are*

$$f_1(x, a) = 1 + \lceil x \leq b \rfloor$$
$$f_0(x, a) = (2/\bar{a}) \min(\bar{a}, 1/2 - x).$$

Proof We use the fact that the length of C_n is $\bar{a}$, and the length of $\text{arc}(w_n, w_{n+1})$ containing $\hat{w}_{n+1}$ is at most $1/2$. For case $J_n = 1$, w_n is at either ends of C_n. So to achieve $X_n = x$, when $x + \bar{a} \leq 1/2$, two places for w_{n+1} are possible, and when $x + \bar{a} > 1/2$, only one place for w_{n+1} is possible. For case $J_n = 0$, w_n is uniformly distributed on C_n. So to achieve $X_n = x$, when $x + \bar{a} \leq 1/2$, exactly $\min(1/2, \bar{a})/\bar{a}$ fraction of C_n can be used for w_n to realize one of two sites for w_{n+1}. The fraction decreases linearly to 0 at $x = 1/2$. ∎

With this lemma, we can evaluate the effective learning rate

$$\Delta A_n \stackrel{\text{def}}{=} A_n - A_{n+1}$$

by conditioning it on the variables $\lceil A_n \geq 1/2 \rfloor$ and J_n. To simplify, let

$$E_{i,j}(a) \stackrel{\text{def}}{=} \mathbf{E}(\Delta A_n \mid A_n = a, \lceil A_n \geq 1/2 \rfloor = i, J_n = j).$$

Lemma 14

$$\begin{aligned} E_{1,1}(a) &= 1/8 + b^2/2 \\ E_{1,0}(a) &= (1/8 - b^3)/(3\bar{a}) \\ E_{0,1}(a) &= a\bar{a}/2 \\ E_{0,0}(a) &= (a^2(1/2 - 2a/3) + ab^2)/\bar{a}. \end{aligned}$$

Proof We use the fact that $\Delta A_n = \min(X_n, A_n)$ and the previous lemma on X_n to evaluate

$$E_{i,j}(a) = \int_0^{1/2} \min(x, a) \, f_j(x, a) \, dx$$

conditioned on $\lceil a \geq 1/2 \rfloor = i$. In particular, we have

$$E_{1,1} = \left(\int_0^b 2 + \int_b^{1/2} 1 \right) x \, dx$$

$$E_{1,0} = \left(\int_0^b 2 + \int_b^{1/2} (2/\bar{a})(1/2 - x) \right) x \, dx$$

$$E_{0,1} = \left(\int_0^a x + \int_a^{1/2} a \right) 1 \, dx$$

$$E_{0,0} = \left(\int_0^a x + \int_a^{1/2} a \right) (2/\bar{a})(1/2 - x) \, dx.$$

The rest is just simple computation. ∎

Theorem 15 *For* Modapt *under* $\mathcal{E}_1$ *to obtain, with confidence* $1 - \delta$, *a* Fast-planner *that can solve all but ϵ fraction of the tasks, at most $n_t \leq \ln_{19/24}(\epsilon\delta)$ steps of training are necessary.*

Proof We show that $\mathbf{E}^{(n)}\Delta A_n \geq 5A_n/24$, which implies the upper bound of $\mathbf{E}A_n \leq (19/24)^n$ in the application of Theorem 4. For $J_n = 1$, w_n must be at the boundary of C_n, an event which occurs with probability $A_{n-1} \geq A_n$. Using the previous lemma, we bound $\mathbf{E}^{(n)}\Delta A_n$ by considering whether $A_n \geq 1/2$. Thus, let $a = A_n$, $x = \Pr(J_n = 1) \geq a$, and $i = \lceil a \geq 1/2 \rfloor$. Then

$$\mathbf{E}^{(n)}\Delta A_n = xE_{i,1}(a) + (1-x)E_{i,0}(a)$$

can be bounded below by setting $x = a$ because $E_{i,1}(a) - E_{i,0}(a)$, the coefficient of $x \geq a$, is nonnegative in the respective ranges of a depending on i. Specifically,

$$E_{1,1}(a) - E_{1,0}(a) = \bar{a}^2/6,$$
$$E_{0,1}(a) - E_{0,0}(a) = a(a^2/3 - b)/(2\bar{a}).$$

Finally, the derived lower bound $aE_{i,1}(a) + \bar{a}E_{i,0}(a)$ simplifies to $g_i(a)\,a$ with

$$g_1(a) = b^2/3 + (1 + 1/(3a))(1/8 + b^2)$$
$$g_0(a) = (1/2 - a^2/3)/2,$$

where $g_i(a)$ has a minimum of $5/24$ at $a = 1/2$ regardless of i. ∎

6. Particular Analysis

So far we have presented in the general analysis section a list of general relationships between important parameters of the learning process such as time cost, space cost, capability, and learning rate. We have also presented in previous section a specific case analysis of the learning algorithm applied to environment $\mathcal{E}_1$. Although the theorems of Table 2 are applicable to all environments, they are too general to yield immediate, useful results. On the other hand, results concerning one specific environment cannot generally be extended to another. It is thus the objective of this section to bridge the gap between general and specific case analysis.

In this section, we present an approach to analyzing speedup learning,[31] which is what the algorithm is doing – seeking to improve program efficiency through learning. We first formalize the concept of improvability, and derive general conditions for such improvements. Next, we introduce two models with additional simplifying assumptions and parameters. Using these models, we then derive sharp bounds on planning cost and capability in terms of training time. Finally, we characterize the improvable situations in terms of the model parameters, and prescribe the amount of training required.

Definition 1 *Let A be a speedup learning algorithm designed to improve the efficiency of another algorithm A'. We say that*

```
        ┌─────────┬─────────┐
        │         │         │
        │    A    │    B    │
        │         │         │
        └─────────┴─────────┘
```

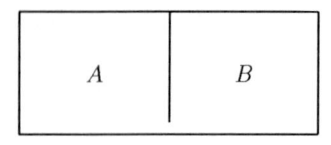

Fig. 4. An environment with two traps.

1. A can *(cost-wise)* improve A' with average failure probability p *iff* A can perform the same task as A' with average probability at least $1 - p$, while costing less on average.
2. A can *effectively* improve A' *iff* A can improve A' with failure probability no greater than that of A'.
3. A can *effectively* replace A' *iff* A can effectively improve A' without relying on A'.

Using the variables of Table 1, we can immediately characterize the conditions under which improvements can be achieved. (Proof is clear.)

Lemma 16 *After n steps of training,*

1. Fast can improve slow with failure probability $\mathbf{E}A_n$ *iff* $\mathbf{E}E_n < 1$.
2. Adapt can effectively improve slow *iff* $\mathbf{E}\Delta F_n < 1$.
3. Fast can effectively replace slow *iff* $\mathbf{E}E_n < 1$ and $\mathbf{E}A_n \leq 1 - \sigma$.

To express these conditions in more useful terms of training time, we need to have further information such as the specification of the vertex ordering function h and the incremental learning strategy Learn. Thus, we introduce two models, one pessimistic (on task complexity), the other randomized (on experience utility), each with different applicability and additional simplifying assumptions. Using these models and the theorems of Table 2, we derive sharper bounds on the variables of Table 1, and explore the ramifications of Lemma 16.

In the pessimistic model, we study the worst-case consequence of learning in environments in which the strategy of Learn is specified, and the connectivity of S under fast is characterized. To motivate, consider a point robot in a planar polygonal environment shown in Figure 4, and let fast be 'go-straight'. Since we are dealing with a point robot, the C-space and the work space are the same. Clearly, the C-space is well-connected locally in the sense that each feasible (configuration) point is connectable (visible) to at least half of the entire C-space under fast. However, this environment may be difficult for learning algorithms with no teachers[18,25] to handle in that the points randomly sampled will tend to form two disconnected components (traps) in A and B, and will not help in solving problems that require reaching B from A. In contrast, with the help of slow, our algorithm will adapt to this environment efficiently as long as the number of components induced by fast is not too large. Thus, using the following definition of the pessimistic model, we explore the consequences when the complexity of the C-space S relative to fast is measured by m in that S is m-coverable. Incidentally, the environment in Figure 4 is 2-coverable under fast, with the two components being A and B.

Definition 2 *Under pessimistic model $\mathcal{M}_p$,*

1. *Learn adopts the minimal memory strategy of adding all necessary edges to take in only the minimal subchain sufficient to Fast-solve the current problem.*

2. *C-space S is m-coverable for some m in that S can be covered by m components (not necessarily disjoint), with the initial configuration in S_1 and each component S_i ($1 \leq i \leq m$) being connected under fast, i.e., every pair of points u, v in S_i satisfies $\mathsf{fast}(u,v)$.*

In the randomized model, we study the average-case consequence of learning in environments in which the number of new trail-markers acquired by Learn and the power of fast are randomized. Thus, we are interested in the average behavior for a class of environments instead of a fixed environment. Using the following definition of the randomized model, we explore the consequences when the utilities of the learned trail-markers are measured by $\bar{\mu}$ and $\tilde{\mu}$. While $\mathcal{M}_r$ may not be physically realizable as opposed to $\mathcal{M}_p$, it does simplify the corresponding results for $\mathcal{M}_p$, and provide reasonable estimation tools as demonstrated in the applications section.

Definition 3 *Under randomized model $\mathcal{M}_r$,*

1. *The number of new trail-markers acquired by Learn, λ, is an independent random variable.*

2. *$\mathsf{fast}(u,w)$ is 1 for any established edge (u,w) in V, and is 1 otherwise with independent probability $\bar{\mu}$.*

Table 3 summarizes and compares the consequences of the two models. It shows that the reliance of Adapt on slow is at most inversely proportional to the training time n; it also shows that the reliance is at most proportional to m, the complexity of S under $\mathcal{M}_p$, and is correspondingly, at most proportional to $1/\tilde{\mu}$, the power of fast under $\mathcal{M}_r$. In fact, the correspondence between these model parameters: m with $\tilde{\mu}$ and $\bar{\mu}$ is indicated throughout the table. The situation in which learning is useful (effective improvement or replacement of slow) is discerned by weighing $1/r$, the speed of fast, against m and correspondingly $1/\bar{\mu}$. To achieve this usefulness, the necessary training time is also prescribed. Finally, the speedup performance of Adapt as measured by the ratio of the planning cost of Adapt to that of slow is presented. The rest of this section is devoted to the details of these results. Readers uninterested in these details may skip to the next section on applications.

6.1. Pessimistic Model

Under the pessimistic model, the complexity of the C-space S relative to fast is measured by m in that S is m-coverable. Using this complexity parameter, the

Table 3. Summary of particular analysis.

prop.	pessimistic model	randomized model
param.	m	$\bar{\mu},\ \lambda,\ \tilde{\mu} \stackrel{\text{def}}{=} 1 - \mathbf{E}(1-\bar{\mu})^{\lambda}$
capab.	$\mathbf{E}A_n \leq \dfrac{m-1}{\sigma n}$	$\mathbf{E}A_n \leq \dfrac{1}{\sigma\tilde{\mu}(n+1)}$
cost	$\mathbf{E}E_n \leq r(3m-2)$	$\mathbf{E}E_n = \dfrac{r}{\mu}(1 - \mathbf{E}A_n)$
improve time	$r < 1/(3m-2)$ $n \geq \dfrac{(m-1)(1+\sigma c)}{\sigma(1 - r(3m-2))}$	$r < \bar{\mu}$ $n \geq \dfrac{1}{\sigma\tilde{\mu}}\left(1 + \dfrac{\sigma c}{1 - r/\bar{\mu}}\right)$
replace time	$r < 1/(3m-2),\ \sigma < 1$ $n \geq \dfrac{m-1}{\sigma(1-\sigma)}$	$r < \bar{\mu},\ \sigma < 1$ $n \geq \dfrac{1}{\sigma(1-\sigma)\tilde{\mu}}$
cost ratio	$\dfrac{\mathbf{E}F_n}{n} \leq r(3m-2) + \dfrac{(m-1)(1+\sigma c)}{\sigma n}$	$\dfrac{\mathbf{E}F_n}{n} \leq \dfrac{r}{\mu} + \left(1+\sigma c - \dfrac{r}{\bar{\mu}}\right)\dfrac{\ln(eA_0\sigma\tilde{\mu}n)}{\sigma\tilde{\mu}n}$

following theorem says that the failure probability of **Adapt** is at most proportional to m and inversely proportional to the amount of training n, and that this bound is tight up to some constant factor.

Theorem 17 *Under $\mathcal{M}_p$, the expected **Fast**-failure probability of **Adapt** after n steps of training has a upper bound of*

$$\mathbf{E}A_n \leq \frac{m-1}{\sigma n}, \tag{5}$$

and a lower bound of

$$\mathbf{E}A_n \geq \frac{m-1}{e\sigma(n+1)} \tag{6}$$

for $n \geq (m-1)/\sigma$ and some environment dependent on n.

Proof (5) Let w_j be the jth random problem. Let $X_{i,j}$ be the 0-1 random variable indicating that $w_{j+1} \in S_i$ and w_{j+1} is not **Fast**-solvable using V_j. Since the S_i's cover S, we have $\mathbf{E}A_n \leq \mathbf{E}\sum_{i>1} X_{i,n}$. Also, since **Adapt** never forgets, we have $V_j \subseteq V_{j+1}$ for all j, which implies that $\mathbf{E}X_{i,j} \geq \mathbf{E}X_{i,j+1}$ because $\mathsf{Fast}(w; V_j, h) \implies \mathsf{Fast}(w; V_{j+1}, h)$ for all w. Finally, if $X_{i,j} = 1$ then **Slow** will be called. If it is successful, w_{j+1} will be remembered in V_{j+1}, causing $X_{i,j'} = 0$ for $j' > j$. Consequently, for any i, $\mathbf{E}\sum_j X_{i,j} \leq 1/\sigma$, which is the expected number of times that **Slow** will be called to reach some points in S_i before one is remembered. Combining all three inequalities,

we have
$$\mathbf{E}A_n \leq \sum_{i>1} \mathbf{E}X_{i,n} \leq \sum_{i>1} \sum_{0<j\leq n} \frac{\mathbf{E}X_{i,j}}{n} \leq \sum_{i>1} \frac{1}{\sigma n} = \frac{m-1}{\sigma n}.$$

(6) Let S be composed of exactly m non-overlapping components, with every component disconnected from each other except S_1 under **fast**. Let the distribution be uniform within each component and have total probability $p = 1/(\sigma(n+1))$ on S_i for $i > 1$, and probability $1 - (m-1)p$ on S_1. Then for $i > 1$, the probability that Adapt will be given j training problems from S_i after n steps and failed to learn from them is $\binom{n}{j}p^j(1-p)^{n-j}(1-\sigma)^j$. Thus, with this probability summed over j, Adapt will **Fast**-fail on S_i. Summing up each $i > 1$, we have

$$\mathbf{E}A_n \geq (m-1)p \sum_j \binom{n}{j}(p(1-\sigma))^j(1-p)^{n-j} = (m-1)p(1-p\sigma)^n,$$

yielding the desired lower bound. ∎

Using m again as a complexity measure of the C-space, the following theorem says that the **Fast**-planning cost of Adapt is at most linear in r and m, and that this bound is tight up to a constant factor. Further, the number of times **slow** will be needed is at most proportional to m and inversely proportional to its capability σ, and that this bound is tight with sufficient amount of training.

Theorem 18 *Under $\mathcal{M}_p$, the expected* **Fast**-*planning cost of* Adapt *after n steps of training has an upper bound of*

$$\mathbf{E}E_n \leq r\left(3m - 2 - (m-1)\left(1 - \frac{\sigma}{m-1}\right)^n\right). \tag{7}$$

The expected number of calls to Slow *has an upper bound of*

$$\mathbf{E}K_n \leq \left(\frac{m-1}{\sigma}\right)\left(1 - \left(1 - \frac{\sigma}{m-1}\right)^n\right). \tag{8}$$

Conversely, there exists an environment in which $\mathbf{E}E_n = r(2m-1)$, and an environment in which the equality of (8) is reached.

Proof Let J_n be the number of times **slow** is successful. Classify the trail-markers of ρ into two types: type 1 sharing the same S_i for some i with the current V, and type 2 does not. According to the 'memory minimizing' strategy of Learn, at most one marker of the subchain can be of type 1, and at most two markers per component not 'occupied' by the current V can be of type 2. (Otherwise, an edge can be introduced to shorten the subchain.) Thus, counting all subchains, the total number of type 1 and 2 markers are at most J_n and $2(m-1)$, respectively. Hence, $\|V_n\| \leq 1 + 2(m-1) + J_n$.

For (7), it suffices to show that $\mathbf{E}J_n \leq (m-1)\left(1 - \left(1 - \frac{\sigma}{m-1}\right)^n\right)$. Partition S into m disjoint components with the ith component being $S'_i = S_i \setminus \bigcup_{j<i} S_j$. Let X_i be the 0-1 random variable indicating that one of the n training problems is both

in S'_i and solvable by Slow. Then $J_n \leq \sum_{i>1} X_i$ because there can be at most one successful call of Slow for each $i > 1$. Let p_i be the probability that a random problem is both in S'_i and solvable by Slow. Then $\mathbf{E} X_i = 1 - (1 - p_i)^n$ and $\sum_{i>1} p_i = \sigma - p_1$. Using the fact that $(p_2, \ldots, p_m)$ majorizes[23] $(\frac{\sigma - p_1}{m-1}, \ldots, \frac{\sigma - p_1}{m-1})$, we have $\sum_{i>1}(1 - p_i)^n \geq (m-1)(1 - \frac{\sigma - p_1}{m-1})^n \geq (m-1)(1 - \frac{\sigma}{m-1})^n$, as desired. For (8), simply notice that $\sigma \mathbf{E} K_n = \mathbf{E} J_n$.

For the lower bound on $\mathbf{E} E_n$, let S be composed of exactly m non-overlapping components, with the first $m - 1$ components consisting of exactly 2 points, $s_{i,1}$ and $s_{i,2}$. Let the only inter-component connections under fast be between $s_{i,2}$ and $s_{i+1,1}$. Let $s_{1,1}$ be the initial configuration, and let the distribution be 0 on the first $m - 1$ components. Let h select the markers in the increasing order of the component index. Then upon solving the first problem, a path of $2m - 1$ markers connecting $s_{1,1}$ to a point in S_m will be incorporated into V. Consequently, a Fast-planning cost of $r(2m - 1)$ is required for latter problems.

For the lower bound on $\mathbf{E} K_n$, let S be composed of exactly m non-overlapping components, with every component disconnected from each other except S_1 under fast. For each $i > 1$, let the distribution have equal total probability $1/(m-1)$ on S_i. Then the learning process becomes effectively a coupon collector's problem[11] with $m - 1$ types of coupons. Thus, $\mathbf{E} J_n = \sum_{i>0} \mathbf{E} X_i$, where X_i is the 0-1 random variable indicating that one of the n training problems is both in S_i and solved by Slow. The bound now follows since $\mathbf{E} X_i = 1 - (1 - \frac{\sigma}{m-1})^n$. ∎

Using the estimates provided above, the following theorem discerns the situations in which Adapt is useful by weighing $1/r$, the speed of fast, against m, the complexity of S. For those situations in which Adapt can be useful, it also prescribes the amount of training required.

Theorem 19 *Under $\mathcal{M}_p$, if $1/r > (3m - 2)$, then Adapt can effectively improve slow with*

$$n \geq \frac{(m-1)(1 + \sigma c)}{\sigma(1 - r(3m - 2))} \qquad (9)$$

steps of training. If slow is also not complete, then Fast can effectively replace slow with

$$n \geq \frac{m - 1}{\sigma(1 - \sigma)} \qquad (10)$$

steps of training. If $2m - 1 < 1/r \leq 3m - 2$, then after

$$n < \ln_{\left(\frac{m-1}{m-2}\right)} \left(\frac{m - 1}{3m - 2 - 1/r} \right) \qquad (11)$$

steps of training, Fast can still improve slow with average failure probability no greater than $(m-1)/(\sigma n)$. If $1/r \leq 2m - 1$, then there exist environments in which neither Adapt nor Fast can improve slow.

Proof From Theorem 18, we have $\mathbf{E}E_n < r(3m-2)$ for any finite n. If (9) holds, then from Theorem 17 and Theorem 2, we have $\mathbf{E}\Delta F_n \leq 1 - r(3m-2) + \mathbf{E}E_n < 1$, as prescribed by 2 of Lemma 16. Further, if $\sigma < 1$ and (10) holds, then $\mathbf{E}A_n \leq 1-\sigma$ and $\mathbf{E}E_n < 1$, as prescribed by 3 of Lemma 16. If $2m - 1 < 1/r \leq 3m - 2$ and (11) holds, then from Theorem 18, $\mathbf{E}E_n < r(2m-1+(m-1)(1-(3m-2-1/r)/(m-1))) = 1$, as prescribed by 1 of Lemma 16. If $2m - 1 \geq 1/r$, then by Theorem 2 and Theorem 18, there exists an environment in which $\mathbf{E}F_n \geq \mathbf{E}E_n = r(2m-1) \geq 1$, contrary to what is required for improvement in Lemma 16. ∎

Finally, the speedup performance of Adapt is provided by the following.

Theorem 20 *Under $\mathcal{M}_p$, the ratio of the average cost of Adapt to that of slow is bounded above by*

$$\frac{\mathbf{E}F_n}{n} \leq r(3m-2) + \frac{(m-1)(1+\sigma c)}{\sigma n}. \tag{12}$$

Hence, it is bounded asymptotically by $r(3m-2)$ as the number of training problems approaches infinity.

Proof The upper bound follows immediately from Theorem 2 and Theorem 18 with the simplications that $\mathbf{E}K_n \leq r(3m-2)$ and $\mathbf{E}K_n \leq (m-1)/\sigma$. ∎

6.2. Randomized Model

Under the randomized model, the utilities of the trail-markers learned are measured by $\bar{\mu}$ and $\widetilde{\mu}$, with their reciprocals measuring the capability of fast. In parallel with the previous subsection, the following theorems estimate the key variables of the learning process with the capability of fast essentially replacing the role of the C-space complexity m.

Theorem 21 *Under $\mathcal{M}_r$,*

$$\mathbf{E}A_n \leq \frac{1}{\sigma\widetilde{\mu}(n+1)}, \tag{13}$$

where $\widetilde{\mu} \stackrel{\text{def}}{=} 1 - \mathbf{E}(1-\bar{\mu})^\lambda$ denotes the average utility of a learned chain ρ.

Proof Let $L_n = A_n - A_{n+1}$ be the additional probability of problems learned through the incorporation of ρ_n into V_n. We call $\mathbf{E}(L_n \mid A_n)$ the expected learning rate, which evaluates to $A_n \sigma \widetilde{\mu}$.

It now suffices to show that for $n > 0$, $\mathbf{E}A_n \leq 1/(\alpha(n+1))$ for an expected learning of $\mathbf{E}(L_n \mid A_n) = \alpha A_n$, $\alpha < 1$. For $n = 0$, $\mathbf{E}A_0 \leq 1 \leq 1/\alpha$. For $n = 1$, $\mathbf{E}A_1 = \mathbf{E}A_0(1 - \alpha A_0)$ has maximum value $1/(4\alpha)$, which is less than the desired upper bound of $1/(2\alpha)$. For $n \geq 2$, we have from Theorem 9 that $\mathbf{E}A_n \leq (\alpha(n+1))^{-1} \exp((0.52 - \ln n)/(2n))$, which implies the desired upper bound. ∎

Theorem 22 *Under $\mathcal{M}_r$, the expected cost of* **Fast** *after n steps of training is*

$$\mathbf{E}E_n = \frac{r}{\bar{\mu}}(1 - \mathbf{E}A_n). \tag{14}$$

Consequently, the expected cost of **Adapt** *per problem after n steps of training is*

$$\mathbf{E}\Delta F_n = r/\bar{\mu} + (1 + \sigma c - r/\bar{\mu})\mathbf{E}A_n. \tag{15}$$

Proof Let N be the number of markers in V_n, and Q_i be the probability that problem $n+1$ cannot be reduced by **fast** to any of the first i markers. Then

$$\begin{aligned}\mathbf{E}E_n = r\mathbf{E}\sum_{0\le i<N}\mathbf{E}(Q_i\mid N) &= r\sum_{0\le i<N}(1-\bar{\mu})^i\\ &= \frac{r}{\bar{\mu}}(1-\mathbf{E}(Q_N\mid N))\\ &= \frac{r}{\bar{\mu}}(1-\mathbf{E}A_n).\end{aligned}$$

The formula for $\mathbf{E}\Delta F_n$ follows from Theorem 2. ∎

Theorem 23 *Under $\mathcal{M}_r$,* **Adapt** *can effectively improve* **slow** *with sufficient training iff $r < \bar{\mu}$. Sufficient training can be achieved with*

$$n \ge \frac{1}{\sigma\widetilde{\mu}}\left(1 + \frac{\sigma c}{1 - r/\bar{\mu}}\right). \tag{16}$$

number of examples. If **slow** *is also not complete, then* **Fast** *can effectively replace* **slow** *with*

$$n \ge \frac{1}{\sigma(1-\sigma)\widetilde{\mu}} \tag{17}$$

steps of training. If $r \ge \bar{\mu}$, then **Fast** *may still improve* **slow***, but only with minimum failure probability $\mathbf{E}A_n \ge 1 - \bar{\mu}/r$.*

Proof From Theorem 2 and Theorem 22, we have $\mathbf{E}\Delta F_n = r/\bar{\mu} + (1+\sigma c - r/\bar{\mu})\mathbf{E}A_n$. Combining Lemma 16, Theorem 21, and this formula yields the desired theorem. ∎

Theorem 24 *Under $\mathcal{M}_r$, the ratio of the average cost of* **Adapt** *to that of* **slow** *is bounded asymptotically by $r/\bar{\mu}$ as the number of training problems approaches infinity. More globally, the behavior is*

$$\frac{\mathbf{E}F_n}{n} \le \frac{r}{\bar{\mu}} + \left(1 + \sigma c - \frac{r}{\bar{\mu}}\right)\begin{cases}\frac{\ln(eA_0\sigma\widetilde{\mu}n)}{(\sigma\widetilde{\mu}n)} & \text{if } A_0\sigma\widetilde{\mu}n > 1;\\ A_0 & \text{otherwise.}\end{cases} \tag{18}$$

Accordingly, the maximum value that the ratio can attain at any n is at most

$$\mathbf{E}F_n/n \le r/\bar{\mu} + (1 + \sigma c - r/\bar{\mu})A_0. \tag{19}$$

Proof Let $\alpha = \sigma\widetilde{\mu}$. From Theorem 2 and Theorem 22, it suffices to prove that $\mathbf{E}K_n \leq \ln(eA_0\alpha n)/\alpha$ if $A_0\alpha n > 1$; and $\mathbf{E}K_n \leq A_0 n$ otherwise. Since $\mathbf{E}K_n = \sum_{j<n} \mathbf{E}A_j$, and $\mathbf{E}A_n \leq \min(A_0, (\alpha(n+1))^{-1})$, we must have $\mathbf{E}K_n \leq A_0 x + (H_n - H_x)/\alpha$, for all positive integers $x \leq n$. Since $H_n - H_x \leq \ln(n/x)$, we may extend the domain of x to the reals and obtain $\mathbf{E}K_n \leq A_0 x + \ln(n/x)/\alpha$, which yields the theorem when minimized at $x = \min(n, 1/(\alpha A_0))$. ∎

7. Application and Verification

We now demonstrate the applicability of **Adapt** and the fidelity of our theory on a variety of robot environments. First, we investigate the simplest environment possible using the pessimistic model.

7.1. Pessimistic Model

Going back to the example in Figure 4 of a point robot in a 2-coverable workspace, we see from Theorem 17 that the expected failure probability of **Adapt** can be no greater than $1/(\sigma n)$ with n being the number of training problems. Notice that this result does not depend on what the problem distribution is, as long as it is fixed for both training and subsequent problem-solving. More generally, we have the following theorem for a point robot in simple planar environments.

Theorem 25 *Consider a point robot in a planar simple polygonal workspace filled with b simple polygonal obstacles and having a total of c corners. Let* **fast** *be "go-straight". Then the workspace is $(2b+c-2)$-coverable. Consequently, after n steps of training, the expected probability that* **Adapt** *will succeed in reaching the next random goal using only "go-straight" via the learned markers V_n is at least $1-(2b+c-3)/(\sigma n)$. Further, the expected cost of* **Fast** *is at most $3r(2b+c-2)$.*

Proof It suffices to show that the feasible workspace (a simple polygon with holes) can be triangulated into $m = 2b + c - 2$ triangles, since each triangle is convex, hence connected under "go-straight". It is known that every simple polygon with holes has a valid constrained triangulation[28] whose edges are a superset of the input edges, and whose vertices are the input vertices (no added vertices are necessary). Let m and d be the number of triangles and edges in such a triangulation. Then by Euler's formula,[28] we have $(m+b+1) - d + c = 2$. Also, counting each edge of every triangle yields $3m = 2d - c$. Hence, $m + b + c - (3m+c)/2 = 1$ implies the desired result of $m = 2b + c - 2$. ∎

Beyond immediate applications to point robots, the theories thus developed can also help us make plausible performance predictions for more complicated robots. Figure 5a shows a 10-dof robot in a planar environment, which has been studied by others.[18] Let **fast** implement the following procedure:

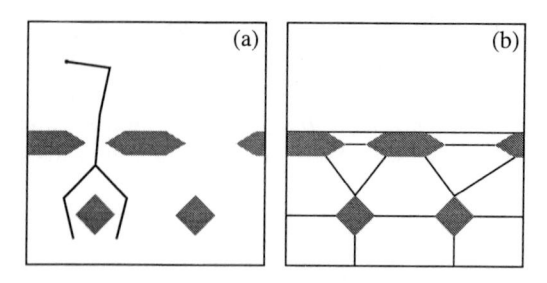

Fig. 5. A 10-dof robot environment.

1. move one end of the robot straight to the desired location with the rest of the robot complying;
2. with the first end point fixed, move the other end of the robot straight to the its desired location with the rest of the robot complying;
3. with both end points fixed, move the rest of the robot to their desired configuration using standard potential field approach.

Since the robot is snake-like with high dof, it is likely that **fast** will succeed if both end points are visible from their desired locations and if there is indeed a solution. Given that **fast** succeeds under this condition, we can bound the number of **fast**-connected components necessary to cover the 10-dimensional C-space. From Figure 5b, we see that the workspace is 11-coverable for each end point under visibility. Also, from visual inspection, we see that there are at most 12 topologically distinct inverse-kinematic solutions for a given pair of end points. Hence, the C-space is at most $11 \cdot 11 \cdot 12 = 1452$-coverable. Consequently, if the teacher **slow** is a complete planner, it will take at most 145100 training problems for **Adapt** to attain a 99% expected capability.

7.2. Randomized Model

Although the results from the pessimistic model can give us worst-case bounds on important quantities such as training time and capability, they are often too loose to predict the actual behaviors well. To complement and address this deficiency, we now demonstrate the randomized model by explaining data and making performance predictions on two separate experiments. Figure 6a shows a planar 2-link robot environment in which **Adapt** is applied. The environment exemplifies the planar component of a typical robot workcell in a SCARA configuration[10] with the z-component decoupled. In this experiment, **slow** implements an incomplete but fairly effective planner,[4] and **fast** implements a simple potential-field based hill-climb. There are 5 polygonal obstacles in the fixed workcell, and a goal set consisting of 9 preselected goal positions. Starting at home position 0, the robot is to go through a sequence of

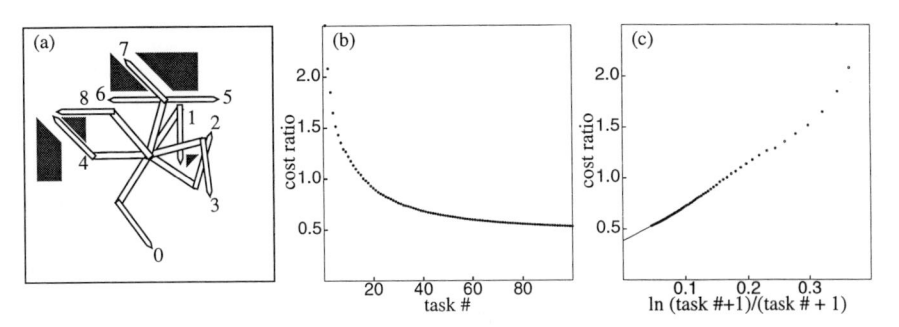

Fig. 6. Time improvement on a planar 2-link robot environment.

100 goals randomly selected from the goal set. In Figure 6b, the ratio of the cumulative planning cost of **Adapt** to that of **slow** only is plotted against problem number n. The planning costs are averaged over 100 runs and are measured by the number of robot-to-obstacle distance evaluations, which is the dominating factor in the computing cost of each planner. Figure 6c plots the ratio against $(\ln(n+1))/(n+1)$ to show their asymptotic linear relationship, hinted at by Theorem 24.

The experiment shows that **Adapt** is able to increase its performance relative to **slow** from 150% slower (ratio $\doteq 2.5$) to 50% faster at the end of 100 training examples. It also shows that **Adapt** needs about 16 training tasks before becoming competitive with **slow**, a fact attributable to both the task simplicity for **slow** and the significant learning costs incurred by **Adapt** during solution abstraction. We can use Theorem 24 to predict the maximum speedup achievable. If we believe that (18) is also an asymptotic lower bound, then the plot implies that $r/\bar{\mu} \doteq 0.38$ is the minimum achievable cost ratio, equivalent to a maximum speedup of 62%. From other empirical observations, we estimate that $r \doteq 0.1$, $c \doteq 1$, $\sigma \doteq 1$, and $A_0 \doteq 0.9$. Hence, $\bar{\mu} \doteq 0.26$. Since $\lambda \doteq 2$, we also estimate $\tilde{\mu} \doteq 0.45$. To see how consistent these numbers are, we estimate the number of training problems required by **Adapt** to have its cumulative cost first become less than that of **slow**. Using (18), we have $\sigma\tilde{\mu} \doteq 0.45$ and $\frac{\ln(eA_0\sigma\tilde{\mu}n)}{eA_0\sigma\tilde{\mu}n} \doteq \frac{1+\sigma c - r/\bar{\mu}}{eA_0(1-r/\bar{\mu})} \doteq 0.156$. giving us $eA_0\sigma\tilde{\mu}n \doteq 19$, or $n \doteq 17.2$, which is very close to the observed $n = 17$ in the plot.

We use our theory to explain and predict another experiment in which **Adapt** is applied on a 3-dimensional 6-dof gantry robot environment. The same **slow** and **fast** used for the planar case are also used here. In this environment (left side of Figure 7), there are 4 obstacles: a $(16+2)$-sided polyhedral approximation of a cylindrical cask, two cask stands, and a floor. Motivated by problems in radiation survey,[16] the goal positions are chosen randomly, and correspond to the robot end effector touching the cask surface in a prescribed orientation. The tasks are sufficiently difficult that the original planner, **slow**, fails to reach 7 out of a sequence of 100 random goals. In contrast, **Adapt** is able to accomplish all but 1 task during the exercise, thereby

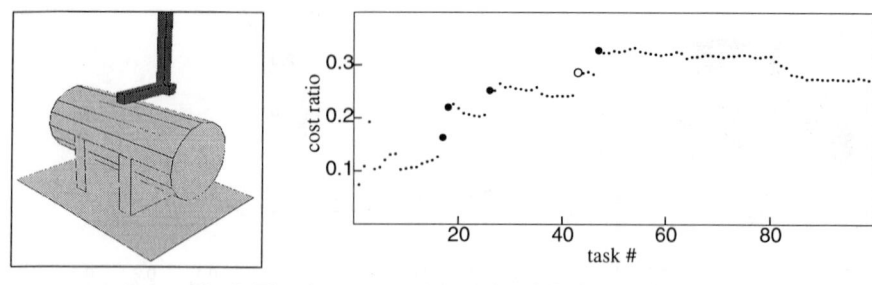

Fig. 7. Time improvement on a 3-d, 6-dof robot problem.

increasing the capability of the original planner. Moreover, **Adapt** calls **slow** only 5 times, and stores only 11 trail-markers in addition to the initial robot configuration. Figure 7 plots the task number against the the ratio of the cumulative effort expended by **Adapt** to that expended by **slow** only. Efforts are again measured by the number of robot-to-obstacle distance evaluations. The 5 large points indicate **Adapt**'s calling of **slow**, and the single white point indicates the only failure of **Adapt**. Initially, **Adapt** is able to plan without **slow** because the tasks are relatively easy. Later, **Adapt** starts to learn as indicated by the jumps of the cost ratio. When the task number reaches 50, **Adapt** has basically learned the environment as shown by the gradual decline of the cost ratio.

Using the data, we estimate $\sigma \doteq 93\%$ because of the 7 failures; $\tilde{\mu} \doteq 1 - 0.01^{1/5} \doteq 80\%$ because only 5 chains are involved. Using (17) of Theorem 23, we then estimate $n \doteq 1/(0.93 \cdot 0.07 \cdot 0.8) \doteq 19.2$ to be the number of training tasks n required for **Fast** to improve both the speed and the capability of **slow**. This estimate means that **fast** is already very powerful, and that roughly only 2 calls (#17 and #18 in the plot) to **slow** are necessary for **Fast** to catch up with **slow** in task solving capability.

With Theorem 24, we can predict the maximum speedup achievable. We estimate $A_0 \doteq 1/17 \doteq 6\%$ because **Adapt** first failed at task #17. We also estimate $c \doteq 0.1$ from empirical observation. Again, if we believe that (18) is also a lower bound, then the maximum cost ratio is $r/\bar{\mu} + (1 + 0.1 - r/\bar{\mu})0.06 \doteq 0.32$ from the plot, which implies that $r/\bar{\mu} \doteq 0.27$, which is incidentally very close to the cost ratio at the end of task #100. Consequently, we do not anticipate **Adapt** will do much better with more training. Overall, **Adapt** is able to reduce the planning time of **slow** by a factor of $\bar{\mu}/r \doteq 4$, and increase the task solving capability of **slow** by $1/\sigma - 1 \doteq 7.5\%$.

8. Extension to Changing Environments

After presenting and analyzing **Adapt** for the fundamental stationary case, we now extend it to handle incrementally changing environments. In this more general environment, we assume that for each robot task, the obstacles are stationary, but

---

Algorithm Gen-Adapt(Fast, Slow; Trace)
   $u \leftarrow$ current position; $v \leftarrow u$; $G \leftarrow (\{v\}, \emptyset)$;
  do forever
      $\boxed{\text{Repair}(G);}$
      $w \leftarrow$ goal;
      if (**not** Fast$(u, w; G, h)$) **then**
         if (**not** Slow$(u, w; G, h)$) **then continue**;
         $\boxed{\rho \leftarrow \text{Obj-Abstract}(\text{Slow}[u, w; G, h]);}$
         $G \leftarrow$ Learn(G, ρ);
      endif
      $\boxed{\text{if Trace}(u, w; G, h) \text{ then}}$
         execute(Fast$[u, w; G, h]$); $u \leftarrow w$;
  enddo
end.

Fig. 8. An adaptive path planning algorithm in changing environment.

may slowly change their configuration or shape over the course of the robot performing many tasks. In other words, we assume that the environmental change is both occasional and localized. By occasional, we mean that the interval between workcell changes is large compared to the amount of time spent on each task. By localized, we mean that the workcell change involves only a few objects in a relative small area of the workspace. Both conditions are prevalent in many applications and have their intuitive implications: Occasional implies that old experience may be useful for a significant amount of time, and localized implies that old experience may have salvage value.

Our algorithmic extension consists of two experience manipulation schemes: For minor environmental adjustments, we use an object-attached experience abstraction scheme to increase the flexibility of the learned experience; for major environmental modifications, we use an on-demand experience repair scheme to retain those experiences that remain valid and useful. In addition to presenting this extension, we also compare it with three other variant strategies for using old experiences in new environments. Formally, the generalized learning algorithm **Gen-Adapt** is shown in Figure 8. It is the same as the algorithm for stationary environments except for the three boxed fragments. The second boxed fragment replaces **Abstract** with **Obj-Abstract**, the object-attached experience abstraction scheme. The third boxed fragment introduces **Trace**, the on-demand experience repair scheme. The first boxed fragment, which introduces **Repair**, is not part of the algorithm, but is included for

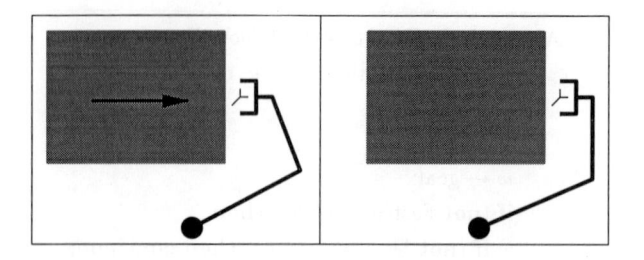

Fig. 9. Object-attached experience using critical tag-points.

later discussion of other variants of the algorithm that use it.

8.1. Object-attached Experience Abstraction

Recall in the specification of **Adapt** for stationary environments that the procedure **Abstract** is used to condense a work experience slow$[v, w]$ into a short chain ρ of trail-markers for **fast** to traverse. In fixed environments, it does not matter how we choose to represent markers because they always correspond to fixed robot configurations. However, to increase the flexibility of their use in changing environments, we now require that the markers returned by **Obj-Abstract** (second boxed fragment in Figure 8) be relative robot configurations associated with nearby objects, rather than the absolute positions in the stationary case. That is, instead of remembering the robot positions as points in absolute space, we now remember each of them as an offset from some nearby object serving as a landmark.

One way to implement this strategy is to create a tag-point (a 6 degrees-of-freedom coordinate frame for the robot tool point) for each critical robot position, and affix the tag-point to the local coordinate of a nearby object. Then, as this nearby object changes its location or orientation, the tag-point can be adjusted accordingly so that the robot tool point can maintain its distance to the object under change. Figure 9 shows an example. In the left frame, the robot position is recorded via the tag-point of the robot tool, and is attached to the rectangular object. As the object moves toward the right, the tag-point moves along with it, enabling the robot to comply with the change. If the tag-point had not been attached to the object, the corresponding robot position would have become invalid in the new environment.

One potential drawback of this tag-point method is that solving the inverse-kinematics for the tag-point will be necessary to recompute the robot configuration for the trail-marker. Thus, multiple solutions may arise not all of which may be feasible. Solutions may also disappear for tag-points whose attached objects have moved too much. Nevertheless, under this object-attached experience abstraction scheme, we can adjust to any minor environmental change without expensive experience repair.

8.2. On-Demand Experience Repair

Of course, if the environment changes significantly, the validity of G will deteriorate. The edges that used to be valid in previous environments may not stay valid in the same environment in the sense that **fast** may no longer be powerful enough to traverse them. How much deterioration G will suffer depends on how drastically the environment changes. If the change is major and extensive, then it may be better to start over with no experience (G reinitialized), rather than to work with the old impaired experience. In the more interesting case where the change may be major (e.g., introducing a new object) but not extensive (e.g., the rest of the workcell is undisturbed), the right choice is not as clear. Therefore, we introduce an on-demand repair scheme (third boxed fragment in Figure 8) to retain those experiences that remain valid and useful.

In this scheme, we plan as if G is connected, until $\mathsf{Fast}(\cdot)$ succeeds and we actually need to produce a path. Then, to generate $\mathsf{Fast}[\cdot]$, we require the success of $\mathsf{Trace}(\cdot)$ to provide a connected sequence from $\hat{u}$ to $\hat{w}$. As $\mathsf{Trace}(\cdot)$ searches for and verifies such a sequence, it may come across invalid edges, which it simply deletes. If $\hat{u}$ is already connected to $\hat{w}$ in G, then no repair need take place. If, however, $\hat{u}$ and $\hat{w}$ do not belong to the same (connected) component due to the deterioration of G, then **slow** is called to reestablish their connectivity. It is of course possible that connectivity cannot be reestablished due to the environmental change. In this case, the portion of G connected to $\hat{w}$ is deemed useless, and hence discarded. The procedure for $\mathsf{Trace}(\cdot)$ is as follows:

1. While there exists a sequence Γ of vertices in G connecting $\hat{u} = \Gamma_1$ to $\hat{w} = \Gamma_k$ for some $k \geq 1$ do
   (a) If $\mathsf{fast}(\Gamma_i, \Gamma_{i+1})$ for all $1 \leq i < k$ then return success;
   (b) Else remove edge (Γ_i, Γ_{i+1}) with smallest i such that $\neg\mathsf{fast}(\Gamma_i, \Gamma_{i+1})$.
2. If $\mathsf{slow}(\hat{u}, \hat{w})$ succeeds then augment G with $\mathsf{Abstract}(\mathsf{slow}[\hat{u}, \hat{w}])$; return success;
3. Else remove the (connected) component of $\hat{w}$ from G, and return failure.

8.3. Other Repair Strategies

It is also possible to cope with major environmental change using other variants of the on-demand repairing strategy. One trivial strategy as stated above is simply to forget the old experience and start over (with G reinitialized) whenever there is a change in the environment. The corresponding algorithm, A_0, can be obtained from Figure 1 by skipping the boxed condition, and defining $\mathsf{Repair}(G)$ to be the reinitialization procedure.

Another less trivial strategy is to verify each edge of G first whenever there is a change. Then with the time investment, we can initialize G to the home component that contains the current robot position. The corresponding algorithm, A_1, can again be obtained from Figure 1 by skipping the boxed condition, and defining Repair(G) to be the above home-component extraction procedure.

Notice that both strategies above only update G according to environmental change, and do not really repair old experience. In contrast, a third strategy that repairs actively is to first apply Trace to attempt to reach every vertex of G from home before taking on any new task. The corresponding algorithm, A_2, can be obtained from Figure 1 by skipping the boxed condition, and defining Repair(G) to be the above repair-all procedure.

All of the suggested algorithms (including the repair-on-demand algorithm A_3) have their advantages and disadvantages. Intuitively, if the environment undergoes a major and extensive change, then starting over with A_0 may be the best choice. On the other hand, if slow costs much more than fast, then using A_1 to save some old experience may be better. Alternatively, if the change is only local, then repairing old experience with A_2 or A_3 may be more beneficial. Which algorithm to use thus depends on the particular application.

8.4. Solution Quality and Redundancy

So far we have focused on task solvability but not solution quality. If solution quality is not important, then in Fast[·] we can simply produce the solution of going through Γ with fast. In this situation, the experience graph will always be a tree. However, if solution quality is important, then it may be worthwhile to locally optimize Γ by seeking to "cut corners" whenever possible. The result of this compression is that G may be augmented with additional edges to enable shorter sequences in the future. Also, the redundancy introduced may be useful in combating experience deterioration. In fact, for environments with a shrinking set of obstacles such as those encountered in waste-remediation, compressing known solutions would be a simple and cost-effective approach to improving solution quality.

8.5. Example

We illustrate the generalized learning algorithm with a simple example involving a point robot in a 2D workspace. Let fast implement a go-straight procedure, with fast(u, w) returning success iff w is visible from u, and fast[u, w] returning the line segment $\overline{uw}$. Let slow implement a greedy 2-step go-straight procedure, with slow(u, w) returning success iff the two points are connectable by at most 2 line segments, and slow[u, w] returning the shortest such connecting path. To complete the algorithmic specification, let the heuristic used in Fast and Slow be $h = h_1$, with h_1 ordering the

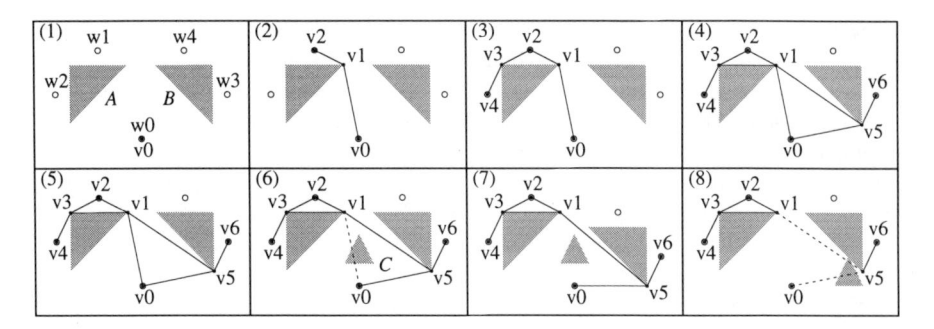

Fig. 10. Snapshots of Gen-Adapt under environmental change.

vertices of G according to the distance to w, starting with the closest point first.

Figure 10 illustrates Gen-Adapt with a series of snapshots. Frame (1) shows the initial setting with the robot at home $u = w_0$ amongst two objects A and B. The robot's initial tasks are to inspect both A from w_1 and w_2, and B from w_3 and w_4. To begin, the experience graph G is initialized to the single vertex $v_0 = w_0$.

The first goal indicated by w_1 is shown in Frame (2). Since Fast is unable to plan using only fast and G, Gen-Adapt then calls Slow. Using h, Slow chooses to extend from v_0 to w_1, since v_0 is the only vertex in G. The path produced by slow(v_0, w_1) consists of the line segments $\overline{v_0v_1}$ and $\overline{v_1v_2}$. This path is then abstracted into the chain connecting v_0 to v_1 and v_1 to v_2. The result of augmenting G is that G now becomes the 3-vertex chain. Using this augmented G, Fast is now able to produce a path from $u = w_0$ to w_1, which consists of the segments $\overline{uv_0}$, $\overline{v_0v_1}$, $\overline{v_1v_2}$, and $\overline{v_2w_1}$, with $\overline{uv_0}$ and $\overline{v_2w_1}$ being null segments.

With the first task accomplished, the next task is to go to w_2 shown in Frame (3). Since Fast is again unable to plan using only fast and G, Gen-Adapt then calls Slow. Using h, Slow chooses to extend from v_2 to w_2, and produces the line segments $\overline{v_2v_3}$ and $\overline{v_3v_4}$. The result of augmenting G is that G now becomes the 5-vertex chain with new vertices v_3 and v_4.

With this G, Fast is still unable to succeed in reaching w_3 in Frame (4). Consequently, Slow chooses to extend from v_0 and produces 2 more segments $\overline{v_0v_5}$ and $\overline{v_5v_6}$. Thus, before calling Fast[·], G is a 7-vertex chain with new vertices v_5 and v_6. After calling Fast[·], however, G becomes cyclic due to the addition of edges (v_3, v_1) and (v_1, v_5) as a result of locally optimizing the solution path $(v_4, v_3, v_2, v_1, v_0, v_5, v_6)$.

Frame (5) shows that Fast is now capable of reaching w_4, with $\hat{w}_4 = v_1$. Consequently, for the first time, Slow is not called and G is not modified.

So far, the workcell has been stationary. In Frame (6), we return the robot to its home and introduce a new object C. With A_0 using the start-over strategy, we would lose the entire G and not retain anything from the 3 previous calls to Slow. With A_1, we would verify all 8 edges in G with fast, remove the only broken edge (v_0, v_1), and retain the rest of G since it remains connected. If fast costs much less than slow, then the return on the initial time investment is certainly justifiable compared to that of A_0. This case demonstrates that improving solution quality can also increase experience redundancy, which in turn decreases experience deterioration under change. (The introduction of edge (v_1, v_5) in locally optimizing $(v_4, v_3, v_2, v_1, v_0, v_5, v_6)$ alleviates the damage of edge (v_0, v_1) caused by object C.) With A_2 using the active-repair scheme, we would also just remove edge (v_0, v_1) from G at the end of Repair(G). With A_3 using the repair-on-demand strategy, we simply do nothing.

Frame (7) shows what happens if we introduce some minor change by moving object B and its object-attached goals w_3 and w_4. Because of the object-attached abstraction scheme, v_5 and v_6 also move along with B. Consequently, if the robot were to go back to w_3, it would again succeed by simply reaching toward v_5 and v_6.

Frame (8) shows what happens if we move object C to a corner and decide not to inspect object B anymore. In this case, A_1 would be identical to A_0 in reducing G back to the single vertex v_0, except that A_1 would also have to spend time verifying all 7 edges of G before removing them. With A_2, G would be actively repaired, which means that it would call slow twice to reestablish the connectivity of the 2 components to v_0. With A_3, we again do nothing until the need arises. If we choose not to inspect B anymore, then only one component needs to be reconnected to v_0, which means only one additional call to slow would be required in the future. This case demonstrates a situation where using A_3 is better than using A_2.

8.6. Computational Experience

We have applied Gen-Adapt on the simple 2-link planar robot environment of Figure 6a again, this time with an environment change. Recall the initial setup: the workcell has 5 polygonal obstacles and the goal set consists of 9 preselected robot configurations. Starting at home 0, the robot is to go through a sequence of goals randomly selected from the goal set. During the exercise, we introduce an incremental environmental change by adding a new obstacle to the workcell and a new goal position to the goal set, as shown in Figure 11.

The result of this experiment, with Gen-Adapt using all 4 different repairing strategies, is shown in Figure 11. Here, the ratio of the cumulative planning cost required by Gen-Adapt to that required by slow only is plotted against the task number. The planning costs are averaged over 100 runs and are again measured by the number of robot-to-obstacle distance evaluations, which is the dominating factor in the computing cost of each planner. The environment change is introduced after task 40. To emphasize the important features of the result, the initial portion of the curve corresponding to ratios greater than 1 is not plotted. The unplotted portion actually decreases monotonically from 2.5 at task number 1 to 1.0 at task number 16. The

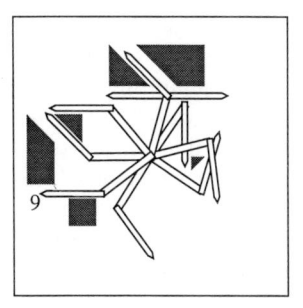

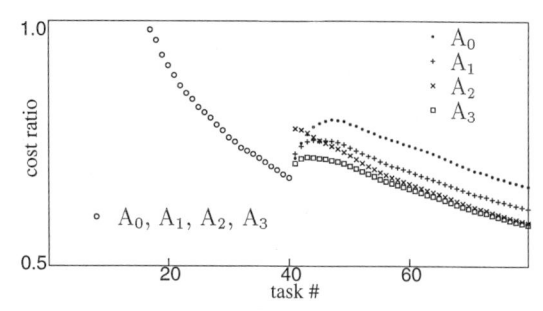

Fig. 11. Workcell change and time improvement of Gen-Adapt over slow.

experiment shows that before the environmental change, **Gen-Adapt** is able to learn and speed up its performance relative to **slow** from 150% slower to 33% faster. (As shown in Figure 6b, this speedup can be increased to 50% faster if the environment stays fixed.)

After the environmental change, the performance curve for **Gen-Adapt** splits up into 4 curves, each corresponding to a different experience repairing strategy. The curves for A_0, A_1, and A_3 exhibit similar behaviors in that they all gradually increase and then decrease at roughly the same rate, with A_3 being clearly better than A_1, which is in turn clearly better than A_0. The curve for A_2 is different in that it first jumps to a high point and then comes down rapidly to approach the curve for A_3. The jump is due to the high initial cost of active repair, and the rapid decrease is due to the benefit of the repair. Overall, the relative performance of the repairing strategy is as expected, since the environmental change is incremental, involving only local and occasional change. In fact, one can devise an experiential cost/benefit model to formalize the concept of local and occasional change, and prove the optimality of the on-demand repair strategy A_3 relative to the other variants A_0, A_1, and A_2 under such change.[7]

9. Conclusion

We have presented a learning algorithm that can improve path planning performance. The algorithm adapts to its working environment by maintaining an experience graph with vertices corresponding to useful robot configurations. It can both reduce time cost and increase task solving capability of existing planners.

We have demonstrated the effectiveness of the algorithm via both mathematical analysis and experimentation. Our analysis involves uncovering general quantitative relationships between important variables such as capability, planning cost, memory requirement, and training time. It also includes studying the implications of these relationships under two stochastic models: pessimistic $\mathcal{M}_p$ and randomized $\mathcal{M}_r$. The models have different assumptions and applications: $\mathcal{M}_p$ quantifies C-space complexity while $\mathcal{M}_r$ quantifies experience utility. Using these models, we characterize the situations in which learning will yield a speedup, and provide global quantitative bounds on planning cost and capability in terms of training cost. Empirically, we have also demonstrated the applicability of the algorithm and the fidelity of its theory on several robot path planning environments. In particular, we have illustrated a technique for predicting the maximum achievable speedup. Finally, we have extended our algorithm to handle changing environments with object-attached experience abstraction and on-demand experience repair. The performance of the generalized algorithm is characterized and compared with three other plausible experience repair strategies.

Although our algorithm is presented in the context of robot path planning, there is no restriction that it be applied only to robotics. In fact, at a higher level of abstraction, the algorithm may even be applicable for an intelligent agent navigating in a space of information subject to potential incremental change. In this sense, our elementary theoretical results and techniques should be useful for studying other types of probabilistic learning as well.

10. Acknowledgements

This work is supported by the Laboratory-Directed Research and Development Program at Sandia National Laboratories and by the U.S. Department of Energy under Contract DE-AC04-94AL85000. The author thanks Yong Hwang, Sharon Stansfield, David Strip, and Patrick Xavier for their help in reviewing and improving this work throughout the years.

11. References

1. Barbehenn, M., Chen, P.C., Hutchinson, S., "An Efficient Hybrid Planner in Changing Environments," *Proc. of IEEE Int. Conf. on Robotics and Automation,* pp. 2755–2760, 1994.
2. Barraquand, J., Latombe, J., "A Monte-Carlo algorithm for path planning with many degrees of freedom," *Proc. of IEEE Int. Conf. on Robotics and Automation,* pp. 1712–1717, 1990.
3. Bessiere, P., Ahuactzin, J.M., et al, "The 'Ariadne's Clew' algorithm: Global planning with local methods," *Proc. of IEEE/RSJ Conf. on Intelligent Robots and Systems*, 1993.
4. Chen, P.C., "Effective Path Planning through Task Restrictions," Sandia Report SAND91-1964, 1992.
5. Chen, P.C., "Improving Path Planning with Learning," *Machine Learning: Proc. of the Ninth Int. Conf.,* pp. 55–61, 1992.
6. Chen, P.C., Hwang, Y.K., "SANDROS: A Motion Planner with Performance Proportional to Task Difficulty," *Proc. of IEEE Int. Conf. on Robotics and*

Automation, pp. 2346–2353, 1992.
7. Chen, P.C., "Adaptive Path Planning in Changing Environments," Sandia Report SAND92-2744, 1993.
8. Chen, P.C., "Adaptive Path Planning for Flexible Manufacturing," *Proc. of Fourth Int. Conf. on Computer Integrated Manufacturing and Automation Technology,* Oct., 1994.
9. Chen, P.C., "Adaptive Path Planning: Algorithm and Analysis," *Proc. of IEEE Int. Conf. on Robotics and Automation,* 1995.
10. Craig, J., *Introduction to Robotics: Mechanics and Control,* pp. 268–269, 1989.
11. Feller, W., *An Introduction to Probability Theory and Its Application,* 3rd edition, v. 1, John Wiley & Sons, 1968.
12. Fujimura, K., Samet, H., "Time Minimal Paths among Moving Obstacles," *Proc. of IEEE Int. Conf. on Robotics and Automation,* pp. 1110–1115, 1989.
13. Glavina, B., "Solving Findpath by Combination of Goal-Directed and Randomized Search," *Proc. of IEEE Int. Conf. on Robotics and Automation,* pp. 1718–1723, 1990.
14. Goel, A., Callantine, T., Donnelian, M., Vazquez, N., "An intergrated experience-based approach to navigational path planning for autonomous mobile robot," *Proc. of IEEE Int. Conf. on Robotics and Automation,* pp. 818–825, 1993.
15. Gupta, K.K., Zhu, X., "Practical Global Motion Planning for Many Degrees of Freedom: A Novel Approach within Sequential Framework," *Proc. of IEEE Int. Conf. on Robotics and Automation,* pp. 2038–2043, 1994.
16. Harrigan, R.W., Sanders, T.L., "A Robotic System to Conduct Radiation and Contamination Surveys on Nuclear Waste Transport Casks," Sandia Report SAND89-0017, 1990.
17. Hwang, Y.K., Ahuja, N., "Gross Motion Planning – A Survey," *ACM Computing Surveys,* **24**, no. 3, pp. 219–292, Sept. 1992.
18. Kavraki, L., Latombe, J.-C, "Randomized preprocessing of configuration space for fast path planning," *Proc. of IEEE Int. Conf. on Robotics and Automation,* pp. 2138–2145, 1994.
19. Kant, K., Zucker, S.W., "Planning Collision-free Trajectories in Time-Varying Environments: a Two-level Hierarchy," *Proc. of IEEE Int. Conf. on Robotics and Automation,* pp. 1644–1649, 1988.
20. Kondo, K., "Motion Planning with Six Degrees of Freedom by Multistrategic Bidirectional Heuristic Free-Space Enumeration", *IEEE Tran. on Robotics and Automation,* **7**, no. 3, pp. 267–277, June 1991.
21. Latombe, J., *Robot Motion Planning,* Kluwer Academic Publishers, 1991.
22. Lozano-Pérez, T., "A Simple Motion-Planning Algorithm for General Robot Manipulators," *IEEE J. of Robotics and Automation,* **RA-3**, no. 3, pp. 224–238, June 1987.

23. Marshall, A.W., Olkin, I., *Inequalities: Theory of Majorization and Its Applications,* Academic Press, 1979.
24. Natarajan, B.K., *Machine Learning: A Theoretical Approach,* Morgan Kaufmann, 1991.
25. Overmars, M.H., Švestka, P., "Probabilistic Learning Approach to Motion Planning," *Workshop on the Algorithmic Foundations of Robotics,* Feb. 1994.
26. Pandya, S., Hutchinson, S., "A Case-based Approach to Robot Motion Planning," *Proc. of IEEE Int. Conf. on Systems Man and Cybernetics,* pp. 492–497, 1992.
27. Paden, B., Mees, A., Fisher, M., "Path Planning Using a Jacobian-Based Freespace Generation Algorithm," *Proc. of IEEE Int. Conf. on Robotics and Automation,* pp. 1732–1737, 1989.
28. Preparata, F.P., Shamos, M.I., *Computational Geometry: An Introduction,* Springer-Verlag, 1988.
29. Reif, J.H., Sharir, M., "Motion Planning in the Presence of Moving Obstacles," *Proc. of 26th FOCS,* pp. 144–154, 1985.
30. Shiryayev, A.N., *Probability,* Springer-Verlag, 1984.
31. Tadepalli, P., "A Theory of Unsupervised Speedup Learning," *Proc. of AAAI,* pp. 229–234, 1992.

Incremental Adaptation as a Method to Improve Reactive Behavior

A.J. HENDRIKS and D.M. LYONS

Philips Laboratories,
Philips Electronics North America Corporation,
345 Scarborough Road,
Briarcliff Manor, N.Y. 10510

ABSTRACT

The brittle nature and the computational complexity of the state of the art "classical" planning systems such as MOLGEN and SIPE when applied to 'real-world' problem domains such as robotics has become evident. Uncertainty about the effects of actions and the state of the environment, in addition to dynamic effects caused by other active agents made these approaches wholly insufficient. Reactive systems, an approach that started to remedy this, generates robust and responsive behavior, but does not support a priori deliberation and globally optimal behavior.

In this chapter, we introduce incremental adaptation as a method to improve reactive behavior. This methodology aims to get the best of both worlds, the fast response of the reactive systems and the deliberative capability of planning systems. Unlike in the latter, a plan does not have to be generated all at once. Rather, the reactive control strategy can be incrementally adapted to better suit the current environment and the possibly changing objectives. We discuss the advantages of this methodology and present a survey of systems using incremental adaptation.

Keywords: planning, adaptation, reaction, reactive systems, reactive behavior.

1. Introduction

Planning is used in AI to refer to a body of techniques to automate the process of selecting and carrying out actions to bring an environment to some desired state. In *Classical Planning* (e.g., STRIPS [1], NONLIN [2], MOLGEN [3], DEVISER [4], SIPE [5], TWEAK[6]) an application domain is encoded in terms of a set of primitive actions and their preconditions and effects characterized as state predicates. A planning problem consists of this domain description plus an initial and final (or goal) state. Planning in this framework consists of searching through the space of action orderings to find one which takes the initial state to the goal state. This set of ordered actions is called the plan.

Plan execution is achieved by indexing the plan (figure 1). The current action is performed, and, depending on its results, the next action is indexed. If a failure is encountered, plan execution is stopped and control is handed back to the planner to replan, starting from the failure situation. The 'blocksworld' is the archetypal domain for classical planning: a world consisting of a number of blocks which the agent (usually thought of as a robot) can stack upon each other (the configurations of the blocks are precisely known at all times). The agent is the only entity active in the environment.

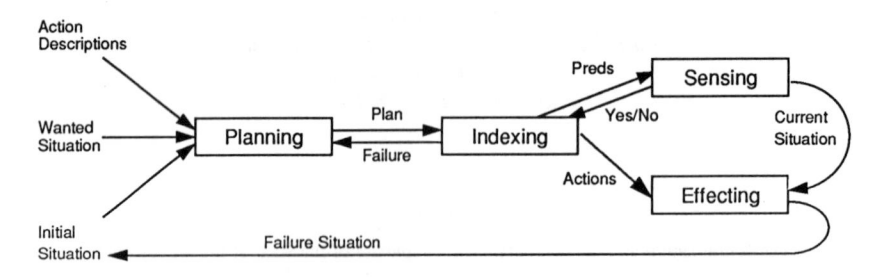

Fig. 1. Typical plan execution strategy in classical planning approaches.

Many real-world application domains, particularly robotic applications, do not have the static characteristics of blocksworld. Schoppers [7] describes an example domain that contrasts strongly with the blocksworld; his 'baby' world. This domain is similar to the blocksworld, except it is inhabited by a "mischievous baby who will flatten block towers, snatch blocks out of the robot's hand, and even throw blocks at the robot." The crucial new ingredients in this problem domain are (1) the agent cannot be certain of the effects of its actions, (2) it cannot make the assumption that the world remains static except when the agent carries out an action, and (3) it cannot assume that it knows everything about the world. This is exactly the problem faced by a robotic machine operating in the 'real world' — the same uncertain and dynamic environment that humans inhabit in everyday life.

Classical planning becomes too brittle in these application domains (Chapman [6], Kaelbling [8], Brooks [9], McDermott[10], Agre & Chapman [11]) for two main reasons: (1) the world can change while planning is in progress, and partial plans may be thus rendered useless; and, (2) there is uncertainty about the effects of actions, hence 'correct' plans may actually fail to achieve their goal. Chapman [6] has summarized the state of the art in classical planning. He has shown that classical planning in general is undecidable, and even in simpler cases can be computationally intractable.

The concept of reactive systems (e.g., Brooks [9], Agre & Chapman[11]) was introduced to address this problem. Such a system doesn't plan in the classical sense, but rather reacts directly to its environment. A reactive machine is a system in which the choice of next action is based on hard-wired response to sensory input and built-in goals. It differs from a classical 'plan executor' in that it is always ready to carry out actions; it isn't waiting for a plan to be loaded. These systems are surprisingly robust and versatile. However, they do not support *a priori* deliberation about action sequences.

In many domains such as the kitting robot in industrial automation [14], medical emergency advice, telescope scheduling, telecommunication, both reaction and deliberation are needed. Naturally, researchers began looking at ways to integrate reaction and deliberation. The emerging methodology, *incremental adaptation*, views plans not as programs, but rather as resources which a reactive system can execute. Concurrently and in parallel, this plan can be incrementally modified, or adapted, by a planning or learning system. Thus, the incremental adaptation methodology aims to get the best of both worlds: fast, real-time response characteristics from the reactive systems work, and the global optimal action patterns from the planning work.

In this chapter, the incremental adaptation methodology is introduced. For completeness, first a short overviews is given of the reactive systems work. Next, the incremental adaptation methodology is introduced and explained. An examination of several systems applying this methodology follows and a summary concludes this chapter.

2. Reactive behavior

Chapman concluded his TWEAK (Chapman [6]) paper by suggesting that improvisation might be a better paradigm than planning. Agre & Chapman [11] went on to carry out the first steps in building systems which can exhibit intelligent behavior in uncertain and dynamic domains. They argued that "before and beneath any activity of plan following, life is a continual improvisation, a matter of deciding what to do *now* based on how the world is *now*."

2.1. Pengi: intelligence through interaction.

Agre and Chapman built a program, Pengi [15], based on these concepts. Pengi played a video arcade game called Pengo. In a typical Pengo game, the computer appeared to hunt down targets, build traps, escape ambushes, take advantage of opportunities and act in a timely fashion. However, the intelligent behavior of Pengi was the result of the interaction of some relatively simple opportunistic strategies with a complex, structured environment. Two key ideas developed for Pengi were the concept of routines and the concept of indexical-functional representation.

Routines. A routine is a pattern of interaction between the agent and the environment. A routine need not be explicitly represented by the agent, but can simply be a property of the regularity in the interaction between the environment and the agent. When guiding a group of tourists for example, these tour groups exhibit a certain behavioral regularity; they listen to the guide, they ask questions, they wander off, etc. The tour guide will typically explain some landmark, accept and respond to questions as they arise. This highly interactive pattern is an example of a routine. Internally, the tour guide may simply have rules that say "answer any asked question" and "describe current landmark". The routine is created by the interleaved effects of these rules and the behavior of the environment (the tour group). Recognizing and exploiting regularity in the environment allows for the construction of simpler and more robust agents that, nonetheless, can effectively cope with uncertainty.

Indexical-Functional Representation. In an indexical-functional representation, properties of the immediate environment are only represented in terms of the impact they have on the objectives of the agent. For instance, a tour guide may have to deal with many people and many tours in any given day. A classical planning approach would involve uniquely naming each individual encountered, e.g., **PERSON-23**. This introduces combinatorics, because the space of persons must now be searched for appropriate instances. In Pengi, every person encountered is not uniquely named. Instead it is represented in terms of their impact on the objectives, e.g., `person-now-asking-question`, or `person-in-danger-of-being-lost`. The perceptual system directly matches the environment with such indexical-functional entities, and no search is required. Pengi used a similar mechanism to directly measure from the environment clues as to what actions to take next; these are called indexical-functional *aspects*.

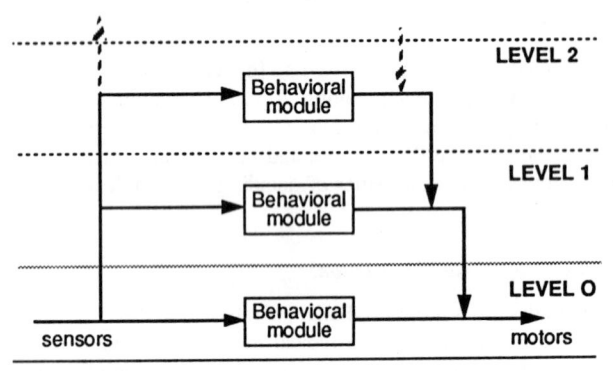

Fig. 2. The subsumption architecture.

2.2. The subsumption architecture

Brooks [9] made the point that the standard view of intelligent robot control architectures as a pipelined collection of functional modules caused problems with robustness, buildability, and testability. He suggested a novel architecture, called the *subsumption* architecture, that emphasized building intelligent control from layers of task-achieving behaviors (figure 2), much in the spirit of Braitenberg[16]. He built a number of robot system (Brooks [9, 17]) that exhibited similar behavior to Pengi. His robots consisted of simple opportunistic control rules, like Pengi, and appeared to act in a robust intelligent way in complex environments.

In the subsumption architecture behaviors are organized in a strict hierarchy. Each level has a competence and can function without the higher level layers present. A few example competencies are avoiding objects, wandering around, exploring areas, and building maps. Each level of competence is allowed to inspect data from lower levels, and also to inject data into the internal interfaces of the lower levels, thereby suppressing the normal data-flow. The lower level continues to run, unaware of the layers above, which sometimes "interfere" with its operation. Individual behaviors are designed as finite-state machines and run on a loosely coupled network of simple processors.

2.3. Designing reactive machines

Reactive machines, machines which produce intelligent but improvised behavior, were a major step forward in addressing the problem of acting in uncertain and dynamic environments. The advantages of such machines are that they respond quickly and robustly, and can be surprisingly intelligent given their lack of internal models. However, much of the power of a Brooks-style robot, for example, comes from the skill with which the behaviors have been designed and composed together in the subsumption architecture. The next major step was to remove this manual element. This was done by developing techniques to design formally correct machines, and by providing tools for automatically generating a reactive machine to suit a set of design criteria.

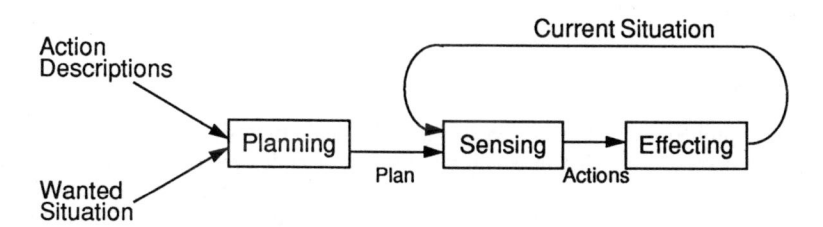

Fig. 3. The Universal Plan approach to execution.

<u>Universal Plans: the autogeneration of reactive machines.</u> The Universal Plan work (Schoppers [7, 18]) is an approach to building reactive machines as highly conditional plans. A universal plan is a plan that can achieve a goal given any possible initial state of the environment. In this approach, actions are selected through the classification of the actual situation encountered at execution time, as opposed to the classical approach of selecting actions based on a single initial situation at planning time. Thus in a universal plan, the failure of an action to achieve its predicted effects does not necessitate replanning. It only requires the selection of a new initial point from which to execute the plan (figure 3). The task of a planner has, therefore, changed from generating a sequence of actions, to anticipating all possible situations and concomitant appropriate responses. At execution time, a sensing module recognizes the actual situation and selects the appropriate action to perform.

Situated Automata: Formally correct reactive machines. The *situated automata* model of Rosenschein & Kaelbling [20, 21, 22] was developed to address the issues of real-time performance and *provably* correct behavior. A program in this model is a finite machine with internal state that transforms inputs to outputs within a small time period. This model, therefore, provides a good base on which to built reactive systems. A Lisp-like language, called Rex, was developed to specify such machines. The output of a Rex program is a description of a digital circuit for a machine that meets the Rex specification.

A key characteristic of the situated automata model is that the correctness of the machine's potential behavior is addressed by considering the *knowledge* that the machine contains, where knowledge is analyzed in terms of the relationship between a machine and its environment. A machine is defined to know a proposition φ in machine state s, if whenever the machine is in state s, φ is satisfied. Given a Rex program, a background theory describing facts about the environment, and a description of the epistemic meaning of the machine inputs, the situated automata model provides a methodology for attacking the problem of verifying the meaning of the machine outputs.

Kaelbling [22] describes a compiler, Gapps, that maps a goal specification and a set of goal reduction rules to a Rex program, for a machine to achieve those goals. The goal reduction rules capture domain specific information. In this fashion, a correct reactive system can be automatically generated off-line.

3. Incremental Adaptation

A crucial problem with any one reactive machine (built or designed) is that should the environment diverge from that in which the machine was designed to operate, then the machine may produce inappropriate behavior. The best a reactive machine can do in the face of such a change is to degrade gracefully. That is, its responses may not be as efficient or appropriate in the new environment. However, these responses will not result in self-destruction. Designing a reactive machine that can handle any encountered situation would of course obviate this problem. It is very difficult to anticipate all possible interactions and conflicts at design time however, and it may be computationally prohibitive to achieve real-time control. In addition, changing objectives then force a (lengthy) off-line redesign.

To address the latter problem completely, it is necessary to integrate the concept of the reactive machine with the concept of *a priori planning* of some kind. Note that these two concepts have some complementary characteristics. A reactive machine is not able to deal with the need to look ahead and plan, while an on-line time-constrained planner cannot react sufficiently quickly to deal with unexpected hazards. The role of the planning system thus has changed. Instead of the a priori generation of a plan that is followed to the letter by an executor, the planner continually advises a reactive system so that its behavior becomes more goal directed. If a change is warranted, the planner can *incrementally adapt* the reactive system such that its behavior is better suited for the current environment.

3.1. The incremental adaptation methodology

The incremental adaptation methodology is depicted in figure 4. A controller (the reactive system) is acting upon the world according to the given goals. It is perceiving the effects through its sensors. The adaptor (the planning system) operates in parallel. It examines the actual effects of the actions in the world, and the effects predicted by the world model. If the two diverge, then the adaptor can modify its world model to better predict the actual world. Concurrently, the adaptor can modify the controller by issuing adaptations to improve the control strategy with respect to the goals and environment as it currently manifests itself.

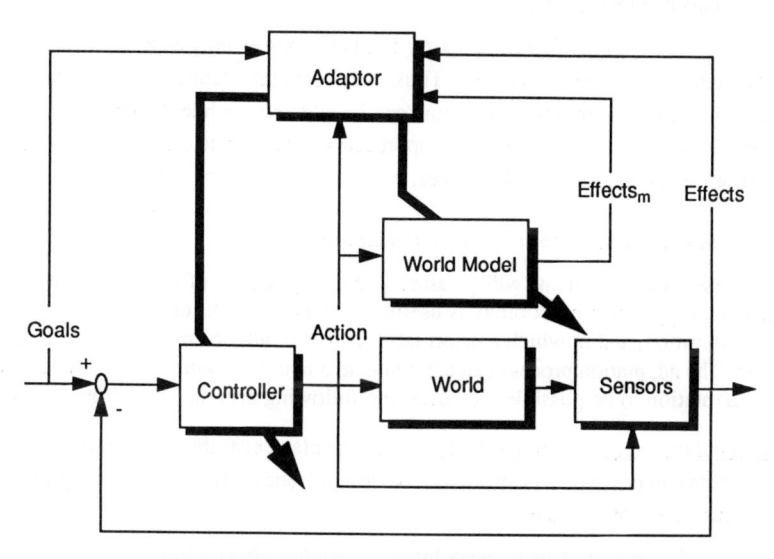

Fig. 4. The incremental adaptation methodology.

Without the presence of an adaptor, the controller acts as a stand-alone reactive system. With the adaptor present, both the controller and the model of the world can be fine tuned. In addition, long term, not readily perceptible objectives can be taken into account. Thus, the incremental adaptation methodology aims to preserve all the advantages of the reactive systems approaches, while addressing their disadvantages. Incremental adaptation offers the following additional advantages:

(i) <u>Shorter lead-time to production of useful behavior.</u> The control strategy can be planned for incrementally. Thus the controller can start acting, while the planner is still working out action details for later parts of the control strategy.

(ii) Flexibility to suit changing objectives. When some objectives change during operation, the adaptor can alter the control strategies in use incrementally, with the system continuing to function. Halting, off-line redesign, and reboot are not necessary.

(iii) Increased robustness. Incremental adaptation opens the way for alternative strategies in response to failures. A purely reactive system would be limited to retrying the same strategy.

(iv) Improved efficiency. Incremental adaptation allows controller specialization with respect to the current environment. Hence, special simplifications for cooperative environments can be exploited.

(v) Amortization of planning complexity. The planner need only address those issues that the actual environment presents. Thus, only relevant situations, as indicated by the world, need to be planned for. Some can be addressed while the system is in operation. In contrast, an off-line design approach would have to consider all potential situations, even the most unlikely ones, before the system could start to act.

3.2. Domain requirements for incremental adaptation.

Of course, not all domains will be amenable to application of the incremental adaptation methodology. The methodology is useful for repetitive tasks or tasks that are relatively long in duration, and in which a priori information about the environment is scarce and uncertain. The adaptation process can add value to a reactive controller in those situations. For the adaptation to be possible, however, the following criteria have to be satisfied:

(i) Redundancy in options or methods. If only one plan exists that satisfies the goals, then this plan can be given a priori to the controller. Hence incremental adaptation cannot increase the robustness of the system.

(ii) No "hard" time dead-lines, prohibiting reasoning efforts. Reasoning about actions takes time and in general cannot be expected to be a real-time process. Thus, the tasks must be such that some time is available for the adaptor to reason about changes to the control strategy. The control strategies themselves can be real-time though, e.g. visual tracking.

(iii) Breakdown of problem and objectives is possible. If the only plan that works is monolithic in nature, then the incremental algorithms have no use. Instead the planning is a fall back to off-line design. If separate steps can be distinguished, perhaps corresponding to an abstract plan, then these steps can be planned for incrementally and the control system adapted accordingly.

(iv) Information can be extracted to guide adaptation. The power of the incremental adaptation methodology is rooted in the fact that the environment can help the planner decide which parts of the control strategy need change and/or refinement. This information needs to be extracted and made available to the planner. Otherwise a Universal Plans approach is required.

(v) <u>Soft failures</u>. If potential system failures are not recoverable, then incremental adaptation cannot increase the robustness of the system. Hard catastrophic failures should be guarded against and a definite strategy should be in place at the start of operations for failure modes of this kind.

4. Incremental adaptation architectures

In this section we discuss several architectures based on the incremental adaptation methodology. The two major streams in this field are divided by the following choices of action selection mechanism:

(i) Continual redeciding which action to select: THEO, DYNA, ERE.

(ii) Updating (partial) control strategies: XFRM, ADAPT.

The first action selection mechanism restricts action execution to one at the time. The latter allows concurrent and parallel action selection and execution. However, it requires a more elaborate adaptation mechanism to safely switch over from old to new behavior patterns. We now examine each of these systems in turn.

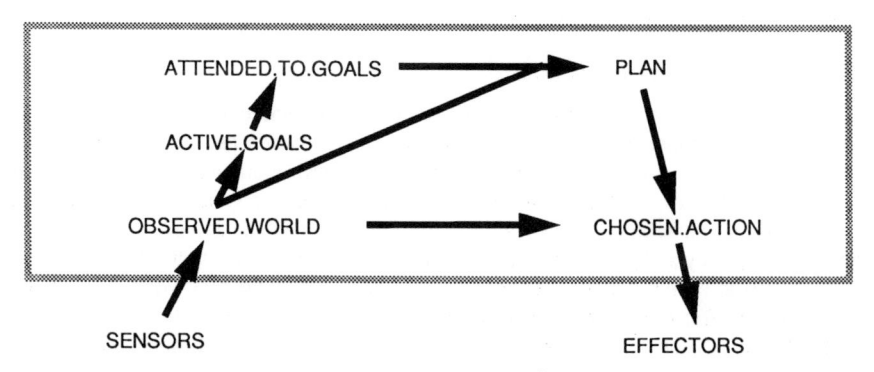

Fig. 5. THEO's strategy.

4.1. THEO: planning on demand.

Theo-Agent (Mitchell [22]) was one of the first architectures designed to combine the advantages of reactive and search-based systems. The adopted control policy in Theo-Agent is a continual reassessment of the next action to perform, followed by execution of that action, and subsequent updating of its world model based on sensor inputs.

Theo's action selection strategy is depicted in figure 5. Action selection is based on *stimulus-response* rules, whenever possible. These stimulus-response rules constitute THEO's reactive component. If one of these rules applies to the current sensed inputs, the corresponding action (the `chosen.action`) can be taken immediately. In this way, THEO can react quickly to situations in which it knows what to do. If no rules apply, its planner is invoked to determine an appropriate action.

Whenever Theo is forced to plan, it acquires a new stimulus-response rule. Based on the observed world and the active goals which Theo infers from the `observed.world`, it chooses the `attended.to.goals`, the goals it will try to achieve next. The planner generates a plan that achieves those goals given the current observed world. The first action of this plan will be the `chosen.action` to execute next. Explanation-based learning is used to generalize this plan into a stimulus-response rule. This new rule recommends the action which the planner has recommended, in the same situations (i.e. those world states in which the same plan justification would apply), but can be invoked much more efficiently. The learning of new stimulus-response rules provides a demand-driven incremental compilation of the planner's knowledge into an equivalent reactive strategy, guided by the agent's experiences.

Theo's on-demand rule acquisition strategy allows it to become increasingly reactive. In a new, and unfamiliar situation, much time will be expended in planning, resulting in a slow reaction time. Over time however, the continual acquisition of stimulus-response rules allows its response time to drop from a few minutes to a mere second as it encounters more and more familiar situations.

4.2. DYNA: learning control policies.

The Dyna architecture (Sutton [23, 24]) presents an effective application of Dynamic Programming for those domains in which a numeric evaluation can be applied to world states. Dyna aims to combine planning, reaction, and learning into a single, integrated architecture. In Dyna (figure 6), actions are generated by a reactive system, the *policy*, which maps states onto an action. The *world*, representing the task to be solved, receives an action from the policy and produces a next state output and a reward output. The *world model* is intended to mimic the one-step input-output behavior of the real world. Finally, the *evaluation function* rapidly maps states onto values, much as the policy rapidly maps states onto actions. The evaluation function, the policy, and the world model are each updated by separate learning processes.

Dyna is based on the concept that planning is like trial-and-error learning from hypothetical experience. For a fixed policy, Dyna is simply a reactive system. However, the policy is continually adjusted by an integrated planning/learning process. The policy is a plan in a sense, but one that is completely conditional on current input. The planning process is incremental. It can be interrupted and resumed at any time, and *does not intervene between situation and action*. Instead, the reactive system, the policy, never waits for the planner and the planner is not restricted to plan only for the current situation. The choice of trial-and-error learning or planning is simply the choice to apply the policy's action to the real world or the world model.

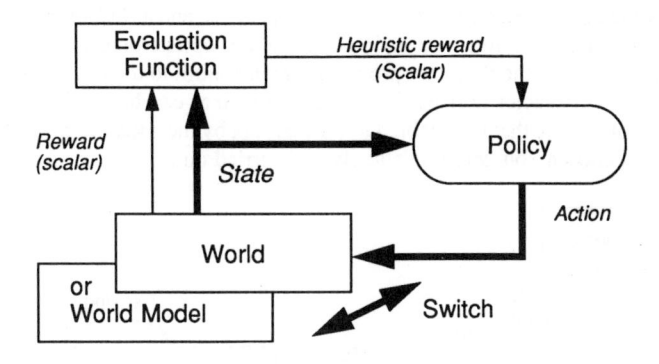

Fig. 6. DYNA.

Learning takes place by continually readjusting the evaluation function in such a way that credit is propagated to the appropriate steps in an action sequence. A real or a hypothetical experience is chosen at the start of every cycle within Dyna's control loop. In the case of a real experience, the current state is chosen. For a hypothetical experience, *any* state suffices. The policy is consulted for the chosen state to select an action. That action is executed and the next state and reward are obtained. If it was a real experience, the world model is updated to reflect the outcome of the action. In any case, the evaluation function is updated such that the evaluation of the old state is more like the evaluation of the new state plus the cost of performing the action.

By approximating Dynamic Programming with machine learning, Dyna offers considerable flexibility with respect to domain knowledge. Neither the structure nor the dynamics of the world have to be known a priori. Even so, Dyna will eventually arrive at an optimal policy. The choice of the learning function — the function updating the evaluation of states — can be tailored to suit the domain. Dyna is simple and powerful, only limited by the need to recognize and numerically evaluate world states. For those domains where this is possible, Dyna offers an integrated architecture for incremental planning, learning, and reaction.

4.3. ERE: generation of situated control rules for action selection.

Bresina & Drummond [25] introduce an alternate approach to systems that have to produce intelligent action under time constraints. Their domain of application is photoelectric telescope scheduling problems. They suggest an architecture, called ERE, that combines the ability to react to the current environment with the ability to plan ahead.

Their system consists of three independent and concurrent components (figure 7); a *Reactor*, which is based on Drummond's plan-net formalism and reacts to the current environment; a *Projector* that determines the future effect of possible next actions, and advises the Reactor on which ones best satisfy the system's objectives, and a *Reductor* that advises the Projector about which possible next actions are the appropriate ones to explore given the systems objectives (i.e., the Reductor provides search control to the Projector). In any given situation, s, the Reactor derives the set of actions that can be carried out. These are simply the set of actions that are enabled in the plan-net by the situation s: enabled(s). One of these is chosen non-deterministically and carried out.

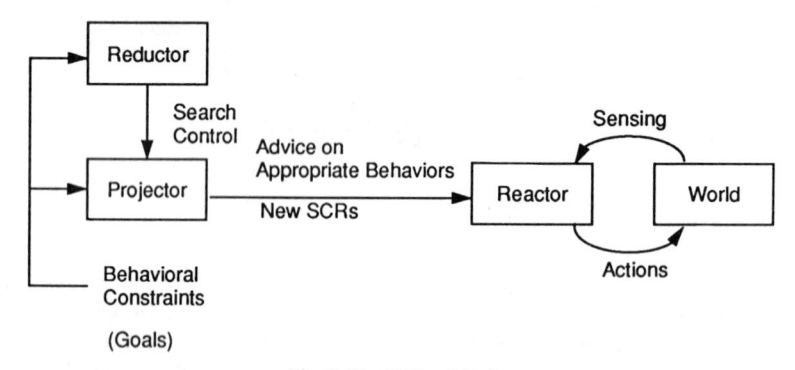

Fig. 7. The ERE architecture.

Using estimates of the probability of conditions and the utility of actions, the Projector component projects forward in time the effects of action sequences that could be selected from the Reactor's plan-net. The theory of temporal projection based on these plan-nets is described in [26]. Forward projection of actions, starting from the current state, is used to find a path satisfying behavioral constraints. Two main search control strategies are employed (figure 8). The *Cut-and-Commit* strategy performs a beam-search. When it finds the first sequence that obeys the behavioral constraint (say in state s3 in figure 8), it compiles a set of *Situated Control Rules* (SCRs) and sends them to the Reactor. These are if-then rules that offer the Reactor advice on the best subset of enabled(s) to carry out in order to keep in line with the behavioral constraint. The strategy then restarts from the state in which the first behavioral constraint is satisfied (s3) to try and find a path for other behavioral constraints. In this way, those SCRs that are generated first most likely satisfy the current needs of the reactor.

Subsequently, the *Deviate-and-Recover* strategy explores side paths. These are paths that could be caused by either external events happening or the agent performing other actions. The strategy aims to recover from transitions in situations for which there exists no satisfactory path yet. If a recovery path is found (s6-s7-s3 in figure 8), new SCRs are generated and downloaded. The Reactor can use these to help it deal with other possible situations in which it may find itself.

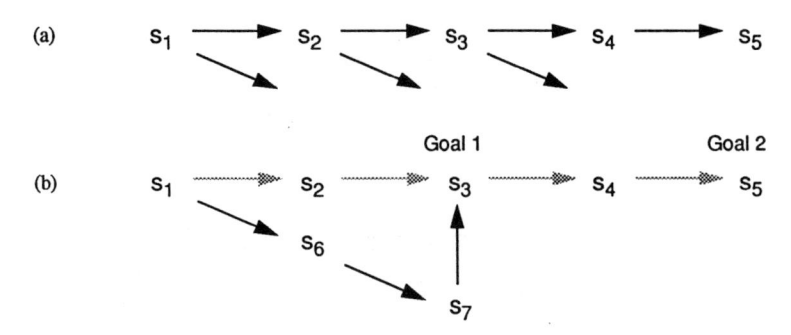

```
(a)   S1 ──► S2 ──► S3 ──► S4 ──► S5
         ╲     ╲     ╲
          ╲     ╲     ╲

                   Goal 1           Goal 2
(b)   S1 ┈┈► S2 ┈┈► S3 ┈┈► S4 ┈┈► S5
         ╲          ▲
          ╲         │
           S6       │
            ╲       │
             ╲      │
              S7
```

Fig. 8. ERE's two main Situation Control Rule generation strategies:. Cut-and Commit (a) and Deviate-and-Recover strategy (b).

The nature of Projector-Reactor interaction is as follows: without any planning, the Reactor should be able to give some acceptable though very myopic behavior; with some planning the Reactor should be able to produce more intelligent behavior in limited circumstances; and with lots of planning the Reactor should be able to produce intelligent robust behavior in various environmental circumstances.

4.4. XFRM: planning by transformation.

XFRM (McDermott [27]) starts with the assumption that an agent *always* has a plan. The plans may be simple, or abstract, but they do have *default methods* for carrying them out. Therefore the agent's plan is always executable, even though those default methods have no guarantee of succeeding. Under this view planning is the operation of *improving* the agent's existing plans, rather than generating programs.

XFRM's architecture is shown in figure 9. The plan is a central resource that is manipulated by both a transformational planner and a reactive executor. The reactive executor executes the current plan. The transformational planner concurrently examines this plan and looks for improvements. When the planner has found one it swaps the current plan for the new one and the reactive executor works off this new version.

Transformational planning starts by projecting the plan forward in time. *Plan critics* then look for *bugs*: detectable problems in the forward projection of the plan. These plan critics are specialized to look for specific bugs, e.g., an overload bug detector detects whether a resource such as a robot hand is over committed. Apart from detecting bugs, the critics evaluate the bug's *penalty* (the cost for the plan to have this bug) and return the bug's signature (a symbolic description of the bug). The critics also provide a *comparer* (a procedure for comparing bug signatures) and most importantly the *transformation* (a procedure that will try to fix the bug by making changes to the plan).

```
        Transformational
            Planner
               |
             Plan
               |
           Reactive
          Interpreter
               |
             World
```

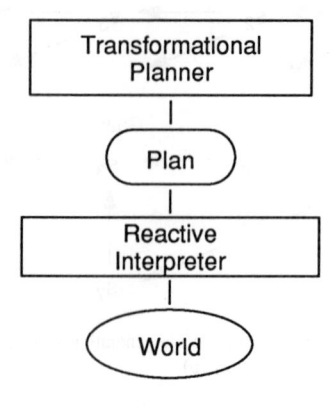

Fig. 9. XFRM.

XFRM operates by keeping a queue of buggy plans. Each plan has several bugs, but the worst bug is special in that it is this one whose transformation will be run if the plan is selected for future search. XFRM selects the most promising plan from this queue, and runs its transformation to generate one or more new plans. These new plans will each be projected forward in time. Then the critics are applied to find remaining bugs and score the resulting plan. Finally, the new plans are merged into the plan queue which is then re-sorted. If a new plan is better than the one that the reactive executor is currently executing, a swap takes place and the new plan becomes the current plan.

The use of plan critics allows XFRM to capitalize on domain specific knowledge. The plan language, however, should be transparent enough to be interpreted and debugged. The design of these plan critics becomes a major issue for a successful application in a selected domain. The set of critics should be sufficient and complete, i.e. they should be able to detect all relevant bugs, and the combined effect of their transformations should result in a correct plan for the encountered environment.

4.5. ADAPT: Planning and Reacting

Lyons & Hendriks et al.[28, 29] address the problem of planning for reaction by considering a Planner and a Reactor as two elements of a concurrent, cooperating system. The Planner interacts with, and is influenced by, the Reactor, in the same fashion that the Reactor interacts with, and is influenced by, the world. The architecture is shown in figure 10.

Their Reactor component is based on a special purpose model of computation developed for representing highly conditional robot plans. The model, called $\mathcal{RS}$, is an extension of the *Robot Schemas* model of Lyons & Arbib [30]. This model inherits the philosophy of Arbib's schema theory in seeing behavior as the result of the cooperation and competition of a set of interacting schema instances. The model is process-based, and process-algebra methods[31] are used to analyze the process network behaviors. Processes can be defined in terms of networks of other processes, grounding out with a set of atomic, predefined, processes. This provides a powerful mechanism for specifying flexible, hierarchical structures. They make the point that to analyze the behavior of reactive machines it is very necessary to know in what sort of environment the machine will be situated. Thus, in reactive machine analysis, the $\mathcal{RS}$ model is used to represent *both* the plan (or controller) and the environment in which the plan will be carried out [32].

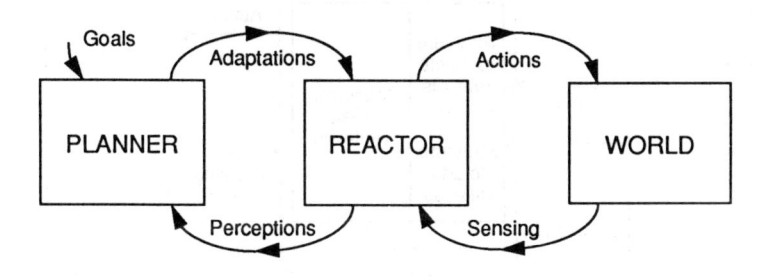

Fig. 10. The Planner-Reactor approach.

The key property of the Reactor is that it can react at *any* time. Unlike a plan executor or hierarchical system of planner and reactive system, the Reactor acts asynchronously and independently of the planner. An $\mathcal{RS}$ Reactor is specified by a set of recursive process equations, where processes can be coupled in networks in two ways: they can communicate messages to each other via *communication ports*, or they can be composed together using several kinds of *process composition operators*.

The Reactor has special structure built-in to facilitate and safeguard adaptations, i.e. the change-over of control strategies in the Reactor, during its operation (figure 11). The safe adaptation mechanism [33] does not interrupt unaffected reactions. It is guaranteed to finish the change-over and does not produce intermediate behavior (i.e a combination of old and new behavior).

The Planner 'tunes' the Reactor so that the concurrent composition of Reactor and the world model continue to produce appropriate behavior. The full system is then a concurrent composition of the Planner and the Reactor-World system. By considering a Reactor to be a set of concurrent situation-labeled actions, Lyons & Hendriks define formally what it means for the Planner to *improve* the behavior of the Reactor over time.

At start-up, the Planner constructs the Reactor incrementally. Adaptations can be sent to the Reactor to achieve either new or changed goals. When the system is given new goals, the reactor structure is at first adapted to reflect the new abstract plans. These abstract steps are incrementally fleshed out based on *time ordering* and *relevancy in the world* (as indicated by perception monitors).

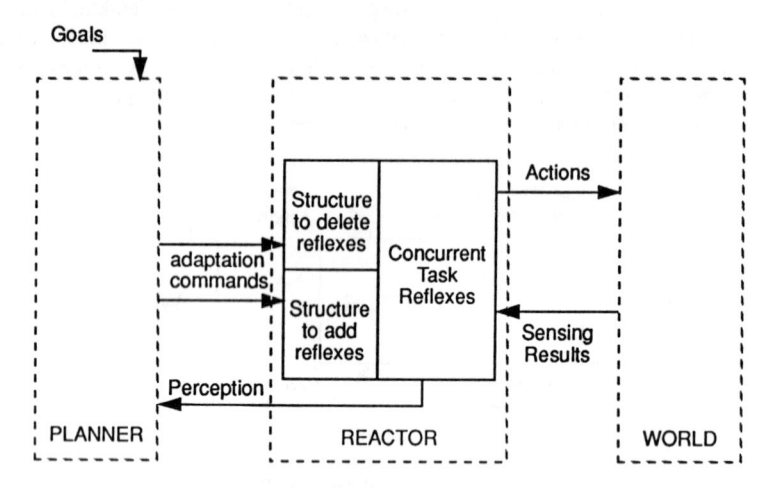

Fig. 11. Planner-Reactor-World interactions in ADAPT.

Once a first complete reactor is in place, the Planner aims to incrementally improve it by widening the conditions under which the reactor operates. This is achieved by relaxation of the assumptions ω which were used initially to speed up the incremental generation of the Reactor. The Planner's ultimate target is the "Ideal Reactor" R^*, a Reactor that achieves the objectives in all possible environments and relies on no assumptions. The incremental assumption relaxation strategy is summarized as follows:

$$R^{\omega 1} \stackrel{\alpha 1}{\Rightarrow} R^{\omega 2} \stackrel{\alpha 2}{\Rightarrow} R^{\omega 3} \stackrel{\alpha 3}{\Rightarrow} \ldots \stackrel{\alpha v}{\Rightarrow} R^*$$

$$\text{for} \quad \omega 1 > \omega 2 > \omega 3 > \ldots > \emptyset$$

with R^ω an ω-ideal Reactor (an ideal Reactor if the assumptions ω hold) and αi are the adaptation commands generated by the planner.

Thus, assumptions are removed incrementally while *at all times* a working Reactor remains. The choice of which assumption to relax can be either planner-directed (based on priorities) or environment-directed (when assumptions do not hold in the environment).

The Planner does not have access to the complete internals of the Reactor. Its observations are restricted to the output of *perception processes* that it can embed into the Reactor. Thus, the Planner can ignore the bulk of the sensing carried out by the Reactor, and concentrate on a few important pieces of information. Perception processes are used to 1) determine when some part of the Reactor is in danger of failing (e.g. an assumption not holding), 2) to reflect on-going resource usage, and 3) to determine whether certain goals or subgoals have been met. In this fashion, the Planner can use the Reactor to focus its reasoning, and in turn the Reactor is guided by the Planner to improve its behavior.

5. Summary

In this chapter, we have introduced incremental adaptation as a method to improve reactive behavior. This methodology builds on both the reactive systems and classical planning work, and aims to incorporate their strengths. The weaknesses of purely reactive machines are clear: they can be very myopic and there are limits to their robustness. It has become evident that some form of additional, longer-term reasoning is necessary. On the other hand, purely deliberative "classical planning" systems cannot cope with the dynamics of an uncertain and changing environments.

Incremental adaptation is a realistic method for achieving intelligent adaptable behavior in uncertain and dynamic environments. It offers the benefit of integration of deliberation into reactive systems, while maintaining the reactive system's capability for a fast real-time response. Several architectures are presented here; their applicability and relative merits depends on the domain of interest. Some architectures favor "experimental behavior", while others are more conservative in their strategies.

There are still a lot of open questions in this area, and of course, much work in other fields of planning, non-monotonic reasoning, robotics, etc., also impacts the incremental adaptation work. We list here just a few of the open areas:

(i) There has been little work in integration with the extensive real-time computation field [34].

(ii) Resources play a key role in many domains —the concept of 'making do' with what is available — but as yet there is no unique viewpoint on resources.

(iii) The long term effects of incremental, iterated adaptations need to be studied. In basic design, the incremental adaptation methodology falls into the class of what in adaptive control are called self-tuning regulators [35]. Stability of a the control strategy under adaptations and convergence to an optimal control strategy need to be studied to ascertain the long term performance of systems implementing the methodology.

(iv) Improvement under efficiency is rarely considered up till now. Yet, the ability to specialize control strategies to suit the current environment it is one of the great benefits incremental adaptation can offer.

6. References

1. R.E. Fikes & N.J. Nilsson, *Strips: A new approach to the application of theorem proving to problem solving*, Artificial Intelligence **2**, (1971), 189-201.
2. A. Tate, *Generating Action Networks*, Proc. IJCAI-5, (1977), 888-893.
3. M. Stefik, *Planning with Constraints (Molgen: part 1), Planning and Meta-Planning (Molgen: part 2)*, Artificial Intelligence **6** (1981), 111-170.
4. S. Vere, *Planning in Time: Windows and Durations for Activities and Goals*, IEEE Transactions on PAMI **5** (1983), 246-267.
5. D. E. Wilkins, *Domain-independent Planning: Representation and plan generation*, Artificial Intelligence **22** (1984), 269-301.
6. D. Chapman, *Planning for Conjunctive Goals*, Artificial Intelligence **32** (1987), 333-377.
7. M. Schoppers, *Universal Plans for Reactive Robots in Unpredictable Environments*, Proc. IJCAI-87 (1987), 1039-1046.
8. L.P. Kaelbling, *An Architecture for Intelligent Reactive Systems*, Workshop on Planning and Reasoning about Action, Timberline OR (1986), 235-250.
9. R. Brooks, *A Robust Layered Control System for a Mobile Robot*, IEEE Journal Robotics & Automation, **2** (1986), 4-22.
10. D. McDermott, *Robot Planning*, Tech. Report YALEU/CSD/RR #861, Yale University, (1991).
11. P. Agre & D. Chapman, *What are plans for?* Robotics and Autonomous Systems **6** (1990), 17-34.
12. P. Agre & D. Chapman, *From reaction to participation*, DARPA planning Workshop, Santa Cruz, CA, (1987), 123-154.
13. D. Chapman & P. Agre, *Abstract Reasoning as Emergent from Concrete Activity*, Workshop on Planning and Reasoning about Action, Timberline, OR (1986).
14. C.J. Sellers & S.Y. Nof, *Performance Analysis of robotic kitting systems*, Rob. & Comp. Integ. Manuf. **6** (1989), 15-24.
15. P. Agre & D. Chapman, *Pengi: An implementation of a theory of action*, Proc. AAAI-6, Santa Cruz, CA (1987), 268-272.
16. V. Braitenberg, *Vehicles*, MIT Press, Cambridge MA (1984).
17. R. Brooks, *Intelligence without Representation*, Artificial Intelligence **32** (1991), 139-160.
18. M. Schoppers, *Representation and Automatic Synthesis of Reaction Plans*, Ph.D. Thesis, Tech. Rep. uiucdcs-r-89-1546, CS dept., University of Illinois at Urbana-Champaign (1989).
19. L.P. Kaelbling, *Goals as Parallel Program Specifications*, AAAI-88 (1988), 60-65.
20. S. Rosenschein and L.P. Kaelbling, *The synthesis of digital machines with provable epistemic properties*, Tech. Note 412, SRI International, Menlo Park, CA (1987).

21. S. Rosenschein, 1st Conf. on Principles of Knowledge Representation and Reasoning, Toronto, Canada (1989), 386-393.
22. T. Mitchell, *Becoming Increasingly Reactive*, Proc. AAAI-90 (1990), 1051-1058.
23. R. Sutton, *Integrated Architectures for Learning, Planning, and Reacting Based on Approximating Dynamic Programming*, Proc. 7th Int'l Conference on Machine Learning, 1990, 216-222.
24. R. Sutton, *Planning by Incremental Dynamic Programming*, Proc. 9th Int'l Workshop on Machine Learning, 1991, 353-357.
25. J. Bresina & M. Drummond, *Integrating Planning and Reaction*, in: AAAI Spring Workshop on Planning in Uncertain, Unpredictable or Changing Environments, Stanford, CA, J. Hendler (ed.) Systems Research Center, Univ. of Maryland (1990), 24-28.
26. M. Drummond & J. Bresina, *Anytime Synthetic Projection: maximizing the probability of goal satisfaction*, Proceedings AAAI-8. Anaheim, CA (1990), 138-144.
27. D. McDermott, *Planning Reactive Behavior: A progress report*, DARPA Workshop on innovative approaches to Planning, Scheduling and Control, San Diego, CA (1990).
28. D.M. Lyons, *A Process-Based Approach to Task-Plan Representation*, IEEE Int. Conf. on Robotics & Automation, Cincinnati, Ohio (1990), 2142-2147.
29. D.M. Lyons, A.J. Hendriks & S. Mehta, *Achieving Robustness by Casting Planning as Adaptation of a Reactive System*, IEEE Int. Conf. on Robotics & Automation, Sacramento, CA (1991), 198-203.
30. D.M. Lyons & M.A. Arbib, *A Formal Model of Computation for Sensory-Based Robotics*, IEEE Trans. on Rob. & Automation **5** (1989), 280-293.
31. C.A.R. Hoare, *Communicating Sequential Processes*, Prentice Hall Int. Series in Computer Science, (1986).
32. D.M. Lyons, *A Formal Model for Reactive Robot Plans*, 2nd Int. conf. on Computer Integrated Manufacturing, RPI, Troy, NY, (1990).
33. D.M. Lyons & A.J. Hendriks, *Safely Adapting a Hierarchical Reactive System*, SPIE Intelligent Robotics & Computer Vision Boston, MA, (1993).
34. P. Gopinath, *Programming and Execution of Object-Based, Parallel, Hard Real-Time Applications*, Ph.D. thesis, Dept. of Computer and Information Sciences, Ohio State University, 1988.
35. K.J. Åström, *Adaptive Feedback Control*, Proc. of the IEEE **75** (1987), 185-217.

AN SPN-NEURAL PLANNING METHODOLOGY FOR COORDINATION OF MULTIPLE ROBOTIC ARMS WITH CONSTRAINED PLACEMENT

N. Bourbakis and A. Tascillo
Binghamton University
School of Engr., AAAI Lab
Binghamton NY 13902

ABSTRACT

This paper presents a planning methodology based on Stochastic Petri Nets (SPN) and Neural nets for coordination of multiple robotic arms working a space with constrainted placement. The SPN planning method generates a global plan based on the states of the elements of the Universe of Discourse. The plan includes all the possible conflict-free planning paths to achieve the goals under constraints, such as specific locations on which objects have to be placed, order of placement, etc. An associated neural network is used to search the vectors of markings generated by the SPN reachability graph for the appropriate selection of plans. Moreover, it preserves all the interesting features of the SPN model, such as synchronization, parallelism, concurrency and timing of events. The coordination of two robotic arms is used as an illustrative example for the proposed planning method, in a UD space where the location of objects placement are restricted.

KEYWORDS: Stochastic Petri-nets; Neural Nets Applications; Planning Strategies;

1. INTRODUCTION

There is a variety of planning methodologies developed the last twenty years [1-13]. Most of these methodologies (NOAH, NONLIN, DEVISER, SIPE, TWEAK, etc) generate a partially ordered plan network for the achievement of conjunctive goals. More specifically, these methods are based on state oriented planning, where each plan is constructed by generating a subplan for each goal state, while the detection and resolution of conflicts among these subplans take place in the state space. In addition, a well defined planning methodology, called RPP [8], has been proposed for the resolution of parallel plans. This method considers resources as active elements in constructing plans and generates conflict-free subplans by controlling the flow of a particular resource, while synthesizes a complete plan in cooperation with the other subplans. The RPP methodology constructs conflict free subplans prior to the expansion of the current plan network. Another methodology, uses heuristic search on a Petri-net framework and a metric space for efficient planning [9].

Most of the methods discussed above lack in timing and stochastic synchronization of the events (actions) which take place during the execution of a plan. Moreover, most of them they cannot handle efficiently cases, where the number of elements increases significantly.

In this paper, a planning methodology based on the combination of stochastic Petri-nets and neural nets is proposed. More specifically, the plan network is expressed as a SPN network where all the states of the elements and the actions of the universe of discourse are expressed as *Petri-net places P and transitions T* respectively, in various levels of abstraction. A neural network model searches the SPN framework for the appropriate selection of plans. Decomposition and synthesis of subplans can be easily achieved on the SPN model. A neural net model is used to search the state space in oredr to achieve a desirable path. An illustrative example is provided by using the SPN network methodology, for the coordination of two robotic arms in a space with constraints such as specific locations for placement of objects.

This paper is organized into five sections. Section 2 presents the important notations and definitions. Section 3 deals with the SPN modeling of the planning methodology. Section 4 presents the results of the illustrative example and section 5 summarizes the overall work.

2. NOTATIONS AND DEFINITIONS

Notation 1: The set (UD) of all the elements Ei, $i \in Z$, and the knowledge related with them for a particular domain:
UD = {Ei/ Ei is an element with properties {$F(Ei)$}, and {$R(Ei,Ej)$} relationships among elements}

Notation 2: A state $S(k,t)$, with $k \in Z$ and t=time, is a the representation of the set UD (or a subset of it) at a given time t after an action A.

Notation 3: An action Am, $m \in Z$, represents the mapping of UD (or a subset of it) from the state $S(i,t)$ into state $S(j,t')$:
$$Am : \Sigma \times S \longrightarrow S$$
where Σ ={Am/ Am is a primitive action performed on Ei}, and S = {$S(k,t)$/ $S(k,t)$ is any state of UD}

A plan would be considered as a sequence of actions performed on the elements of UD at a state for the achievement of the goal state. More specifically,

Definition 1: A plan PLi = {{Am},{Ei},{$S(i,t)$}} is a tri-tuple, where {$S(i,t) \in S$} is a sequence of states achieved by performing a sequence of actions {$Am \in \Sigma$, $m \in Z$} on the set of elements at {$Ei \in UD$, $i \in Z$} starting from a state $S(j,t0)$ for the accomplishment of the goal states {$S(k,tn), n \in Z$}.
$$PLi : UD[Ex\{S(k,t)\}] \xrightarrow{\{Am\}} UD[Ey\{S(j,t')\}]$$

Notation 4: A plan is successful *[PLi]* if and only if it can achieve the goal state.

Notation 5: The set of all the successful plans represents the direct graph (or network) G.
It is the synthesis of all *[PLi]*, $i \in Z$:
$$G = \$_i \{[PLi]\}$$

Notation 6: The set G can also expressed as the union of all transformations of the elements of UD which participate in the plan $PLi(En)$
$$G = \bigcup_n G\{PLi(En)\}$$

3. SPN PLANNING METHODOLOGY

In this section the SPN planning methodology is presented.

3.1. WHY STOCHASTIC PETRI-NETS

There is a variety of methodologies used for planning, such as formal languages, directed graphs, classical mathematical models, queuing models, Petri-nets, etc. In this paper, a modified version of the Petri-net will be used as the modeling framework of an SPN Planning strategy for two robot hands. The major reasons for the use of the Petri-net are [9,13-17]:
. *SPN is an efficient modeling tool for the functional description and analysis of complex systems.*
. *SPN is able to describe simultaneously concurrency, paralelism and synchronization of events that take place in a complex system, especially when other methodologies lack of adequate results.*
. *SPN can be used as a modeling tool for hierarchical and abstracted (top-down or bottom-up) processes.*
. *SPN provides timing during the execution of various events*
. *SPN presents compatibility with neural nets*
. *SPN is an efficient interface for control and communication*
. etc.

Notation: The Petri-net used in our case is a Stochastic (SPN) one with time capabilities.

Notation: RT(Mk,NO,FS) represents the reachability tree of SPN.
Notation: RT(Mr) is the set of all marking which are reachable from Mr, $r \in Z$.
Notation: NO represents the number of moving objects
Notation: FS represents the number of open corridors

Note that the SPN used here is a k-bounded one. Also, the association of random time with the transitions will be used. In particular, the SPN transitions will be classified into categories: 1) immediate transitions (which fire in "zero" time once they are enabled) and 2) the timed transitions (which fire after a random, exponentially distributed, enabling time). This assumption will reduce the complexity of SPN when it is needed.

3.2. SPN MODEL

Definition 2: The structure of the SPN planning model is based on a stochastic Petri-net model. Thus, its formal definition is a 11-tuple :

$$SPN(P) = \{P,T,I,O,M,L,X,MC,Q,R,F\}$$

where
. **P** ={P1,P2,P3,...,Pn} is a finite set of places. Each place Pj represents a particular state of an element Ei;
. **T** ={T1,T2,T3,...,Tm} is a finite set of transitions. A transition Ti, $i \in Z$, represents an **action** performed on a set of elements at a state $S(k,t)$;
. **Ii** C (**PxT**) is the input function and
. **Oj** C (**TxP**) is the output function;
. **Mi** = {mi1,mi2,mi3,...,min}, $ij \in Z$, is the vector of marking (tokens) mij, which represent the status of the places; for i=0 m0j, j=0,1,...,n denotes the number of tokens in place Pj in the initial marking M0.
. **Q** = {t1,t2,t3,...,tm} is the vector of time values related with the time required by an action to be performed.
. **R** = {r1,r2,r3,...,rk} is the set of the relationships among elements;
. **F** = {f1,f2,f3,...,fn} is the set of the properties of the elements.
. **MC** = { a1,a2,...,ak }. $k \in Z$, is the alphabet of communication
. **L** = { l1,l2,...lm } is the set of possibly marking-dependent firing rates associated with the Perti-net transitions.
. **X** = { x1,x2,...,xm } is the set of delays associated with the Petri-net transitions

Notation: The Petri-net used in our case is a Stochastic (SPN) one with time capabilities.

Notation: RT(Mk) represents the reachability tree of SPN.

Notation: RT(Mr) is the set of all marking which are reachable from Mr, $r \in Z$.

Notation: The SPN used here is a k-bounded one.

Notation: The association of random time with the transitions will be used:
    1) immediate transitions (which fire in "zero" time once they are enabled)
    2) the timed transitions (which fire after a random, exponentially distributed, enabling time).

Notation: A marking Mki e RT represents a node of the reachability tree.

3.3. THEORETICAL ASPECTS OF THE SPN PLANNING MODEL

Definition: An SPN path (SPNPH) is defined as a sequence of markings {Mk,$k \in Z$} on the SPN reachability tree RT.

Definition: An SPN plan (SPNPL) = {Mk,Tm} is defined as a sequence of paths on the SPN reachability tree RT associated with the appropriate set of transitions, so that the following conditions are satisfied:

1) There is an marking Mk0, k0∈Z, which is the starting "point" of SPNPL;
2) Every marking Mki ∈ SPNPL with ki≠k0, is reachable by another marking Mkj ∈ SPNPL in **κ** steps or transitions.

Notation: The transition rate from the Mi to Mj is

$$q(ij) = \sum_{k \in Hij} l(k) \quad (1)$$

where Hij is the set of transitions enabled by marking Mi, whose firing generates marking Mj.

Notation: The **goodness** of a plan will be defined by searching various "paths" on the reachability tree with minimum cost under the existing constraints.

Definition : An SPN plan is **k-bounded** if the number of tokens in each place is less or equal to **k** for all markings in RT(Mi).

Definition : An SPN plan is **safe** if the number of tokens in each place is less or equal to 1, for all markings in RT(Mi).

Definition : The length **v** of an SPN path is the number of markings Mk, k∈Z, which compose the path.

Definition : The length **ls** of an SPN plan is the number of transitions required, from an initial marking Mk0, to reach the goal markings Mk ∈ SPNPL

Definition : An SPN path is **empty if v=0** and **unique if v=1**.

Definition : An SPN plan is **empty if ls=0** and **unique** if the number of paths **Lp**, which compose the plan is **Lp=1**.

Definition: An SPNPLi plan is closed if there is no transition from any state of SPNPLi to any state of ˆSPNPLi, where ˆSPNPLi,i∈Z represents the complement plan.

Proposition: Two SPN plans are equivalent, SPNPLi = SPNPLj if lsi = lsj and have the same cost.

Proposition: The synthesis (@) of two paths requires the union of the marking sequences

Proposition : The synthesis of two SPN plans requires the union of the paths and the appropriate adjustment of the transitions associated with the markings

Proposition : SPNPHi @ SPNPHj ? SPNPHj @ SPNPHi; and SPNPLi @ SPNPLj ? SPNPLj @ SPNPLi

3.4. SPN PLANNING IN UD USING TWO ROBOTIC ARMS

Let's consider the set UD = {a,b,c, {**on,table, available, clear**}, {**above,left,right**}, {**L1,L2,L3**} } as shown in figure 1, where Li, i=1,2,3, represents a particular location on the table. Also the set of actions **Σ** = {start(**sr**), stop(**sp**), move(**mv**), grab(**gr**),pick-up(**pu**), release(**re**), stack(**st**), unstack(**us**), put-down(**pd**), wait(**w**)}. Then, the plan-network **G** for two successful plans is shown in figure 2. The decomposition of **G** into three subplans G{P(a)}, G{P(b)}, G{P(c)} is shown in figure 3, under the condition of parallel execution by three independent units [8]. In the case where only two units (robot hands) are available, then the subplans are presented in figure

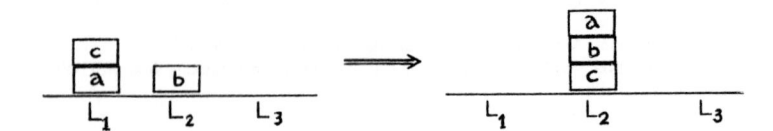

Figure 1

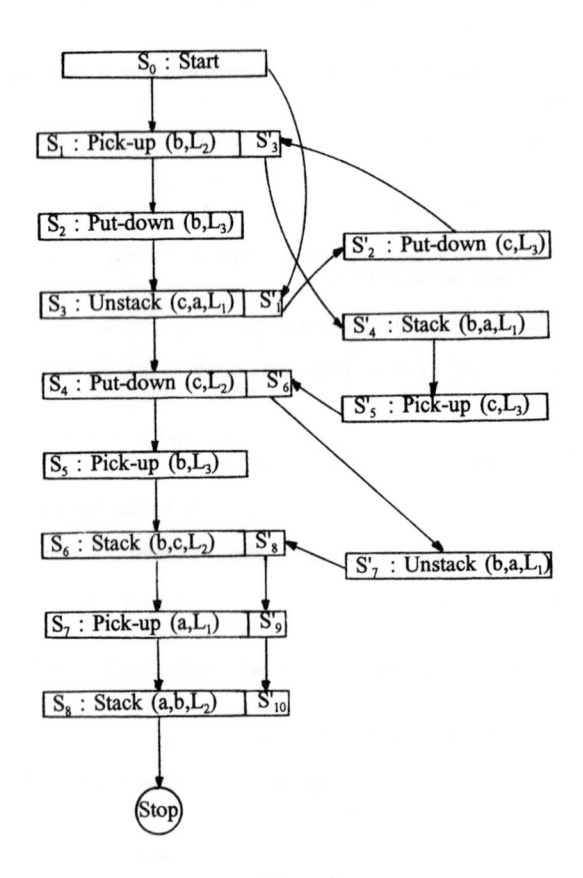

Figure 2

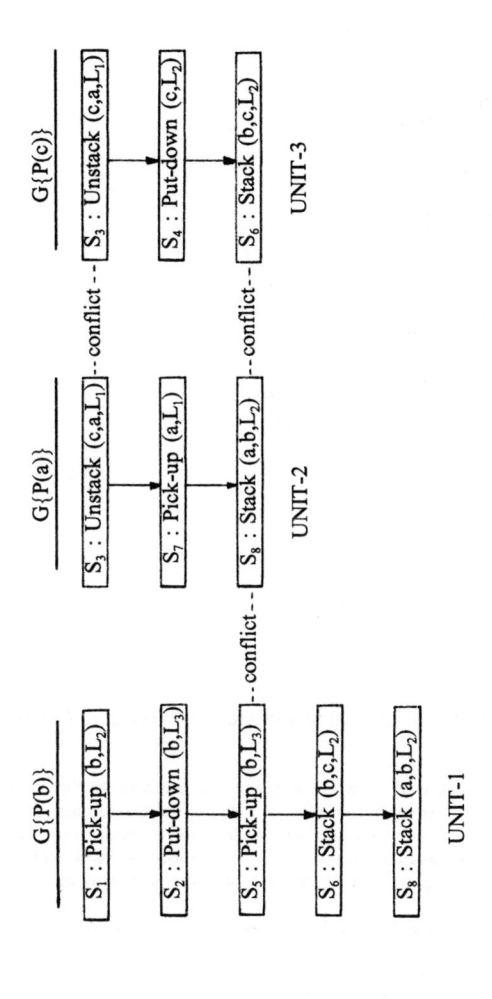

G{P(b)}
- S₁ : Pick-up (b,L₂)
- S₂ : Put-down (b,L₃)
- S₅ : Pick-up (b,L₃) --conflict--
- S₆ : Stack (b,c,L₂)
- S₈ : Stack (a,b,L₂)

UNIT-1

G{P(a)}
- S₃ : Unstack (c,a,L₁) --conflict--
- S₇ : Pick-up (a,L₁)
- S₈ : Stack (a,b,L₂) --conflict--

UNIT-2

G{P(c)}
- S₃ : Unstack (c,a,L₁)
- S₄ : Put-down (c,L₂)
- S₆ : Stack (b,c,L₂)

UNIT-3

Figure 3: Parallel execution of the three subplans [8]

527

4. It is understandable that in all these cases above there is some conflict during the parallel execution of some actions. The avoidance of these kind of conflicts can be done by developing the SPN planning models as shown in figures 5 and 6. Figure 5 shows a generic SPN plan model for two robots hands by synchronizing any possible conflict of actions during the parallel and concurrent execution of the plans. Note that the thick transitions Ti, represent stochastic time delays due to conflicts, while the thin ones represent direct transition with zero delay for the execution of a particular action.

SPN's Main Places and Transitions (actions)

P0 : *starting place of SPN;*
P1 : *one or more objects are available;*
P2 : *assignment of the locations status;*
P3 : *robot hand H2;*
P4 : *3-D workspace for execution of the plans;*
P5 : *robot hand H1;*
P6 : *free state of H2 with probability* **p1**;
P7 : *busy state of H2 with probability* **q1=1-p1**;
P8 : *availability of the 3-D workspace with probability* **p2**;
P9 : *unavailability of the 3-D workspace with probability* **q2=1-p2**;
P10 : *free state of H1 with probability* **p3**;
P11 : *busy state of H1 with probability* **q3=1-p3**;
P16 : *object grabed with probability* **p4**;
P17 : *object blocked with probability* **q4**;
P18 : *object available at the top of a stack with probability* **r4**;
P19 : *object available at the table with probability* **u4**;
P20 : *location free with probability* **p5**;
P21 : *location unavailable with probability* **q5=1-p5**;

Tmv : *move a robot hand;*
Tre : *release the object;*
Tgr : *grab the object;*
Tpu : *pick-up the object;*
Tun : *unstack the object;*
Tpd : *put down the object;*
Tst : *stack the object;*

4. AN ILLUSTRATIVE EXAMPLE OF THE SPN PLANNING WITH A SELF-ORGANIZING NEURAL NETWORK

In this section a neural network model is used for the efficient search of the SPN framework by using the simple example of the previous section. In particular, an ART1 self-organizing network for categorizing binary vectors [11,12] is employed to extract an efficient strategy for a blocks manipulation problem by using two robotic arms (hands), three block and three blocks' positions. The used ART1 network is trained to always achieve the same goal, **a** above **b** above **c** on **L2**, within a minimum number of time steps and from any starting configuration.

As a strategy supervisor, the neural network decomposes a larger strategy that may not be intuitive into a hierarchy of smaller goals. The most obvious example of the benefit of strategy decomposition is when Hand 1 grasps its first block and must decide to either wait or proceed toward one of the two other positions.

Sixteen states and nine previous actions were arranged in a vector and coded as "1" for valid and "0" for not valid for a particular time step. Foe a given end goal, the network was trained with four starting states:
(**c** above **a** above **L**1, **b** above **L**2),
(**a** above **b** above **L**2, **c** above **L**3),

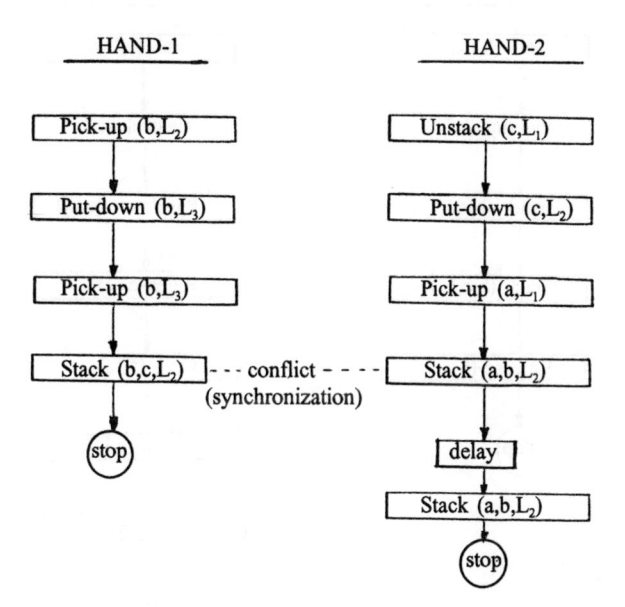

```
        HAND-1                    HAND-2

  ┌──────────────────┐      ┌──────────────────┐
  │ Pick-up (b,L₂)   │      │ Unstack (c,L₁)   │
  └──────────────────┘      └──────────────────┘
           │                         │
  ┌──────────────────┐      ┌──────────────────┐
  │ Put-down (b,L₃)  │      │ Put-down (c,L₂)  │
  └──────────────────┘      └──────────────────┘
           │                         │
  ┌──────────────────┐      ┌──────────────────┐
  │ Pick-up (b,L₃)   │      │ Pick-up (a,L₁)   │
  └──────────────────┘      └──────────────────┘
           │                         │
  ┌──────────────┐                ┌──────────────────┐
  │ Stack (b,c,L₂)│--- conflict ---│ Stack (a,b,L₂)   │
  └──────────────┘  (synchronization) └──────────────┘
           │                         │
        (stop)                   ┌────────┐
                                 │ delay  │
                                 └────────┘
                                     │
                            ┌──────────────────┐
                            │ Stack (a,b,L₂)   │
                            └──────────────────┘
                                     │
                                  (stop)
```

Figure 4: Two robot hands: execution of a plan in a parallel manner with conflict

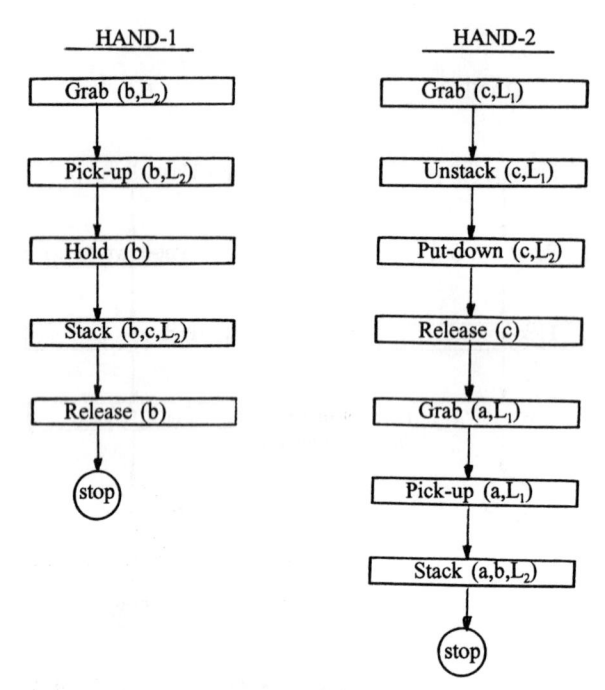

```
     HAND-1                HAND-2

  ┌─────────────┐      ┌──────────────┐
  │ Grab (b,L₂) │      │ Grab (c,L₁)  │
  └─────────────┘      └──────────────┘
         │                    │
  ┌─────────────┐      ┌──────────────┐
  │Pick-up (b,L₂)│     │Unstack (c,L₁)│
  └─────────────┘      └──────────────┘
         │                    │
  ┌─────────────┐      ┌──────────────┐
  │  Hold (b)   │      │Put-down (c,L₂)│
  └─────────────┘      └──────────────┘
         │                    │
  ┌─────────────┐      ┌──────────────┐
  │Stack (b,c,L₂)│     │  Release (c) │
  └─────────────┘      └──────────────┘
         │                    │
  ┌─────────────┐      ┌──────────────┐
  │ Release (b) │      │ Grab (a,L₁)  │
  └─────────────┘      └──────────────┘
         │                    │
       (stop)           ┌──────────────┐
                        │Pick-up (a,L₁)│
                        └──────────────┘
                               │
                        ┌──────────────┐
                        │Stack (a,b,L₂)│
                        └──────────────┘
                               │
                             (stop)
```

Figure 5: Two robot hands executing of a parallel plan without conflict

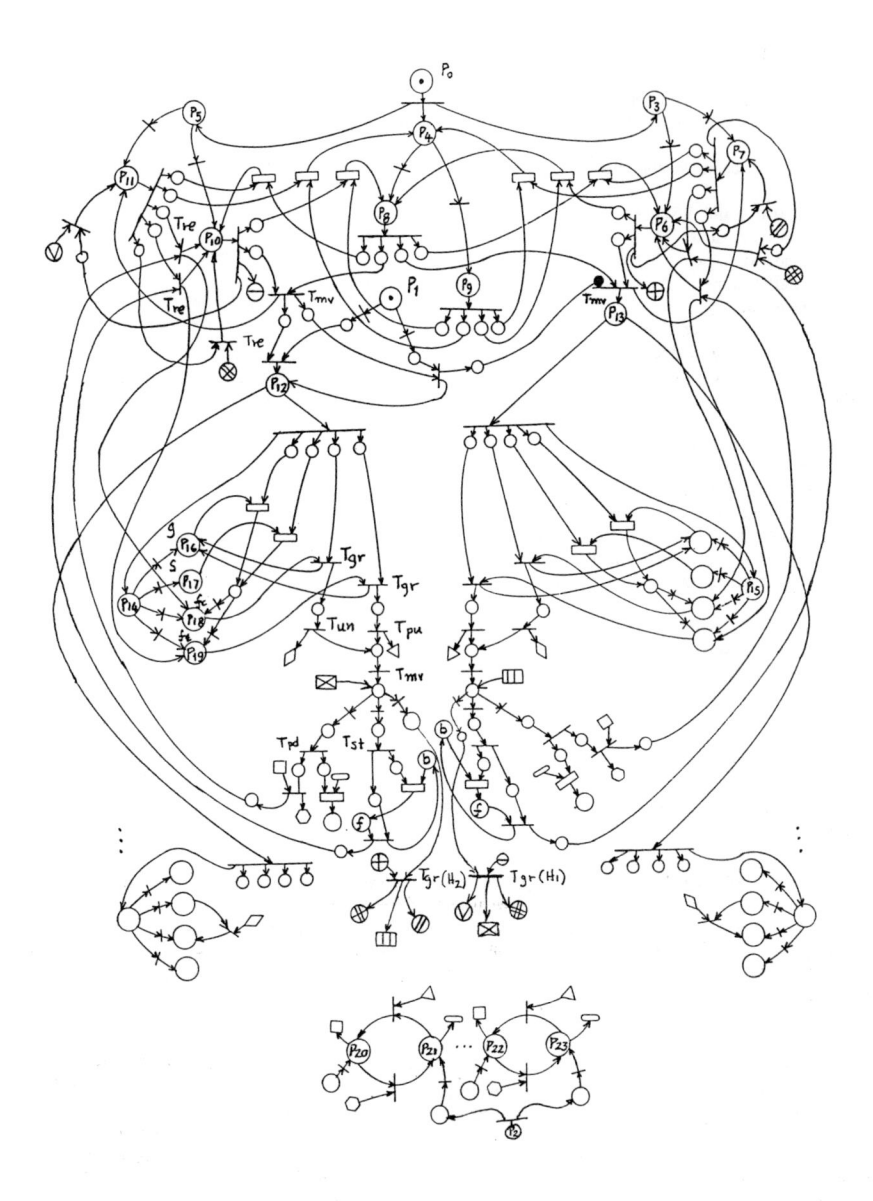

Figure 6: The two robot hands SPN model

(a above c above b above L2), and
(b above a above c above L3),

The goal state for the Test scenario 1 was the network to achieve (a above b above c above L3), and the goal for Test Scenario 2 was (a above c above L1, b above L2), to be completed directly after Senario 2.

The states and actions vectors are:

Test Scenario 1 and 2 States:
S(1) Hand1 available
S(2) Hand2 available
S(3) Hand1 waiting
S(4) Hand2 waiting
S(5) L1 Position clear above
S(6) L1 Position available
S(7) L2 Position clear above
S(8) L2 Position available
S(9) L3 Position clear above
S(10) L3 Position available
S(11) Block **a** clear and available
S(12) Block **b** clear and available
S(13) Block **c** clear and available
S(14) Hand grasping or releasing Block **a**
S(15) Hand grasping or releasing Block **b**
S(16) Hand grasping or releasing Block **c**

Test Scenario 1 Possible Actions:
A(1) move c to Not L2
A(2) move Block (**a** or **b**) to Not above c
A(3) move **a** to Not L2
A(4) Move **b** to Not L2
A(5) move c to L2
A(6) move **a** to Not above **b**
A(7) move **b** to above c
A(8) move **a** to above **b**
A(9) wait

A block is considered unavailable if grasped by a hand or covered by another block. A grasp requires movement into the vicinity of the target position through a possibly busy workspace or to grasp an object that is not free to be grasped. Releasing, similar to grasping, includes movement through the workspace away from the immediate vicinity of the block. The act of waiting implies that the active hand moves out of the immediate workspace for benefit of the other hand. After a hand waits a time step, it then reviews its recommended substrategies (actions) and chooses one that has no conflict.

Determination of Possible Actions

A minimum number of possible actions was chosen to attain the next possible subgoals. Actions 1 and 2 are designed to move blocks away from an initial configuration. A common strategy in a stacking "situation" for a next-to-bottom object is to place it in a location other than its goal state to clear space for the bottom object. Actions 1,3 and 4 have this logic designed in for each block, as there is a very good chance that a block must end up on top of another in the desired final configuration. Actions 5 through 8 provide the robot arms an opportunity to stack blocks in relation to other blocks or locations as referenced to their final desired configuration. Action 9 is the action of waiting, mandatory for multiple hand scenarios.

Note that the actions are not specific as to location other than to their final destinated location. This provision should be able to accomodate more or fewer spare locations in a future scenario.

Logic Supervisor Choice of Actions

A vector representing the current state of the scenario is introduced to the network, which produces a recommended action category. After a next action has been chosen by the neural net and supervisor, a new vector of active states and actions is created and added to the training set. The next action category is suggested by the network given the new vector of states. A maximum depth of only the most recent state/action vectors is used to maintain an update time of less than 0.25 seconds in language C. Each associated substrategy, or action, was checked for conflict (Block or Position Busy) and feasibility (Block is Clear Above). If no option could be executed, the robotic hand would wait. Choices were additionally restricted to avoid backtracking, that is, if an option had already been executed for a given block scenario, or an action returned a block to its previous position, this substrategy was given a lower priority than those who had not been executed previously.

The categories suggested by the neural network (7 for each hand) were compared to a logic supervisor's suggested actions for those steps within the training scenario. If more than one action is found within the category, the actions are prioritized following the same logic tree used by the supervisor. For the Test Scenario 1, the neural network categories and their associated actions for each robotic hand are:

Test Scenario 1

Neural Network Category	Recommended Hand 1 Actions	Recommended Hand 2 Actions
1	3,4,7,8	5
2	3,4,7	1,8
3	6,9	2,3,5,7,8
4	2,7	1,3
5	4,5	5
6	6,7	8
7	2,8,9	9

It is interesting to note that Hand 2, which always follows Hand 1 at each time step, acquired a category 7 devoted to waiting.

Test Scenario 1 Resulting Actions:

Robot Hand and Step	Resulting Category	Recommended Actions	Chosen Actions
H1-1	4	2,7	2
H2-1	3	2,3,5,7,8	2 -> 9
H1-2	-	(2,7)	2
H2-2	-	(2,3,5,7,8)	2
H1-3	5	4,5	5
H2-3	-	(2,3,5,7,8)	2
H1-4	-	(4,5)	5
H2-4	3	3,5,7,8,2	7
H1-5	7	8,9,2	8
H2-5	-	(3,5,7,8,2)	7
H1-6	-	(8,9,2)	8

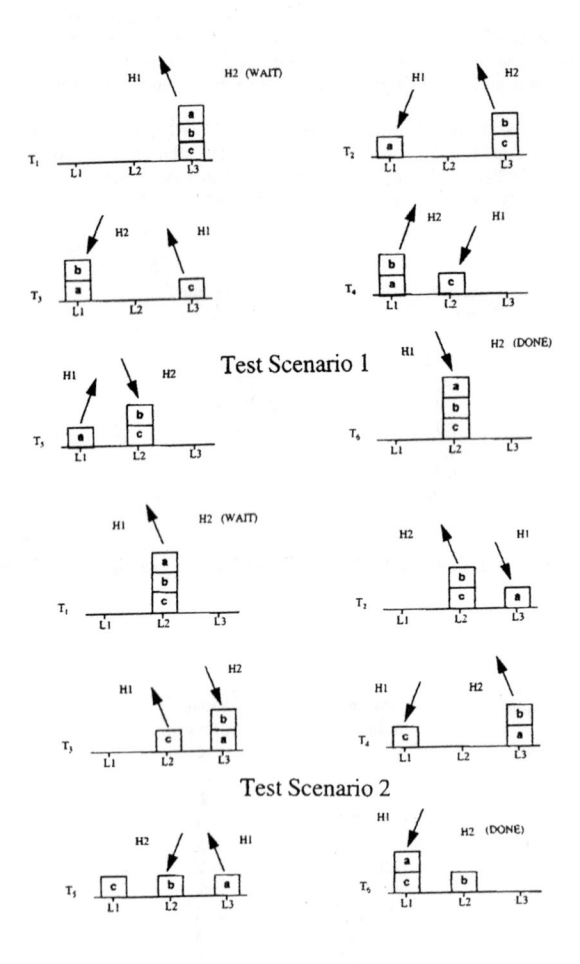

Test Scenario 1

Test Scenario 2

Robotic Hand and Step	Logic Followed
H1-1	**a**, above **c**, is available
H2-1	L3 is busy with Hand 1
H1-2	**a** placed on L1 (Not L2)
H2-2	**b** is now grasped
H1-3	**b** is busy, **c** grasped
H2-3	**b** is placed on L1
H1-4	**c** is placed on L2
H2-4	**a** not clear, **c** busy, **b** clear
H1-5	**a** clear
H2-5	**b** is placed on **c**
H1-6	**a** is placed on **b**

Test Scenario 2

The previous example is extended now to move to a new end state. The available actions are modified to accomodate the new goal states and their desired locations.

The new goal state is to be achieved by moving from the previous goal state (**a** above **b** above **c** above L3) to the new goal (**a** above **c** above L1, **b** above L2).

Test Scenario 2 Possible Actions

A(1) move **c** to Not L1
A(2) move (**a** or **b**) to Not above **c**
A(3) move **a** to Not L1
A(4) move **b** to Not L2
A(5) move **c** to L1
A(6) move **a** to Not above **c**
A(7) move **b** to above L2
A(8) move **a** to above **c**
A(9) wait

Test Scenario 2 Resulting Actions

Robotic Hand and Step	Chosen Actions	Human Actions	Logic Followed
H1-1	3	3	**a** is directed away from destination
H2-1	2->9	2->9	Hand 2 must wait for Hand 1
H1-2	3	3	**a** is placed on L3
H2-2	2	2	**b** begins toward default L1
H1-3	5	5	**c** is directed toward its destination
H2-3	2	2->9	**b** is redirected by Hand 1 with **c**
H1-4	5	5	**c** is placed on L1
H2-4	7	7	**b** is directed toward its destination
H1-5	8	8	**a** is directed toward its destination
H2-5	7	Done	**b** is placed on L2
H1-6	8	8	**a** is placed on **c**

5. CONCLUSION

In this paper a stochastic Petri-net planning model associated witha a neural network were used in a synergetic way for the selection of the appropriate actions to achieve the goal state for the blocks manipulation problem. More specifically, the SPN plan model was generated for the development of the appropriate framework on which the neural network was used as an efficient search planning strategy.

In the scenarios used as illustrative examples it was observed that, in scenario 2, Hand 2 (at Test Step 2) had more than one destination option. Given no other options the default is set at (**L1** before **L2** before **L3**).

However, if Hand 2 is in the process of placing a block on a space that Hand 1 intends as a target location for the block it is grasping, Hand 2 will go to another location which is still acceptable to the action constraints (time step T3).

This planning methodology (at the neural network search level) has a preoccupation with completing a subgoal task, which it has been given, interrupting with a wait step only when a location is being occupied by the other hand. An extension of this work is to change plans or paths midway through a task execution as the state of the process changes, although in the example above if a human's attempts to save steps does not decrease the overall time to complete the task. Another future extension of this work is the use of feedforward diagnostic nodes, which may be added to aid the next step strategy of a fuzzy supervisor.

REFERENCES

[1] E.D.Sacerdoti, **A structure for plans and behavior**, American Elsevier, 1977
[2] A.Tate, **Generating Project Networks**, Proc.IJCAI,1977, 888-893
[3] S.A.Vere, **Planning in time: windows and patterns for activities and goals**,IEEE T-PAMI,1983, 246-267
[4] D.E.Wilkins, **Domain independent planning: representation and plan generation**, AI,22, 1984, 269-301
[5] D.Chapman, **Planning for conjunctive goals**, AI,32, 1987, 333-377
[6] D.Corkill, **Hierarchical planning in distributed environments**, AI, 1979, 168-175
[7] E.D.Sacerdoti, **Planning in hierarchy of abstraction space**, AI,5,1974,115-135
[8] S.Lee and K.Chung, **Resource oriented parallel planning**, IJAIT,1,1,1992,85-115
[9] A.Passino and P.Antsaklis, **Planning via Heuristic Search in a Petri-net framework**, Proc.IEEE Conf. on Intelligent Control, 1988,350-356
[10] M.Genesereth and N.Nilson, **Logical Foundation of AI**, M.Kauffmann Pub.,1987
[11] S.Grossberg, **Studies of the Mind and Brain**, Drodrecht, Holland: Reidel Press, 1982.
[12] T.Kohonen, **Self-Organization and Associative Memory**, 2nd Edition, Berlin: Springer-Verlag, 1987.
[13] N.Bourbakis, **A neural knowledge base for high performance processing and planning**, IJAIT,3, 4,1995
[14] C.A.Petri, **Communication with Automata**, PhD thesis RADC-TR-65377,NY Jan.1966
[15] L.J.Peterson, **Petri-net theory and modeling of systems**, Prentice Hall 1981
[16] G.Sarridis and F.Wang,**A model for coordination of intelligent control using Petri-nets**, IEEE Symp. on Intelligent Control, VA 1988,pp.28-33
[17] F.Wang,K.Kyriakopoulos, T.Tsolkas and G.Sarridis, **A Petri-net coordination model of intelligent mobile robots**, IEEE Trans on SMC, 21,4,1991,777-789